Springer

Berlin
Heidelberg
New York
Hong Kong
London
Milan
Paris
Tokyo

Christian Körner

Alpine Plant Life

Functional Plant Ecology
of High Mountain Ecosystems

2nd Edition

With 218 Figures, 4 Color Plates, and 47 Tables

Springer

Professor Dr. CHRISTIAN KÖRNER
Institute of Botany
University of Basel
Schönbeinstraße 6
4056 Basel
Switzerland

ch.koerner@unibas.ch

Cover illustation: Variation in slope, exposure, relief and substrate type over a short distance are keys to understanding alpine plant life (Central Swiss Alps with Finsterahorn, 4200 m in the background) Left: *Culcitium nivale*, 4600 m, Cayambe, Ecuador

ISBN 3-540-00347-9 2nd ed. Springer-Verlag Berlin Heidelberg New York
ISBN 3-540-65054-7   1st ed. Springer-Verlag Berlin Heidelberg New York

Library of Congress Cataloging-in-Publication Data
Körner, Christian, 1949–
    Alpine plant life : functional plant ecology of high mountain ecosystems / Christian
Körner. – 2nd ed.
      p.   cm.
    Includes bibliographical references (p.).
    ISBN 3-540-00347-9 (alk. paper)
    1. Mountain plants – Ecology.   2. Mountain plants – Ecophysiology.   I. Title.

Springer-Verlag Berlin Heidelberg New York
a member of BertelsmannSpringer Science+Business Media GmbH
http://www.springer.de

© Springer-Verlag Berlin Heidelberg 1999, 2003
Printed in Germany

Cover design: Design & Production, Heidelberg
Typesetting: SNP Best-set Typesetter Ltd., Hong Kong
31/3150WI – 5 4 3 2 1 0 – Printed on acid-free paper

Preface to the second edition

Recent years have seen renewed interest in the fragile alpine biota. The International Year of Mountains in 2002 and numerous international programs and initiatives have contributed to this. Since nearly half of mankind depends on water supplies originating in mountain catchments, the integrity and functional significance of the upland biota is a key to human welfare and will receive even more attention as water becomes an increasingly limited resource. Intact alpine vegetation, as the safeguard of the water towers of the world, is worth being well understood. This new edition of *Alpine Plant Life* is an update with over 100 new references, new diagrams, revised and extended chapters (particularly 7, 10, 11, 12, 16, 17) and now also offers a geographic index. My thanks go to the many careful readers of the first edition for their most valuable comments, in particular to Vicente I. Deltoro (Valencia) and Johanna Wagner (Innsbruck).

Basel, April 2003   Christian Körner

Preface to the first edition

One of the largest natural biological experiments, perhaps the only one replicated across all latitudes and all climatic regions, is uplift of the landscape and exposure of organisms to dramatic climatic gradients over a very short distance, otherwise only seen over thousands of kilometers of poleward traveling. Generations of plant scientists have been fascinated by these natural test areas, and have explored plant and ecosystem responses to alpine life conditions. *Alpine Plant Life* is an attempt at a synthesis.

This book has roots in a century of research into alpine ecology at the Insitute of Botany in Innsbruck, Austria. Anton Kerner pioneered the field in the late 19th century. Arthur Pisek founded modern comparative and experimental ecology of alpine plants, and was the first to systematically combine field with controlled environment studies. Walter Larcher introduced the ecosystem approach, the question of scale. During my doctoral project with him on alpine plant water relations, he stimulated my interest in links between plant structure and function and in plant developmental processes. My former senior colleague Alexander Cernusca introduced me to environmental physics and thanks to him I began to think in terms of fluxes and pools. This text could not have been written without these influences.

Privileged to grow up in a green environment, my interest in biology was stimulated by my parent's fascination in plants and landscape gardening, their painting and photography, and their strong feel for natural aesthetics. Alpine vegetation is often like a garden, a mosaic of beauty, a small-scale multitude of ways of coping with life, attractive to both the naturalist and the scientist. Perhaps the reader will also find some morsels of this fascination between the lines of this scientific text.

Essential contributions to this volume were the patience and help by my wife Raingard and the graphical work, help with literature and laboratory analysis by Susanna Pelaez-Riedl. I am grateful to J Arnone, E Beck, MM Caldwell, T Callaghan, FS Chapin, M Diemer, B Holmgren, S Pelaez-Riedl, F Schweingruber, J Stöcklin and H Veit for commenting on drafts of various parts of the text. A number of colleagues helped with information, plant samples or unpublished data, namely WD Bowman, J Gonzalez, S Halloy, W Larcher, G Miehe, J Paulsen, H Reisigl, R Siegwolf, M Sonesson, RC Sundriyal, U Tappeiner and P Volko. R Guggenheim and his team produced the scanning electron micrographs in Chap. 9 and 16. H Schneider assisted with electronic treatment of photographs. The University of Basel provided an ideal environment for research and teaching alpine ecology to students and permitted the needed work-leave. I thank the Abisko Research Station in N-Sweden for hosting me during much of the literature and text work.

I also gratefully acknowledge the fruitful cooperation with the Springer team throughout this project. In particular, I would like to thank S Bunker, D Czeschlik, A Schlitzberger, K Matthias, and K-H Winter.

Alpine Plant Life was written for a broad readership. This has made it necessary to start several chapters with rather general introductions. On the other hand, I have tried to cover the bulk of scientific findings. In trying to cover as many relevant topics as possible, the reader will often only be given a reference to find answers elsewhere. There is no way to treat this field of science exhaustively in a single volume like this. Hence, the product is a compromise, which hopefully will interest the specialist as well as a wider audience.

Basel, February 1999   Christian Körner

Contents

1   Plant ecology at high elevations

Plants respond to the harsh alpine environment with a high degree of specialization, the structural and functional aspects of which this book aims to explore. Palaeorecords suggest that life on land started out in sheltered, warm and moist environments, and gradually expanded into more demanding habitats where water is rare, thermal energy is either low or overabundant or where mechanical disturbance is high. More than 100 million years ago, when the large, hot deserts of the Cretaceous period where formed, coping with drought became a matter of survival in higher plants. Why is this of relevance here? Because survival of both drought and freezing temperatures requires cell membranes which can tolerate dehydration. When plant tissues freeze, ice is first formed in gaps between cells, which draws water from protoplasts (see Chap. 8). A link between the ultrastructural and molecular basis for freezing tolerance and the **evolution** of dehydration tolerance of biomembranes has therefore been suggested (Larcher 1981). Plant survival in cold as well as hot "deserts" – the two thermal extremes on the globe – thus may have common evolutionary roots, although life under such contrasting thermal conditions requires many additional, rather different metabolic and developmental adaptations. The ability to survive **low temperature extremes** opened the highlands of the earth to plants. The Tertiary (and still ongoing) uplift of mountain ranges strongly accelerated the evolution of alpine taxa (Billings 1974; Agakhanyantz and Breckle 1995). Embedded in different floras of the world, high mountains became both highly fragmented refugia and corridors of cross-continental migration, and often bear plant diversities richer than those in their surrounding lowlands (Körner 1995a; Barthlott et al. 1996; Chap. 2 and color Plates 1–3 at the end of the book).

The concept of limitation

Life in high mountains is mainly constrained by physical components of the environment (Fig. 1.1) and some high altitude plant specialists can survive incredible "extremes", for example dipping in liquid nitrogen. Some species manage to grow at altitudes of 6000 m (Webster 1961; Grabherr et al. 1995; see Chap. 2). The study of traits which enable plants to live in such climatic extremes has fascinated generations of biologists, but what is an extreme? Once the ability to cope with environmental extremes has evolved, such extremes become elements of "normal" life. If we move genetically adapted plants to what – from our human perspective – might be less extreme, most of these specialist plants would either die or would be suppressed by species native to the new habitat. Hence, in an ecological context, the concept of **"limitation"** becomes problematic (Körner 1998b). The term emerged from agronomy, where limitation was defined as a limitation of biomass production when compared with some maximum value that might be achieved with all resource limitations and environmental perturbations eliminated. However, in nature mass production only matters if it contributes to survival and reproduction, hence **fitness**. In so-called limiting environments, the enhancement of "limiting" resources or the removal of physical constraints

Fig. 1.1. Fit to survive and reproduce under demanding environmental conditions (*Ranunculus glacialis*, Alps 2600 m)

(stress) may stimulate (in the short term) growth and reproduction, but in the long term, may eliminate an organism from its previously more "limiting" habitat by competitive replacement. Environments are only limiting to those which are not fit.

The ability to cope with specific environmental demands can be achieved in three ways: (1) by evolutionary (phylogenetic) **adaptation**, (2) by ontogenetic **modifications**, which are non-reversible during the life of an individual (or its modules such as leaves or tillers) but are not inherent, or (3) by reversible adjustment, often termed "acclimation" or **modulation**. If, by any of these adaptive mechanisms, a plant achieves the ability to cope with the demands of its environment and successfully reproduces, it is fit – which by itself says nothing about the adaptive mode employed. Natural selection usually sieves for genotypic fitness. Populations of species with particular fitness for life in a particular environment are called "ecotypes" (Turesson 1925; Clements et al. 1950; Hiesey and Milner 1965). The history of the ecotype concept is closely linked with high mountain plant ecology (Billings 1957). High and low

altitude provenances of plants of the same species were the first for which clear ecotypic differentiation was demonstrated (e.g. Engler 1913; Turesson 1931; Clausen et al. 1948; Clements et al. 1950), and evidence for altitude ecotypes dates back to the last century (see Langlet 1971).

Altitude specific **ecotypes**, however, are only halfway to speciation. Though ecotypic differences within a species may be larger in some cases than differences among certain species, in the long term, it is the higher taxon difference at the species or even genus level that sets the strongest contrast between alpine and lowland specialists. Specialists exclusively found at high altitudes will more likely reflect a high degree of "adaptation" in their characteristics, and hence can be expected to behave as more typically "alpine" than plants which radiate from lower elevation centers to high elevation outposts (Gjaerevoll 1990). However, it will be demonstrated in this book that even specialist species with a narrow high altitude range are weak indicators of life zone specific behavior. The reason for this is the large structural and functional diversity that is found among plant species even at highest altitudes (Körner 1991). It is the habitat (altitude) specific community of species and the relative frequency of traits among those species that bears the most solid message with respect to **life zone specific adaptive responses** (cf. Billings 1957). Provenances or ecotypes of single species from a wide altitudinal range, extending far beyond the zone of greatest abundance, have the advantage of close taxonomic relatedness, but may be "Jacks of all seasons", and hence are less likely to bear the most characteristic features of the highest life zone of plants (Fig. 1.2).

These views of limitation, adaptation and life zone specific responses governed this synthesis, which adopts a **comparative approach** across large geographic and altitudinal scales and large groups of species. Wherever possible, consideration of **frequency** distributions of traits among species or community means rather than single species data are given priority. Given the taxonomic and micro-habitat diversity found in the alpine zone, monospecific studies are at risk to reflect

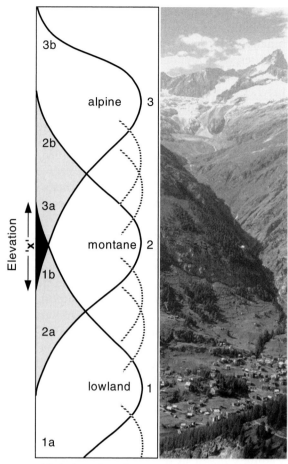

Fig. 1.2. Abundance ranges of different plant species or groups of plant species (*1, 2, 3*) or ecotypes of species (*1a,b; 2a,b; 3a,b*) along an altitudinal gradient. Ranges of maximum abundance do not necessarily reflect physiological optima but are simply centers of distribution. The comparative study of *1b* vs *1a*, *2b* vs *2a* and *3b* vs *3a* has the advantage of taxonomic (and possibly functional) relatedness and the disadvantage of comparing distribution tails (ecotypes from marginal habitats) rather than plants from the center of their range. In contrast, comparisons of plants with abundance peaks in the center of environmental antipodes (e.g. *3* versus *1*) are more likely to hit strong, life zone specific attributes, but taxonomic, and possibly functional diversity needs to be covered by comparing data for whole communities. Both approaches have been used in the past. For some questions, comparisons within overlapping zones, in particular the "*x*" zone bear a lot of scientific potential, but this has been explored to a lesser extent

curiosity rather than more general principles, even though the curiosity by itself may be a scientific jewel.

A regional and historical account

Research on functional ecology of alpine plants has a century-long **history**, and has its roots in comparative plant geography. By 1997 I had recorded approximately thousand publications dealing with functional aspects of alpine plant life and treeline biology (the phytogeographic and taxonomic literature on alpine plants is at least twice as large). The following brief and necessarily incomplete historical overview may assist readers in spotting some key references for geographic regions of interest. The later chapters are not structured by geography but by ecological topics. Where adequate, most recent examples have been used, which allow tracing references to earlier work. Figure 1.3 illustrates the geographical distribution of alpine plant research as reflected by the number of publications.

The pre-World War II research in alpine plants was almost exclusively conducted in the **temperate zone of Europe**, more specifically the Alps and the southern Scandes. The earliest scientific description of the elevational change in vegetation, the first mountain monograph of the world, is *Descriptio Montis Fracti* by K Gessner, who climbed Pilatus (Luzern, Switzerland) in 1555 and drew a rather precise picture, still valid today (Grabherr 1997, Zoller 2000). Possibly the first experimental attempts were those with transplants by Naegeli (mid 19th century from the Alps to Munich) and by Kerner (1869) from low altitude to alpine altitudes in Tirol. Interestingly, both these tests were not very successful. Most of Naegeli's alpine plants died at the "more favorable climate" at low altitude and also Kerner's lowland plants had obvious difficulties at the treeline, which led him to conclude that there must be an inherent (genetic) component associated with plant adaptation. From his famous reciprocal transplant experiments in the French Alps and

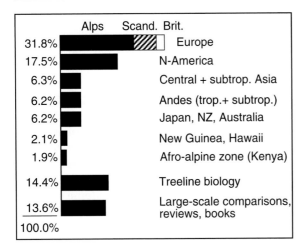

Fig. 1.3. Statistics of the geographical distribution of published research on the functional ecology of alpine plant ecology (evaluation of 800 publications). Publications on treeline biology are given in bulk. This survey does not include: geobotanical or floristic literature, papers on high altitude meteorology and climatology, soils, and conservation issues. Also work documented in the form of PhD or masters theses, institutional reports and publications in regional print media have been disregarded, because their inclusion would have biased the assessment toward the scientific environment of the author

Pyrenees, Bonnier (1890a, 1895) concluded instead that the environment has an overwhelming morphogenetic influence. Bonnier was the first to note with surprise (and by using incredibly simple devices) that the photosynthetic capacity of alpine plants is comparatively high. This conclusion matched well with the results of a broad leaf anatomical survey in alpine plants conducted during the same period by Wagner (1892) in the mountains around Innsbruck.

Following a first synthesis of knowledge by Schröter (1908, 1926), Swiss and Austrian researchers then took a lead in the more physiologically oriented alpine plant ecology for the first half of the 20th century. It was Senn (1922) and his student Henrici (1918) in Basel, who performed (in the Swiss Alps) the first reliable measurements of photosynthesis, transpiration and growth in alpine plants. Later, the Innsbruck group of Pisek (see his review in 1960) produced a large body of

evidence on temperature and drought resistance and many aspects of gas exchange in the major functional groups of plants at and above treeline. This research found a post-war continuation in treeline and subalpine research (e.g. Pisek and Larcher 1954; Friedel 1961; Tranquillini 1964, 1979) and in studies of a wide range of aspects of physiological ecology of alpine plants including high altitude extremes (Cernusca 1976; Moser et al. 1977 and the reviews by Larcher 1980, 1994, and Körner and Larcher 1988). As part of the International Biological Program, Larcher's group conducted the first process oriented, ecosystem research in the alpine zone (Larcher 1977) with follow up projects in the Man and Biosphere Program (Cernusca and Seeber 1981; Cernusca 1989). Alpine research in Scandinavia, which includes the transition to the arctic, and in Scotland, Europe's most humid alpine outpost, contributed almost one third of the current European literature on alpine plant life (e.g. Turesson 1925; Callaghan 1976; Wielgolaski 1975; Gauslaa 1984; Dahl 1986, Sonesson et al. 1991 for Scandinavia; Woodward 1983; Friend and Woodward 1990; Grace 1987 as examples of the work in Scotland).

After the genecological experiments with Californian alpine plants (Clausen et al. 1948), physiological ecology of alpine plants became a leading domain in the **temperate zone of North America** as well. Decker (1959) and Billings et al. (1961) were the first to test alpine plants under contrasting CO_2 concentrations, and, among other aspects, the physiological characterization of alpine ecotypes by Billings and Mooney (see their review in 1968) became a classic in plant ecology. Many other functional aspects of alpine plant ecology including reproductive biology have been studied in the following years by students of Billings (see his summary in 1987) and Bliss (1971, 1985; review by Campbell 1997). Caldwell (1968) pioneered research on solar radiation effects, including UV, water relations were studied by Ehleringer and Miller (1975) and others. At and above the Rocky Mountains treeline, a number of studies on plant nutrition and gas exchange have been conducted more recently (e.g. Bowman et al. 1993; Hamerlynck and Smith 1994, a synthesis by Bowman et al. 2001).

Beginning in the 1930s, CO_2-gas exchange and water relations of alpine plants were studied in **temperate central Asia** (primarily in the Pamir and mostly published in Russian) by Blagowestschenski (1935), Zalenskij (1955), Semichatova (1965), Sveshnikova (1973), Izmailova (1977); later extended to questions of assimilate allocation, plant nutrition and developmental aspects by Agakhanyantz and Lopatin (1978). Most recently Pyankov et al. (1992) surveyed C4-species in high altitudes. In the 1970s eco-physiological and eco-climatological work was taken up in the Central Caucasus (e.g. Nakhutsrishvili 1976; Nakhutsrishvili and Gamzemlidze 1984) followed by the project by Rabotnov (1987) in the northwest Caucasus. For more recent work from the Caucasus, see Onipchenko and Blinnikov (1994), Tappeiner and Cernusca (1996) and Nakhutsrishvili 1999. Shibata (1985) reviewed work done in the **Japanese alpine zone** and examples of more recent work are the papers by Masuzawa (1987), Shibata and Nishida (1993) and Kikuzawa and Kudo (1995). Alpine plant research in the **temperate zone of the Southern Hemisphere** is exemplified by the work of Mark (1975) in New Zealand, Costin (e.g. 1966) and Slatyer (e.g. 1976, 1978) in the Snowy Mountains of southeastern Australia.

Functional ecology of alpine plants in the **subtropics and tropics** is largely underrepresented in the literature (Fig. 1.3.), and a substantial fraction of observational material collected did not find its way into easily accessible publications. In part this had to do with the difficulties for field work in tropical alpine conditions, and the resulting often incomplete data sets which did not meet the standards of (temperate zone) journals. Furthermore, priority of alpine research in these regions, had to be given to documentary work on the flora and life conditions, since most often not even the simplest base data were available. Examples of early observational material date back to Schimper (1898; see also Schimper and von Faber 1935), with the first detailed functional analysis of a **tropical** alpine vegetation by Hedberg (1964) and Hedberg and Hedberg (1979). Approaches of more geobotanical emphasis are those by Vareschi (1951); Troll and Lauer (e.g. 1978) and Walter and Breckle (1991–1994 volumes); for new references for Africa (Kenya, Tanzania) see Hemp (2002) and Beck et al. (2002). Experimental work in the modern sense remained restricted largely to three truly alpine areas in the tropics: (1) The Páramos of Venezuela, with studies on gas exchange, freezing tolerance and growth forms (e.g. Larcher 1975; Baruch 1979; Goldstein et al. 1985; Meinzer et al. 1985; Rada et al. 1987; Smith and Young 1987; Monasterio and Sarmiento 1991; and comparative work in tropical Northern Chile (Arroyo et al. 1990; Squeo et al. 1991), (2) the afro-alpine vegetation on Mt. Kenya (e.g. Schulze et al. 1985, Beck 1994), and (3) Mt. Wilhelm in New Guinea (e.g. Walker 1968; Hnatiuk 1978; Körner et al. 1983). Syntheses of tropical-alpine plant ecology with individual contributions from all major tropical high mountain regions were edited by Vuilleumier and Monasterio (1986) and Rundel et al. (1994).

In the **subtropics**, again most research is from three areas, the alpine zone of the northwest Argentinan Andes (e.g. Ruthsatz 1977; Halloy 1982, 1991; Geyger 1985; Gonzalez et al. 1993), the southern Himalayas (e.g. Purohit et al. 1988; Pangtey et al. 1990; Sundriyal and Joshi 1992; Terashima et al. 1993) and the special island situation of Hawaii (Ziska et al. 1992; Sullivan et al. 1992; Lipp et al. 1994).

Introductions to alpine plant ecology **with a global perspective** are contained in Ives and Barry (1974); Franz (1979; a general overview, covering plants, animals as well as soils), and in the relevant chapters of the Walter and Breckle series (1991–1994). Klotz's (1990) and Archibold's (1995) books contain comparative overviews on alpine vegetation. Reviews on alpine plant diversity and its causes are contained in Chapin and Körner (1995). Functional aspects of the treeline problem have been reviewed by Wardle (1974); Tranquillini (1979); Grace (1989) and Körner (1998). The ecophysiology of alpine plants was reviewed in the above mentioned articles by Billings and Mooney, Bliss, Larcher, and Körner

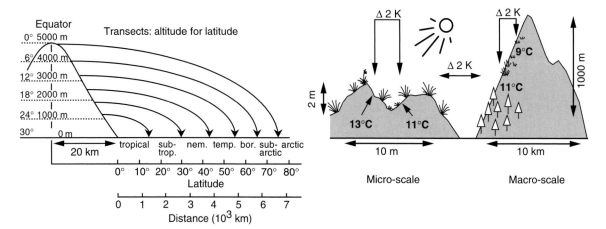

Fig. 1.4. Altitude for latitude: over short elevational distances thermal gradients represent the climate across vast latitudinal distances. Space for time: thermal contrasts across the relief "simulate" differences in temperature along elevational gradients. These are "natural" experiments, ideally suited for comparative functional ecology

Fig. 1.5. The compression of climatic life zones over short distances makes mountains hot spots of biological diversity (Bathlott et al. 1996) and offers unsurpassed possibilities for comparative ecological research. This photograph shows the peaks of Langtang (>7000 m, Nepal), a broad alpine belt (*center*) and the upper montane forest dominated by *Quercus semecarpifolia* and *Tsuga dumosa* (*foreground*)

and Larcher. Billings (1988), Campbell (1997) and Ozenda (1988) provide geobotanical reviews on the North American alpine flora and the flora of the Alps, respectively, and consider functional aspects as well. More specific articles will be discussed in the relevant chapters.

The challenge of alpine plant research

The literature selected for this brief geographical overview illustrates how plants cope with the environmental demands at high altitudes. Generally, it is not a simple physiological weakness related to stress tolerance, carbon assimilation, or nutrient acquisition, but a more complex, partly inherent restriction of growth that causes alpine plants to remain small (Larcher 1983; Körner and Pelaez Menendez-Riedl 1989). Alpine plant species were selected for small size, and thereby do not fully experience some of the climatic stresses presumed to dominate life at high elevation (see Chaps. 4 and 8), but there are ultimate physical limits: the end of the tails in Figure 1.2. The upslope thinning of those abundance tails reflects the gradual increase of environmental constraints. Hence, it is the interplay of both adaptation AND increasing climatic limitation that governs alpine plant life. The content of the following chapters is an attempt to draw together the currently available knowledge about the mechanisms which contribute to fitness of alpine plants and assess the constraints they experience from their environment, despite the manifold adjustments to life at high altitudes.

The alpine life zone is one of the most fascinating regions for scientific research. Nowhere else on land do we find steeper environmental gradients than in alpine terrain. Over a distance of few meters, we may find snow-bed communities in wet cold soil and hot "desert" micro-habitats on rocky outcrops with succulent plants utilizing crassulacean acid metabolism (see color Plate 3 at the end of the book). Within a half-hour drive or a short cable car ride we can move across life zones otherwise 3000 km distant in geographical latitude at low elevation (Fig. 1.4). This compression of life zones (Fig. 1.5) and the small scale patterns of life conditions in steep alpine terrain represent **natural experiments** which provide unbeaten opportunities for comparative ecological research, the study of plant adaptation and the mechanisms for survival of physical stress (Larcher 1967, 1970; Billings 1979).

These gradients not only permit "elevation for latitude" tests in order to separate thermal from other mountain climate influences (Körner et al. 1991; Prock and Körner 1996), but also "space for time" studies in the context of global warming. At very small scales thermal gradients do occur in an otherwise similar macro-climatic, geological and florisic matrix. In contrast to short-term experimental manipulations (e.g. warming treatments), vegetation at such natural temperature gradients had time to adjust, and is thus more likely to reflect a realistic picture of long-term trends.

2 The alpine life zone

What does "alpine" mean? One common explanation is that the term is of Latin origin and means "white" or "snow–covered" (from "albus" = white, with reference to the North Italian peaks of the Alps as seen by the Romans; Löve 1970). However, today linguists consider this as purely coincidental and the term is most likely of pre-Roman origin, with "alp" or "alb" standing for "mountain" in general. Even the Basques use "alpo" for mountain flanks in their non-Indo-Germanic, ancient language. Traditional terms as Alpe/Alpes (Romanic languages), Alp (Swiss and other Allemannic regions) or Alm (Austrian or Bavarian) used by farmers in the Alps refer to man-made pastures for summer grazing at or below the upper treeline. In today's common language "alpine" is often applied to whole mountain regions including valleys and townships or is used as a general substitute for mountains. These uses of the word "alpine" do not match the meaning of "alpine zone" in the current context.

In the plant-geographic sense relevant here, the "alpine" life zone exclusively encompasses vegetation above the natural high altitude treeline (see Chap. 7). The term is applied to any low stature vegetation above the climatic treeline worldwide, and is not restricted to the European Alps from where the name originated. For the Andes, the term "high-andean" is customarily used instead of alpine, and "afro-alpine" is often used in the African mountains.

While the alpine life zone may be reasonably well defined (see also below) this is not the case for **"alpine species"**. As was discussed in Chapter 1 and illustrated in Figure 1.2, the ranges of some lowland species extend into the treeline ecotone or above (1), other species may spread from their montane center to both low and high altitudes (2), still others may be centered in the alpine zone but may also be found at lower altitudes (3), and finally, there are those which are restricted to the alpine zone (4). For comparative purposes, the consideration of all four types of species in the alpine zone may reveal a lot of insight into the adaptive mechanisms of plant life at high altitudes. However, in the strict sense, I agree with Gjaerevoll (1990) that only the last two categories (3 and 4) should be treated as "alpine plant species".

Altitudinal boundaries

The lower boundary of the alpine life zone, the **treeline**, is often fragmented over several hundred meters of altitude. Where an upper treeline is missing, as in many arid mountain regions or because of deforestation, the position of the nearest live remnants of highest altitude tree growth are often taken as a rough guideline. Alpine treelines vary in altitude between a few hundred meters above sea level at subarctic latitudes to about 4000m in the tropics and subtropics (Figure 2.1). The zone between the closed upper montane forest and the uppermost limits of small individuals of tree species is often termed "subalpine". However, these boundaries do not fit a straight forward definition and "subalpine" has not become a widely used term outside Europe. In order to avoid confusion, I refrain from using this term (see the discussion by Löve 1970 and Chap. 7), and address the forest-alpine transition zone by

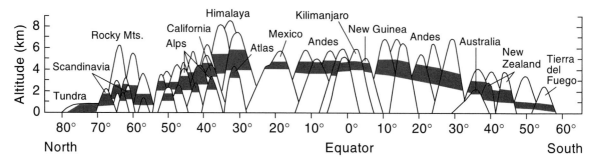

Fig. 2.1. A schematic presentation of the altitudinal position of the alpine life zone from arctic to antarctic latitudes

"treeline-ecotone" (or simply "near the treeline"). A more precise definition seems impractical.

The altitudinal limit for tree growth is one of the most striking phenomena in vegetation science. It obviously marks a biological threshold for plant life, an abrupt transition between rather different life forms. Because of the importance of understanding treeline formation in order to understand life beyond the treeline, this topic will be addressed separately in more detail in Chapter 7.

Closer to the **poleward end** of plant growth it becomes increasingly difficult to distinguish between "arctic" and "arctic-alpine". Ultimately, the two life zones merge. At these high latitudes, a meaningful distinction needs to account less for altitude than for relief. The steeper the slopes, the more topographically structured the terrain, the more abundant are the typical mosaics of alpine microhabitats with all their biological richness (Körner 1995a). Even within the arctic tundra region, rocky outcrops, south-facing slopes or mounts such as the so-called pingos, bear vegetation which (as rated by physiognomy, taxonomy and habitat conditions) has more in common with alpine than flat tundra vegetation (Walker 1995). In contrast, some high altitude "arctic-alpine" plateaus with extensive peat land may better fit the tundra concept, despite their more southerly position. Further south, the alpine zone becomes more distinctively different from the arctic, which is why the term "tundra" is best avoided when talking

about alpine vegetation (see discussion and references in Löve 1970, Billings 1973, 1988, Körner 1995a).

The highly fragmented plant cover above the belt of closed alpine vegetation, (including the summit region with only isolated plant occurrence) is termed "subnival" or "nival" (i.e. vegetation just below or above the permanent snowline, Schröter 1926, Reisigl and Pitschmann 1958, Landolt 1983, Ozenda 1988). This zonal distinction remained largely confined to the central European literature, perhaps (as Reisigl and Pitschmann, 1958 pointed out), because the so-called snowline is not visible in the landscape, and relief driven mosaics of both closed alpine mats and open scree or rock field vegetation make an altitudinal separation difficult. In the global survey attempted here, the term "alpine" does also include these highest ranging outposts of higher plant life.

What are the **upper altitudinal limits** of higher plant life? Various altitude records have been traded over the years (Webster 1961, Grabherr et al. 1995). The current record is held by *Saussurea gnaphalodes* which was found by E Shipton in 1938 on a scree slope at 6400 m altitude on the north flank of Mt. Everest in the central Himalayas, i.e. at subtropical latitudes (Miehe 1991; General Herbarium, British Museum, London). This species is a perennial herbaceous rosette plant of very compact growth habit, covered by a thick felt of white hairs, and it belongs to the family Asteraceae. The genus is widespread at high altitude in

Table 2.1. Altitudinal limits of higher plants (after a list compiled by Grabherr et al. 1995)

	Individuals	Communities	Closed vegetation
Tropical mountains (maximum height)			
Mt. Kenya (5190 m)	5190	–	4400
Kilimanjaro (5896 m)	5760	5700	4300
Ruwenzori (5119 m)	5119	–	4500
Chimborazzo (6310 m)	5100	–	4600
Subtropical mountains			
Himalayas (8846 m)	6400	5960	5500
Andes (in this part, 7084 m)	5800	–	4600
Temperate zone mountains			
Alps (4607 m)	4450	3970	3480

the Himalayas and does also occur in northern temperate and subarctic mountains. Several other spot-findings of higher plants between 5800 and 6200 m in the Andes and Himalayas have been reported (Table 2.1). Genera found above 6000 m are *Arenaria*, *Stellaria* and *Ermania* (all Caryophyllaceae).

As will be pointed out in Chapter 4, altitude per se becomes a doubtful criterion for estimating conditions for life in high mountains, because the microenvironment can overrule the macroclimate during sunshine hours. It is likely that these isolated plant individuals inhabit microhabitats whose thermal regime is similar to the more common situation at more than 1000 m lower alpine altitudes. However, these plants live with only about half the CO_2-partial pressure measured at sea level (an aspect discussed later in Chaps. 3 and 11) and are certainly exposed to extremely low night temperatures and frozen soil for most of the year (Rawat and Pangtey 1987). It is unlikely that they reproduce at these extreme altitudes, but recruit from diaspores brought in from somewhat lower altitudes. Patches of plant communities have been recorded at altitudes as high as 5700–6000 m (East Africa, Himalayas), and the upper limits for closed swards of vegetation in these mountains is somewhere between 4600 and 5500 m altitude (Miehe 1991).

According to Anchisi (1985) the altitude record for the Alps appears to be held by *Saxifraga biflora* at Dom du Mischabel in the Swiss Alps of canton Valaise at 4450 m (47°N lat.). Several other members of the genus *Saxifraga* and species of *Androsace* (Primulaceae), *Gentiana*, *Achillea* and *Ranunculus* have been reported for Finsterahorn, 4270 m, in the same part of the western Alps (Werner 1988). Again, the limit for communities and closed vegetation are much lower. According to the data compiled by Grabherr et al (1995) micro-communities have been found in the central Alps (where the treeline is between 2100 and 2300 m) at altitudes between 3800 and 4000 m, and closed swards of higher plants can be found up to 3500 m. At 40°N lat. in the Sierra Nevada of California, lush plant communities can be found at summits close to 4000 m altitude (e.g. on Mt. Dana, in the Tioga Pass area).

The altitudinal ranges of lichens and mosses exceed those of higher plants. These "lower plants" which tolerate complete desiccation are found at the highest mountain peaks, for instance at the summit of Kilimanjaro, 5900 m (Beck 1988), the high Andes, 6700 m (Halloy 1991) and the absolute record again comes from the Himalayas for lichens at 7400 m (Kunvar, cited in Miehe 1989). At least 50 species of lichens have been reported by Gams (1963) to occur at altitudes above 4000 m in the Alps. Snow algae are reported between 5100–5700 m on Langtang, Nepal (Yoshimura and Koshima 1997). Small animals (e.g. a Salticidae spider) have been found living above the limits of higher plants, depending on a food web based on wind blown debris (Swan 1992). Bacteria have no altitudinal limits as long they find some organic dust and short spells with liquid water and a number of taxa were isolated from substrate collected at 8400 m altitude on Mt. Everest, not including any cyanobacteria but a curious member of the Actinomycetales called *Geodermatophila obscurus* ssp. *everestii* (Swan 1992). Swan thinks that this locality is the environment on Earth most comparable with that of Mars.

Global alpine land area

Alpine vegetation, as defined above, represents the only biogeographic unit on land with a global distribution. Depending on altitude, it can be found at all latitudes (Figure 2.1.). This remarkable fact was first documented by Alexander von Humbolt (cf. Troll 1961), who pioneered the concept of comparative mountain ecology.

The global land area covered by alpine vegetation is fragmented into many mountain regions (Figure 2.2). A geographical data base for alpine vegetation does not exist. Rather, this type of land cover is amalgamated with arctic tundra or is treated as mountain pasture (where cattle grazing is still possible) or as non-vegetated bare land. The more open higher altitude vegetation usually falls in the same category as rocks and glaciers (*sensu* useless, i.e. "unproductive"). An assessment of the discrepancy between actual plant cover and official statistics for the central Alps (Körner 1989a) revealed that 50% of the land officially rated as bare and useless bears some of the biologically richest alpine ecosystems. Based on this analysis for the Alps, approximately 30% of the area of alpine life zone falls in the "bare of any vegetation" category of landscape inventories. This 30% estimate does not account for some isolated plants and cryptogams.

How large is the alpine area of the globe? Taking the altitudinal ranges of the alpine life zone from published records (Hermes 1955, Troll 1973, Wardle 1974; see also chapter 7 and the more detailed analysis presented in Figure 7.1) and personal observations from many parts of the world, and combining these with a global geographic information system (thanks to W Cramer, Potsdam for his help) yields the approximate aerial extent of the alpine life zone (Figure 2.3). Alternatively, using the "tundra" notation in currently available digital land cover maps (in which "alpine" is often incorrectly nested) and discriminating altitudinal ranges that fall below treeline according to Figure 2.3, resulted in completely unrealistic estimates of the aerial extent and regional distribution of true alpine vegetation.

The total area of the alpine life zone (Figure 2.3.) overestimates the vegetated alpine area, because bare land surfaces are included, altitudinal limits are compressed in oceanic areas compared with latitude specific global means, and because the depression of altitudinal limits near the equator compared with the subtropics (Halloy 1989) is not accounted for. Another uncertainty comes from vast mountain ranges with so-called mountain-tundra in eastern Siberia, the ranking of which as alpine or arctic is subjective (50% included here). With these caveats, and the fraction of land surface completely bare of any vegetation in the alpine life zone of the Alps (30% of total alpine area) applied globally, the total vegetated alpine area of the earth amounts to ca. 4 million km² or nearly 3% of the terrestrial surface.

Although this area appears relatively small, it is inhabited by roughly 10 000 species of higher plants (see below). Due to their remote position, many of these high altitude ecosystems have retained a relatively natural vegetation, and thus represent some of the last wilderness reserves in highly populated parts of the world. Alpine ecosystems also strongly influence life (including human) down–slope. About 10% of the world's population live in mountain regions, and more than 40% depend in some way on mountain resources, in particular on drinking and irrigation

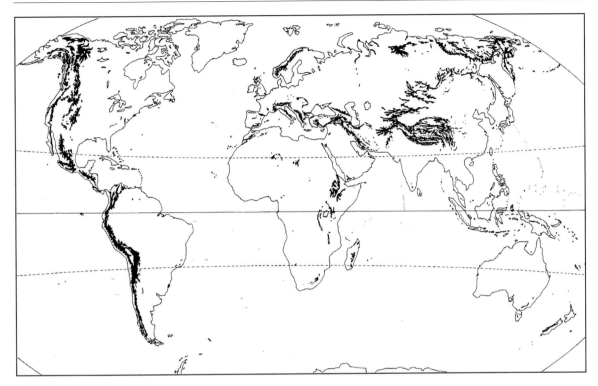

Fig. 2.2. The global distribution of the alpine life zone. Nearly 3% of the terrestrial land area is covered by true alpine vegetation, which includes approximately 10 000 species or 4% of all known higher plant species. (Körner 1995a)

water yielded by high altitude catchments (Messerli 1983; Chap. 17).

Alpine plant diversity

The alpine floras of the world are nested within a great variety of regional floras, partly explaining the great overall species diversity among alpine vegetation (Körner and Spehn 2002). Even within a single mountain region such as the Caucasus or the Venezuelan Andes, plant species diversity in the alpine zone alone may approximate that of the whole arctic tundra (5% of the global land area and ca. 1500 species). Geographic isolation, tectonic uplift, climatic changes, glaciation, strong micro habitat differentiation and a varied history of migration and/or evolution lead to a high degree of taxonomic richness (Packer 1974;

Agakhanyanz and Breckle 1995; Körner 1995a). Many alpine regions are biodiversity hotspots and are inhabited by substantial numbers of endemic species. For instance, ca. 830 species out of 1500 listed by Polunin and Stainton (1988) for the western part of the central Himalayas alone grow in the alpine zone above 3900 m. The total alpine flora of the mountains of Central Asia may be two to three times as large. More than 1000 species are found in the alpine zones of the Great Caucasus or the Venezuelan páramos (Agakhanyanz and Breckle 1995; Vareschi 1970). The European Alps, New Zealand Alps and the southern Rocky Mountains region have alpine floras of 600–650 truly alpine species each (Mark and Adams 1979; Hadley 1987; Ozenda 1993).

After checking several floras and trying various types of compilation, it appears justified to assume a total **alpine flora of the world** of 8000 to 10 000

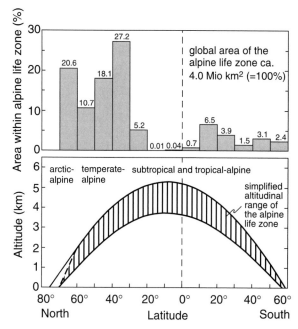

Fig. 2.3. The latitudinal distribution of the alpine life zone. The *lower* diagram illustrates the mean altitudinal limits of the alpine life zone, the *dashed line* on the *left* accounts for the uncertainty between arctic-alpine vegetation *versus* so-called mountain tundra (see the text). The *upper* diagram illustrates the relative contribution of each 10° range in latitude to the total global area falling in the alpine life zone (as defined by the *lower* diagram). Note that 82% of the alpine life zone is situated in the Northern Hemisphere. Arctic and Subantarctic alpine areas represent 23%, cool-temperate alpine 32%, warm-temperate to nemoral alpine 29% and the subtropical and tropical alpine life zone together cover 16% of the global alpine land area. (Körner 1995a)

species of higher plants. These belong to approximately 100 families and about 2000 genera and represent ca. 4% of all known higher plant species. Hence about one-fourth to one third, depending on the family key, of all plant families of higher plants have alpine representatives. Some families are particularly abundant. Most important are Asteraceae and Poaceae, with Brassicaceae, Caryophyllaceae, Cyperaceae, Rosaceae and Ranunculaceae forming further prominent families. Commonly of poor cover, but regularly found across the world, are species of Gentianaceae, Apiaceae, Lamiaceae, and, somewhat less regu-

larly, Primulaceae, Campanulaceae and Polygonaceae. The alpine shrub vegetation is dominated by Ericaceae and Asteraceae in most parts of the world, but regionally representatives of other families, for instance Scrophulariaceae (New Zealand), Epacridaceae (Tasmania), Proteaceae (Australia) or Hypericaceae (Andes), may become important. Families with otherwise prominent positions in the list of plant species of the world such as Fabaceae and Orchidaceae are strongly under-represented. Though present in nearly all alpine floras, they are not among the specialists of highest altitudes (notable exceptions are *Astragalus* and *Oxytropis* species in Central Asia, some of which thrive above 5000 m, cf. Miehe 1991). Typical tropical families such as Arecaceae, Araceae, Moraceae and Piperaceae are absent from truly alpine floras, but some taxa of families which are very prominent in the humid tropics such as Rubiaceae and Melastomataceae (e.g. *Coprosma* sp. in the Australian Snowy Mountains and *Brachyotum* sp. in the equatorial Andes) or in humid or semiarid subtropics such as Cactaceae and Bromeliaceae are present (e.g. *Tephrocactus* sp. in northwest Argentina, *Opuntia* sp. in Peru and *Puya* sp. in Ecuador).

As a rule of thumb, **regional floras** of the alpine zone extending over a few to about a hundred km² include in the order of 300 species, and numbers do not increase substantially until whole mountain ranges are considered, and such regional floras appear to represent about half of the total number of alpine species in the larger mountain systems. This is the surprising result of surveys in the Alps summarized by Wohlgemuth (1993; Figure 2.4). Similar numbers (ca. 200–280 species) are reported for several other regional alpine floras (Table 2.2). On average, these regional floras are composed of about 40 families of higher plant species. I list these numbers because they indicate the sort of sample size one would have to consider for taxonomically representative data sets of functional attributes of alpine plant species or plant families.

The alpine flora often seems to represent about one fifth to one fourth of the total regional flora

Table 2.2. The number of higher plant species in various distinct mountain regions

Region	No. of species (No. of families)	References
Afro-alpine (E Africa)	278 (39)	Hedberg (1957)
Beartooth Plateau (Montana, USA)	210 (–)	Johnson and Billings (1962)
Cumbres Calchaquies (NW Argentina)	200 (49)	Halloy (1982)
Hokkaido alpine zone (Japan)	225 (42)	Tatewaki (1968)
Ruby Range (Colorado, USA)	220 (35)	Hartman and Rottman (1987)
Scandes (Scandinavia)	250 (29)	Nilsson (1986)
Snowy Mountains (Australia)	250 (40)	Costin et al. (1979)
Teton Range (Wyoming, USA)	260 (36)	Spence and Shaw (1981)
Upper Walker River (California, USA)	280 (–)	Lavin (1983)
Mean	241 (39)	

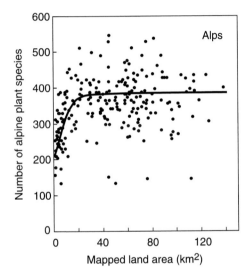

Fig. 2.4. Numbers of plant species (higher plants only) per inventoried area of regional alpine flora in the Alps. (Wohlgemuth 1993)

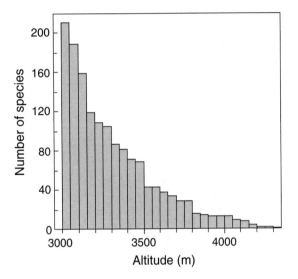

Fig. 2.5. The reduction of plant species number with altitude in steps of 50 m in the Alps. Note that the treeline is at ca. 2000 m and the upper limit of closed vegetation is mostly found around 2800 m altitude. (Grabherr et al. 1995)

including the plains. This is the result of a comparison of ten different Asiatic mountain regions (extremes 5–35%, mean 23 ± 9%) reported by Agakhanyanz and Breckle (1995), and matches fairly well with the situation in the Alps where approximately 650 alpine taxa are part of an overall flora that includes the immediate forelands, of ca. 3000 species (i.e. ca. 22%).

Alpine plant diversity decreases with increasing altitude, but, as Grabherr et al. (1995) suggested for the eastern central Alps, often not gradually, but in steps, with rather stable numbers over 200 to 400 m of altitude. This wave-like reduction of species richness reflects the zonation of alpine vegetation. The uppermost, largest "wave" is illustrated in Figure 2.5 which shows the exponential reduction of species numbers between 3000 and 4450 m altitude. Figures 2.6 to 2.8 show the elevational trends in species richness for higher plants and bryophytes along discrete elevational tran-

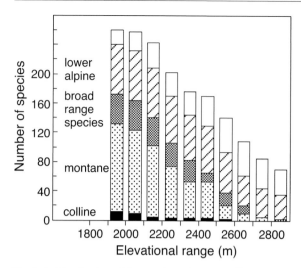

Fig. 2.6. Distribution of the number of species of higher plants in altitudinal steps of 100 m along a transect in Belalp, Swiss central Alps. Different *shades* refer to groups of species of a common elevational range of distribution (as noted on the *left*). (Theurillat and Schlüssel 1996)

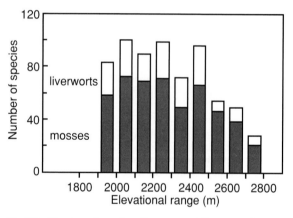

Fig. 2.7. Number of species of mosses and liverworts in altitudinal steps of 100 m along a transect in Belalp, Swiss central Alps. (Geissler and Velluti 1996)

sects. The reduction in species number above the treeline is ca. 40/100 m of altitude and follows the altitudinal reduction in land area (Körner 2000). The close to linear diversity/area relationship suggests continued proportionality between species numbers and the total remaining land area as the alpine archipelago narrows with altitude.

Origin of alpine floras

A large body of literature on the possible historical **origins of today's alpine taxa** contains evidence for a mix of ancestral, mostly Tertiary elements, immigrants from various source floras and new evolutionary lines (e.g. Jerosch 1903; Braun-Blanquet 1923; Gams 1933; Merxmüller 1952–1954; Troll 1968; Hedberg 1970; Vuilleumier 1971; Raven 1973; Packer 1974; Billings 1974, 1978, 1979; Hübl 1985; Salgado-Labouriau 1986; Van der Hammen and Cleef 1986; Dahl 1990; Simpson and Todzia 1990; Agakhanyanz and Breckle 1995). As mentioned above, geographical fragmentation, the speed of tectonic uplift and glaciation represent the major selective events at a continental scale.

The east-west orientation, greater fragmentation and separation from the arctic flora of the Eurasian mountain ranges compared to the north-south stretching Cordilleras, causes them to show greater regional variation and speciation (e.g. Gupta 1972). Tectonics, in the case of "block-orogenesis" i.e. the simultaneous uplift of large mountain "blocks" in contrast to volcanic orogenesis, can indeed induce speciation. Agakhanyanz and Breckle (1995) report annual uplifts of the Pamir and Caucasus between 7 and 10 mm (extremes 20 mm), movements which create new (cooler) habitats at rates comparable to the speed of evolutionary plant adaptation and species formation.

Glaciation does not remove the entire alpine flora of an area, but restricts it to microclimatically favorable high altitude refugia above or between the ice stream net. From these refugia ("nunatakker") species may re-invade habitats released by retreating ice shields (Merxmüller and Poelt 1954, Gjaerevoll and Ryvarden 1977, Pignatti and Pignatti 1983, Dahl 1987, Chomes and Kadereit 1998; Hungerer and Kadereit 1998; recent review by Stehlik et al. 2000). During the earliest post-glacial period, cool low altitude corridors

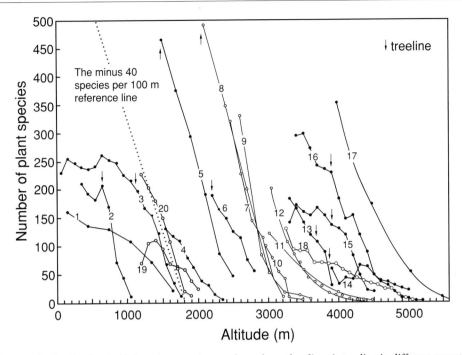

Fig. 2.8. The altitudinal reduction in higher plant species numbers above the climatic treeline in different mountain regions (the *dashed line* is the 40 sp/100 m reference; from various sources in Körner 2002). Data for Kilimanjaro by A. Hemp (pers. commun.; loss of 20 sp/100 m) and for one of the two regions in New Zealand by Mark et al. (2000; loss of 12 sp/100 m on Mt. Armstrong) deviate from the general pattern. Both have comparatively small floras. Remarkably, the Mt. Eyre profile in New Zealand is three times steeper than the Mt. Armstrong profile (a loss of 30 sp/100 m), but also starts from a two to three times richer flora at the treeline (235–245 instead of 70–115 species). The lower the species number at the climatic treeline, the less steep the elevational decline, which reflects a proportional rather than an absolute loss of species with elevation. Note the overall sigmoid trend, with loss rates of <10 sp/100 m asymptotically approaching zero (the upper limit of higher plants)

for species migration were opened. This is how the edelweiss, a member of a central Asian genus (*Leontopodium*) made its way to the Alps, a rather recent addition to the flora. Another example is the occurrence of completely disjunct, small patches (relicts) of *Carlina bibersteinii* near the treeline in the Alps, with the main area of this species' distribution today in Siberia (Körner and Meusel 1986). Migration corridors strongly influence floristic richness and endemism. In part, the flora of Nepal is so rich because it lies midway along a several 1000 km bow of mountains which span at least 20° of latitude between the east and west ends and the most southern middle part. For references on the postglacial history of high altitude vegetation consult the above references on the origin of alpine plants, Lauer (1988) and Flenley (1992) for tropical mountains, and Ammann (1995) for temperate zone mountains. Endemic species, i.e. those restricted to a certain mountain region, are usually less frequent in the summit floras, but reach greatest abundance in the lower alpine zone (mostly less than 10% of the alpine flora in Asia; Agakhanyanz and Breckle 1995). Extreme high altitudes are commonly dominated by otherwise widespread species. Long-term isolation of alpine floras does not necessarily cause a depauperation in species, as is illustrated by the richness of the alpine flora of New Zealand compared with the species pool in Australia (Table 2.2.) from where most of these New Zealand alpines descended (Raven 1973).

In addition to species diversity, a number of authors considered within-species genetic diversity of alpine plants (see review by Packer 1974; Albott et al. 1995; Gugerli et al. 1999; Till-Bottraud and Gaudeul 2002). There appears to be no real consensus about altitudinal trends, but it may be concluded that genetic diversity is generally high (perhaps except for refugia species (cf Dahl 1990; Bauert et al. 1998), despite a substantial fraction of apomictic or clonally expanding species. For instance Steinger et al. (1996) illustrated, by applying modern DNA analysis, that the obligatory clonal species *Carex curvula* in late succession alpine turf exhibits a high degree of genetic differentiation among intermingling clones. Similarly, Bauert (1993) identified (by studying isoenzymes) substantial genetic variation in populations of the bulbil-forming *Polygonum viviparum*, also in the Alps. It does not seem that the increased polyploidy that was found in arctic vegetation is paralleled in the alpine zone. Favarger (1961) reports that roughly half of all alpine species tested were polyploid, just as many as in the adjacent lowland flora. However, Packer (1974) refers to some studies where slightly higher percentages were documented.

Despite the sheer overwhelming multitude of species and their floristic and evolutionary relatedness, there are some common patterns. Plant life forms and morphotypes exhibit striking convergences throughout the world's mountains, providing a coherent basis for comparative alpine plant ecology (Halloy 1990).

Alpine growth forms

Growth form is a "plan", life form is the result of the interplay between the plan and the environment (Rauh 1978). Life form may differ from growth form. The environment can shape the growth form tree into the life form shrub. However, in most cases this distinction is not particularly important for the undisturbed

Table 2.3. Life form spectra (relative fraction of total flora, %) of alpine species in the Hindu Kush Mts. above 4000 m altitude. (Breckle, cited in Agakhanyantz and Breckle 1995)

Altitude range	Ph	Ch	H	G	B	A
>5400 m	–	10	70	10	10	–
5200–5400	–	16	74	5	5	–
5000–5200	–	12	71	4	8	6
4800–5000	–	12	72	4	6	5
4500–4800	2	12	69	7	4	7
4000–4500	3	17	61	7	2	9

Ph, small phanerophytes (shrubs); Ch, chamaephytes (dwarf shrubs, thorny cushions); H, hemicryptophytes (perennial herbs and graminoids); G, geophytes (bulb-, rhizome- or tuber-geophytes); B, biennial plants (mostly rosette plants); A, annuals. Note, sums are not always 100% because of rounding errors.

native alpine vegetation, were life form and realized growth form largely converge. A classic example is cushion plants. Alpine cushion plants are not shaped by the harsh alpine environment, as is often believed, but are "genetic" cushions i.e. they retain this growth form even when exposed to a milder climate (Rauh 1940, Spomer 1964). In contrast, grazing pressure in mediterranean, dry continental or subtropical mountains may lead to cushion shaped shrubs (or trees) which would immediately "abandon" this life form when relieved from their consumers.

The vegetation found in the alpine life zone is composed of ten principal groups of growth forms, eight of higher plants, two of cryptogams, irrespective of whether individuals perform clonal growth (Chap. 16). The first four groups are most important (see color Plate 1 at the end of the book):

- Low stature or prostrate woody shrubs,
- Graminoids such as grasses and sedges, mostly forming tussocks,
- Herbaceous perennials, often forming rosettes, and
- Cushion plants of various types.

Less common or of more regional importance are:

- Giant rosettes mostly in tropical mountains,
- Geophytes, mainly confined to seasonal mountains,
- Succulents, with both stem and leaf succulence, and
- Annuals (sometimes biannuals), which become quite rare at high altitudes.

Cryptogams, i.e. mostly desiccation-tolerant, non-flowering plants:

- Bryophytes ("mosses"), in some areas also ferns and lycopodioids, and
- Lichens (fruticous and crustaceous subsumed).

These growth forms, in mixtures of varying abundance of each, both clonal and non-clonal (Chap. 16), and of regionally rather different evolutionary (taxonomic) origin, compose the "alpine vegetation" (for plausible subdivisions see Billings 1988). In addition algae and fungi play an important role, but are not considered as "vegetation" here. As an example Table 2.3 illustrates growth form spectra for the Hindukush Mountains above 4000 m altitude. The dominance of perennial herbaceous species, in this case including graminoids, which together represent the "hemicryptophyte" category, is obvious and is characteristic for all alpine floras.

A number of alpine plant species are succulents and many of them perform, at least temporarily, the crassulacean acid metabolism (CAM, common in semi–arid lowlands). They inhabit dry niches in the alpine zone at all latitudes including the subarctic (for example *Sempervivum montanum* found at altitudes up to 3250 m in the Alps, Larcher and Wagner 1983; *Sedum lanceolatum* at 3730 m in the Rocky Mountains, Jolls and Bock 1983; *Echeveria columbiana* at 4100 m in Venezuela, Medina and Delgado 1976; *Opuntia lagopus* at 4500 m in Peru, Troll 1968; see also Chap. 11).

A very characteristic growth form of the tropical alpine life zone are the giant rosettes of the genera *Espeletia*, *Puya*, (Andes), *Senecio*, *Lobelia*, (Africa), *Cyathea* (a tree-fern, New Guinea), *Argyroxiphium* (Hawaii) and *Echium* (Tenerife) of subtropical mountains. The tropical giant rosettes in Africa and South America, reaching sizes of more than 3 m, strongly contrast with the surrounding low stature grass and shrub vegetation and attracted biologists for many years (review by Cuatrecasas 1986, Mabberley 1986, Smith 1994 and other contributions in Rundel et al. 1994). Hedberg (1964) concludes that giant rosettes show very little resemblance to a tree, and indeed, the patterns of dry matter allocation documented by Monasterio (1986) with ca. 26% of total dry matter of life biomass in leaves (leaf mass ratio, LMR) places them among particularly leafy herbaceous perennials (mean LMR 21% compared to > 5% in trees; Körner 1994). Giant rosettes are not restricted to tropical alpine floras, but also occur in warm temperate and subtropical semi-deserts (species of *Yucca*, *Xanthorrea*, *Aloe*, *Dracena*, *Aeonium*), in the lowland and montane humid tropics (Arecaceae and Pandanaceae) and in temperate climates (*Cordyline australis* and tree ferns). Specimen measuring more than 1.5 m exist even on the subantarctic Kerguel islands ("Kerguel cabbage", *Pringlea antiscorbutica*, Brassicaceae, Rauh 1978). One of the most impressive giant rosettes, reaching sizes of 3–4 m, is found near the treeline in Tasmania (*Richea pandanifolia*, Epacridaceae).

Annuals commonly do not represent more than 2% of the total alpine flora, and become increasingly rare with increasing altitude (Billings 1974, Jackson and Bliss 1982, Agakhanyanz and Breckle 1995). On average, two to three annual species reach the open scree zone at high altitudes in the temperate zone (often Brassicaceae and Caryophyllaceae).

For more detailed information on alpine plant life forms I refer to works by Troll (1968), Troll and Lauer (1978), Hedberg and Hedberg (1979), Klötzli (1991a) and Smith (1994). Halloy (1990) developed a (quite sophisticated) key to the morphologies of alpine plants, which proved to be useful for cross-continental comparisons. Consequences of plant structure for microclimate are a topic of Chapter 4. Because of the large physiognomic convergence of alpine floras, they represent a natural worldwide network of comparable low temperature vegetation for basic research in functional ecology and for monitoring changes of global dimensions.

3   Alpine climate

Which alpine climate?

The climate of a steep equator facing slope? The climate in a poleward exposed avalanche track? The climate of an alpine observatory on a windswept ridge? The climate where alpine life occurs, e.g. in the surface of a cushion plant ? In winter or in summer? Under overcast or clear sky conditions? In the tropics or in Alaska? Obviously THE alpine climate does not exist and meteorologists or biologists, a mountaineer or a nematode in the soil will have different views on the subject, and a number of biased assumptions about the alpine climate can be found in the literature. Here are some classics:

- "Solar radiation increases with altitude". If one traces this assumption back to the original texts such as the one by Sauberer and Dirmhirn (1958) one finds with surprise an explicit statement that this conclusion only holds for standardized weather conditions such as a clear sky. When full data sets, including all weather conditions, were considered, most often no such trend was found because of altitudinally increasing cloudiness (Fig. 3.1).
- "The alpine life zone is characterized by strong winds". This is a typical temperate zone perspective and wrong in this general form. There are strong latitude and exposure effects, with some isolated mountains in temperate and polar latitudes being truely wind beaten, while some of the highest mountains in the world in the subtropics and tropics and the inner parts of larger mountain systems are commonly quite calm (Grace 1977, Barry 1981).

- "Precipitation increases with altitude". While true in some, mainly humid temperate zone mountains, this is certainly not true for many subtropical and tropical mountains which show a mid altitude maximum and reductions at higher elevations (Flohn 1974, Lauscher 1976).
- "The alpine life zone is characterized by short growing periods", clearly, again a view valid for extratropical mountains only. If this were a key determinant of alpine plant characteristics, we should not find such characteristics in tropical mountains. For instance, if the general phenomenon of treeline formation were linked to season length we should not find upper treelines in the rather constant climates of the tropics.

Common features of alpine climates

Quantitative descriptions of the specific features of the alpine climate date back more than 200 years (see the impressive account on De Saussure's work by Barry 1978) and a number of reviews on mountain climatology are available today (e.g. Schimper 1898, Mörikhofer 1932, Sauberer and Dirmhirn 1958, Ives and Barry 1974, Fliri 1975, Troll and Lauer 1978, Franz 1979, Barry 1981, Marchand 1991, Rundel 1994).

To explain plant characteristics in the alpine zone we first need to ask: what are the truly **common attributes** of an alpine climate? Answering this question is of key importance, because it must be these attributes that matter if we want to explain those features of plants common to all alpine zones of the world, while in contrast, those more variable climate elements addressed above

Fig. 3.1. While valleys receive full sunshine, mountain tops are often cloud covered, which dampens or may even reverse the well known clear sky increase of total solar radiation with elevation (Swiss central Alps, upper Valais)

Table 3.1. The standard free atmosphere

Altitude (m)	Pressure (hPa)	Mean temperature (°C)	Air density (kg m^{-3})	Saturation vapor pressure (hPa = mbar)
0	1013	15.0	1.2250	17.1
1000	899	8.5	1.1117	11.1
2000	795	2.0	1.0581	7.1
3000	701	−4.5	0.9093	4.1
4000	616	−11.0	0.8194	2.4
5000	541	−17.5	0.7864	1.3
6000	472	−24.0	0.6601	0.7

Numbers represent a hypothetical state, approximating mean annual conditions at mid–latitudes (Barry 1981). Note that in the tropical atmosphere pressure is about 15 hPa higher at 3000 m and 20 hPa higher at 5000 m as compared to mid-latitudes (i.e. equal pressures are found at ca. 200–300 m higher altitudes in the tropics than assumed in this list).

are likely to be responsible for the differences in morphology and physiology which might be found among alpine floras. The global variability of some and the constancy of other attributes of an alpine climate represent a gigantic natural experiment, with "differential treatments" imposed for so long that selective processes can be expected to have evolved life zone specific biolog-ical answers, provided natural vegetation is considered.

Atmospheric pressure declines almost linearly with increasing altitude (in the lower few km, Table 3.1, Fig. 3.2). For example, at 2600 m altitude the total pressure is ca. 750 hPa (= mbar) compared with 1013 hPa at sea level, i.e. it is 26% lower at the higher altitude. At the upper limit of higher plant

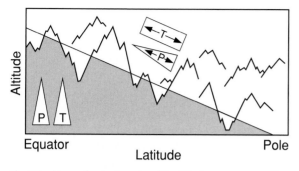

Fig. 3.2. Atmospheric changes with altitude common to all mountains of the world. Pressure (*P*) decreases with altitude and so does mean atmospheric temperature (*T*). Within the growing season, temperatures are similar in the alpine zone across latitudes (boundary of *shading*), whereas atmospheric pressure, and therefore the partial pressure of CO_2 (and oxygen) in the alpine life zone generally increases with latitude, because of latitudinally decreasing elevation of the alpine zone. (Körner et al. 1991)

life (5500 to 6000 m in the Himalayas and Andes) plants and animals live at about half the pressure found at sea level. The partial pressures of CO_2 and oxygen are reduced exactly by the same proportion, because the atmosphere is well mixed in the range of altitudes of interest here, so that the mixing ratio of dry gases does not significantly change with altitude (e.g. Zumbrunn et al. 1983). In other words, the mean CO_2 concentration expressed in units like ppm, $\mu l\,l^{-1}$, $\mu mol\,mol^{-1}$, $\mu bar\,bar^{-1}$ or $Pa\,MPa^{-1}$ is always the same, in 1996 ca. 360 ppm (currently rising by ca. 1.4 ppm a^{-1}). At sea level this value corresponds to a partial pressure of 36 Pa (= 360 μbar). The difference between mixing ratios (which do not significantly change with altitude in the range of interest here) and partial pressures (which do change) often causes confusion. The following two examples further underline the significance of this distinction for high altitude ecology.

Above, I selected 2600 m as an example because at this altitude the reduction of the partial pressure of CO_2 corresponds approximately to the difference in partial pressure of CO_2 at sea level between the year 1800 (ca 28.5 Pa) and 1996 (ca. 36 Pa). This means that plants living at 2600 m today experience a partial pressure of CO_2 that

lowland plants had experienced at the beginning of the industrial revolution. Another, more extreme example: it is known that plants lived on equator facing slopes of mountain peaks ("nunatakkers") which protruded through the ice shields during the last ice age when the partial pressure of CO_2 was 18–19 Pa at sea level (ca. 18 000 years before the present, according to ice core data). If they lived at that time at the altitudes they inhabit today (ca. 5000 m), such plants would have had to cope with a partial pressure of CO_2 of ca. 9–9.5 Pa (or four times less than found at sea level today). Would this have been enough for their survival?

Pressure reduction also increases molecular diffusivity (less likelyhood of molecular collisions). De Saussure (1779–1796 in Barry 1978), concludes from his experiments on Mt Blanc: "other things being equal, a decrease of about one third in air density causes more than a doubling of the evaporative amount" (not to be referred to quantitatively). Of course "other things" are not equal as noted already by De Saussure. Temperature is much lower at high altitude, which reduces the rate of molecular diffusion, and the capacity of the air to take up moisture is drastically diminished (see Table 3.1 and below) and the effect requires absence of any convective transport (wind) to become fully expressed, as in the interior of leaves (Gale 1972), through egg shells (Rahn et al. 1977) or in any undisturbed aerodynamic boundary layer of an evaporating surface. Thus, it depends on the situation as to how much the pressure reduction will actually stimulate evaporation or other gas fluxes.

The so-called adiabatic lapse rate (i.e. the reduction) of **atmospheric temperature** with altitude varies between annual means of ca. 0.8 K per 100 m of altitude in coastal areas or on island mountains and ca. 0.4 K per 100 m in more continental areas, but extremes of close to 1 and zero, and even negative lapse rates under conditions of winterly temperature inversions in valleys are found during certain periods of the year. During the growing season of temperate zone mountains the gradient is usually steeper than during winter.

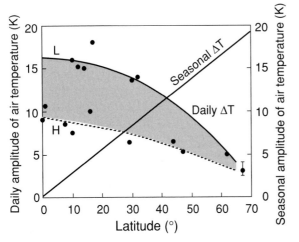

Fig. 3.3. The latitudinal variations in diurnal and seasonal amplitudes of air temperature at high elevation stations, based on data compiled by Lauscher (1966), Rundel (1994) and from the Abisko Station in North Sweden. Diurnal amplitudes are annual means, seasonal amplitudes are calculated from monthly means. *L* Low humidity (low cloudiness), *H* high humidity (high cloudiness)

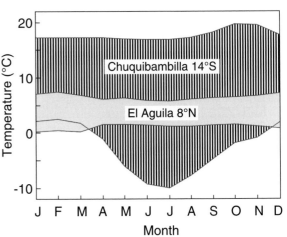

Fig. 3.4. The seasonal variation of minimum and maximum monthly mean air temperatures (amplitude *shaded*) for two high Andean stations of contrasting continentality. El Aguila (4120 m) experiences frequent cloudiness, in contrast to dry continental Chuquibambilla (3910 m). Note that regional climate effects can surpass latitudinal effects. (Sarmiento 1986)

A lapse rate of 0.60 K* per 100 m seems to be the best estimate for both the temperate zone summer (e.g. the central Alps) and year round in the tropics. Annual means for 2400 to 4500 m in New Guinea (extremely humid) and 2700 to 4800 m in tropical Chile (arid) are almost identical: 0.61 and 0.60 K per 100 m (data compiled by Körner et al. 1983; Lauscher 1976/1977). By pooling data from various tropical regions Rundel (1994) also arrived at a mean of 0.60 K per 100 m.

The altitudinal lapse rate of temperature also undergoes substantial diurnal fluctuations because the diurnal amplitude of air temperature at low altitude is greater than at high altitude (Lauscher 1966, Barry 1981, Fig. 3.3). However, diurnal amplitudes in air temperature in tropical mountains are not always much greater than in the temperate zone, as is often stated ("winter and summer every day"; Fig. 3.4). This view emerged

from experience in relatively dry tropical and subtropical areas were amplitudes of more than 15 K are observed at alpine altitudes (Rundel 1994). In humid tropical mountains (and on island vulcanoes) the amplitudes differ relatively little from those in mountains from higher latitudes. Furthermore, data from the temperate zone are mostly taken from wind-beaten summit stations such as Zugspitze in Germany or Sonnblick in Austria, whereas those from the tropics are commonly from less "ventilated" high valleys (Lauscher 1966). Finally, and particularly relevant to plant biology, the periods considered for averaging exert a great influence on such comparisons. While mean monthly amplitudes for the free atmosphere in temperate zone mountains are around 5 K at 3000 m throughout the year (Lauscher 1966), amplitudes under mid-growing season weather conditions comparable to those prevailing in the drier tropics year round (bright weather with clear nights) can be quite similar to those reported for some tropical mountains. More than 15 K amplitudes in temperate zone mountains can occur, but the frequency of such condi-

---

* In accordance with the International System of Units "K", the "Kelvin", is used to describe *differences* of temperature, independently of the practical use of a relative "grade" scale (°C) for temperature.

tions is very much lower, which is, however, of little relevance when biologically decisive singularities are concerned. Such signals are drowned in means which are dominated by the more frequent bad weather conditions in many higher latitude stations. Of course, the seasonal amplitude of monthly mean air temperature increases almost linearly with latitude (Fig. 3.3).

A reduction of atmospheric temperature automatically reduces the **moisture content** of vapour saturated air (the saturation vapour pressure, Table 3.1) and thereby reduces the vapour pressure deficit at any given relative humidity. As a consequence, half of the total atmospheric moisture of the earth is contained in the lower 2000 m of the atmosphere. Unless transpiring surfaces heat up during periods of strong radiation (see Chap. 4) or are warmer than the air for other reasons (e.g. human skin) the driving forces for evaporation of water are reduced at high altitude, despite increased diffusivity.

A word on relative humidity: this common descriptor of air humidity (the ratio of actual to saturating vapour pressure of air at a given temperature times 100) can be particularly misleading in high mountains. Based on this measure "low humidity at high altitude"(under certain weather conditions or in semi-arid mountains) has often been suggested to enhance evaporation. However, at low temperatures the saturation vapour pressure is low (see above) and relative humidity may be as low as 10% while the actual vapour pressure deficit (i.e. the capacity of air to take up more moisture) may be negligible. Vice versa, the prevailing "high humidities in high mountains" (which meteorological statistics indicate for most mountain areas) represent little absolute moisture content compared with lower altitudes for the same reason. As will be discussed later, the specific evaporative demand in the alpine zone can only be assessed by accounting for surface heating due to solar radiation (see Chap. 4).

The **radiation climate** exhibits some features common to all mountains (Sauberer and Dirmhirn 1958, Dirmhirn 1964, Barry 1981): areal fluxes of solar radiation increase with altitude under both a clear sky (reduced atmospheric tur-

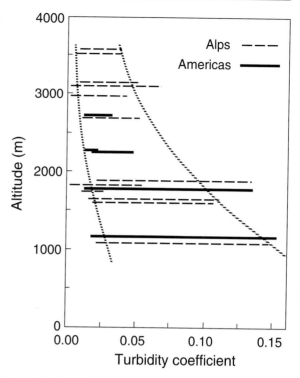

Fig. 3.5. A global comparison of altitudinal ranges in atmospheric turbidity. The dotted boundary illustrates the overall trend. (Volko 1971)

bidity, Figs. 3.5 and 3.6) and overcast conditions (thinner cloud layer) and so do high radiation extremes, but also the relative frequency of local cloud cover increases with altitude (Fig. 3.6), counteracting these trends (see below). At any given global radiation, the UV fraction is increased at high altitude (Caldwell 1968, Caldwell et al. 1989, Blumthaler et al. 1993) with gradients, depending on region, season and solar angle, ranging from almost similar to twice as steep as for global radiation. The annual or seasonal dose may increase less with altitude (or not) when the greater cloudiness at high elevation is accounted for. Another common feature of the radiation climate at high altitudes, again a consequence of reduced turbidity, is the reduced fraction of diffuse radiation under clear sky conditions, which enhances the contrast between sun exposed and shaded surfaces (Sauberer and Dirmhirn 1958). In contrast, thinner cloud layers at high altitude enhance

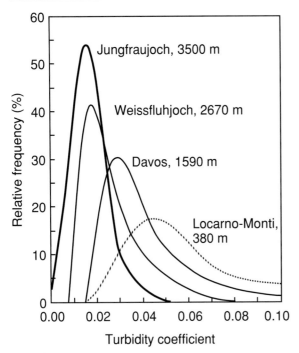

Fig. 3.6. Relative frequency of atmospheric turbidity (expressed as the extinction coefficient at 500 nm) for different altitudes in the Alps. (Volko 1971)

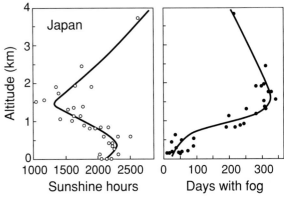

Fig. 3.7. Long-term sums of global radiation or, as here, numbers of sunshine hours do not follow any uniform pattern along altitudinal gradients in the mountain systems of the world. They may increase, or decrease with altitude or, as exemplified here for the Japanese mountains, decline for the first 2 km and then increase, following the patterns of cloudiness or fog. (Yoshino 1975)

diffuse radiation under overcast conditions. Low turbidity also enhances radiative heat losses during clear nights which may have a significant impact on plants (see Chap. 8). Common to all extratropical alpine regions is the strong influence of snow on plant available solar radiation, either by snow cover (reduction) or by reflection (enhancement; see Chap. 6).

Regional features of alpine climates

A number of climate characteristics in alpine areas reflect regional rather than global patterns, which complicates functional explanations of attributes of alpine plants. Besides the obvious latitudinal differences in seasonality, altitudinal patterns of long term integrals of solar radiation, precipitation and wind vary substantially among different mountain systems. For instance, means or sums of

radiation may not change with altitude or may even decline. The latter is evident for instance for the mountains of Scotland (Barry 1981, p. 265), mountains in Japan (Fig. 3.7) and most extremely, the humid tropical mountains of New Guinea (possibly also true for the Ruwenzori Mountains in Africa) where the alpine zone receives only one third of the radiation measured at low altitude due to almost permanent cloud cover (cf. Körner et al. 1983).

No significant altitudinal changes in growing season means of solar radiation are found in the Alps (Tranquillini 1960; Figs. 3.8 and 3.9), the Rocky Mts. (Caldwell 1968) and the mountains of New Zealand (Table 3.2). Altitudinally increasing radiation sums can be found in arid regions. For the desert mountains of Chile and northwest Argentina (21–22°S) a 6% increase per 1000 m in the 2–5 km elevation range is estimated by Lauscher (1976/1977).

It is a common observation, that radiation sums do not change as much with latitude as midday maxima (related to solar angle) do, because the daylength increases with latitude. Comparative measurements of quantum flux density (QFD) in

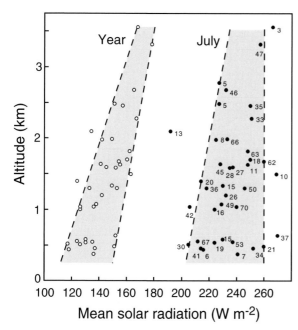

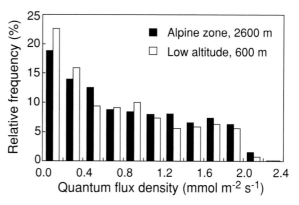

Fig. 3.9. Frequency distribution of quantum flux density (QFD, 400–700 nm, only hours with QFD >30) at 2600 and 600 m altitude in the Austrian central Alps near Innsbruck for the main part of the growing season at each altitude, i.e. April to mid-June at low and mid-June to August at high altitude. Both the frequency distribution and the sums of QFD (not shown) are similar, but higher maxima of QFD occur at high altitude. (Körner and Diemer 1987)

Fig. 3.8. Solar radiation in the Swiss Alps for 37 stations between 273 and 3580 m altitude (47°N), possibly the densest network of stations measuring solar radiation in a mountainous region (Swiss "A-net"). Only stations close to or within the Alps, but from both the northern and southern ranges, and from the center were used. *Left* Annual means; *right* July means (calculated from daily sums for the 10-year period between 1983 and 1992, the *shaded area* indicates the range for ca. 90% of the data). Note the absence of an altitudinal trend in July, but a 10% increase per 1000 m in annual means, because of more frequent fog and greater screening of the horizon by mountains at low altitude during winter. (Meteotest 1995)

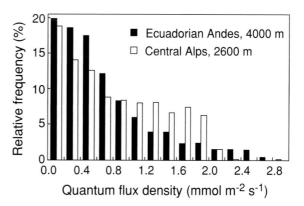

Fig. 3.10. The frequency distribution of quantum flux density (400–700 nm) at biologically comparable altitudes in the Ecuadorian Andes (0° latitude) and the central Alps (47°N): pooled data for 5 months of measurements on Guagua Pichincha (4600 m) and Páramo de la Virgen (4000 m) and growing season data (mid June to end of August) for the Central Alps, Mt. Glungezer 2600 m. (Diemer 1996)

the photosyntheticly active part of the spectrum (400–700 nm) over the whole alpine growing season indicate only small differences in QFD sums between the Alps and an arctic-alpine site at the Scandinavian polar circle (Prock and Körner 1996): 14 h day length (only hours with QFD >30) with a mean of $750 \mu mol\, m^{-2} s^{-1}$ in the Alps versus a >21 h day length with a mean of $415 \mu mol\, m^{-2} s^{-1}$. The greater number of hours ith weak light in the north partly outweighs greater intensities in the south. There is little difference with respect to sums of QFD measured at 4000 to 4600 m in the equatorial Andes (which are much

drier, and thus less cloudy than humid tropical mountains) and at a biologically comparable altitude in the European central Alps (2600 m, Fig. 3.10).

Table 3.2. Summary of daily means of global radiation at lowland and alpine altitudes for the summer period in New Zealand (December–February, 43°S) and the Alps (June–August, 47°N). Compiled from different sources by Körner et al. 1986)

Site	Altitude (m)	Daily mean global radiation (MJ m^{-2} d^{-1})	Observation period
Invercargill	0	20.1	1977–1980
Christchurch	30	22.0	1977–1980
Mount John	1027	23.1	1977–1980
Old Man Range	1220	21.5	1976–1977
Craigieburn Range	1550	20.7	1969–1974
Central Alps (Tirol)	2000	20.0	long term

Note: treeline is situated at 1200 m in this part of New Zealand and at 2000 m in the central Alps of Tirol.

While integrated values of solar radiation are important for photosynthetic CO_2 fixation, maximum values or higher intensities in shorter wavelengths may have substantial influence on morphogenesis and many other life processes. Under certain sky conditions the solar radiation received by alpine vegetation may exceed the "solar constant", i.e. the intensity measured in the direction of the sun, outside the atmosphere (ca. 1360 W m^{-2}, Barry 1981). This can happen when a thin layer of clouds (strato-cirrus) leaves a gap in the direction of midday sun (Fig. 3.11). In this case, a plane surface receives the direct solar radiation of e.g. 1000 W m^{-2} or a QFD of ca. 2000 μmole quanta m^{-2} s^{-1} plus the intense diffuse radiation from the thin clouds surrounding the gap, e.g. 500 W m^{-2} or 1000 μmol m^{-2} s^{-1}, summing up to 1500 W m^{-2} or 3000 μmol m^{-2} s^{-1}, values actually measured by the author in midsummer at 2600 m altitude near Innsbruck. Turner (1958a) measured 18% more than maximum clear sky radiation under such conditions near the treeline. Reflections from surrounding snow may further enhance the radiative load in certain areas. Alpine plants have to cope with such short term extremes, unlikely to occur at lower altitudes.

Similar to the situation with solar radiation sums, there is also no general pattern for changes in **precipitation** sums with altitude. The most common observation at lower altitudes is an elevational increase. In most areas of the temperate zone, such as in the Scottish mountains, the Alps, but also in the high mountains of central

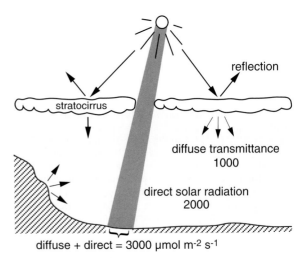

Fig. 3.11. Leaves and flowers of alpine plants may receive short term extremes of solar radiation exceeding the extra-terrestrial solar constant when direct beam and high intensity diffuse radiation overlap as it happens when the sun shines through a gap in a thin cloud layer. (Unpublished QFD measurements in the Alps by the author). Adjacent snow banks may further enhance the radiation

Asia, Japan, New Zealand and Australia, this trend continues to the highest altitudes (Flohn 1974). However, in many subtropical and tropical mountains, precipitation tends to decrease above a mid-altitude maximum (Lauscher 1976, Rundel 1994; Hemp and Beck 2001). Intuitively one would expect the reduced atmospheric carrying capacity for moisture at high altitudes (see above) also to reduce precipitation as it does in many tropical mountains. Flohn explains this latitudinal difference by the overriding advective moisture deposition in the windier high latitudes compared with the more pronounced local convective moisture transport in equatorial latitudes. However all sorts of exceptions may be found (see Lauscher's 1976 review). The tropical cordilliera of New Guinea is soaked in moisture brought in by the trade winds up to the highest peaks (see references in Körner et al. 1983). Some peaks in the Andes in northwest Argentina receive significant moisture only above 4500 m. Except for truly arid mountains, these differences are of relatively little relevance for plant water supply (in the strict sense), since the most common situation in the alpine zone is an overabundance of moisture as rated by evaporative demand even in subtropical mountains with as little as 300 mm of annual precipitation (see Chap. 9). Consequences of snow pack, snow duration, influences on the nutrient cycle (which is severely impaired by low top soil moisture) and radiation deficits due to clouds or fog appear to be the biologically more significant effects of these different precipitation patterns.

With respect to the **wind** climate, the reader is referred to the extensive treatments of the subject by Grace (1977) and Barry (1981). The latter publication in particular makes it quite clear that the general notion of high mountain environments being particularly windy does not match the data. As Barry states "the most important characteristics of wind velocity over mountains are related to their topographic, rather than their altitudinal effects". Of course, wind speeds up to $200\,km\,h^{-1}$ are measured at some mountain stations on isolated peaks, and plants living on exposed ridges in certain mountain regions are facing severe mechanical wind stress, just as many coastal plants do. However, the friction caused by mountain topography affects the flow of air masses, causing the climate within mountainous areas to be less windy than in the plains (data compiled for the Alps by Fliri 1975; examples for the Andes in Lauscher 1976/1977), and at smaller scales the relief creates additional shelter for most of the alpine vegetation. Whoever has visited the vast alpine areas in the mainland tropics and subtropics will have noticed the relatively calm situation. Mountain islands in the Roaring Forties are not representative for the majority of alpine land area, but they exemplify an interesting extreme situation that adds to the multitude of "alpine climates" in the world.

Table 3.3. Comparison of climatic conditions in arctic-, temperate- and tropical-alpine areas that support similar short stature alpine vegetation. Data compiled from various sources by Körner and Larcher (1988)

Area[a]	arctic	temperate			tropical	
		oceanic	semi-oceanic	continental	humid	mesic-dry
	A	B	C	D	E	F
Latitude	71°N	44°N	47°N	40°N	0°/5°S	8°N/10°S
Elevation (m)	5	1600	2600	3500	4400	4100
Elevation of upper forest line (m) (4100)	–	1200	2000	3400	3500	3200
Mean length of growing season (days)	70	70–100	70–80	80	365	365
Mean daily global radiation in July (A,B,C,D) and total year (E,F) ($10^6\,Jm^2$)	18.1	ca.20	20.0	20.2	ca. 15	ca. 21.5
Growing season photoperiod (hd^{-1})	24	16	15.5	15	ca. 12	ca. 12
Mean air temperature in the warmest month (°C)	+4	+4	+5	+8.5	+3	+3/+4
Mean soil temperature in the warmest month (°C; 10–25 cm depth)	+2.4	+8/+10	+7	+13	ca. +5	ca. +5

[a] A, Alaska (Barrow); B, Southern Alps of New Zealand; C, Austrian Central Alps; D, Rocky Mountains (Niwot Ridge); E, Mt. Wilhelm, Papua New Guinea or Izombamba (3050 m, radiation only), Ecuador; F, Andes in Peru and Venezuela.

Other aspects of the alpine climate, particularly those relating to snow distribution, biologically effective temperature, radiative cooling or heating and evaporative forcing need to be discussed in a microclimatological context (Chaps. 4, 6, 7 and 9), because the atmospheric conditions measured above the vegetation in the alpine zone most often have little in common with those at vegetation level.

In summary, the alpine (macro-) climate above the vegetation shows a number of common, but also a number of different features across the globe. The most important common components are reduced pressure and reduced atmospheric temperature with the associated reduction of vapour pressure deficits. High maximum solar radiation and a greater short wave contribution, but otherwise rather similar radiation doses across altitudes and latitudes (with a few exceptions) are another common characteristic. In a global perspective, neither precipitation nor wind exhibit altitudinal patterns which can be called as typically alpine. Table 3.3 summarizes approximate values for some key atmospheric variables together with soil temperature data (which lead to the next chapter) obtained from altitudes with comparable alpine vegetation for the major climatic zones.

4   The climate plants experience

Walking across an alpine fell field on a bright day, the mountaineer may find it still chilly and windy, and "dry" air may desiccate the skin. A nearby meteorological station with all its weather masts confirms: +4 °C air temperature, 5 ms^{-1} wind speed, 40% relative humidity. While our mountaineer experiences the harsh life in the mountains, the world looks different for those who stay close to ground. Micrometeorological research in alpine vegetation has shown that, under the conditions described above, the climate within a compact leaf canopy 1–2 cm above the ground, for instance among cushion plants, may be +27 °C, 98% relative humidity and no convective air movement – life in a humid tropical forest? Seed beds of alpine plants on dark humic soils exposed to full sunlight have been shown to heat up to temperatures of around 80 °C (Alps of Tirol, Turner 1958b; Australian Snowy Mountains, Körner and Cochrane 1983). Such temperatures are sterilizing the topsoil – life in hot deserts? How cold is the cold climate of alpine plants?

It all depends on three components:

· Solar radiation
· Slope and exposure
· Plant stature

Additional components which modulate the interplay of the three main drivers are wind velocity, ambient air temperature above the vegetation and soil properties such as surface structure, moisture and thermal conductivity.

Solar radiation in the alpine zone, as discussed above, is stronger both under clear and overcast conditions, but the greater abundance of overcast conditions may cut the overall dose substantially (Chap. 3). The above examples of ground warming are for a clear sky. Solar radiation of less than 20% of full midday intensity, such as under moderately thick clouds, has only a small heating effect and below 10%, the radiant warming effect vanishes (the situation under dense clouds or during early morning and evening hours). Almost no links exist between air temperature and temperature below snow. During clear nights, plant surfaces may cool several K below the air temperature due to radiative heat loss. The spectrum ranges from extremely positive to substantially negative deviations in vegetation temperature from "met-station" temperature. However, a large, often overlooked, fraction of alpine plant life occurs at temperatures almost identical to air temperature, the periods of rain, fog, dense cloud cover and all overcast nights. Hence, besides weather, the differences between the temperatures which plants experience and air temperature depend on the time of year, time of day, type of plant organ and the particular plant function considered.

Interactions of relief, wind and sun

Though trivial, one of the key elements of alpine plant life are slopes which can cause, anywhere in the world, equatorial solar incidence angles or permanent shade, with all possible intermediate sun-surface angles and their seasonal and diurnal variation (Fig. 4.1). Direction of exposure adds to the multitude of possibilities and may enhance diurnal patterns (for extensive reviews see Geiger 1965 and Barry 1981). However, sun is not the only vector interacting with slope or exposure. Wind

Fig. 4.1. Above the treeline relief, i.e. slope and exposure, become major determinants of the climatic conditions experienced by plants

and gravity are the two other ones. While effects of gravity are more important for soil processes, wind – in interaction with the relief – influences the microclimate directly by affecting the aerodynamic boundary layer, and thus affecting convective heat loss, evaporative cooling and the distribution of precipitation, snow in particular. Of course, these topographic effects are not restricted to high mountains, but their influence is more pronounced because slopes tend to be steeper, climatic vectors are stronger, and, most importantly, because plant life becomes more dependent on decoupling from a "hostile" atmosphere the higher the altitude. In the following I will briefly discuss two topographic effects: (1) the direct influence of exposure on microclimate during the growing season and (2) the more complex indirect effects of relief on the combination of snow distribution and slope specific interception of radiation.

All relevant texts stress the importance of **slope exposure** for microclimate (e.g. Geiger 1965; Isard 1986), but data for truly alpine situations are not very abundant. This situation may change with the wider use of the new generation of micro-dataloggers (Fig. 4.3). When this text was written I had knowledge of only one published data set of full season plant-temperature measurements in the alpine zone, collected from the same altitude but at steep slopes of contrasting orientation, the work by Moser et al. (1977) discussed below. The data by Bliss (1956) in the Medicine Bow Mountains include slope exposure as a variable, but these were rather gentle slopes, which exerted no pronounced climatological differentiation. However, Bliss's data revealed a rather important interaction between exposure and local weather. His north-facing slope had slightly warmer rather than cooler ground temperatures than the south-facing slope. This unexpected phenomenon resulted from common bright mornings and cloudy afternoons, as observed in many mountains, which caused the north slope to profit from very early sun, whereas the south slope had little advantage from afternoon insolation.

Moser's data, collected at 3184 m altitude in the Alps, represent a situation near the upper limit of plant life (mean air temperature, measured at 2 m above the surface during the warmest month of the investigation was 0.2 °C). It is important that Moser measured temperatures on leaves of the same plant species (*Ranunculus glacialis*), but at contrasting slope directions. Leaf temperatures during the growing seasons between 1968 and 1972 varied between the absolute extremes of −12 and +44 °C. Rather complete data sets for south slope, north slope and ridge for the three warmest months of 1968 (Table 4.1) indicate that south slope and ridge plants experienced 60% more hours in the 0 to 15 °C range than the north slope plants, which in turn experienced 75% more hours in the −5 to 0 °C range. While about 3% of all hours at the ridge and south slope fell in the 15–30 °C range (even 0.5% of all hours >30 °C), such temperatures never occurred at the north slope. The −6 °C temperature, which is critical for freezing damage in summer-active leaves (see Chap. 8), was measured repeatedly on the north slope and on the ridge (radiation freezing) but never occurred on the south slope. It needs to be mentioned that the "ridge" in fact was a narrow rocky plateau with plants growing in sandy patches between the rocks. The horizontal distance between the plants studied on the north and south slopes was 10 to 15 m.

Clear summer days are quite rare in the Alps and even rarer on high peaks such as Hoher Nebelkogel. Since temperatures are very low during the prevailing periods with weak radiation (70% of all daylight hours with less than one third of clear sky midday radiation) slope effects become particularly important during the short remaining periods of high radiation. Figure 4.2 illustrates leaf temperatures of *Ranunculus glacialis* during a bright day in July (the warmest month) – the thermal deficit on the north slope is obvious.

The link between slope inclination or slope direction and the interception of solar radiation is so obvious that topographic maps have been used to predict incoming radiation per unit ground area

Table 4.1. The influence of micro-habitat exposure on leaf temperature in *Ranunculus glacialis* on Hoher Nebelkogel (3184 m, Tirolian Alps). Numbers are the relative frequencies (%) of hours (100% = ca. 1800 h) in each of five temperature classes during the summer of 1968 (Moser et al. 1977)

Leaf temperature classes (°C)	<−5	−5 to 0	0 to 15	15 to 30	>30
North slope	3.1	57.9	38.5	0.5	0
Ridge	1.2	30.4	63.9	3.5	0.5
South slope	0	35.4	61.5	2.9	0

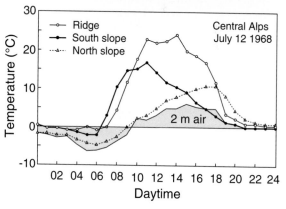

Fig. 4.2. Life conditions of *Ranunculus glacialis* of 3184 m altitude in the Alps (Hoher Nebelkogel, Tirol). *Above* Snow in midsummer; *below* leaf temperatures at microhabitats differing in slope direction during a clear day (July 12th, 1968; Moser et al. 1977)

and to draw "radiation maps" (see Garnier and Ohmura 1968, for principles; field data sets e.g. by Williams et al. 1972, for the peak area of Mt. Wilhelm in New Guinea, and Isard 1983, for the

Indian Peaks in the Rocky Mts.). With the availability of rather precise geographical information systems for mountains (Price and Heywood 1994) digitized topographic maps can now be converted to radiation maps which also allow estimation of the distribution of latent and sensible heat flux and surface temperatures on slopes. Remote sensing in combination with radiative transfer models permits mapping of the radiation balance, and from that, or directly by scanning thermal bands, the topographic effects on surface temperatures on landscape scales can be derived. By using data from the Landsat Thematic Mapper, Duguay

(1994) produced energy balance maps of Niwot Ridge in the Rocky Mountains which show strong contrasts between north- and south-facing slopes. Ground data, such as those shown in Fig. 4.3 underline the large thermal contrasts plants may experience at equal elevation but with different exposure.

The patterns of **snow distribution** in structured alpine terrain are the most visible consequences of relief and its interaction with climatic vectors. Particularly in spring, snow melt figures on slopes document the conservative nature of these patterns, which have even been given names by local

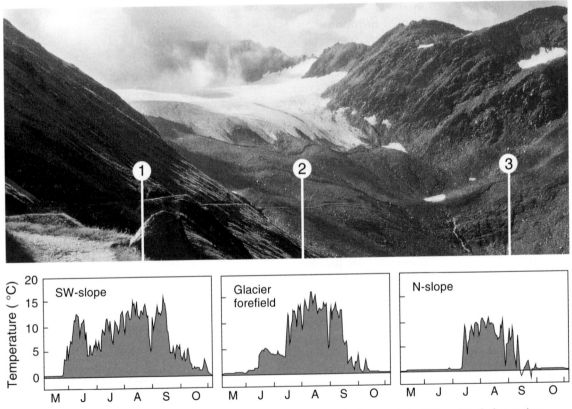

Fig. 4.3. The annual course of root zone temperature (10 cm below the surface) at precisely 2500 m altitude, but at sites contrasting in slope exposure. *1*, SW slope with tall alpine grassland; *2*, almost horizontal glacier forefield with sparse pioneer vegetation; *3*, steep N slope with fellfield vegetation. Unpublished data from 1996/1997 near the Furka Pass, Swiss Alps, ca. 300 m above the climatic treeline.

Fig. 4.4. Snowmelt figures in alpine terrain are spatially conservative and determine plant distribution (location as Fig. 4.3)

this field by his classical mapping and monitoring work. It is this sort of research where results may have been assumed to be self-evident, but where someone finally did the very detailed, cumbersome analysis, which has replaced assumption, intuitive knowledge or common belief by hard rock evidence. Unfortunately the work was not published in a prominent place and did not receive the attention it deserved. Re-investigating his plots in the Ötztal valley of Tirol half a century later might yield valuable long-term data.

Friedel (1961) carefully mapped vegetation mosaics, snow depth and the spatial dynamics of snow melt on strongly structured steep terrain covered by various alpine, mainly dwarf shrub, associations and found very close correlations between vegetation boundaries and contour lines of mean dates for snow disappearance (Fig. 4.5). The patterns of these contour lines did not change over the years, but the dates when they were reached did vary.

A number of conclusions emerged from Friedel's work which apply to plant distribution in orographically structured alpine terrain in general, and which confirm what earlier researchers have intuitively assumed. Friedel noted:

- Boundaries between units of low stature alpine vegetation are mostly sharp rather than gradual and can be defined with an accuracy better than half a meter.
- When relief comes into play, climatic isoline distances become narrower and so do the boundary lines between different vegetation units.
- Wind- and relief-controlled snow distribution can cause the dates of snow disappearance in spring to vary by 4 months, and the inter-annual variation of these patterns can be one month. Note: Friedel worked at the treeline ecotone with a 6 month growing season; much larger or ecologically even more significant temporal variations my occur at higher altitudes where Moser et al. (1977) found that some plants may miss out one or two complete summers due to snow pack.

people (Fig. 4.4). However, while these melting patterns, particularly in later stages are quite consistent over the years, winter snow depth is rather variable and the timing of melting also changes from year to year and also across topographic profiles, depending on wind direction during late snowfalls. A substantial literature documents the fundamental influence of these snow distribution patterns on alpine vegetation (for example Gjaerevoll 1956; Billings and Bliss 1959; Nägeli 1971; Douglas and Bliss 1977; Helm 1982; Williams 1987; Ozenda 1988; Barbour et al. 1991; Holtmeier and Broll 1992; Kudo and Ito 1992; Komarkova 1993; Walker et al. 1993; Stanton et al. 1994). One of the most detailed analyses is that by Friedel (1961). He formulated the law of "constancy of spatial patterns despite varying temporal patterns" (by extending what researchers before him have called "the conservative distribution of snow", cf. Gjaerevoll 1956) and certainly pioneered

Snowmelt patterns   Vegetation zones

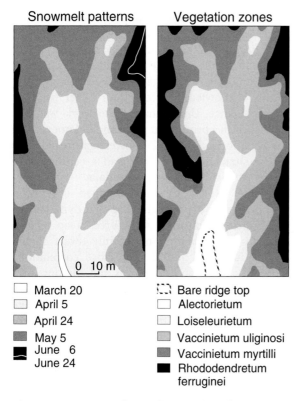

March 20
April 5
April 24
May 5
June 6
June 24

Bare ridge top
Alectorietum
Loiseleurietum
Vaccinietum uliginosi
Vaccinietum myrtilli
Rhododendretum ferruginei

Fig. 4.5. Convergence of vegetation mosaics and contour lines of snowmelt in spring on steep terrain in the Ötztal Alps of Tirol. The plant sociological units indicate the importance of certain species in each zone. (After Friedel 1961)

- Despite some inter-annual fluctuations, the overall pattern of snow distribution is "rigidly "engraved" into the alpine plant cover ("Darum liegt die räumliche Variation des Isolinienfeldes als eisernes Gesetz im Gelände eingeschrieben").
- He re-emphasizes that, at high altitudes, the spatial variability in life conditions becomes more important than the temporal variability (the latter being the usual focus of climatology), and that
- elevation per se becomes a less useful predictor of life conditions the higher one ascends in the mountains.

The last two and most important statements have been made repeatedly over the years before Friedel did his careful analysis, and have found support in many later studies (see the above references; most impressively by Gottfried et al. 1998). They are also fully in line with Moser's data from the upper limit of higher plant life discussed earlier.

Plants exhibit rather specific micro-environmental requirements that are affected by relief, and duration of snow cover is one of the most important ones. Some species profit while others suffer under late lying snow, and these requirements determine the spatial distribution of species, their vigor of growth after snow melt and their reproductive behavior (e.g. Billings and Bliss 1959; Eddelman and Ward 1984; Galen and Stanton 1995 and references therein). Some of the mechanisms linking snow and plant distribution will be discussed in Chapter 5. One micrometeorological study which directly relates to Friedel's work, because it considers the same type of toposequence of ericaceous dwarf shrub communities in the lower alpine zone, is the analysis by Cernusca (1976) – possibly the most detailed ever conducted on alpine plants.

Cernusca worked along a 70 m catena (at the treeline in the central Alps near Innsbruck, 1950 m) ranging from a *Rhododendron ferrugineum* gully, packed by up to 3 m of winter snow, to a windswept ridge with prostrate mats of *Loiseleuria procumbens* which rarely see any snow. In contrast to Moser's work, Cernusca measured temperatures (and other parameters) in very different plant canopies (tall versus prostrate) on the same slope, but at positions differing in wind exposure. Taken together, these two studies from the upper and lower ends of the alpine life zone illustrate the topography-induced amplitude of deviations of plant temperature from air temperature above the treeline.

The data shown in Figure 4.6 (similar to those in Fig. 4.3) substantiate the well-known protective effect of snow in winter (Sakai and Larcher 1987; see Chap. 5). Almost perfectly stable temperatures of 0 °C persist over 7 months in the canopy of evergreen *Rhododendron*, the tallest of all species, and known to be unable to survive without such protection (Larcher and Siegwolf 1985). Life at freezing point persists for 6 months in stands of summer-green (and half as tall) *Vaccinium myrtillus* and is reduced to a period of only one month at the wind exposed end of the catena, where carpets of the evergreen *Loiseleuria* cover the ground. As a consequence of the lack of snow cover, and thus full exposure to solar radiation during the day and night-time cooling, *Loiseleuria* experiences canopy temperatures between January and March of −10 to +30 °C, in harsh contrast to 2 m air temperatures which never exceeded +6 °C during this period. *Loiseleuria* is well adjusted to cope with these life conditions by maintaining deep physiological dormancy irrespective of periodic canopy warming (Körner 1976; Grabherr 1976; Larcher 1977). Clearly, wind exposure and relief exert enormous influences on conditions for life, answered by nature by stepwise replacement of vegetation types, such as the ones mapped by Friedel (1961; Fig. 4.5).

The growing season data in Figure 4.6 contain a big surprise: one might have expected for sunny weather that topographic shelter from wind favors canopy warming, and wind exposure dampens it, whereas the opposite was found. *Rhododendron* experiences a much cooler summer than *Loiseleuria* does, even though both grow at similar altitude and overall slope orientation to the sun (mean 2 m air temperature of the warmest month 7.5 °C). Here relief is no longer able to explain microclimate. What these data illustrate is the overruling influence of plant stature and canopy structure – a topic considered in more detail later in this chapter.

Many other aspects of the alpine climate as experienced by plants are driven by relief. Mechanical effects of wind on wind edges may influence virtually all aspects of plant growth (Grace 1977; Bell and Bliss 1979; Biddington 1985). The abrasive and shaping effects of wind, as described for trees at the treelines of some temperate zone mountains or on islands are well known (Tranquillini 1979; Marchand 1991), but plants of lower stature growing on exposed sites

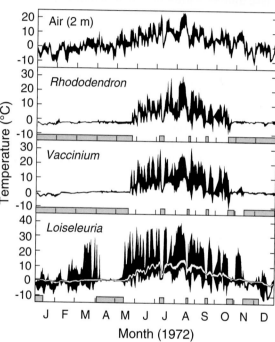

Fig. 4.6. Plant canopy temperatures along a profile of increasing wind exposure at the central Alp's treeline in Tirol. The site with *Rhododendron* is sheltered and deeply covered by snow in winter. Carpets of *Loiseleuria* are found at the wind – blown end of the transect. *Vaccinium* stands hold an intermediate position. The *black area* indicates the amplitude between daily minimum and maximum temperatures, the *white strip* in the lowest diagram shows the amplitude in −10 cm soil temperature. *Grey bars* mark the duration of snow cover. Further details in the text. The photograph on *top* shows the dwarf shrub heath in this area in autumn. Note the dark patches of *Rhododendron* in sheltered depressions. (Cernusca 1976)

are also affected (Caldwell 1970). The relative position of plants between ridges and valleys exposes them to different local wind systems (Franz 1979; Barry 1981) which can substantially alter plant temperatures. Relief – either directly or via snow cover – also strongly affects the soil climate. Soil heat flux, root zone temperatures, phenomena like frost heaving or needle ice formation, all vary with exposure. Finally, meltwater affects local moisture and nutrient supply (e.g. Stanton et al. 1994) and these influence growth, leaf area index, litter accumulation, canopy structure and transpiration, which in turn feed back on microhabitat climate.

The small-scale multitude of microclimates and "stresses" together with edaphic patterns created by relief explain the high biological diversity in the alpine life zone (Aulitzky's 1963 "wind-snow ecogram"; Fox 1981; Körner 1995a). It is impossible to draw conclusions about life conditions of alpine plants from common meteorological sources. The reason that *Saussurea gnaphalodes* is able to grow at 6400 m altitude in the Himalayan (see Chap. 2) is simply that its microhabitat exposure (and its stature) create thermal conditions that may otherwise be found at 4000 m elevation. Meters of altitude have limited ecological meaning once steep slopes, solar radiation and wind are interacting on low stature alpine vegetation.

How alpine plants influence their climate

In addition to effects of solar radiation and relief discussed above, the third important determinant of the climate experienced by alpine plants are the plants themselves. Stature, leaf arrangements, height above ground, and surface roughness of the plant canopy exert strong influences on aerodynamic coupling to the free atmosphere (Huber 1956; Geiger 1965). The data in Figure 4.6 illustrate the effect of canopy structure on plant climate. The prostrate dwarf shrub, despite its greater wind exposure, had the highest leaf temperatures under otherwise similar atmospheric temperatures. The favorable thermal climate of small or prostrate

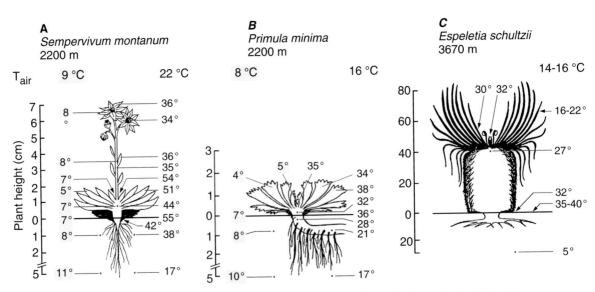

Fig. 4.7. Temperatures in alpine rosette plants of different stature, all exposed to full sunlight (in **A** and **B**, minimum temperatures measured early in the morning, numbers on the *left*). **A, B** central Alps; **C** Venezuelan Andes. **A**, from Larcher (1977 and Larcher cited in Franz 1979; **B, C**, from Larcher 1980; see also Fetene et al. 1998)

alpine plants under sunny conditions had already been noticed by scientists in the last century and a vast literature, impossible to be exhaustively presented here, has accumulated in the meanwhile (for references see Körner and Cochrane 1983; Gauslaa 1984 and the most recent account for the tropics by Meinzer et al. 1994). Besides heat accumulation in the leaf canopy (provided there is sufficient radiation energy) alpine plants are exposed to extremely steep thermal gradients within their plant body. While leaves may heat up to 30 °C, roots may still rest on frozen soil layers (Billings et al. 1976). Figure 4.7 provides examples for temperature profiles in plants of different stature from temperate- and tropical-alpine habitats.

Obviously, no uniform plant temperature exists under such field conditions, and tissue specific thermal adjustments and/or requirements are to be expected (see Chap. 11). Plants experimentally grown in a climate uniform for all plant parts may, in fact, experience a functionally rather unequal treatment of above- and belowground organs. The natural "climate" of the different parts of an alpine plant may differ as much across a 5 cm distance as it does in trees between lowlands and treeline, or over even greater climatic gradients. Figures 4.8 and 4.9 illustrate the influence of **plant life form** on leaf temperature. For comparison, in Figure 4.8, data for similar life forms but from a low altitude are presented as well.

The more plants attach to the ground, the more they decouple their climate from the ambient, and the higher the heat accumulation in the leaf canopy. Without evaporative cooling (dry moss or dry soil surface) temperatures can become lethal for active tissue even at high altitudes. Turner (1958b) and Körner and Cochrane (1983) report **soil surface temperatures** on bare spots as high as 80 °C near the alpine treeline. In herbaceous plants it makes a big difference whether leaves have petioles or are sessile, the latter warming up more (Fig. 4.7). One of the most characteristic growth forms in alpine areas, compact **cushion plants**, are known as particularly efficient heat-traps. Their dense canopy structure is genetically determined

Life form	Temperature (° C) 900 m	Temperature (° C) 2050 m	ΔT (K)
2 m air	33.7	20.0	-13.7
(tree)	40.7	21.7	-19.0
(shrub)	37.3	–	–
(dwarf shrub)	–	40.8	–
(rosette with petioles)	46.8	31.2	-15.6
(rosette without petioles)	47.7	44.0	-3.7
(tussock grass)	48.8	44.0	-4.8
(dry moss cushion)	–	49.6	–
(dry grey sand)	68.0	61.9	-6.1
(dry dark humus surface)	81.0	81.9	+0.9

Fig. 4.8. Maximum temperatures on clear days of fully sunlit leaves in coexisting plants of contrasting life form or soil surfaces in the Australian Snowy Mountains. For comparison data for alpine altitudes (upper treeline at Mt. Perisher, 2040 m) and low altitude (Lake Jinderbyne, 940 m) are presented. Note the convergence of temperatures between altitudes (difference alpine-lowland) with decreasing plant size and the effect of soil darkness. 1 *Eucalyptus* tree, 2 shrub 50 cm, 3 dwarf shrub 5 cm, 4 and 5 rosettes of leaves with or without petioles, 6 tussock grass, 7 dry moss cushion, 8 dry grey sand, 9 dry dark humus surface. (Körner and Cochrane 1983)

and not simply a phenotypic response to the environment (Rauh 1939, 1940). By means of a hand held infrared thermometer, the diurnal course of the surface temperature of a dome-shaped specimen of *Silene acaulis,* one of the most widespread alpine cushion plant species in the Northern Hemisphere, was documented throughout a bright day for distinct positions on the cushion (Fig. 4.10). Temperatures regularly deviated from the air temperature by 15 K. Gauslaa (1984) reports a maximum leaf/air temperature gradient of 24.5 K

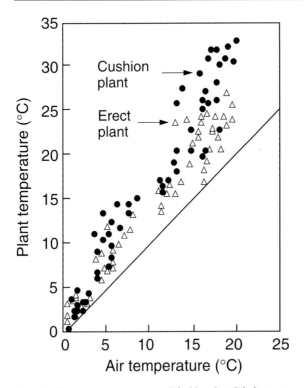

Fig. 4.9. Plant temperature as modified by plant life form at 3800 m altitude in the Rocky Mountains of Colorado. (Salisbury and Spomer 1964)

for *Silene acaulis* cushions in central Norway (45.5 °C at an air temperature of 21 °C). He suggested in accordance with Dahl (1951) that such prostrate plants are unable to survive at lower altitudes, because of these pronounced decoupling effects from atmospheric temperature conditions.

However, all these data are for bright sunshine and the warmest period of the year in the temperate zone, and thus are exaggerations of much less spectacular trends during most of the remaining periods of plant life at such altitudes. The possible benefits of such high temperatures under sunny conditions are questionable. For plant gas exchange such extreme canopy temperatures are more likely to have a negative effect because they stimulate respiratory losses more than photosynthetic gains, particularly if they occur during the dormant season. Even heat damage in such pros-

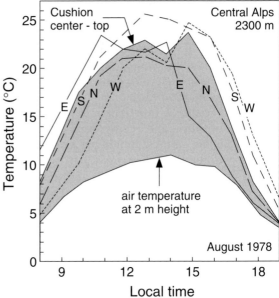

Fig. 4.10. The diurnal variation in surface temperature of a dome shaped specimen of the alpine cushion plant *Silene acaulis ssp. exscapa* (*above*), measured on a bright day at 2300 m altitude in the Alps. Letters mark the geographic orientation of measurement points. Note the large (*shaded*) differences between cushion (*center-top*) and air temperature. (Körner and DeMoraes 1979)

trate alpine plants is a realistic possibility (Larcher and Wagner 1976; Gauslaa 1984; see Chap. 8). Hence, benefits other than canopy warming in bright weather need to be considered when explaining the high abundance of compact life forms in the alpine zone – perhaps the following three advantages are most important:

- During periods of overcast conditions, when photosynthesis is co-limited by low temperature, slightly warmer temperatures of only 2 or 3 K above air temperature may improve the carbon balance significantly.
- There may be benefits of stored heat for growth and development, particularly for reproductive processes which are much more dependent on warmer temperatures than carbon fixation is.
- The benefits of such compact growth forms may be unrelated to thermal effects, but have to do with plant nutrition, with periodic canopy overheating becoming a negative side effect. Compact cushion plants become more abundant the more windy an environment is, and reach their greatest abundance on islands of the cool temperate and Antarctic zone. Long-lived, and commonly inhabiting poorly developed soils, nutrient acquisition and nutrient preservation become a severe problem, when leaf litter is blown off by wind, interrupting the local nutrient cycle. Hence, the cushion growth form must also be seen as an effective litter trap, which closes the microhabitat nutrient cycle (Körner 1993; Chap. 10).

Possibly all three components co-contribute to the advantages of compact life forms in the alpine zone. Soil microbes as well as the soil fauna also profit from these periodically warm, moist (Rauh 1939, Körner and DeMoraes 1979) and always detritus-rich spots in a cold and rocky world (Franz 1979; Schinner 1982).

Tussock grasses are perhaps the most abundant life form on stable, less steep terrain at medium altitudes within the alpine life zone worldwide. Their microclimate is modified in a peculiar way through dead leaves. Except for the youngest leaf cohorts, a large terminal part of leaves is dead, making a rather effective windbreak and causing the lower parts of the canopy to warm up. Decayed, but still standing leaves further add to a calm tussock environment. However, in humid mountains, where tussocks are particularly abun-

dant, their canopies usually warm up much less than is the case in more prostrate growth forms, as illustrated by Fig. 4.11. Unusually low temperature optima for photosynthesis, as were found in tall snow-tussocks of New Zealand, fit this picture (below 10 °C, Mark 1975). In contrast, cores of tussocks in less humid mountains with sufficiently long sunny periods which allow tussock centers to dry, as for instance in the afro-alpine *Festuca pilgeri*, have been found to heat up by 15 to 20 K above air temperature (Beck 1994). The price to be paid by either the compression of leaves to a narrow layer as in cushion plants or by clustering necromass as in tussocks, is a loss in light capture and a low green (!) leaf area index in both life

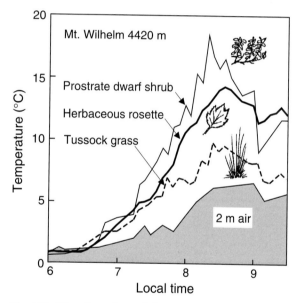

Fig. 4.11. The effect of growth form on leaf temperature in humid tropical-alpine vegetation at 4420 m altitude on Mt. Wilhelm, New Guinea. The prostrate microphyllous dwarf shrub *Styphelia suaveolens* (Epacridaceae) warms up most (max. 13 K) under the short peaks of strong radiation, whereas the tussocks of *Deschampsia klosii* (Poaceae) show the smallest deviation from air temperature (max. 5 K).The herbaceous rosettes of *Ranunculus saruwagedicus* are only slightly less efficient traps for thermal energy than the dwarf shrub. For Quantum flux density see Fig. 4.10. Note that substantial canopy warming persists throughout overcast or foggy periods, with low stature plants measuring ca. 6 K above air temperature. (Körner et al. 1983)

forms. The leaf area index (LAI) of cushion plants is around 1 to 2 m² of leaf area per m² of ground area (Körner and DeMoraes 1979), and values between 2 and 3 are common in tropical and subantarctic tussock vegetation (Hnatiuk 1978) and in the Alps (Cernusca 1977). In both cases, control over nutrient cycling, and perhaps long-term space occupancy or freezing and fire resistance (Beck 1994) may be equally or more important selective drivers for the presence of these life forms than canopy warming observed during sunny periods.

Canopy warming in low stature or otherwise compact (e.g. giant rosette) alpine plants has dramatic **influences on moisture gradients** to the atmosphere, and thus leaf transpiration and whole ecosystem vapor loss (Smith and Geller 1979). Just like human skin, warm plant layers may lose as much or more moisture at high compared with low altitudes, because of the physical conse-

quences of such temperature gradients, and despite low ambient temperatures and almost moisture saturated air. Figure 4.12 helps demonstrate this phenomenon for a 3300 m altitude difference in the tropics. Under full sun, the temperature differences between leaves of the dominant shrub species and free air at each site range from +3.5 K (mostly +1 K) at low altitude to +13 K (mostly +10 K) at high altitude. The midday vapor pressure deficit (vpd) at low altitude ranges from 20 to 25 hPa and never exceeds 1 hPa at the high altitude site. At low altitude, the actual leaf to air vapor pressure gradient is similar to vpd because there is no significant leaf to air temperature difference for most of the time. At high altitude canopy temperatures of 15–17 °C versus air temperatures of 5–7 °C, in almost saturated air, represent a vapor pressure gradient of ca. 12 hPa (a pure consequence of temperature), i.e. half the gradient of the hot tropical low altitude site with

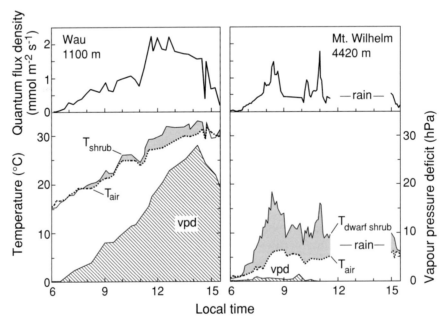

Fig. 4.12. A comparison of leaf/air temperature differences (*shaded areas*) in shrubs across a 3300 m altitudinal gradient in the humid tropics of New Guinea. *Vaccinium albicans*, a 1–2 m shrub near Wau at 1100 m and *Styphelia suaveolens* a 3–4 cm prostrate dwarf shrub at 4420 m near the peak of Mt. Wilhelm. Further explanations in the text. (Körner et al. 1983)

midday temperatures of 25–30 °C. Since stomata of leaves opened twice as much at high compared with low altitudes, the calculated actual water loss becomes similar for both sites (Körner et al. 1983) – a really surprising result in view of the dramatic differences in weather experienced by an experimenter. Even greater leaf-air humidity gradients can build up in dry tropical mountains (Schulze et al. 1985).

Plant morphology also has pronounced effects on **night-time temperature**. Clear skies, particularly at high altitudes, cause horizontal surfaces to lose much more heat by thermal radiation than vertical structures. In addition, upright structures are usually better coupled to atmospheric temperature by enhanced heat convection (Figs. 4.8 and 4.9). Figure 4.13 shows this phenomenon for a plant community at 4510 m altitude in the Ecuadorian Andes. It is important to note that both these plant species have their apex 2–3 cm below the ground, hence meristem temperatures may not be so different.

In addition to these passive effects of plant structure on microclimate, alpine plants may actively influence their microclimate by leaf movements – a phenomenon also well-known from hot desert environments. The most prominent example is the night-time closure of giant rosettes in the tropic-alpine zone (Fig. 4.14). In addition to screening the sensitive young core of the rosette, some of these species bath their most actively growing center in exudate water, the heat capacity of which delays the night-time decline in bud temperature (Beck et al. 1982).

Root zone temperatures are reported much less frequently, although it is not obvious why these should be less important than temperatures for aerial parts of plants (see also Chap. 7). Temperature usually has a greater effect on dark respiration than on photosynthesis, and most alpine plants have more than half of their respiring plant mass below the ground (Chap. 12). Even though a large fraction of the roots tends to accumulate in upper soil horizons in more humid mountains, plants in both humid and drier mountains also have very deep roots persistently exposed to rela-

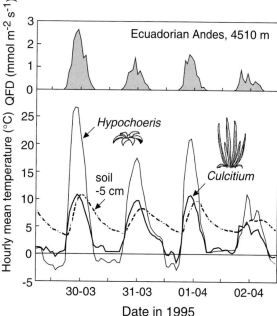

Fig. 4.13. Diurnal course of soil temperature and leaf temperatures of two herbaceous perennial plants (*Culcitium nivale* and *Hypochoeris sonchoides*) at Guagua Pichincha, 4510 m, Ecuador (as shown *above*). *Hypochoeris* forms flat dark rosettes, *Culcitium* has pubescent (*white*) and erect leaves. The *upper diagram* shows quantum flux density. (Diemer 1996)

tively cool temperatures, whatever the oscillations of temperature in the leaf canopy. On the other hand, upper soil layers may be warmer than air temperatures during periods of bright weather, but this is strongly influenced by the type of vegetation cover.

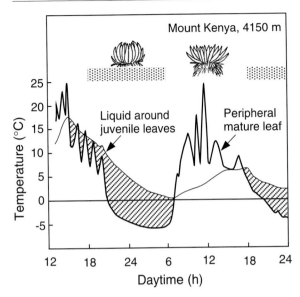

Fig. 4.14. Diurnal variations of temperature in mature outer parts and juvenile inner parts of the giant tropical rosettes of *Lobelia keniensis* (Mt. Kenya 4150 m) which show nocturnal "bud" closure. Note, the temperatures in the core of the rosette are for liquid "cisternas" developing from exudate in the most actively growing part. (Beck et al. 1982)

Lüdi (1938) for instance, compared −2 and −10 cm soil temperatures measured synchronously at three altitudes above the treeline in the central Alps on a clear midsummer day. While daily mean air temperature decreased by −0.57 K/100 m, the −2 cm soil temperature *increased* by +1 K/100 m and the −10 cm temperature by +2.5 K/100 m. This resulted partly from a reduction in plant canopy density from a closed dwarf shrub heath near the treeline to open sedge communities at the highest altitude. The less steep gradient for the −2 cm daily means resulted from the more pronounced night-time heat loss at high altitude – again a consequence of poor plant cover. The daily amplitude at Lüdi's lowest (2140 m, mean air temperature 13.5 °C) and highest (2740 m, 9.5 °C) site was 9 and 7.5 K for air temperature, 5.5 and 11.5 K for −2 cm and 0.5 and 3.5 K for −10 cm at the respective altitudes. Note that in contrast to a decreasing amplitude in air temperature, the amplitude in soil temperature increases, similar to the situation for plant canopy temperatures, as discussed earlier in this chapter. Mean temperatures in −2 cm depth of alpine ground, bare of any vegetation linearly follow the sum of daily global radiation (Mahringer 1964), but temperatures at greater depth may lag substantially behind atmospheric temperatures. For instance Bliss (1956) noted a lag of maximum temperatures by 4–6 hours at 10 cm depth.

The remarkable life conditions under which roots may operate near the upper limits of higher plant life are illustrated by data of Moser et al. (1977). At 3184 m altitude in the Alps, monthly mean root zone temperatures of *Ranunculus glacialis* in 10 cm soil depth exceeded 0 °C only during 3 months per year, reaching +2.8 °C in July (+0.7 and +0.6 °C in August and September). The mean maximum root zone temperature during July was +6.1 °C – in harsh contrast to the leaf canopy maxima illustrated in Figure 4.1. For the rest of the year the soil is frozen with temperatures down to −12.5 °C. When plants flush in June, the mean maximum soil temperatures are +0.7 °C. Hence, for a large part of their active life, these roots operate near the freezing point of soil water, a situation also found on permafrost in the arctic (Billings et al. 1976).

Long-term measurements of soil temperatures under contrasting ground cover at high subtropical altitudes in northwest Argentina were conducted by Halloy (1982 and unpublished follow-up work). Figure 4.15 illustrates temperature profiles for summer and winter in the alpine belt at 4250 m altitude, which supports a rich, deep-rooted tussock, cushion and dwarf-shrub flora. Mean temperatures at 30 cm depth during the 3 warmest months of the 6-months season is +8.1 °C (Halloy and Mark 1996). The annual amplitude of topsoil temperature in this winter-dry climate is close to 60 K and is reduced to 8 K at 1 m depth. Soil freezing reaches down to 70 cm depth, but dry top soil thaws regularly even in mid-winter, providing rather peculiar life conditions for plants rooting across such a profile. Figure 4.2 suggests that winter time root zone temperatures in alpine plants of the temperate zone are much higher due to snow cover.

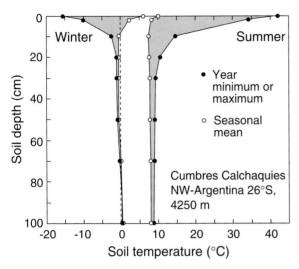

Fig. 4.15. Root-zone temperatures in a subtropical high-plane in northwest Argentina (Cumbres Calchaquies, 4250 m, 26°S; see Fig. 9.10). *Winter* and *summer* refers to mean temperatures measured during a number of days during the warmest (February 1985) and coldest (August 1984) month. According to the author, these profiles are consistent with measurements obtained at one or two soil horizons over several years between 1977 and 1990. Note the narrow annual amplitude of means between zero and ca. +8 °C. Data available for depths down to nearly 2 m suggest a merging of profiles below 3 m at an annual mean of ca. 5 °C. (S. Halloy, pers. comm.)

The geographic variation of alpine climate

Is what people consider "alpine", inferred from similar plant formations, similar in terms of climate? A Europe-wide assessment of ground temperatures under alpine grassland provides a positive answer (Fig. 4.16). There are differences between the far north, the extremely oceanic sites (Scotland) and the Mediterranean, but seasonal mean temperatures are quite similar, and local snow distribution rather than latitude controls thermal minima experienced by roots and below-ground apical meristems in winter.

In summary, plant stature and canopy structure are key determinants of plant temperatures in alpine vegetation. Unlike trees or larger shrubs, low stature vegetation, as found in the alpine life zone, is effectively decoupled from atmospheric conditions, causing heat to accumulate in the canopy and top soil during all periods of moderate to high radiation. The degree of uncoupling decreases in the sequence cushion plant – prostrate dwarf shrub – herbaceous rosette – tussock grass – isolated taller shrubs or tall herbs – Krummholz vegetation – treeline trees. The relevance of standard meteorological data for plant life increases in this sequence, although treeline trees have been shown to undergo some shoot warming as well (Baig and Tranquillini 1980; Smith and Carter 1988). Except for ridge tops, root-zone temperatures near the treeline ecotone are consistently higher under low stature vegetation than under forest trees (as was documented for instance for a forest versus grassland altitudinal transect in New Zealand by Körner et al. 1986; Körner et al. 2003; see Chap. 7). Deep roots operate at significantly lower temperatures than shoots for most of the time.

An exception to the above life-form related sequence of plant size are large-leafed tall herbs, a curiosity found in sheltered nutritious places in almost all high mountain areas of the world (various *Senecio* species in the Andes, *Rumex alpinus* in the Alps, *Rheum nobile* in the Himalayas, *Ranunculus anemoneus*, in the Australian Alps) or the giant rosettes in the afro-alpine zone and the equatorial Andes. In these plants substantial leaf/air temperature differences result from large leaf size in combination with low to moderate wind speeds (Grace 1977).

In conclusion, cold alpine climates are not always cold for plants, particularly not for leaves when photon supply for photosynthesis is high. Growing season root zone temperatures in the upper soil profile may be higher, but temperatures in deep roots may often be lower than under forests near the treeline (see Chap. 7). The steep spatial temperature gradients within a single plant and the strong temporal variability complicate simulations of alpine plant climates. There is no such thing as a uniform temperature across the body of alpine plants.

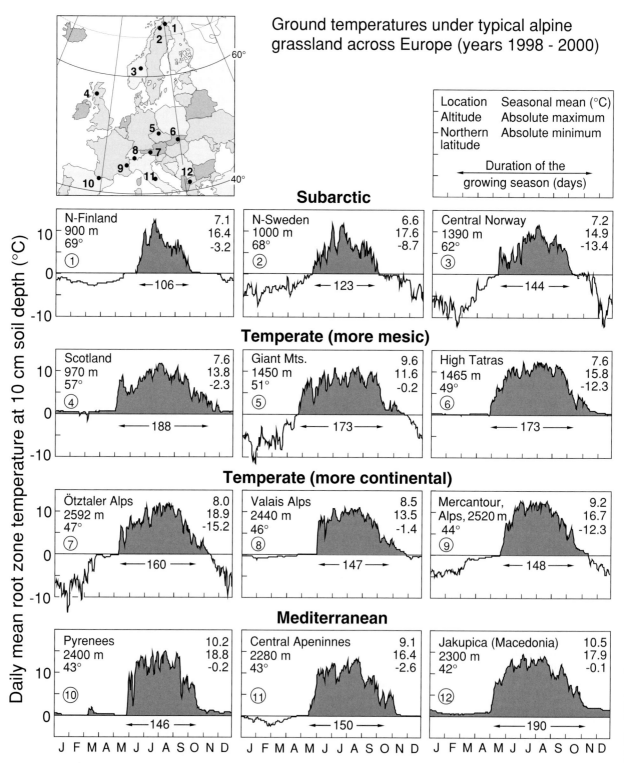

Fig. 4.16. Alpine climates across Europe: the seasonal variation of root zone temperature (−10 cm depth) in horizontal patches of closed alpine grassland ca. 250 m above the regional climatic treeline. (Körner et al. 2003)

5   Life under snow: protection and limitation

In the alpine zone snow can fall and cover vegetation at any time of the year at all latitudes. Higher plants can be found in habitats of mean snow duration of anywhere between 330 and a few days (in the tropics) per year. In some cases, at higher latitudes, alpine plants may even have to (and do) survive one or two years of permanent snowpack, as can happen due to yearly variable snow distribution by wind (Moser et al. 1977). Snow cover strongly determines the distribution of plant species in the alpine zone as was discussed in Chapter 4. Some profit, others suffer. This chapter deals with the conditions for life and plant responses under snow at high altitudes. For more extensive treatments of snow physics and the climate under snow I refer to Geiger (1965), Gray and Male (1981), and Marchand (1991).

The protective aspects of snow cover are easily listed: snow prevents plant exposure to low temperature extremes, winter desiccation, ice blast and solar radiation (potentially dangerous to dormant tissue) during the cold season or during and after cold episodes in the alpine tropics. The adverse effects are less clear. Shortening of the length of the growing season is the only obvious limitation. Others, like effects on plant respiration by elevated soil temperatures in winter, effects on microbial activity, nutrient cycling, melt water seepage and water logging, ground ice formation and possible anoxia in and above the soil, mechanical pressure and shearing effects on slopes, snow mold and other pathogen effects or below snow rodent activity, plus effects on soils during thaw–freeze cycles are more difficult to evaluate.

Temperatures under snow

By **thermal insulation** snow dampens temperature oscillations in low vegetation and commonly keeps soil temperatures high compared with above-snow temperatures during winter or during clear nights in general (Figs. 4.3, 4.6). Eckel and Thams (1939) found a snow layer of only 35 cm sufficient to maintain topsoil temperatures near zero when ambient air temperatures reached −33 °C after a series of days with temperatures around −20 °C in December 1937 near Davos, Switzerland. In one of the coldest areas of the world, in eastern Siberia, a 20 cm snow layer was found to delay subsoil freezing by two months, during which ambient temperatures fell to below −40 °C (Zimov et al. 1993).

The lower the **density of snow,** the more it insulates. The thermal conductivity of snow varies between 0.3 and 4 mW cm^{-1} K^{-1} in light fresh and compacted old snow respectively. For comparison: glass wool has a value of 0.4 and concrete one of 12 (data compiled by Marchand 1991). Hence, when fresh fallen snow of a density of 0.1 g cm^{-3} compacts to "old snow" of a density of 0.4 g cm^{-3} thermal conductivity increases more than tenfold. The formation of a crust of ice by thaw-freeze processes (density ca 0.8 g cm^{-3}) causes specific thermal conductivity to increase by another four times (Geiger 1965). Apart from snow quality, **snow depth and duration** are the most important determinants of plant and soil temperatures during cold periods. A number of studies have documented the almost linear negative correlation between the thickness of the snow cover and the depth of **soil freezing** (Sakai and Larcher 1987).

When alpine soils freeze during the early part of the winter because of lack of snow, later snowfalls can cause soil temperatures to rise again and soils may even thaw despite decreasing ambient air temperatures (Larcher 1957). Except for ridges or extremely wind exposed terrain, it can be assumed that soils in the lower alpine belt of the temperate zone remain unfrozen for most if not all of the year, provided autumn snow falls were not delayed for too long. At higher elevations and/or higher latitudes it requires a rather thick and early snow cover for soils not to undergo winter-time freezing (see also Chap. 6).

The complete absence of snow during the cold season may cause soils to freeze to great depth and expose plants to severe physical and physiological stress. An extreme example is parts of the semi-arid subtropical Andes in nothwest Argentina, where plants of a very rich alpine flora have to survive a 6-months winter in altitudes of 4000–5000 m, mostly bare of snow protection (Fig. 9.10).

Snow also plays a protective role in the alpine zone of the equatorial tropics (just as it does under certain conditions in midsummer at higher latitudes) when storms are followed by clear nights. The lowest temperatures actively growing (i.e. not winter-dormant) plants ever experience are those due to radiative cooling during the first clear night after a cold front had passed (see Chap. 8). A thin snow cover may be decisive for the fate of aboveground parts in many species under such conditions.

Relatively warm temperatures below a winter snow cover reduce the need for plants to invest in cryoprotective measures, and indeed, species growing in habitats well protected by snow are less resistant than exposed ones (Larcher 1980, see Chapter 8 and Fig. 5.7). On the other hand, "warm" soils lasting for a 9-month dormant season may not be all that "desirable" for plants, given that this causes metabolic costs by continued respiration while growth is ceased. All plants of cold climates exhibit substantial respiration at 0 °C and it needs temperatures of −5 to −10 °C for rates to approach zero. If, at the same time, solar radiation penetrat-ing the snow facilitates photosynthetic CO_2-fixation, some of these "costs" resulting from snow cover could theoretically be covered. Does this happen?

Solar radiation under snow

Many alpine plants do retain green structures above the ground over winter – not only the so-called evergreens. Overwintering leaves are quite common even in forbs (e.g. in the genera *Geum* and *Potentilla* in the Rosaceae) and grasses such as *Nardus sticta*. The bark of young shoots of dwarf shrubs is also photosynthetically active, most obviously in those with green stems such as in *Vaccinium myrtillus*. Mosses and lichens also retain photosynthetic structures over winter. Two questions arise from this. Is there enough photo-synthetically active radiation passing through snow? Is the photosynthetic machinery able to make any use of it at temperatures close to freezing point? In principle, all alpine plants ever studied are able to photosynthesize at temperatures between 0 and −2 to −6 °C. During the growing period some plants can reach as much as 20–30% of their maximum rates of **photosynthesis** when temperatures are around 0 °C (Pisek et al. 1967; see Chap. 11). During winter, however, photosynthetic capacity is strongly reduced in higher plants, (though not always completely), whereas it is largely retained in lichens (Sonesson 1989; Kappen et al. 1995).

The radiation climate under snow is determined by two processes: reflectance at the surface, and extinction within the snow. Both processes do not follow simple rules and the literature is not univocal. When meteorologists measure **reflectance**, they usually consider global radiation (0.3–3 μm) and call it albedo. Biologists measuring the reflectance of "quantum or photon" flux, which excludes wavelengths >700 nm, i.e. roughly half of the total radiative energy, might better talk about quantum flux reflectance, so as not to get confused, both with established terminology and meaning. This distinction is important because reflectance of solar radiation is larger the shorter the wavelength. For wavelengths longer than ca. 1.5 μm, snow represents almost an ideal black body and hardly reflects. Therefore, reflectance of quantum flux in the 400–700 nm range should always be larger (commonly >90% on clean fresh snow) than albedo (ca. 60–80%), which is biologically important. The next problem is that snow cover, like all natural surfaces, is not an ideal plane, so the cosine correction for non-90° insolation cannot be applied in a straightforward way. This becomes important at lower solar angles and a clear sky at high altitudes, when fresh snow reflectance of visible radiation (including the scattered radiation) can approach almost 100% of incoming quantum flux density (QFD) measured with a horizontal sensor. Were incoming QFD measured in the direction of the sun, the resultant "reflectance" measured by a horizontal sensor would be smaller the lower the solar angle. As part of a study on the effect of red snow algae on reflectance (which was very small) Thomas and Duval (1995) measured "white" snow QFD reflectance in the Sierra Nevada of California of 58 to 65% when compared with QFD measured in the direction of the sun (at noon in June). Accounting for the cosine-effect at their latitude, this would yield a ca. 75% reflectance based on horizontally measured incoming as well as reflected QFD. Hence, measurement conventions co-determine estimates of QFD actually penetrating into snow.

The remaining "net QFD" – in a first approximation – is absorbed exponentially with increasing thickness of the absorbing layer (Geiger 1965, Fig. 5.1). However, radiation that penetrates the snow creates **back-scatter**. Hence, even when considering an idealized, completely homogenous profile, snow does not meet the criteria for the application of Beer-Lambert's extinction law. Two rather different "**transmittance**" values would be obtained under a 10 cm layer of snow by two horizontal sensors if one were placed at the soil surface (almost no back-scatter), and the other one at 10 cm below the surface of an otherwise deeper snow layer. The reading of the latter sensor may be somewhere around twice that of the first,

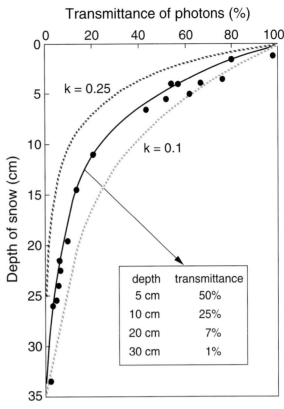

Fig. 5.1. Solar radiation under snow. *Dotted lines* Theoretical transmittance of radiation calculated from Beer-Lambert's extinction law for two different extinction coefficients (0.25 and 0.1 cm^{-1}). *Solid line* Actual transmittance measured for wet spring snow in Abisko, northern Sweden (k = 0.14; original data). With QFD = 1000 μmol m^{-2} s^{-1} above the snow and 80% QFD reflectance the actual QFD measured on a dark plain (e.g. the soil surface) under 10 cm of snow would be ca. 50 μmol m^{-2} s^{-1}

even though the absorbing snow layers above the sensors were identical in thickness and quality. Thus, measurements with a horizontal "point" sensor (a conventional QFD sensor) placed at various depths within a given snow profile, do not produce the same transmittance/thickness relationship as the same sensor placed at the soil surface under layers of snow of different thickness. This does not mean that one of the two ways of measuring transmittance or extinction is wrong, rather they have a different meaning. One extinction coefficient includes, the other one largely excludes back scatter effects. I explain this because of the obvious biological relevance if one compares the light regime of a moss or lichen thallus at the ground or leaves or branches of upright plants surrounded by snow at otherwise equal distance from the snow surface. Without accounting for back scatter effects, results of QFD measurements below snow can create substantial confusion.

Commonly, less than 10% of QFD reach the ground when snow depth is 10 cm (depending on snow quality). Only 2–3% of QFD may be left after radiation has passed through a 20 cm snowpack. QFD capture by plant structures protruding into snow at these same distances from the snow surface may be twice as high because of back scatter effects (see above). Below a 30 cm snow cover, the environment is essentially "dark" for photosynthesis, because 1% of full midday QFD in summer is the minimum usually required for a net carbon gain by leaves. In midwinter, under overcast conditions and earlier or later in the day, snow

layers of only 5–15 cm thickness may have the same "dark blanket" effect.

Fresh, low density snow not only reflects more, it also tends to absorb more transmitted light per cm of depth than old or melting snow. However, the relationships between **snow quality** and transmittance are again not straight forward. The same high **density** of snow reached by mechanical compaction or by melting processes has opposite effects on transmittance, and settling of snow under dry, cold conditions decreases transmittance on a per cm depth basis. I refer the reader to the detailed considerations of the fate of solar radiation in snow of varying quality by Marchand (1991). For plants living under snow it is important that more radiation penetrates the snow during melting because reflectance is reduced and transmittance is increased and at the same time the wetting plus the melting process rapidly cause the thickness of the snow layer to shrink (Fig. 5.2). At temperate or subpolar latitudes, all this usually happens when solar radiation is close to its annual peak in late May or June, which further enhances

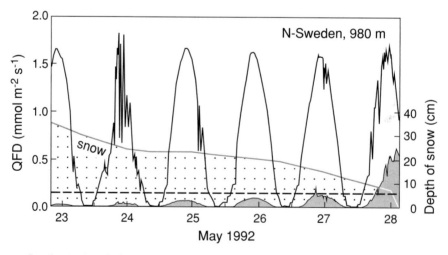

Fig. 5.2. Quantum flux density (QFD) above and below (*shaded area*) melting snow under bright arctic-alpine spring conditions at Latna Jaure near Abisko, northern Sweden (68°N, 980 m; M. Sommerkorn, pers. comm.; below-snow QFD is enhanced due to backscatter effects because the sensor's diffusor cap was about 5 cm above the ground and surrounded by snow, a situation not dissimilar to the one experienced by fruticose lichens). Temperatures in actively photosynthesizing lichens on the ground were around +0.5 °C throughout the observation period, while ambient air temperatures varied between +13 and –3 °C. The level of saturation of lichen photosynthesis by QFD is 160 μmol m^{-2} s^{-1} at this thallus temperature (Kappen et al. 1995), wich is indicated by the *dashed line*

the radiation reaching the ground before the snow finally disappears. This is the period when photosynthetic profits from light penetrating a thinned snow cover are most likely.

Within the photosynthetically active part of the spectrum (400–700 nm) the composition of solar radiation is changed relatively little by snow, but a slightly better transmittance in the photosynthetically least useful part between 480 and 580 nm has been reported by Richardson and Salisbury (1977) and Kappen et al. (1995) whereas other authors found no clear **spectral trends** (cf. Geiger 1965), but perhaps they did not study deep enough snow. The deeper the snow, the more the spectrum is skewed to the blue-green range, with the biologically important red fraction becoming less and less. In the infrared part of the spectrum (>700 nm), absorptance, as mentioned above increases rapidly (Curl et al. 1972).

In summary, both intensities and the spectral quality of radiation potentially permit plants to achieve net photosynthetic **carbon gains under snow**, provided the snow cover is shallow, the ambient radiation is high and plants are physiologically active. Figure 5.2 illustrates an arctic-alpine snow-melt situation in northern Sweden where fruticose lichens were shown to exhibit substantial carbon gains, while still covered by a 10–25 cm layer of wet spring snow (Kappen et al. 1995). Up to 250 µmol photons $m^{-2} s^{-1}$ reached the lichen layer during midday, which substantially exceeds the light saturation of photosynthesis at snow-melt temperatures. Hamerlynck and Smith (1994) measured 500 µmol $m^{-2} s^{-1}$ at the edge of a snowbed when snow depth became less than 5 cm.

However, radiation plays three different roles in plant life. In addition to influences on microclimate or heat budget and photochemical effects discussed above, a third and most important role has not yet been touched on: radiation carries information, either via **spectral quality** or **photoperiod**. Plants are able to perceive minute amounts of radiation and learn from these the time of day, the time of year and whether it is appropriate (from evolutionary experience) to initiate certain developmental phases of their life.

Whatever the photochemical gain from radiation penetrating snow may be, there will almost always be enough radiation to carry such information, no matter how deep snow is, as has been demonstrated so elegantly by Richardson and Salisbury (1977) in their famous snow-tunnel experiment in the mountains of Utah. Using a combination of photomultiplier studies and bioassays they proved that signals penetrating snow down to 2 m depth are strong enough to induce seed germination, sprouting and morphological responses of test plants. As an example of their results, Figure 5.3 shows that in seeds of lettuce (certainly not a snowbed specialist), which were kept in complete darkness, but were then exposed for only 10 minutes to light penetrating 77 cm of snow, germination rates correlated with the time of exposure and/or wavelength perceived. Richardson and Salisbury concluded that this response was due to diurnal spectral shifts below snow (which

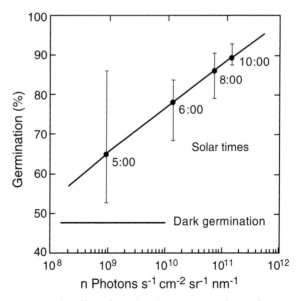

Fig. 5.3. The effect of very low, but increasing doses of spectrally altered solar radiation penetrating 77 cm of snow on the germination of lettuce seeds exposed to this radiation for only 10 min in a special snow tunnel on 11 May, 1974. The time of day during the exposure is indicated. Note that sunrise was at 5.30 solar time. (Richardson and Salisbury 1977)

they documented). How much more must we expect plants other than lettuce to respond to such signals, given their success depends on such messages?

These radiation signals, plus melting water and increasing temperatures, cause many alpine plants to initiate flowering, to sprout or germinate before snow disappears (see color Plate 3 at the end of the book). Though the specific mechanisms await to be understood, the advantages for the effective utilization of the often extremely short snow-free period for reproduction and carbon gain are obvious.

Gas concentrations under snow

As discussed above, snow effectively inhibits heat diffusion and thereby creates conditions predominantly favorable for plant survival and plant metabolism during winter, which in turn depend on gas diffusion. Since sinks and sources of reactive gases are separated from the atmosphere by snow as a resistor, concentrations under snow will always differ from ambient ones. The thickness and quality of the snow layer and the biological activity in the soil (soil temperature) co-determine the magnitude of such differences. The **diffusion of gases** through snow follows Fick's law of diffusion. For CO_2, a gas of particular interest here, the presence of snow with common porosities between 50 and 60% reduces the diffusion coefficient for free molecular diffusion of ca. $0.135\,cm^2\,s^{-1}$ at 0 °C and sea level pressure to values of around $0.052\,cm^2\,s^{-1}$ (Mariko et al. 1994). With increasing altitude the diffusivity increases and is ca. 30% higher at 3000 m elevation, as a result of reduced atmospheric pressure (see Chap. 3).

Under dry, soft snow of 20 cm thickness long term means of 12 to 36 ppm CO_2 enrichment near the ground have been measured in the extremely cold winter climate of subarctic Siberia (Zimov et al. 1993), similar to the <30 ppm observed by Kelley et al. (1968) in Alaska. In both studies the highest CO_2 enrichments were noted early in the winter and a few weeks before the disappearance

of snow. Under wet snow of similar thickness Kappen et al. (1995) found means of ca. +70 ppm at their arctic-alpine site, but short-term excursions of **CO_2 concentrations** up to 200 ppm above ambient have been noted. The temporal variation and depth dependency of CO_2 concentration under snow in the tundra of northern Sweden is illustrated in Figure 5.4. Below 50 cm of snow, CO_2 concentrations were found to vary between 400 and 900 ppm above ambient, and even higher concentrations were measured near the soil surface under 90 cm of snow (800–1700 ppm). Under deep snow (1 m) and on biologically very active ground such as under montane forests, Mariko et al. (1994) found CO_2 concentrations of ca. 2000 ppm in Japan (Fig. 5.5).

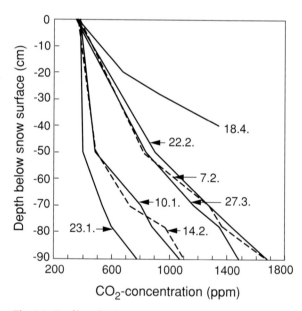

Fig. 5.4. Profiles of CO_2 concentration under snow in the subarctic birch-tundra near Abisko (northern Sweden), *arrows* indicate dates. Because snow depth varied, the snow surface has been used as a reference for depth. By April 18, snow was compacted by melting to 40 cm. Note snow quality varied across the profile, causing the CO_2-gradients to be non-linear. The ground vegetation is a rich dwarf shrub/moss community and ambient air temperatures varied between +5 and −20 °C (mostly around −8 °C) during the measurement period. (M. Sonesson and M. Tijus, unpubl. data for 1989)

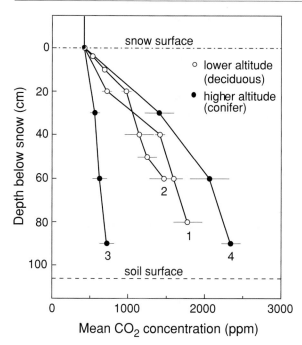

Fig. 5.5. Profiles of CO_2 concentration in >0.8 m deep snow in Japanese montane forests at 1320 (*1*), 1500 (*2*), 1980 (*3*) and 2200 (*4*) m altitude measured between March and May. Soil temperature between 0 and −10 cm soil depth was between 0 and +2 °C at all sites. Snow depth was 82–98 cm at sites *1, 2* and *4*, and 190 cm at site *3*. Note the nearly linear gradients. Different slopes reflect differences in winter-time CO_2-evolution from soil in the range of 20 to 75 mg CO_2 m^{-2} h^{-1} (means and SD for 5 profiles each). (Mariko et al. 1994)

Since concentrations of CO_2 in soil normally exceed 1000 ppm, the above numbers for further enhancement of concentrations below snow represent only a moderate spatial excursion of soil air into the soil-snow boundary. A significant photosynthetic benefit (in absolute terms) from CO_2 enrichment under snow is very unlikely because of low light and because low temperatures drastically reduce CO_2 sensitivity of photosynthesis in this range of concentrations.

Strongly compacted snow and solid ice may reduce diffusivity so much that oxygen consumption due to heterotrophic respiration (CO_2 production) may lead to periodic **anoxia** (Andrews 1996). Is there a realistic possibility of **oxygen shortage** under snow? Mitochondrial respiration of plants is not significantly affected by reductions of oxygen unless concentrations of less than 5% (compared with 21% in the free atmosphere) are reached, but certain micro-organisms in the soil may be more sensitive. Underneath an ice-crust above well developed organic soils, Newesely et al. (1994) measured a depletion of oxygen concentrations down to and below 5%. No such depletion was found under similar conditions, but on degraded substrate of low biological activity. The situation becomes more severe if **soils** become **waterlogged** during snowmelt, particularly when drainage is blocked by an ice table in deeper soil layers. Soils respond to such repeated anoxia by the formation of blue-gray horizons of gley on otherwise well drained high alpine grassland (alpine "pseudogley", Burger and Franz 1969, see Chap. 6). Snowbed communities are particularly exposed to such stress situations. It is not yet known how sensitive alpine snowbed species are to oxygen depletion, but northern populations of some common arctic plant species (in contrast to more southern populations) have been found to be extremely robust (Crawford et al. 1994). Anoxia situations are generally more tolerable to plants at low temperatures (Crawford 1992). Given the fact that highest microbial activities have been observed immediately after thawing (Schinner 1983), soil micro-organisms as a whole do not seem to suffer, but certain groups of microbes may.

Besides these direct microclimatic and atmospheric effects of snow cover on the sub-snow environment, snow also exerts **mechanical forces** on vegetation (Larcher 1985a; Marchand 1991). It distorts or even breaks upright plants, allowing only those shrubs and Krummholz trees to survive which can tolerate tons of shear forces by elastic stems and robust buds (Fig. 5.6). On steep slopes and under certain types of spring weather, snow frozen to the vegetation may tear sods of alpine turf off the ground when sliding downhill. This problem is known to cause damage to abandoned alpine pasture land which enters winter with long standing-dead crops of grass. The movement of rocks and loose ground by sliding snow or

Fig. 5.6. Ice blast, snow pack and freezing may brake stems and branches and damage overwintering leaves. Tons of wet snow were packed to these trees by an antarctic blizzard followed by freezing. This occurring once every 20 to 50 years would be enough to turn trees into multistemmed shrub. One of the highest growing trees of *Eucalyptus pauciflora* in the Australian Snowy Mountains at 2050 m altitude on Mt. Perisher at the end of June 1989

avalanches with subsequent mud or debris deposition exert massive mechanical loads (Fig. 5.7). In addition, snowpack affects soil erosion and slope stability in various ways (Billings 1973; Franz 1979; Neuwinger 1980; Fox 1981; Stanton et al. 1994; see also Chap. 6).

Plant responses to snowpack

As discussed in Chapter 4, patterns of snow distribution and snow duration in alpine areas create rather stable mosaics of vegetation types which follow the isolines of snow melt, but underlying mechanisms are not always clear. There is a close relationship between snow duration, plant cover and plant productivity (Billings and Bliss 1959; Klug-Pümpel 1982; Ostler et al. 1982; Galen and Stanton 1995; Fig. 5.8). The longer the duration of snow cover, the more pronounced effects become and the fewer species survive. In the following I will, therefore, discuss the snowpack problem with

Fig. 5.7. Avalanches create a special alpine habitat with regular mechanical stress, long snow cover, a supply of plant debris and a fine substrate. They also cause the alpine flora to extend below the treeline. (Sellrain, Tirol)

special emphasis on long lasting snow cover in snowbeds.

There are three fundamental pre-requisites ("filters") for life under prolonged snow cover:

- Resistance to physico-chemical stresses and snowbed pathogens,
- Plant phenorhythmics have to match the timing of snow coverage, and
- Annual carbon gains must suffice to allow the completion of the life cycle and support persistence by reserve formation and clonal or reproductive propagation (the latter may not be

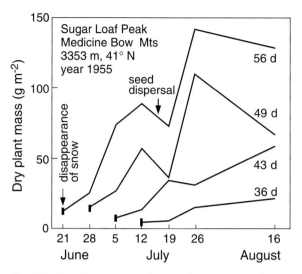

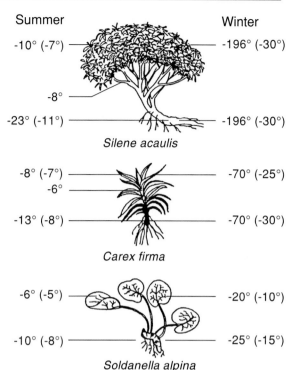

Fig. 5.8. Plant dry matter production along a snow melt gradient in the Rocky Mountains. Total above ground live biomass has repeatedly been determined in plots situated on isolines of dates of snow disappearance. The reductions in biomass in the plots released earliest (56 d isoline data) are due to seed dispersal and late-July senescence of some older leaves. (Drawn from data by Billings and Bliss 1959)

Fig. 5.9. Freezing resistance of the alpine snowbed species *Soldanella alpina* – a species of southern European distribution, in comparison to species commonly overwintering at more exposed (snow-deprived) habitats. The arctic-alpine *Silene acaulis* often grows on wind-swept ridges, *Carex firma* holds an intermediate position with respect to snow cover. *Numbers* indicate maximum frost resistance, and in *brackets*, resistance after 3 to 5 days of warm temperature pretreatment, indicating the potential reduction of resistance during warm spells. (Larcher 1980)

essential when diaspores are imported from "better" surrounding habitats).

The first depends more on the snow cover period and the length and nature of thaw-freeze cycles at snow melt, the last two are also critically dependent on the duration of the snow-free period. The carbon gain is co-determined by weather, soil moisture, nutrient availability and herbivory during the growing season, all general aspects of alpine plant life in seasonal climates, which will be treated in several later chapters. In the following, topics of immediate relevance to the snowpack question will be discussed, but thematic boundaries are of a practical rather than scientific nature.

First, the existential "filters" must be passed. What are the **stress**-physiological barriers for plant survival in areas of prolonged snowpack? Nowhere else in the alpine terrain do we find plants so well protected from frost, and thus requiring so little resistance, than in places of reli-

able winter snow cover (Larcher 1980; see also Körner and Larcher 1988; Fig. 5.9). However, freeze-thaw situations during snowmelt bear a risk of frost damage in these commonly less resistant or less hardened plants (Larcher 1985b). Given the high soil moisture during the peak growing season (e.g. Billings and Bliss 1959), heat and drought effects are unlikely to affect snowbed plants (but see below).

However, drought stress may come into play in two indirect ways. Firstly, through a reduction of **nutrient availability** during periods of top soil desiccation (not necessarily affecting turgor because of sufficient moisture in deeper horizons). Snow beds released from snow very late are particularly endangered because the water holding capacity of the substrate is weak due to poor humus content (Billings and Bliss 1959). Secondly, drought may become a problem to snowbed plants when **snow** is abruptly **removed** in winter by strong winds in an "unconventional" direction, by avalanches, animals or human influence (skiers). Under such conditions winter drought can kill plants in few days, as was demonstrated by an excavation experiment by Larcher and Siegwolf (1985; Fig. 5.10). These authors showed that *Rhododendron ferrugineum*, an evergreen dwarf

Fig. 5.10. Plant species requiring deep snow cover for survival can become active at any time in winter when exposed to sun, which is commonly fatal. Here, an in situ examination of photosynthesis of *Rhododendron ferrugineum* on Mt. Patscherkofel near Innsbruck. (cf. Larcher and Siegwolf 1985)

shrub known to require safe snow cover (Friedel 1961), dies within 3 days by leaf desiccation when snow is removed and when exposed stems freeze. Unlike the wind edge species *Loiseleuria procumbens* (Körner 1976, Grabherr 1976), *Rhododendron* is able to photosynthesize significantly within 2 hours after snow removal and reaches a relatively high stomatal conductance for water vapor – a death sentence when its snow shelter is removed in winter.

Another critical factor in snowbed plant life is **radiation stress at emergence** from snow. Within one or two days, even within few hours, when snow rapidly melts, plant tissue may become exposed to the highest intensities of solar radiation measured on earth (see Fig. 3.11), in particular when reflecting snowflecks are still around. While mature and hardened leaves of alpine plants, and snow algae have been shown to be very well equipped to screen damaging radiation from the photosynthetic apparatus (Caldwell 1968; Robberecht et al. 1980; Bergweiler 1987; Caldwell et al. 1989; Thomas and Duval 1995; Wildi and Lütz 1996), still dormant or premature tissue suddenly released from snow is in danger. The yellow or reddish color of freshly emerged shoots or leaves at the edge of snowbeds is a well-known phenomenon (color Plate 3 at the end of the book). Similar to flushing tropical trees, the lack of green chloroplasts during this phase reduces short wave absorptance and allows cells to expand, vacuolize and establish a protective screen before the photosynthetic machinery is fully installed. In some species, particularly the most rapidly expanding ones, this may take several days. Frost events during this period can interrupt or even terminate this greening process, leaving life-long chlorotic leaf scars (C. Lütz, pers. comm.). Caldwell (1968) found a rapid reduction of UV transmittance in extracts of stems of *Ranunculus adoneus* released at the edge of a melting snowbed (see Chap. 8).

Other plants emerge from snow with fully active, green leaves, (e.g. *Soldanella alpina*), which immediately exhibit high electron transport capacity in their photosystem (Bergweiler 1987), and thus channel potentially damaging solar energy into biochemical energy even before snow fully disappears. Kume and Ino (1993) were surprised that *Aucuba japonica*, which is an overwintering green under heavy snowpack in Hokaido, maintained full pigmentation and photosynthetic activity over 110 days of storage on ice, and similarly, Lösch et al. (1983) found a snowbed moss *Anthelia sp.* to maintain its photosynthetic capacity for a 9-month period of cold storage, not tolerated by another snowbed moss, *Polytrichum* sp. Evergreen Ericaceae (but apparently not all, as discussed above for *Rhododendron ferrugineum*) shut down or reduce photosynthetic capacity in the dormant stage under snow (depending on exposure) and it may take two weeks or more to fully recover under favorable conditions (Grabherr 1976, Karlsson 1985). In contrast to snow covered leaves, exposed winter green leaves are often found discolored, which is considered as a protective mechanism against superfluous radiant energy (see color Plate 3 at the end of the book). Suddenly released, fully green, but physiologically deactivated leaves may thus be more vulnerable to high radiation (Larcher 1985b).

Yellow birth, pink flush, active green, passive green or bleached survival, or other modes of passing through this critical phase of life in snowbed plants may be grouped into at least six different **types of "strategies"** (with some examples discussed later):

1. Remain green and retain full photosynthetic capacity over winter (e.g. *Soldanella*).
2. Remain green, but deactivate photosynthesis, with a period of up to 2 weeks required to reach full capacity after snow melt (some evergreen Ericaceae *Vaccinium vitis idea, Loiseleuria procumbens*).
3. Flush "conventionally", i.e. some time after snowmelt, with new leaves emerging green (e.g. Krummholz species of *Alnus* or *Betula*, deciduous alpine dwarf shrubs, *Carex* sp.).
4. Initiate leaf expansion before snow melt and commence greening and activation of photosynthesis immediately when released (*Erythronium grandiflorum*, snowbed *Ranunculus* sp.)
5. Initiate leaf expansion before or at snow melt, but delay greening and activation of photosynthesis (fast expanding tall herbfield species from snowbeds in genera like *Geranium, Rumex, Angelica, Luzula*).
6. Survive as seed and germinate at or after snowmelt and thereby circumvent some of the problems of sudden release from snow (few rare snowbed annuals).

Strategies 2, 3, 5 and 6 are commonly found at lower alpine altitudes, near the treeline, with a shorter snowy period. Strategies 1 and 4 dominate sites with longest snow duration at higher altitudes, but all 6 types may co-occur. A systematic, comparative analysis of the snow-release physiology of such functional types still needs to be done, but it seems, there exist various solutions to the same problem. In any case, these response types strongly relate to species specific phenorhythmics (see below).

Besides the potentially harmful physicochemical stressors associated with snowpack discussed above, the most decisive component of plant resistance in snowbeds is of a biological nature: the impact by **microbial pathogens** and in particular **snow mold**. Prolonged temperatures around 0 °C and wetness provide ideal life conditions for certain fungi (e.g. *Herpotrichia* sp., *Phacidium* sp.), and damage can be regularly found after long snow cover in spring (e.g. Watson et al. 1966; Smith 1975; Frey 1977; Aulitzky 1984; Larcher 1985b; Sturges 1989). It may be the most critical factor for tree regeneration above the treeline, and possibly explains the predominance of plant species which overwinter completely below ground in late snowbeds. Such habitats often appear almost bare at snow melt. Even some lichens do not tolerate prolonged snow cover (Benedict 1990). If a species has not acquired true snow mold resistance, belowground regeneration buds avoid the infestation. There appears to be an interaction between snow mold susceptibility and the nutrient richness of overwintering leaves. The addition of small amounts of full mineral fertilizer to evergreen dwarf shrub vegetation in early summer has ended with a disaster after 5 years: fertilized plots were selectively wiped out by snow mold (Körner 1984).

Developmental aspects to the snowpack problem have already been touched upon above. A classic developmental adjustment of plants to long snow cover and a short growing season is the preformation of leaf- and flowerbuds in the previous season (Larcher 1980; see also Chaps. 13 and 16; Fig. 16.1). In some species, two generations of flower initials can be found preformed. These can all be activated in one season when plants are transplanted to a warmer climate (Prock and Körner 1996). But again, not all species have adopted this **bud-preformation** strategy, which is found in species flowering at or soon after snow melt. Others, like many alpine Caryophyllaceae (e.g. *Cerastium* sp.), produce flowers on newly emerged shoots, and thus flower later, but these are not the most typical snowbed plants. Molau (1993) found a higher abundance of **apomixis and vivipari** in late snow melt species, which also tend to exhibit higher ploidy levels than earlier flowering fellfield species, which in turn, according to Molau, have greater outbreeding rates and low seed : ovule ratios. Whether seedlings can establish may depend on the occurrence of a series of exceptionally long growing seasons and specific disturbance regimes (Chambers et al. 1990; Galen and Stanton 1999), while for the remaining time (which can be centuries) **clonal propagation** will prevail (Bell and Bliss 1979, Steinger et al. 1996).

Plants with different **flowering and leafing phenologies** may respond differently to prolonged snow duration. For instance, Kudo (1992) observed that long snow-free periods caused evergreens to shorten their leaf life span (faster amortization of leaves) whereas deciduous shrubs maintained leaves longer – possibly a consequence of a melt-date induced earlier sprouting but a (fixed), photoperiodically controlled, senescence date. Inouye and McGuire (1991) have shown that in *Delphinium nelsonii*, flowering was delayed in years of lower snow accumulation, which they associated with lower ground temperatures in winter, but it may also indicate some photoperiodic control. Walker et al. (1995), like many others,

have noticed that key phenological events such as the date of maximum leaf length or flower number were closely related to the date of snow disappearance in two Rocky Mts. forbs. However, counter-intuitively, they found greater leaf length in a year with delayed snow melt, which they explain as a beneficial effect of the greater snowpack of that year on soil conditions during midsummer. From this they concluded that the nutritional and moisture situation in snowbeds may be more important than the actual length of the snow-free period (within certain ranges).

There is a lot of evidence for **opportunistic plant responses** at the beginning of the alpine season (Prock and Körner 1996 and Chap. 13). According to Friedel (1961) and the more recent work by Stanton et al. (1994), long-term spatial patterns of snow distribution are fixed, but short-term temporal ones are not, and the disappearance of snow controls the onset of the alpine growing season in habitats with late lying snow. Any delay of development (e.g. by photoperiodic control) would reduce the chances of achieving a positive annual carbon balance and of producing seeds during the remaining short season. Plants are sensitive to physical precursors of snow melt and many species initiate growth or reproductive development even before snow disappears (as discussed above). A survey of phenorhythmics of 184 species of high altitude plants in the central Himalayas by Pangtey et al. (1990) revealed a close synchronization of development with snow melt.

As soon as more than one species is considered in greater detail, answers tend to become rather complex – and alpine plants are no exception to this general rule, as will be illustrated in this book repeatedly. Galen and Stanton's (1995) snowbed study in the Rocky Mts. is a good example. They first mapped the distribution of snowbed forb species along a gradient of 80 to 35 days of annual snow-free period, and then manipulated snow pack duration at either end of the gradient. As illustrated in Figure 5.11, species whose absolute ranges of distribution are the same fall in at least three categories of microhabitat preference along the snow melt gradient: *Geum rosii, Artemisia*

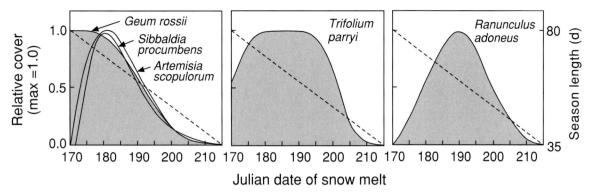

Fig. 5.11. Gradients of relative plant cover within populations of five typical Rocky Mts. snowbed forbs along a line of increasing duration of snow cover. *Left*, Species that reach greater cover (abundance) near the edges of a snowbed; *center* a species of relatively constant cover over a wide range of snow cover; *right* a species reaching greatest cover half way to the center of the snowbed. No species "centers" near the 35-day zone. (After data in Galen and Stanton 1995)

scopulorum and *Sibbaldia procumbens* are more frequently found near the margins of the snowbed, *Ranunculus adoneus* reaches greatest cover half way to the center, and *Trifolium parryi* likes it everywhere. All five snowbed specialists can grow in the center, but there is no higher plant species with a "preference" (sensu greatest relative abundance) for a place with a cool 35 days growing season.

The speed of initial growth and **development after snow melt** differs among these species (Fig. 5.12). *Ranunculus* is the first to flower and initiate leaf expansion, and *Trifolium* is the last, with the species sequence for the two traits staying the same between these two extremes. Once growth is initiated, the duration of the vegetative flush is shorter for those who start late, and longer for those who start early, with one species (*Geum*) holding an intermediate position in all cases. Note that leaf expansion commences over a range of 1 to 8 days after snow melt and shoot expansion takes between 5 and 16 days – periods during which it became 3 K warmer in the snowbed. Galen and Stanton (1995) noted that any manipulation of snowpack (i.e. the timing of snow melt) for a period of three years had more immediate consequences for changes in ground cover in the "fast" initiating species, i.e. those whose development follows snow-melt with minimum delay, whereas

effects in the slower ones were small. The "fast" species may profit from earlier snow melt through maintaining a greater metabolic readiness under snow – which, however, has its energetic trade-off and also requires developmental "readiness" by bud preformation. There may be a threshold under prolonged snow cover, when the continued maintenance of readiness becomes counterproductive, perhaps illustrated by the reduced success of *Ranunculus adoneus* in the center of the snowbed.

I have described this experiment in more detail because it illustrates and quantifies a number of important aspects of developmental plant responses to prolonged snow cover. First, and well known to vegetation ecologists, species have "preferences" for certain ranges across a snow duration gradient, but the width of "preferred" ranges varies among species ("preferred" means maximum abundance, and does not imply an optimum in a growth physiological sense). Second, independently of the zonal position, species differ in the dynamics of their development, even within one type of snowbed habitat. Third, the predictive value of these developmental properties for responsiveness to varying snow duration may be greater than the current position along a snow-melt gradient. And fourth, the metabolism before snow-melt may determine the post-snow-melt development.

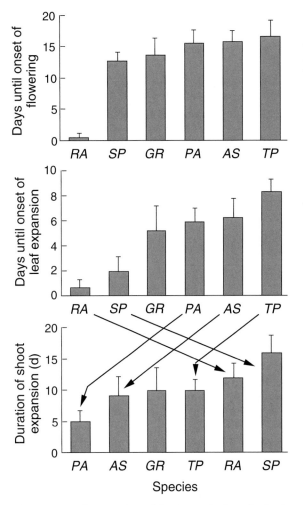

Fig. 5.12. Developmental speed in six snowbed species (as in Fig. 5.11, plus *Sibbaldia procumbens* and *Poa alpina*). Note that the ranking of species by time elapsed from snow melt to leaf intitiation and flowering is the same, but the duration of shoot expansion (*lowest* diagram) tends to be shorter in late initiating species and is longer for early initiating species. Canopy air temperature increased by 3 K during the observation period 3–17 July (from 15.5 to 18.5 °C). (Galen and Stanton 1995)

It is not yet known for sure whether alpine snowbed species evolved development-ecotypes, but it is almost certain that they did. There is evidence from arctic snowbeds that ecotypic, i.e. genetic, differentiation of carbon investment between snowbed and fellfield populations does

exist over short distances. McGraw (1985a, 1995) showed in a pot experiment that snowbed eco-types of *Dryas octopetala* were superior com-petitors and that this had to do with biomass allocation.

Growth and photosynthesis. It seems unlikely that sub-snow photosynthesis contributes signi-ficantly to the annual carbon balance of higher plants exposed to long-lasting snowpacks (see the previous Sect.). The C-gain of a single peak season bright day is probably greater than the sum of below snow photosynthesis of a whole winter. Hamerlynck and Smith (1994) measured CO_2 gas exchange in the snowbed geophyte *Ery-thronium grandiflorum* before and one day after snow melt and found negative values in both cases, despite high sub-snow QFD. But 2 days after snow melt, leaves reached a photochemical steady state and soon after exhibited full photosynthetic capacity. They conclude, in accordance with what Galen and Stanton (see above) found, that it is the speed of development after snow melt and the efficient use of the short season which matter, rather than below snow photosynthetic activity. These authors also found, by comparing snowbeds of different melt dates, that the snowbed plants were always bigger at emergence in snowbanks with early melt dates, most likely a consequence of greater below ground reserves produced during the previous year. Except for a short transition period during snow melt or under generally thin snow cover, sub-snow photosynthesis does also not appear to contribute substantially to the carbon balance of lichens (Sonesson 1989; Kappen et al. 1995). Rather it seems that lichens, but also some mosses, suffer under the mild temperatures and high moisture under long-lasting snowpack, conditions which stimulate respiratory losses (e.g. Benedict 1990 for lichens; Lösch et al. 1983 for mosses).

An important determinant of growth is **leaf construction cost.** Leaves of some herbaceous snowbed species share characteristics with plants from shady environments, such as soft tissue or high chlorophyll content on a dry leaf mass basis as was observed already by Henrici (1918). In competition experiments McGraw (1985b) found that snowbed ecotypes of *Dryas octopetala* were more shade tolerant than fellfield ecotypes. The soft leaves of many deciduous snowbed plants (e.g. Kudo 1992) may reflect at a microscale, what has been observed in alpine plants across a large latitudinal scale, namely that specific leaf area (SLA, the area of leaf produced per amount of dry matter) increases with the poleward shortening of the alpine growing season, with maxima in arctic alpine plants and minima in tropical-alpine plants (Körner 1989, Prock and Körner 1996 (see also Fig. 13.7)). It is known from functional growth analysis that high SLA is a key feature for fast growth also in cold-adapted plants (Atkin and Cummins 1994) – a plausible explanation in view of the extremely short period available for growth in such habitats. Large carbon investments per area in leaves of only 6 weeks life span would significantly reduce returns.

Soil fertility in snowbeds co-determines how fast biomass production can proceed in the short season. Several studies have considered the effects of local snow accumulation on soil nutrients. Walker et al. (1995) suggested that nutritional effects of snow melt are critical for individual species responses in the alpine zone of the Rocky Mts., and similar conclusions were drawn by Squeo et al. (1993) for the alpine belt around 4000 m in Chile at 30°S. As a rule, early melting zones in snowbeds have richer soils than late melting ones (Fig. 5.13), in particular, they have higher humus content, which was found to correlate positively with various parameters related to plant growth (Scott and Billings 1964; Stanton et al. 1994). Figure 5.8 illustrates that the later released plots showed a much smaller relative increase in biomass during the first week of growth (+22%) compared with successively earlier released plots (+71, +80, +104%). Early melting snowbeds are also richer in species than late melting ones (Stanton et al. 1994). Snow distribution also affects cryogenic soil processes (Johnson and Billings 1962), litter decomposition (O'Lear and Seastedt 1994) and microbial activity (Schinner 1983). Snow by itself is a substantial

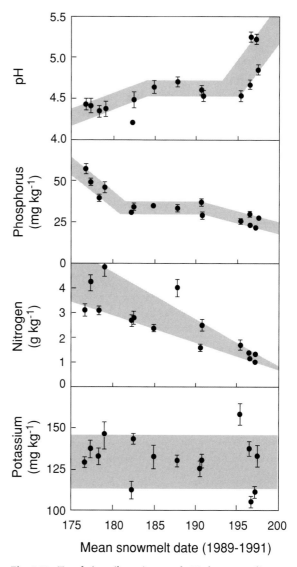

Fig. 5.13. Trends in soil nutrients and pH along a gradient of 15 microsites of differing (Julian) melting date in a large snowbed (2 mm soil fraction, Mosquito Range, Colorado, 3650 m). Note the opposing trends in pH and P (amonium acetat extraction), and the continuous reduction of total N (Kjeldahl) with increasing snow duration. K, Ca (extracted as for P) and soil moisture do not change (the last two not shown). *Shaded areas* cover ca. 90% of the data points. (Stanton et al. 1994)

source of soluble nutrients which accumulate over the winter and are released at once during the melt (Haselwandter et al. 1983; Bowman 1992; Maupetit et al. 1995).

In summary, snowbeds represent a rather specific part of the alpine life zone from subtropical to polar latitudes, with microenvironmental peculiarities and the co-occurrence of a variety of plant response types with respect to stress resistance, development and biomass production. Over very short distances and periods of time we find extreme changes in life conditions – a natural "experiment", which will continue to provide promising opportunities for the study of plant adaptation.

Alpine plants acquire water and mineral resources from substrates which differ in many respects from those common at lower altitudes. Over very short distances one can find a large variety of alpine soil types, ranging from alpine sand drifts to peat bogs, from shallow A/C (e.g. rendzina) soils to the deeply weathered profiles of some alpine grassland soils. Vegetation mosaics closely follow these below-ground patterns as was documented in a first synthesis by Braun-Blanquet and Jenny (1926), and many later studies have substantiated and confirmed the tight link between soils and plant distribution in alpine terrain (see the reviews by Neuwinger 1970, 1980; Gracanin 1972; Retzer 1974; Franz 1979). Here I will not enter the complicated matter of alpine soil typology (I refer to the above texts), but try to summarize some generally important aspects of soil formation and function in the alpine zone. With respect to the geomorphological determinants of alpine soil formation, I refer to the classical text by Troll (1944) and the more recent accounts by Caine (1974), Franz (1979), Harris (1981) and French (1996). I will first discuss some physical effects on soil formation, followed by a section on soil organic matter and, related to this, brief comments on soil organisms and their activities. Aspects of alpine plant nutrition will be considered in Chapter 10.

Physics of alpine soil formation

There are four principal sources or vectors that facilitate the local accumulation of fine mineral substrate in alpine terrain: (1) on site erosion of parent rock, (2) gravity (on and below slopes),

(3) sedimentation by water or snow (avalanches etc.), and (4) sedimentation by wind. Cryogenic processes play a role in all four cases. Apart from sedimentation in alluvial flats, the coarser matrix into which finer sediments fill or attach to initially, is created by landslides, glacial deposits, scree accumulation or solid rock weathering, and in some areas also by volcanic cinders and ashes. In the matrix of such coarse materials, the fine substrate accumulates from the bottom up and may not reach the top for a very long time. This explains the rich plant life often found on coarse, stony fellfields where the fine substrate is buried under rock debris (Fig. 6.1). From visual observation, alpine scree slopes or rock fields are often erroneously considered too dry for plant growth. This may not be true, as illustrated by Pisek's (1956) data on water relations of plants growing on such "hot", equator-facing calcareous scree slopes in the Alps, and a quick check by removal of top layers of coarse material will convince those who have doubts. There is usually plentiful moisture in the ground because of perfect screening of the substrate from direct atmospheric forcing. However, except for sun-exposed slopes or lower latitudes, these substrates are usually rather cold for plants, because roots have to grow deep and radiant soil warming is very much reduced.

The accumulation of fine substrate by **aeolic sedimentation** is a particularly important component of alpine soil formation (for references see Jenny 1926; Franz 1979; Litaor 1988). Dust deposition, commonly associated with drier lowlands, is enhanced in the alpine zone – particularly in geologically "younger" mountains (virtually all the large mountain ranges of the world) for several

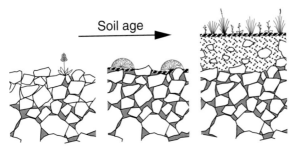

Soil age

Fig. 6.1. Because the filling of coarse substrate with fine material starts at the bottom, "soils" often exist long before they become obvious at the surface. These buried substrates are commonly rather moist and cool even in otherwise dry environments. Wind deposition of dust is an important input

Fig. 6.2. Coarse gravel or scree material often provides the first niches for plants to settle. *Viola cheiranthifolia*, Mt. Teide, 3600 m; Tenerife)

reasons. Most importantly, the presence of a newly eroded landscape provides a steady source. In glaciated mountains, glacial silt is a particularly significant dust source. Regular local wind systems ("mountain winds") contribute to the steady dust allocation, and relief and vegetation enhance its local deposition.

An interesting data set documenting the intensities of aeolic sedimentation was collected by Gruber (1980). He sampled dust from 7–8 month old snow fields at 20 alpine and 2 upper montane locations in the central Alps and also measured dust deposition during the snow-free period from 6 sites in the same area for 2 years. Excluding one site with extremely high deposition rates (see below) he ended up with means of 70–80 g of mineral dust per m^2 during the period of snow cover, and 15–30 g in summer, yielding a total of about $1 t ha^{-1} a^{-1}$. With a given density of $2.55 g cm^{-3}$ and a common pore volume of soils of 50% he calculated mean annual increments of soil profiles in sink areas of 0.08 mm (80 mm in 1000 years). The full range in his data was 0.01 up to 1.45 mm per year. The minimum rate of deposition would be sufficient to accumulate a layer of 10 cm of fine substrate during the whole postglacial period. The extreme of $1.45 mm a^{-1}$ was recorded in the center of the Hohe Tauern region of the Alps, downwind of bare phyllitic scree and the boundaries of a retreating glacier. Even small sand "dunes" are reported for this otherwise rather

moist area. It should be noted that the total dust deposition would be ca. 20% higher if the organic dust component was included. Not surprisingly, such sand and dust drift phenomena are enhanced in mountains of drier regions (cf. Franz 1979).

Retzer (1974) and Franz (1979) also mention a number of examples of long-distance transport of dust into alpine sedimentation zones of different geology, explaining calcium enriched soils in otherwise silicatic areas, and siliceous dust in purely calcareous areas, which further explains some phytosociological peculiarities. Again, the work by Franz's student Gruber (1980) in the central Alps revealed some surprising numbers: the annual addition of $CaCO_3$ in dust to vegetation on non-calcareous parent material ranged from 12 to $1100 kg ha^{-1}$ – the latter high value from an area with nearby sources, but values as high as $100 kg ha^{-1}$ were found in places with the next upwind point source at least 5 km away.

As a result of such wind-driven allocations of fine mineral substrate, alpine soil profiles often contain – counter to expectations – large fractions of very fine grain sizes. Topsoil layers developed under closed alpine grassland on plateaus, not in reach of any significant aqueous sedimentation, may contain several decimeters of stone-less very fine sand or silt. Holtmeier and Broll (1992) illustrated the strong influence of relief and plant cover on such wind-driven soil formations. They found loess accumulation in the leeward vicinity of islands of crippled conifers above the treeline in the Rocky Mts. – whole mini-ecosystems which, according to Benedict (1984), may migrate across the alpine terrain with a speed of 2 cm per year. Alpine loess is reported by Franz (1979) for a number of other mountain regions.

The **grain size distribution** found in alpine soils also depends strongly on the parent material (Fig. 6.3). Siliceous rock (schist, and particularly granite and gneiss) produce coarser initial grain structure

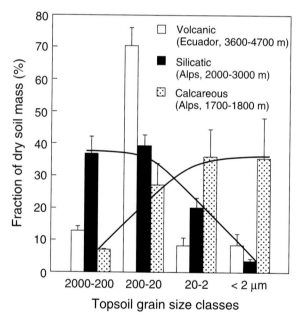

Fig. 6.3. Grain size structure in topsoils of three different mountain areas (means for the upper ca. 3–15 cm. Note the greater fraction of coarse grain sizes in the crystalline central parts of the Alps (nine different locations and soil types in the Hohe Tauern), the extremely high content of finest grain sizes from sites in the northern calcareous ranges of the Alps (three locations and soil types on Rax Alpe), and the narrow spectrum and relatively coarse grain size found in the young volcanic soils of the mountains of Ecuador (four locations on Cotopaxi, Chimborazo and others). (Franz 1979)

than calcareous rock, which, besides humus content, has important consequences for water holding capacity. Apart from substrate translocated as dust, grain size tends to increase with altitude, because more recently weathered material is coarser and alpine soils tend to be younger. Alpine soils formed from recent volcanic depositions show much narrower (and coarser) grain size spectra than those from older depositions.

Among the physical effects on alpine soil formation, **freezing phenomena** play a central role (Troll 1944; Johnson and Billings 1962; Smith 1987). In the short term, the uppermost soil layer may freeze in diurnal cycles, which is often associated with needle ice formation (Fig. 6.4). In the longer term, deeper soil freezing may occur following seasonal cycles. In both cases, soil plus vegetation are heaved and the substrate becomes periodically loose. On slopes, these frost-induced movements in combination with moisture loading in spring lead to secondary processes such as soil creeping (solifluction), and in flat terrain deep frost tables can induce waterlogging. The sorting and shaping effects of frost heaving of the ground are well known, since they occur both in the arctic and in the alpine landscape (Troll 1944; Rieger 1974). Evidence for such cryopedogenetic (or "periglacial") effects in the alpine life zone comes from all parts of the world, including tropical mountains. There appears to be consensus that seasonal frost cycles are far more important for soil processes than diurnal cycles. However, for plants diurnal frost heaving cycles may be crucial. In the following I will give a brief account of both types of freezing phenomena.

Frost heaving of soil results from the development of ice within freezing soil. Diurnal freeze-thaw cycles, mainly in the snow-free period at higher latitudes and at any time of the year in the tropics, penetrate the soil only to a few centimeters, whereas seasonal cycles in higher latitudes may reach a meter deep, largely depending on snow cover. The volumetric increase at the liquid-solid transition of water contributes relatively little to the heaving process. The bulk of the heaving is induced by the growth of ice-needles or ice-lenses at the freezing front which attracts upward migration of water. It requires a special soil texture and a combination of a certain rate of cooling and soil moisture for the development of the most harmful form of topsoil freezing for plant seedlings, the formation of bristle-like **needle-ice** ("Kammeis" in German) which may separate the upper few millimeters or centimeters of soil from the rest by up to 10 cm. Heaving of 1–3 cm is quite common, which uproots seedlings and leaves a disintegrated, soft soil surface after thaw (Fig. 6.4). By spraying paint on frost boils, Johnson and Billings (1962) demonstrated that 10% of all surface particles were turned after three nights of ground freezing. Even closely related plant species may exhibit rather different responses to such diurnal physical disturbances, as was illustrated for two species of Draba at 4250 m altitude in the Venezuelan Páramos by Pfitsch (1988). One species was found in frost heave locations, the other one, although found in the immediate neighborhood, was located in microhabitats without frost heaving.

However, slight diurnal frost heave of soil may also occur without such spectacular needle-ice formation. Smith (1987) recorded about 20 diurnal frost heave events per year in the Canadian Rocky

Mts. (a mean of 10 mm per heave), with a seasonal peak at the end of the growing season (October). Thanks to a worldwide review of the needle-ice phenomenon by Lawler (1988) a global map and altitudinal ranges of its occurrence are available (Fig. 6.5). The lower boundary rises from sea level in the temperate zone up to 3500 m altitude in the tropics, hence it is nearly always far below the treeline. The upper boundary follows more or less the line of alpine (shallow) permafrost, which also roughly marks the upper altitudinal limits of higher plant growth. This means that needle-ice formation can be found throughout the alpine life zone, worldwide. It occurs during snow-free periods, at higher latitudes, peaking early and late in the season.

Deeper penetration of soils by freezing temperatures during the cold season at higher latitudes causes massive **heaving of bulk soil**, leading to all sorts of ground-patterns (Troll 1944; Johnson and Billings 1962). Fahey (1974) observed up to 30 cm vertical movement in the Indian Peaks area of the Rocky Mts. in Colorado. Smith (1987) measured the soil level continuously for 2 years in the Canadian Rocky Mts. at 2400 m altitude and recorded peaks of 23 to 45 mm in two winters respectively. Smaller movements are reported by

Fig. 6.4. Needle- or hair-ice formation in the uppermost soil layer (mostly during clear nights) causes continuous seed bed instability and may uproot seedlings. Photograph taken at 6 a.m. at 4600 m altitude on Mt. Cayambe, Ecuador

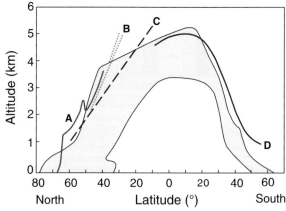

Fig. 6.5. The global distribution of needle-ice formation in the topsoil based on 113 published observations. Note that the upper limit is closely associated with the permafrost lines indicated for various latitudes (*A-D*). (Lawler 1988)

Matsuoka (1994) for a site with poorly vegetated debris at 2900 m in the Akaishi Range of Japan. With soil frost penetrating below 1 m he measured a heave of 20 mm. Matsuoaka explains this comparatively small heave by a very high freezing speed which prevented more water from migrating to the ice front. According to literature reviewed by Smith (1987) any freezing below about 25 cm depth does not contribute much additional soil heave. Hence, it is the uppermost layer of the soil that is lifted. Both authors could quantitatively link their observations to downslope creeping of soil after thaw, which was about 3 cm per year in the uppermost 10 cm of the profile in the example from Japan (slope 29°), and between 0.5 and 0.8 cm a^{-1} in Smith's experiment (slope 15–20°). Mark (1994) reported rates of downhill soil movement of 0.35 cm a^{-1} in solifluction terraces vegetated by cushion plants on Old Man Range, New Zealand. At slopes of only 3–7° hummocks graded into active stripes, underlining the strong effect of slope angle. Gamper (1981) noted a mean 3.7 cm per year drift in mobile soil stripes at only 2400 m altitude on steep westslopes of the Swiss central Alps. According to his observations, plant cover correlates negatively with alpine soil stripe movements, but it remains unclear whether cover is increased because the stripes became inactive (more likely) or whether dense vegetation had stabilized the stripes.

On a yearly basis, it is not possible to attribute such soil movements to freezing phenomena alone. Moisture loading of inclined topsoils plays an additional, in many cases, more important role. Veit et al. (1995) observed several decimeters per year soil creeping, largely due to spring wetting, in the Austrian central Alps (see also Veit and Höfner 1993). The relevant literature contains many more examples.

The magnitude of these displacements is important, both for understanding dynamics of soil processes, and for estimating the forces plants are exposed to by living in such a mobile substrate. There are several possibilities of what could happen to roots during seasonal soil heaving by frost. Cone-shaped ones could be forced out of the soil, as was documented by Perfect et al. (1988). Cylindrical roots well anchored in deeper soil, if solidly frozen would break, whatever the lifting rate around the neck. Reversed cones would become buried over the years. Some possible morphological compromises plants may have adopted to circumvent these problems are illustrated in Figure 6.6. These examples are from the subtropical Andes of northwest Argentina (4250 m), where vegetation experiences most severe frosts almost without snow protection in winter. Most of the tap-rooted species in this area have a contractile hypocotyl/upper root zone, which apparently contributes to retracting the apical bud several centimeters below ground after seedling establishment. Roots tend to be cylindrical, in some cases even somewhat bottle-shaped. Still, the only way not to break when the upper part is heaved seems

Fig. 6.6. Thick roots are required for plants to withstand the vertical forces created by soil freezing at high altitudes when a protective snow cover is missing, as in this case in the northwestern Argentinan Andes at 4250 m elevation (*Lepidium* sp.). By their contractive nature these roots also gradually pull the shoot apex of young plants several centimeters below the ground surface (see also Fig. 12.16)

to be to remain unfrozen, i.e. elastic, both a mechanical and physiological question that awaits testing. Paradoxically, Jonasson and Callaghan (1992) found plants from frost heaved soils in arctic-alpine polygons to have rather weaker roots than those from less disturbed ground, which they found to be in line with their observation of a general decrease in root diameter with the degree of soil disturbance (see also Callaghan et al. 1991).

Once the winter-heaving has been survived, the horizontal drift on slopes after the thaw has to be managed. Soil creeping at least happens during periods when plants are active and may respond by compensatory growth, an interesting area of functional morphology relating to plant life in steep and mobile ground in general.

On slopes, the long-term consequence of these vertical freeze-thaw motions and moisture loading of soil is downslope creeping, called **solifluction**. Creeping rates change with depth, soil structure and moisture content, and emerging ground patterns depend on whether the movement is restricted to the uppermost soil layers, or includes the whole profile. Commonly, upper layers are ahead of bottom layers and, in the case of a compact turf cover, this leads to pocket formation and revolution of the creeping front – a particularly obvious form of solifluction (Figs. 6.7, 6.8 and 6.9). This type of solifluction leads to a systematic mixing of soil strata and creates buried A-horizons. It may also lead to terrace formation which could counteract surface erosion in steep terrain in a similar way to the effect of terracing by hoofed

Fig. 6.7. Solifluction causes soils to creep down the slope (Furka Pass, Switzerland 2500 m). The ground frozen at a deeper level and water saturation, particularly at snowmelt, accelerate the process

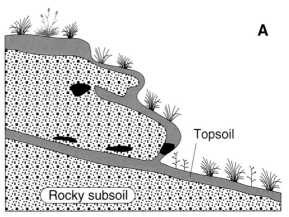

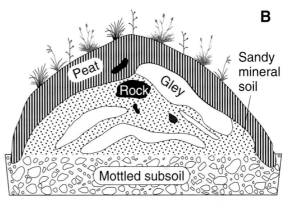

Fig. 6.8. Cross section of a solifluction lobe as shown in Figure 6.7. The *arrow* indicates a buried, dark humus layer (cf. Fig. 6.9A)

Fig. 6.9. Solifluction (*A*) and frost hummock formation (*B*), two typical cryopedogenic phenomena in alpine terrain of higher altitudes or latitudes. (Johnson and Billings 1962)

grazers (both wild and domesticated). The plant cover usually differs consistently between the terrace and its front (the top or the front of a lobe), adding to microhabitat diversity.

On less densely vegetated slopes, the substrate loosened by frost heaving may creep at even higher speeds (see the Japanese example above) leaving only specialist plants a chance of surviving the mechanical tensions. Usually these are plants with very thin and elastic roots and/or rhizomes and with a high potential for regenerating below-ground organs. Traces of roots of some tiny pioneers on such slopes may be found meters uphill of their current position. Burial of plants by debris or sliding rocks is common.

Other cryopedogenetic phenomena associated with frost heaving are polygon soils, garland and stripe formations, frost hummocks (Fig. 6.9) etc., all affecting snow depth, microclimate, soil moisture, nutrient availability and plant exposure to mechanical forces, and all these influence plant species distribution. There are many examples in the literature of vegetation following these ground patterns. Jonasson and Sköld (1983), for instance, noticed basophilous vascular plant species and

cryptogams on frost heaved polygon ground which were absent in the surrounding nutrient-poor shrub heath in the northern Scandes. They could explain this phenomenon by higher soil nutrient contents in the heaved, and thus regularly disturbed polygon ground.

Such cryopedogenetic activities increase with altitude, latitude and precipitation to evaporation ratio, and become more significant the steeper the terrain. Except for diurnal topsoil freezing, these phenomena are less abundant in drier subtropical mountains (locally similar patterns can be created by salt dynamics in the high Andes; H. Veit, pers. comm.), but do exist at higher altitudes under more humid tropical-alpine conditions, for instance in the Mt. Wilhelm area in New Guinea (between 4050 and 4500 m). They are also documented for the afro-alpine life zone above 3500 m (Löffler 1975 and references therein) and above 4300 m in the Andes (Furrer and Graf 1978).

Permafrost is not restricted to the arctic, but is quite commonly found in high mountains as well. Some of it may be fossil, such as in the case of passive rock glaciers, the melting of which under warmer climates could seriously affect high altitude ecosystems by slope destabilization; in other cases it reflects the current thermal balance of land surface and can be continuous or patchy (for references see Péwé 1983; VanTatenhove and Dikau 1990). The line of continuous alpine permafrost commonly matches the line of the upper limit of higher plant distribution (Lawler 1988). High altitude outposts of plants may well be sited on shallow substrate directly overlaying permanently frozen ground, which severely limits their root activity and moisture uptake as described by Rawat and Pangtey (1987) for plants growing at 5600 m in the central Himalayas. At temperate zone latitudes, permanently frozen ground may locally (predominantly on poleward slopes) reach down to treeline altitudes, but at depths which mostly preclude a direct interference with plant growth (>4 m, e.g. VanTatenhove and Dikau 1990; but see Fahey 1974, who reports much shallower frost tables for alpine fellfields in Colorado). In the upper alpine zone, the ground may locally remain frozen at a depth of 0.5 to 1 m as reported by Johnson and Billings (1962) for the Beartooth Plateau of the Rocky Mts. at 3300 m altitude (45°N). At higher latitudes, alpine permafrost is found closer to the ground surface, just as it occurs in the arctic, with all associated effects on soil processes and water relations (Jonasson 1986), and it may even extend into forested land (Kullman 1989). It is important that current alpine soils are also seen from a historical perspective. Many soil properties may be related to a climatic situation several thousand years ago (e.g. to humid periods during the late glacial or the Holocene in the Alps and Andes; Veit 1993, 1996).

The organic compound

The above physical processes create a fine and penetrable substrate for plant roots which in many cases does not develop much further. Large areas of alpine terrain are covered with such **raw and unstructured "soils"** which are regularly disturbed by freezing, erosion and displacement, yet these soils are inhabited by a large number of pioneer plant species and soil organisms. Compact growth forms such as cushion plants and cryptogams (mainly mosses), but also many specialized herbs and graminoids are found on such raw soils. By accumulating "compost" within their dense branchwork, long-lived cushion plants partly decouple themselves from the substrate, and run their "private" nutrient recycling with a rich microflora in an otherwise hostile soil environment (Schinner 1982a, Körner 1993).

As soon as such substrates become more stable and support more extensive and persistent plant growth, dead plant material will accumulate leading to a succession of developmental stages of **biologically driven soil formation**. This may happen over large areas or in crevices of weathered rock. The two important components that now start to interfere with the inorganic substrate are accumulation of organic carbon compounds and a steady supply of protons as the net outcome of photosynthetic reduction of CO_2. Depending on substrate texture, chemical composition, water regime, temperature, plant productivity and ongoing as well as historical influences by sedimentation and freezing phenomena, a multitude of soil types develop over thousands of years (Johnson and Billings 1962; Neuwinger 1970; Retzer 1974; Franz 1979; Veit 1993). There are two questions to be answered here: why and under what conditions does organic material accumulate, and what are the consequences for the inorganic soil matrix?

Per unit land area, **soil organic matter** (SOM), increases with altitude, commonly reaching a peak in the montane forest zone, sometimes extending into the lower alpine zone. At higher altitudes, SOM decreases and reaches close to zero levels in unvegetated substrates. Per unit topsoil dry matter (A-horizon) SOM concentration increases with altitude as well, but remains high at higher elevations, varying more with exposure, age and parent material than elevation (between 5 and 50% C of soil dry matter, according to Rehder 1970). Almost pure organic substrates may be found on some mountain tops, as in the case of some high altitude rendzinas on carbonate rock. This distinction between total pools and layer specific concentrations is important.

The reduction of SOM pools in the upper alpine zone – again disregarding special situations – is a consequence of reduced plant cover, rooted soil depth, reduced annual biomass production, and generally younger age of soils. In essence, it reflects the gradient of snowpack duration in higher latitudes discussed in the previous chapter, where SOM is reduced with increasing length of long-term means of snowcover (Johnson and Billings 1962, Stanton et al. 1994). In the case of snowbeds, SOM concentration decreases toward their center as well.

Unfortunately, SOM pools have rarely been calculated for alpine areas, and without knowing physical profile details, the commonly reported concentration data cannot be accurately converted. Table 6.1 summarizes some estimates for alpine soil profiles in the temperate zone. According to this limited data set, soil carbon pools in mature soils under closed vegetation in the lower and middle alpine zone may vary between 5 and $51 \, kg \, m^{-2}$ (50–510 t C ha^{-1}), with the higher values (31–51 kg C m^{-2}) found in very humic soils under Ericaceae dwarf shrubs at lowest altitudes. Values between 4 and 22 (Webber and Ebert-May 1977 report 5–15) kg C m^{-2} are more typical for closed alpine vegetation at medium alpine altitudes (a mean of 12 ± 6 kg C m^{-2} for the eight locations), which is also the range found in crop fields, pastures and forests at low elevations. A remarkable fraction of 6 to 32% of the total C in these mature alpine soils is accreted in the C horizon at depths of between 0.3 and 1 m. As reported by Retzer (1974) and Franz (1979), profiles with thick peat layers are quite common at the treeline ecotones

around the world, with C-pools possibly even exceeding those reported in Table 6.1.

Since annual net primary production is much lower in alpine compared with lowland ecosystems, the similar range of soil C pools found under closed alpine vegetation is remarkable (see next paragraph). Of the total ecosystem C pool, only 3–5% are bound in life biomass, 2–3% in dead plant mass and litter, and the remaining part of >92% is contained in the SOM. For comparison, in mature forests a common biomass to SOM ratio is 1:1. Roughly similar ratios apply to lowland grasslands, with the important difference that at lower altitudes SOM is usually diluted over larger profiles. According to Table 6.1, SOM concentrations in the A-horizon in these temperate zone alpine soils generally exceed lowland averages by factors of 2 to 5 – not including some extremes of almost pure organic peat. Similar high concentrations of SOM are reported for moist subtropical and tropical soils. Hope (1976) reports 12–19% in the alpine zone >3450 m of Mt. Wilhelm (New Guinea), and Rehder (1994) found between 6 and 13% C in the upper 15 cm of soils on Mt. Kenya at 4150 m. Vitousek et al. (1992) studied soils on lava streams of known age which opens a window to the understanding of long-term dynamics of SOM at high altitudes. A comparison of soil C pools across an age x elevation matrix revealed that C accumulation is initially faster at lower elevations (on young streams), but later (i.e. on older streams) the maximum C pool is found at higher elevations, indicating a negative correlation between SOM accumulation and plant productivity in the long term.

Why does SOM accumulate to such high concentrations in relatively narrow soil horizons, when annual production of new biomass is so much smaller at high altitudes? In young "expanding" soils, SOM accumulates because the mean residence time of C in SOM compared with life biomass is much longer. The life span of a leaf may be one season, e.g. 60 days. The full structural **decomposition** of its remains may take 2 years in high alpine forbs, 5 years in sedge leaves and 10 or more years in evergreen leaves of dwarf shrubs,

Table 6.1. Estimates of soil carbon pools per unit land area at lower and medium alpine altitudes of temperate zone mountains[a]

Location	SOM in A-horizon (% d.m.)	Thickness of soil horizons (cm)		Total pool (kg C m^{-2})	Carbon in C-horizon (% of total)
		(O+) A + B	C-horizon		
A. Rocky Mts.					
(>3500 m)					
Basalt	14	36	58	22	24
Schist	15	25	67	15	14
Granite	4	46	14	4	10
Quartzite	10	56	26	10	6
Shale	7	23	53	7	29
Limestone	8	46	30	8	14
B. Alps (Tirol)					
(Mica schist)					
Dwarf shrub					
heath 1980 m	49	63	35	51	10
Prostrate shrub					
heath 2000 m	13–81	51	25	31	18
as above but at					
2180 m	30	30	22	18	9
C. Alps (Switzerland)					
(Mica schist)					
Sedge turf 2470 m	5–14	30	30	9	32
D. Low altitude soils					
Tilled crop land	1–3			5–20	–
Pastures	2–7		1 m profile	10–30	–
Lowland forests	2–5			10–30	–

[a] Calculations are based on thickness, pore volume, volume density and % SOM for different soil horizons where provided by authors and assuming that 50% of SOM is carbon. Where pore volume and/or density numbers were missing (A, partly B), pore volumes of 65, 50 and 40% for A, B and C horizons respectively, and a solid density of the mineral substrate of 2.5 g cm^{-3} were assumed. Except for C , stone volumes were unknown and remained unaccounted. With the given uncertainties and necessary interpolations these numbers are only approximate and possibly not better than ±30%. For A, type of vegetation and altitude are not specified; data from Retzer (1974 p. 779). B, dwarf shrubs: *Vaccinium myrtillus* at treeline; prostrate shrub: *Loiseleuria procumbens*; wind exposure sharply increases with altitude; from Larcher (1977 p. 312). C, closed turfs of *Carex curvula*, from Körner et al. (1996, Swiss Alps); Danneberg et al. (1980), Austrian Alps – both studies revealed similar numbers). D, ranges for lowland soils were taken from Schachtschabel et al. (1982).

periods substantially longer than those observed at low altitude, where most leaf litter is recycled within a year (Seastedt et al. 2001). For stable situations where annual input of C equals annual output, the net production of SOM per year approaches zero.

Naturally senesced previous season litter of *Carex curvula* exposed at the soil surface for one season at 2470 m altitude in the Swiss Alps lost only 20% of its initial mass during the 10 week season, while low altitude samples collected and exposed in a meadow at 570 m altitude had almost disappeared within a 7 month period of exposure (Arnone and Hirschel 1997). No doubt even this first step of carbon recycling is substantially delayed at high altitude.

However, the delay of litter breakdown at alpine altitudes is not only a consequence of the shorter

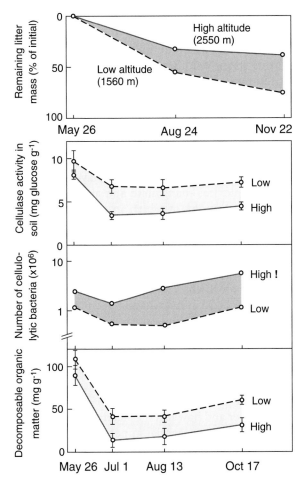

Fig. 6.10. Decomposition of leaf material at contrasting altitudes in the Alps and some associated substrate and microbiological characteristics. Plant samples were exposed in mesh bags 2–3 cm below the soil surface. Note the inverse relationship of cellulase activity and the number of cellulytic bacteria. (Schinner 1982b)

snow-free period. Schinner (1982b) compared the rate of litter decomposition and associated microbial parameters across a 1000 m transect (1560–2550 m) in the Austrian Alps (47°N) within the snow free period and found remarkable differences (Fig. 6.5). At the high altitude site only 46% of the herbaceous leaf material exposed in meshbags 2–3 cm below ground had decomposed between May and October, compared with 86%

at the low altitude site during the same period. Cellulase activity was only half as high but the number of cellulolytic bacteria was several times higher at the upper site. The fraction of decomposable organic matter in the topsoil was much lower in the alpine site. Hence, it appears that **microbial activity** is reduced under alpine field conditions, but microbe numbers are rather increased, suggesting much slower bacterial turnover. This is in line with observations that soil CO_2 evolution decreases with altitude when sieved soil samples are exposed to the same incubator temperatures (Schinner and Gstraunthaler 1981; Schinner 1982b; Davis et al. 1991).

In the next step, carbon gradually enters the SOM pool by further microbial action and forms complex compounds, some of which may take many thousands of years to become "recycled" and ultimately leave the SOM pool as CO_2, CH_4 or some other gaseous or soluble compound. For instance, the mean age of C in humic acids (one of the most resistant group of "humus" compounds) in bogsoil was determined as 5400 years (see Schachtschabel et al. 1982), other compounds such as fulvonic acids may take "only" 2000 or 3000 years for turnover and some low molecular weight carbon compounds may be recycled instantly. Danneberg et al. (1980) analyzed the humus fraction of alpine turf soils in the Austrian Alps at 2300 m. Their data indicate that the 12 kg of C they identified per m^2 of ground, largely consists of colloidal compounds ($7\,kg\,C\,m^{-2}$ in the B-horizon alone) with the highly polymeric, stable, gray humic acid complex dominating below the A-horizon (40% of all "humic compounds").

It depends on the metabolic activity of decomposers whether and how fast SOM will accumulate in such developing soils. For most of the time alpine soil temperatures are below 5 °C, except for a couple of warm weeks during the growing season, when radiant soil warming may occur (see Chap. 4). The same substrate will always take longer to decompose in the alpine zone as compared to lower altitude. Schinner and Gstraunthaler (1981) found a fourfold increase of CO_2 release in sieved soil samples between 4 and 18 °C.

At the same temperature, the alpine samples always exhibited lower rates of release than the lowland samples. Hence, both substrate properties and the climate prolong the mean residence time of C in alpine soils. The system reaches an equilibrium when, under a given constant input of new compounds of the same nature, the mean life span of the oldest components (statistically) expires. The mean life span of SOM compounds determines the time it takes to reach such an equilibrium state. The rate of input is driven by the rate of plant growth. If substrate conditions become more unfavorable for growth and/or nutrient recycling due to SOM accumulation, we have a classic negative feedback loop. This is what commonly happens during alpine zone soil development. The moister and cooler the location, the more pronounced will these negative feedbacks be.

According to Mahaney (cited by Retzer 1974), it takes alpine soils in the Rocky Mts. about 10 000 years to reach a thickness of 30 cm (1000 years for the first 20 cm) and structured three-horizon soils are least 2000 years old – a period during which pH drops by two orders of magnitude while soil organic matter increases from zero to about 6% of soil dry weight (ca. 100–200 t C ha^{-1} as result of a mean net addition of only 5 to 10 g C m^{-2} a^{-1}). Well developed alpine podsols in the Austrian Alps date back to the early Holocene (Veit 1993). Thus, soil formation at high altitudes is a very slow process, underlining the overriding significance of soil conservation in the alpine zone. Once removed, there is no way to replace alpine soils within historical dimensions of time. Since the soil-vegetation complex stabilizes slopes, its disintegration in steep terrain is one of the key components of downslope risk for human life.

The interaction of organic and inorganic compounds

In chemical terms, plant life is a reduction process. Wherever plant products accumulate, protons accumulate as well. Plant wastes acidify the soil. Nowhere else than in cold and moist environments, where recycling of plant products is delayed, is this **acidification** more pronounced. The dynamics of alpine soils are primarily driven by this strong acidification process (pH < 3.5 not uncommon). Acids mobilize cations, which interact with the raw material provided by the physical soil-forming processes discussed in the previous section. Danneberg et al. (1980) quantified the downward allocation of humic acids in alpine turf soils and found a peak early in the season. The seepage of acids into the lower soil layers released, for example, free manganese and aluminum ions, the concentrations of which in the soil solution would be considered toxic by lowland standards (Posch 1980). Moisture availability is the key determinant of such processes. Litaor (1988) concluded from his 5-year studies of the chemistry of soil solutions on Niwot Ridge, Colorado, that precipitation effects per se overshadow any possible effect of acid rain deposition on the ionic composition in soil water in such acid alpine soils. He also found that the concentration of free aluminum and iron in the soil solution was correlated with the total amount of dissolved organic compounds in the soil.

Whether and how fast the steady top-down flow of acids removes essential mineral nutrients from the rooting zone depends on the chemical nature and texture of the inorganic matrix. The input of cations by aeolic depositions or erosion and the intermingling of soil horizons by frost driven perturbations such as solifluction can partly reverse the otherwise unidirectional flow of nutrients. These processes are much more significant for plant nutrition at high compared with low altitudes because other vectors for soil regeneration, most importantly earthworms, are largely missing.

The higher the altitude above the treeline, the less is soil formation co-determined by **animals**. Franz (1980) mentions the negative effects of ground freezing and calcium shortage in acid alpine soils on earthworms, and the range of only a few species appears to include the alpine grassland belt (sporadic observations of *Octolasium* and *Dendrobaena* sp. in the Alps; Franz 1979, p.

257). However, Schinner (1982b) found no difference in litter decomposition whether he included or excluded soil fauna by selecting certain mesh gauges. A bibliography of alpine soil organisms (animals and microbes) was published by Broll (1998).

The diversity of soil **fungi** decreases with altitude and the species composition changes in a characteristic manner (Table 6.2). Also, a lower abundance of plant-mycorrhiza has been found at high altitudes (Haselwandter 1987; see Chap. 10). On the other hand, diversity and abundance of bacteria is high. Schinner et al. (1992) isolated 130 different strains of microorganisms in alpine environments of the western and eastern Alps which can survive and multiply at 0°C. Of these 77% were **bacteria**, 20% yeasts and 3% hyphomycetes. Almost half of the bacterial strains were found to excrete protease into cultivation media when grown at 10°C. Most of the bacterial strains were found to belong to the genus *Pseudomonas*. Mancinelli (1984) conducted a very detailed analysis of bacterial diversity on Niwot Ridge, Colorado, and found *Pseudomonas* and *Bacillus* to be most

abundant and he also detected high proteolytic activity with characteristic seasonal fluctuations. Dinitrogen fixing as well as nitrifying and denitrifying bacteria are abundant in alpine soils (Wojciechowski and Heimbrook 1984, Mancinelli 1984; see Chap. 10).

In summary, the high concentrations of soil organic matter in many alpine soils are the result of both direct negative effects of the climate on organic matter recycling, and the negative feedbacks of SOM accumulation on the soil milieu via acidification. Acidification together with high precipitation/evaporation ratios affect the mineral nutrient balance, and thus nutrient availability for plants (see Chap. 10). Despite these "adverse" effects on alpine soils, we find a large number of plant species, growing under these soil conditions, at comparatively high rates (Table 15.5). When one sees a fully blooming high alpine turf at pH < 4 – (see color Plate 2 at the end of the book) or a colorful fellfield community at the mouth of a glacier, doubts arise about our concepts of "limitation" and our judgement of what is a "good soil", based on lowland standards.

Table 6.2. The altitudinal distribution of certain dominating soil fungi in the Alps. Note the decreasing diversity with increasing altitude; only 1 out of 18 genera appears to be present across the whole 1000 m transect. (Schinner and Gstraunthaler 1981)

Altitude	1560 m	1920 m	2300 m	2550 m
Vegetation	Meadow	Pasture	Alpine turf	Fellfield
soil pH	6.3	4.1	3.4	3.5
Fungal species:				
Cheatomium homopilatum				**
Cladosporium herbarum			*	**
Chrysosporium pannorum			*	**
Monascus sp.			*	
Paecilomyces sp.			*	
Fusarium sp.			*	
Trichocladium opacum		*		
Trichoderma inflatum		*		
Aspergillus versicolor		*		
Aspergillus fumigatus	*	*		
Cylindrocarpon sp.	*	*		
Pseudogymnoascus sp.	**	**		
Volutella sp.	*			
Monilia sp.	*			
Penicillium sp.	**			
Mucor racemosus	*			
Mucor parvisporus	*	**	**	
Mortierella sp.	*	*	**	**

In this section I took a rather general approach, discussing aspects that may apply to the majority of mountain areas with ample water supply and with a focus on areas with closed vegetation on reasonably well developed soils.

In three circumstances alpine plants may grow and thrive in substrates in which the major trends outlined in this chapter do not apply: (1) in unstable substrates where soil formation does not occur, (2) in arid mountains, where moisture shortage limits biomass production, accumulation of soil organic matter and humic-acid dynamics as described above, and (3) in young calcareous or volcanic substrates.

In all three cases we find rather species-rich high altitude floras, but the ground cover and productivity is so low and/or the mineral nutrient status of substrates is so favorable that the above soil constraints do not apply. Further, higher plant life extends to altitudes far beyond the level were "soils" exist. Patches of sand and debris at altitudes above 4000 m in temperate latitudes and above 5000 m in tropical latitudes support plants whose root environment is essentially inorganic. Soil processes in the classical sense do not add to the explanation of the success of these high altitude specialists.

In their recent account of organic matter dynamics in the Rocky Mountains, Seasedt et al. (2001) concluded that the major difference in controls on decomposition processes between alpine soils and soils of the arctic tundra appears to be the relatively large within-ecosystem variability generated in the alpine soils by topography–climate interactions. While arctic soil dynamics are driven by soil freezing, water logging and summer temperatures, alpine soil dynamics depend more on slope and exposure and, thus, snow duration and top soil moisture, which, in turn, reflect exposure-driven wind and radiation regimes.

Treelines, wherever they occur, at thermal, drought, nutritional or salt stress gradients represent an abrupt transition in life-form dominance; lines beyond which massive single stems and tall crowns either can not be developed, become unaffordable or are disadvantageous (Fig. 7.1). Why do trees disappear above a certain elevation, what causes the alpine zone to be treeless, i.e. "alpine"? The answer to this question would also indicate which functional attributes alpine plants must have to be where trees are unable to be. Thus, there is a reciprocal interest in this question, upslope for forest ecology, and downslope (because of its lower boundary) for alpine ecology.

The alpine treeline is possibly the best known and most studied of all distributional boundaries of trees, but still awaits a conclusive functional explanation that withstands testing across the non-arid mountains of the world. In this chapter, which builds on a recent review (Körner 1998), I will first illustrate the global patterns of treeline positions and their climatic relatedness, and will then discuss approaches towards a mechanistic interpretation of alpine treelines. It is only the general pattern which is of interest here, not the regional or local peculiarity.

About trees and lines . . .

What is a "tree"? What is a "line" in plant distribution? Although these two questions sound trivial, confusion about their answers has contributed to a century-long treeline discussion. I will try a short-cut. For practical reasons, a "tree" in the current context is defined as an upright woody plant with a dominant above-ground stem that reaches a height of at least 3 m, independently of whether reproduction occurs or not. This height assures that such a tree would have its crown closely coupled to prevailing atmospheric conditions (in contrast to low stature plants which strongly modify their climate, as was shown in Chap. 4) and would protrude through even deep snow and the height reached by large herbivores foraging in winter. Plant types with tall perennial structures such as bamboo, cacti or giant rosettes, also found at high altitudes, are considered special life forms not included in the term "tree". Whether the few species of tropical-alpine giant rosettes actually do grow above the true climatic treeline may be questioned (see also below).

The definition of a "line" is more delicate, because "any natural boundary is in reality a transition zone, which has its own two boundaries. They are, in turn, also transition zones with their own boundaries, and so on endlessly. So localization of a natural border is in principle inexact and therefore determined by convention". Thanks to A. Armand (1992) for this refreshing end to the debate!

What are useful conventions? There are several in use. The upper "limit" of the closed forest has been called the "**timberline**", but "closure" rarely ends abruptly, nor does it always require logs of "timber" size to establish a forest. A more common situation is a gradual decline of tree size and opening of the canopy. The upper limit of the occurrence of tree *species*, i.e. the uppermost outposts of individuals in the so-called "kampfzone" is another approach, but this "**tree species limit**" may conflict with the definition of "tree" given

Fig. 7.1. The lower boundary of the alpine life zone: the alpine treeline. *Left* on Pico di Orizaba, Mexico, at 4000 m, (*Pinus hartwegii*); *Right* near Haast Pass, South Island, New Zealand, at 1200 m (*Nothofagus menziesii*). The sharp *Nothofagus* treeline reflects a recruitment dilemma: too shady inside the forest, too harsh outside (Wardle 1998)

above or put too much weight on peculiar micro-habitat conditions (in the case of isolated tree outposts). The "**treeline**" (or **forest line**) takes a middle position (preferred here) and roughly marks a line connecting the highest patches of forest within a given slope or series of slopes of similar exposure. This definition corresponds to the one used by Brockmann-Jerosch (1919) and Däniker (1923) in their classical monographs on treeline biology of the Alps, and it is the definition used by Hermes (1955) in his global survey (see below). Since "timberline" and "treeline" are coupled boundaries the fundamental mechanisms causing their general position should be similar. While forest line would terminologically be more adequate, I retain "treeline" because this term is in wide use and has become self-explanatory.

As already mentioned in Chapter 2, the term **subalpine** will not be used here in order to avoid confusion, as Löve (1970) comments: "one can only sadly state that utter confusion reigns, and it is almost necessary to know where, geographically, and to which "school" the discussant belongs in order to make sense out of chaos and misunderstanding". A logical definition would be the tran-

sition zone ("ecocline") between the upper limit of the closed montane forest (the timberline?), and the tree species line (i.e. the beginning of the tree-less alpine zone), but not everybody might agree on that. Subalpine parkland is another plausible term for this transition zone (Rochefort et al. 1994). In central Europe closed forests, several hundreds of meters below the treeline, are often termed subalpine, which would correspond to what is called upper montane forest in other parts of the world.

Current altitudinal positions of climatic treelines

First the distributional pattern of treelines needs to be documented. At which altitudes are climatic treelines situated across the world? Besides the above mentioned classics for the Alps, a large number of authors have described the worldwide positions of treelines in the past (Troll 1973; Wardle 1974, 1993; Franz 1979; Baumgartner 1980; Arno 1984). Historical trends were reviewed recently by Rochefort et al. (1994). For tropical and subtropical mountains, reviews were published by Ohsawa (1990), Miehe and Miehe (1994) and Leuschner (1996). The most extensive quantitative analysis so far has been done by Hermes (1955) who published his results in a local geographical report and in German, the reason why it may not have received the attention it deserves. The following discussion will lean heavily on the Hermes data set, extended by 26 entries from Wardle (1974), a few points from Arno (1984) and my own observations (cf. Körner 1998a). The snowline was included because it is a purely physics-driven reference (Troll 1961); a thermal boundary which connects points, above which the ground remains snow covered throughout most of the year, and precipitation falls as snow (an approximation of the elevation of the 0 °C isotherm of the warmest month). The resultant 120 data pairs for treeline and snowline, plus 30 individual treeline elevations were used for the polynomial regression analysis shown in Figure 7.2. Subsets of data for three north-south transects are shown in Figure 7.3, of which the one for the two Americas is most instructive.

The data in Figures 7.2 and 7.3 quantify trends which in principle have been noted repeatedly in the literature (for references see Troll 1973; Miehe and Miehe 1994). A linear regression of the altitude/latitude relationship in the northern hemisphere between 70 and 45°N yields a change in treeline altitude of 75 m per degree. Over the whole temperate/subtropical transition (50 to 30°N) the slope is 130 m per degree of latitude. The lower slope at higher latitudes fits the analysis by Malyshev (1993), who found a range of between 70 and 90 m per degree for various transects in northern Asia. Adopting a **worldwide** perspective these **patterns** suggest that:

1. There is no strict correlation between the altitude of the alpine treeline and latitude, even if one accounts for the thermal equator being 6–7° north of the geographic equator. There is a steep, almost linear increase in treeline altitude in the temperate zone, with the maximum altitude reached in the subtropics at 32°N and 20°S lat. Over a range of about 50° across the equator, treeline position does not change significantly with latitude.

2. The variation around the latitude specific mean is small in the Southern Hemisphere (north-

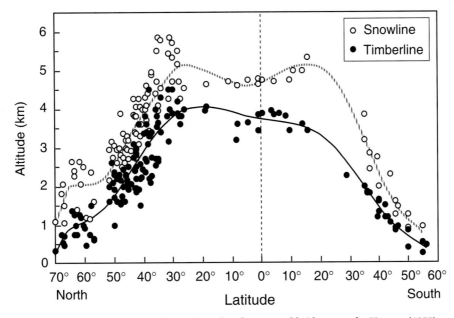

Fig. 7.2. The latitudinal position of treeline and snowline taken from a worldwide survey by Hermes (1955), completed using data from various other sources. Further details in the text

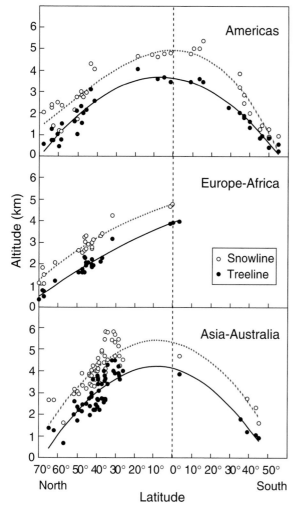

Fig. 7.3. Subsets of data taken from Figure 7.2 for three latitudinal transects. *Top*: this transect is most complete and follows the cordillieras of the two Americas. *Middle*: all European mountains, the Atlas and the equatorial mountains of Africa. *Bottom*: the bulk of the data for Asia (from Caucasus to Japan), connected via Indonesia to the mountains of Australia and New Zealand

south stretching cordilleras of South America, mountains of Indonesia, New Zealand etc.) and is large in the Northern Hemisphere, where the bulk of mountain masses is situated in continental areas. For instance, at 40°N, treeline altitudes vary between 2100 and 3700 m altitude. In the Alps alone, at 47°N, the range is

1600 to 2300 m over short distances. Climatic factors not closely related to latitude must have a strong influence on treeline formation. Hermes (except for Alaska) did not differentiate humid coastal slopes (commonly with lower limits) and drier inland slopes (with higher limits), causing a great deal of the altitudinal scatter seen in Fig. 7.2.

3. The above biological trends run parallel to physical trends illustrated by the snowline. The snowline, like the treeline, does not vary significantly over the same 50° span of lower latitudes. In fact, there is a depression in the snowline altitude around the thermal equator, between 0–10°N, which might have been seen in the treeline as well, if more data had been entered for these latitudes. The combination of these two data sets suggests a common physical driver for snowlines and treelines across the globe.

4. The distance between treeline and snowline, which roughly corresponds to the altitudinal range of the alpine life zone (if plants living above the snowline are disregarded here), varies between 800 and 1600 m, with a mean of about 1100 m (subantarctic Chile and Argentina disregarded, where the range is narrowed to 300 m). The distance reaches a maximum in the subtropics, and there are smaller values in the temperate zone and equatorial tropics. It is important to note that the narrowing trend at low latitudes is not restricted to the humid tropics, but starts in the subtropics.

Treeline-climate relationships

Any discussion of vegetation-climate relationships suffers from three fundamental problems, namely (1) auto-correlation between different climate factors, (2) averaging procedures, and (3) from the uncertainty of whether present or past climates are reflected in the current patterns (see section on palaeo-treelines below). After having explained these problems I will discuss two questions: does the worldwide distribution of treelines follow a common thermal boundary, and if so, which? And what explains the intrazonal variation and the pantropical plateauing of treelines?

Auto-correlation: at the same altitude, lower latitude is associated with higher mean temperatures, lower seasonal amplitudes of temperature, longer duration of the growing period, larger diurnal amplitudes of temperatures (Fig. 3.2), higher solar angles and peak radiation, and often decreasing cloudiness and precipitation/evaporation ratios. In the case of the alpine treeline all these components of the climate may be important, even if one of them, such as a certain temperature mean, shows a reasonable correlation with treeline position.

Means are generally **problematic**, as will be discussed, but often this is the only information available. It is not helpful for an understanding of the treeline phenomenon to state that "everything matters", that all climatic factors must be considered, that climatic means cannot be used etc. and, instead of carefully applying available data, providing verbose elaborates which do not bring us any nearer to solving the problem. Existing data should allow us to rank environmental factors and pinpoint the key ones in a first approximation, even if some other determinants may become important at times as well. When doing this, a distinction needs to be made between gradual influences of a factor (e.g. effects of temperature on growth) and threshold factors (e.g. effects of extreme temperatures on survival).

Past, and not **current climates** may have determined treelines. Because of the long life span of many treeline forming species, and resilience of established forest communities to environmental changes, substantial outphasing between climate and treeline position is to be anticipated. Ives (1978) said: "to relate natural timberlines in Colorado to climatic parameters of the twentieth century will have an air of unreality in face of the trees' ability to persist through perhaps several thousand years". The Methuselahs of treeline trees Ives refers to are quite rare (Klötzli 1991), but a few hundred years of climatic carry-over stored in today's treelines is plausible.

These difficulties severely constrain local explanations of climatic treelines, but on a global scale, they will only increase the noise around an overall relatedness to environmental drivers (which we seek here) and, perhaps, introduce some systematic "error" (lags) with respect to global trends of climate. There is also a practical problem: hardly ever are climate data and treeline positions published together, and climate stations are commonly not where natural treelines are. Adequate data for solar radiation are non-existent for treelines of most parts of the world. Fortunately, temperatures of tree crowns near the treeline are relatively close to air temperatures (Grace 1977; Goldstein et al. 1994), and also soil temperatures tend to be closer to mean air temperatures under trees than under treeless vegetation (Winiger 1981; Körner et al. 1986; Miehe and Miehe 1994). Though usually better coupled to the atmosphere than lowland forests, dense tree canopies at the treeline, may still warm up by 1–2 K (long-term means) compared with the air temperature, and in a few instances larger deviations have been reported (see the discussion in Körner 1998a), this is not specific to treelines but is true everywhere. Finally, at higher latitudes, correlations with temperature (on which I will focus here) do not allow a distinction between direct effects of temperature and effects induced via variable duration of the growing period – both are operating in the same direction.

Back in the nineteenth century, alpine treelines were repeatedly quoted as coinciding with a mean **air temperature of the warmest month** of about 10 °C (for references see Brockmann-Jerosch 1919; Daubenmire 1954; Holtmeier 1974; Grace 1977). From the data I have extracted from various sources and plotted in Figure 7.4 (uppermost line) it seems that this is a temperate zone perspective of the world. For instance, the treeline on Mt. Wilhelm in New Guinea at 3850 m (5°S) lies at a mean air temperature of only 5.6 °C (there is only a 0.3 K "seasonal" variation). Similarly, low mean temperatures apply to climatic treelines in the afro-alpine and tropical-Andean climate (Miehe and Miehe 1994; climate data by Rundel 1994;

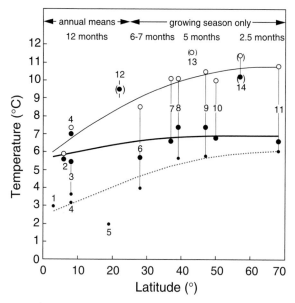

Fig. 7.4. Temperatures at alpine treelines for various latitudes (sites for which the seasonal course of temperature was available). *Upper line* Mean temperature of the warmest month. *Middle line* Mean temperature for the growing season, for sites north of 28° lat calculated from monthly means for May, June, July, August and September. *Lowest line* Temperatures for extremely high altitude patches of forest. Where interpolations were necessary, the local seasonal adiabatic lapse rate was used (between 0.45 and 0.6 K/ 100 m). Treeline sites and altitude (extreme): *1* Kilimanjaro (3950 m), *2* Mt. Wilhelm New Guinea 3850 m (4100 m), *3* Bale Mountains, Ethiopia 4000 m (4100 m, only rough estimates of temperature), *4* Venezuelan Andes 3300 m (4200 m), *5* *Polylepis* record northern Chile 4900 m, *6* Khumbu Himal (Everest region) 4200 m (4420 m, both altitudes for sunny slopes, 200 m less on shade slopes), *7* White Mountains, California 3600 m, *8* Front Range, Rocky Mountains Colorado 3550 m (3900 m), *9* Central Alps 2100 m (2500 m), *10* Rocky Mountains, Alberta 2400 m, *11* northern Scandes, Abisko 680 m (750 m). Island mountains not considered in the lines fitted to the data: *12* Mauna Kea, Hawaii 3000 m, *13* Craigiburn Range New Zealand 1300 m, *14* Cairngorms, Scotland 600 m. Data for the coastal Mt. Rainier near Seattle range halfway between island and mainland data, but were omitted for clarity. Compiled from data by Arno (1984), Aulitzky (1961), Goldstein et al. (1994), Grace (1977), Hermes (1955), Hnatjuk et al. (1976), Lauer (1988), Lauscher (1977), Miehe (1989), Miehe and Miehe (1994), Rundel (1994), Sonesson and Hoogesteger (1983), Troll (1973), Wardle (1971,1974)

Hoch and Körner 2003). According to Wardle (1971, his Table 1) the treeline in the Craigieburn Range in New Zealand (1300 m, 43°S) matches with the 11.6 °C isotherm for the warmest month, i.e. twice the number for New Guinea. In the central, more "continental", part of the Alps the mean air temperature of the warmest month measured right at the treeline at 2100 m (47°N) for 6 consecutive years was 9.5 °C (Aulitzky 1961), close to Daubenmires numbers for the Rocky Mountains. In the Scottish Cairngorms (58°N) *Pinus sylvestris* grows up to 600 m altitude with a July mean of ca. 11.4 °C (Grace 1977). Ohsawa (1990) lists numbers for Hokkaido which suggest a warmest month mean for the 1600 m treeline of 12.4 °C, similar to the 13 °C reported by Takahashi (1944) for the birch treeline at 2350 m in the Hida Mountains. Finally, a subarctic example: the upper limit of the birch forest (not its "krummholz"-form) at ca. 680 m altitude in northern Sweden (68°N) coincides with a mean air temperature of the warmest month of 10.5 °C. The mean for the arctic treeline in Siberia is 11.2 °C (Malyshev 1993).

The "warmest month" rule systematically overestimates the actual temperatures during the growing season at higher latitudes. Since growth is much more sensitive to temperature than photosynthesis (see below) and because most of the expansion growth in this part of the world happens early in the season when temperatures are still low, this temperature is also of limited physiological relevance. Accordingly, Malyshev's (1993) very detailed analysis of the latitudinal variation of the upper treeline in northern Asia showed that temperatures exceeding either 0 or 5 °C are clearly more effective in controling treeline position, than temperatures exceeding 10 °C. Ellenberg (1963) has already suggested counting the days with means of air temperature above 5 °C, and he found that 100 days fit treeline positions much better then warmest month means. Aulitzky (1961) reports that soil at a depth of 5 cm exceeds 5 °C for 128 days at the central alpine treeline of Tirol, but climatic data permitting such an analysis are hard to obtain for other regions.

The few treeline **temperatures for whole growing seasons** I could find illustrate a much smaller discrepancy between treeline temperatures across latitudes (lower thick line in Fig. 7.4). While warmest-month temperatures differ by 5 K (extreme 8 K) between temperate and tropical treelines, seasonal means differ by only 1 K. With more detailed information on diurnal courses, it could possibly be demonstrated that this remaining difference is due to a greater weight of night-time temperatures in the calculation of means for the tropical treelines compared with high latitudes where nights are shorter during the growing period. In other words, the tropical means are slightly depressed relative to the temperate zone ones by the tropical 12 h nights. On a daytime-only basis this remaining latitudinal difference is likely to vanish.

Finally, higher seasonal mean temperatures at higher latitude treelines may be a consequence of averaging. Since most plant-temperature responses are non-linear, means composed of fairly constant temperatures, as in the case of arctic-alpine treelines, versus numerically identical means, resulting from a wide amplitude, as in the tropics, differ in their biological meaning. If metabolic processes, such as mitochondrial respiration, with exponential temperature responses are related to the treeline phenomenon (as suggested by Dahl 1986), one would indeed expect higher arithmetic mean temperatures at treelines of higher latitudes. This discussion illustrates the difficulties in using means when biological processes are encountered. A careful latitudinal comparison of treeline climates based on the frequency distribution of hourly means of temperatures or thermal sums above certain thresholds could clarify the situation. Modelled annual temperature amplitudes (Jobbagy and Jackson 2000) can explain the gap between the actual broad-leaved treelines of the Southern Hemisphere and the potential tree limit (*Pinus contorta* doing well several 100 m above the *Nothofagus* treeline in New Zealand, Ledgard and Baker 1988, with a much narrower gap in southern Chile, Wardle 1998).

Treelines on **islands**, in coastal zones **and forest outposts** don't fit the overall pattern. The season means for treeline in the Cairngorms in Scotland and for Mauna Kea in Hawaii are at least 3 K above those of the mainland mountains, which could be related to the **maritime climate** (Wardle 1974; Leuschner 1996) but more likely reflect a **lack of** appropriate **taxa**: ca. 25-m-tall *Picea abies, Pinus strobus* and *Eucalyptus* sp. thrive several hundred meters above the native *Metrosideros* treeline near the Mauna Kea observatory. Higher wind speeds (tighter atmospheric coupling, Grace 1977,1988) and greater cloudiness may be locally effective, and the use of means of a fairly buffered coastal climate compared with means of oscillating climates (see above) provide another explanation.

Forest patches beyond what is commonly believed to be the "proper" climatic treeline are more difficult to explain (lowest line in Fig. 7.4). They do occur in all parts of the world and at estimated air temperatures of 5 °C or below. For instance, in the Swiss Alps near Zermatt "proper" *Pinus cembra* trees can be found on rocky outcrops (not in sheltered groves) as high as 2500 m, 200–300 m above the "treeline". Holtmeier and Broll (1992) describe **tree islands** ca. 400 m above treeline in the Rocky Mountains. Miehe (1989) reports 20 cm diameter 3.5 m tall *Juniperus recurva* trees for 4420 m in the Mt. Everest region, several hundred meters above the line of continuous tree occurrence in the Central Himalayas. According to Troll (1973) patches of conifer forest reach 4600 m altitude in eastern Tibet, and *Polylepis tomentella* trees are found in northern Chile between 4800 and 4900 m (Hermes 1955, Troll 1973). In Venezuela *Polylepis sericea* stands are found up to 4200 m, i.e. 900 m above the "official" treeline (Goldstein et al. 1994), and form impressive forests at 4100 m (Fig. 7.5).

The *Polylepis* problem. Miehe and Miehe (1994) and Kessler and Hohnwald (1998), provide substantial evidence that the classic "warm slope hypothesis" (Troll 1959), is not supported by available data. This is in line with the analysis by Goldstein et al. (1994) who found the common slight warming of closed canopies compared with free air, but the extent of this is not sufficient to explain the altitudinal difference between the current treeline and these forest outposts on the basis of thermal peculiarities alone (Fig. 7.5). In addition, they report an analysis of 256 *Polylepis* stands in Venezuela by H. Arnal, in which no pref-

Fig. 7.5. Patches of "outpost" forests on block fields, 400 m above the continuous current treeline: *Polylepis sericea* at ca. 4100 m altitude, Merida, Venezuela. Soils under such forests tend to be 1–2 k colder than in adjacent open terrain (Kessler and Hohnwald 1998)

erence of slope orientation to the sun was found, which contrasts with widespread supposition. It is interesting that Beaman (1962) also found no north-south differentiation in treeline altitude in Mexico. On the other hand, Smith (1977, cited by Goldstein et al. 1994) demonstrated (by planting seedlings) a positive sheltering effect of the boulders on which these *Polylepis* stands are usually found. High photosynthetic efficiency and frost tolerance plus the favorable seed-bed conditions are the physiological explanations for the *Polylepis* phenomenon favored by Goldstein et al. (1994).

Miehe and Miehe's global approach to an understanding of the "outpost-problem" appears to support Ellenberg's (1958, 1996) hypothesis, that such forests are remnants in an otherwise deforested landscape due to centuries or millennia of human **land use** by grazing and fire, i.e. not the result of particularly sheltered, warm habitats. **Palaeo-ecological data** (pollen spectra) from high altitude lake sediments in the Andes document (and this holds for other parts of the world as well, see below) that the mean altitude of the treeline was approximately 200 m higher during the warmest postglacial period than today, and it was

formed by *Polylepis* (Lauer 1988). As this author convincingly suggests, *Polylepis* appears to have retained some of its hypsothermic positions, perhaps facilitated by the above favorable factors. In this case, the *Polylepis* treeline is a living "fossil" and does not represent the remnant of a potential treeline at current climates. The peculiar habitat conditions may have favored either the "remnant" or "fossil" position of these forest patches. However, it is remarkable that closely related taxa also belonging to the Sanquisorbeae tribe of the Rosaceae exhibit similar distributional patterns in Africa (see discussion in Troll 1978), which could be interpreted as a taxon-related selection for life at high tropical altitudes.

In the case of the *Polylepis* stands, the (1) "shelter" hypothesis (peculiar habitat) and the (2) "remnant" or (3) "fossil" hypothesis may find a common denominator in the ground structure: the block-fields on which high altitude *Polylepis* is commonly found prevent fire from spreading because of a lack of a continuous fuel cover (dry grass). As grazing land, these block-fields are rather useless. However, for seedlings bolders provide protection from radiative freezing and frost heaving of seedbeds during clear nights. Once established, these forest islands – remnants or fossiles – may create an interior climate which is buffered against climatic extremes (cf. Slatyer and Noble 1992) and resist moderate climatic variation. Depending on whether one favors the "fossil" or the "shelter" hypothesis, the treeline reflecting today's climate, particularly in the subtropics and tropics, might or might not be a few hundred meters higher than we actually see it (the lowest line in Fig. 7.4). Assuming a "true" treeline position at current remnants' positions would, however, only be justified in the case of outpost forests and not on the occurrence of isolated trees (see below). Still other arguments may be required to explain the high altitude forest outposts (or the missing forest in between) in the old world. The important point here, is that seasonal mean ground or air temperatures between 2.5 and 6°C do not per se seem to prevent growth of trees.

The "warmest-month" model is inadequate for predicting treelines worldwide. Growing season means (or, perhaps, some even better descriptions of integrated bio-temperature) are more closely correlated with treeline alitudes across latitudes and range between 5.5 and 7.5°C, which comes close to the 7°C mean for soil temperatures suggested by Walter and Medina (1969), Walter (1973) and others as coinciding with treelines (cf. Miehe and Miehe 1994). Winiger (1981) arrived at a 8–9°C range, but his selection of tropical mountains includes several for which treeline depression by non-climatic factors has been described (E-African vulcanos, Kinabalu in Borneo). However, if the uppermost forest islands indicate the natural climatic treeline, then the critical temperatures for tropical and subtropical treelines would be 2.5 to 5°C, which is less than for such outposts at higher latitudes (5.5–6°C). In most cases, the grassland and shrub vegetation on drained ground found below these isotherms (e.g. most of the Andean "páramos") would then not be truly alpine. The majestic El Angel *Espeletia* "forest" at the border between Ecuador and Colombia at ca. 3600 m altitude may be seen as an example of such a "pseudo-alpine" landscape (Fig. 7.6).

In view of the relatively close cross-latitude relationship between the bulk of the current treeline altitudes and seasonal mean temperatures (independently of the range of local annual temperature extremes!) a direct thermal rather than any other explanation of treeline altitude is the most plausible. This was also Brockmann-Jerosch's (1919) conclusion from his very detailed analysis of the climatic relatedness of treelines across the Swiss Alps. Physiological reasons for the presumed 5.5–7.5°C threshold temperature need to be found. Allowing for the latitudinal bias of means as discussed above, this rather limited data set provides no substance for assuming that **season length** plays a particular role in determining the world-

Fig. 7.6. High elevation giant rosette "forest" dominated by *Espeletia hartwegiana* El Angel, northern Ecuador, 3600 m

wide transition from the high altitude forest to the alpine life zone.

Only long-term micrometeorological studies at subtropical and tropical treelines (including solar radiation and root zone temperatures) such as those available for few temperate zone mountains will allow us to substantiate the temperature relatedness of the treelines of the world. In particular, data for patches of forests above current treelines are needed. It seems that phytogeographic, and partly also physiological knowledge is ahead of a broadly based knowledge of actual life conditions in treeline ecotones of the world.

Intrazonal variations and pantropical plateauing of alpine treelines

Most of the variation seen in Figure 7.2 has to do with regional climate gradients between coastal and inland mountains or between front ranges and central ranges. For instance, in western North America (47 °N), the treeline increases in elevation across the Cascades from ca. 2000 m in the west to ca. 2500 m on the eastern, continental side (Arno 1984). A similar gradient is found between the northern front ranges of the Alps and the central ranges. Front ranges receive almost twice as much precipitation, have greater cloud cover and are exposed to stronger winds (in this case from the northwest). Less moisture, more sunshine and less exposure to frontal weather causes temperatures in the central ranges to be higher, and treeline to rise. This phenomenon is called the "**Massenerhebungseffect**" (mountain mass elevation effect, Schröter 1908/1926, Brockmann-Jerosch 1919) which is discussed in detail by Barry (1981). Figure 7.7 illustrates the consequences of these climatic changes for the frequency of days during the year on which 5 or 10 °C are exceeded. At the treeline (as mentioned above) daily mean air temperature exceeds 5 °C for about 100 days and is above 10 °C for ca. 35 days. The level at which these thermal conditions occur shifts from 1850 m to 2200 m as one moves from the northern to the central Alps (Ellenberg 1963). "Warmest month" means

increase less than do altitudes of treelines, a confusion that irritated Brockmann-Jerosch, but is resolved if one takes Ellenberg's integrating approach. Figure 7.8 illustrates the altitudinal variation of the treeline in Switzerland, based on an assessment by Brockmann-Jerosch (1919). There is a clear north-south zonation following the above thermal gradient.

Thus, there is a relatively close correlation between temperature and treeline at a regional as well as at a global scale. Temperature is closely correlated with sunshine hours or sums of solar radiation, which tend to be more influential on tree photosynthesis than temperature during the growing season (Tranquillini 1979), but it is very uncertain whether the physiology of carbon assimilation is the most critical factor in treeline formation. Developmental and growth processes are likely to be more temperature sensitive (see below). Hence, the elevation of the treeline in the center of larger mountain systems can be either

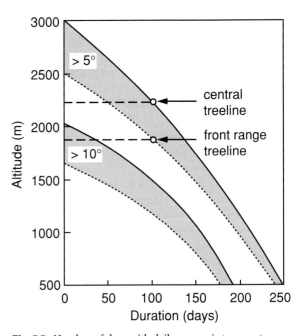

Fig. 7.7. Number of days with daily mean air temperatures above 5 °C or above 10 °C in the Alps. *Solid lines* Central Alps, *dashed lines* northern front ranges. Interpolated from monthly means for 1864–1900. *Arrows* indicate current treeline position on the <5 °C line. (Ellenberg 1963)

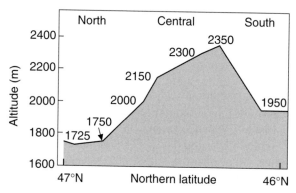

Fig. 7.8. Treeline altitudes along a north-south transect across the Alps, illustrating the "Massenerhebungseffekt". (Brockmann-Jerosch 1919)

warmer temperature or, if photosynthesis matters, increased radiation, or a combination of the two. Except for arid mountains, increased precipitation tends to depress the alpine treeline, possibly via a combination of greater cloudiness and cooler weather, at higher latitudes, increased duration of snowpack. Snow on the ground and associated low soil temperatures inhibit leaf gas exchange and development in branches above the snow even under warm atmospheric conditions as will be discussed later in this chapter.

It has not been conclusively resolved also why treelines peak in the subtropics and are found at comparatively lower altitudes in the humid **equatorial tropics**, but in principle the same arguments that were discussed above for the intrazonal variation of treeline altitudes apply here as well. Increasing precipitation and cloudiness tend to suppress treelines. Since these factors also cause the snowline to be lowered there is a common climatological reason for the parallel flattening of the latitudinal treeline and snowline curves in Figure 7.2. The best evidence for the positive effects of low to moderate moisture supply on treelines are altitudinal records in eastern Tibet and northern Chile where trees are growing not too far from 5000 m altitude under rather dry conditions, compared with the prevailing treeline altitudes in the humid tropics at or substantially below 4000 m. Whether reduced solar radiation for photosynthesis or direct effects of low temperatures due to

enhanced cloudiness are the key is again uncertain. Personally, I favor direct temperature effects, as will be discussed later.

Subzero temperatures occur regularly at both subtropical and tropical treelines, but their predictability is greater in the subtropical climate than in the non-seasonal tropical climate. Hence, the **risk of freezing** damage is greater at the comparatively "milder" equatorial treelines than at the subtropical treelines where the severest low temperature regime falls in the cold season, during which plant activity is reduced. The occurrence of growth-limiting low temperatures due to greater cloudiness (Fig. 7.4) plus a permanent risk of night-time freezing damage at comparatively lower altitudes may be the two most important reasons why equatorial treelines do not reach higher altitudes than those in the subtropics (but see Chap. 8).

Treelines in the past

Altitudes of treelines, like the altitudinal zonation of vegetation in general, underwent large oscillations during the Pleistocene and Holocene periods, and what we see today must also be viewed from this palaeo-ecological viewpoint. During **glacial periods**, temperate zone mountains were deforested, with the alpine life zone reaching down to the lowlands, and in the tropics treelines were depressed by 1000–1600 m (for references see Wijmstra 1978; Lauer 1988 and Flenley 1979; Elias 2001). Snowlines were lowered less, due to the prevailing drier climate in the large mountain systems of the earth during the peak of glaciation. According to Lauer and Wijmstra, treelines in the equatorial Andes were as low as 2000 m. The tropical-alpine life zone can be assumed to have covered a wider altitudinal range and a vastly larger land area than today, though at lower altitude.

Current treelines are depressed only a little compared with the **postglacial** maximum, indicating that the current land area covered by the alpine life zone is close to its postglacial minimum (Elias 2001). Evidence for this comes from fossil tree stumps and pollen records, suggesting that temperate zone treelines varied in altitude by no more than 100–200 m while temperaturs varied by several K. (e.g. LaMarche and Mooney 1972; Zukrigl 1975; Bortenschlager 1977; Ives and Hansen-Bristow 1983; Burga 1988; Rochefort et al. 1994; Graumlich and Brubaker 1995). For instance, the pollen record from the highest known mire in the eastern central Alps at 2700 m altitude shows almost constant pollen abundance of tree species dominant today over the last 8000 years (Bortenschlager 1993). But gradual shifts in subdominant species in the northern calcareous Alps were observed, which are clearly associated with human interference (increase of *Larix*) and date back 5000 years (Zukrigl 1975). Also, for the equatorial and outer tropical Andes, available evidence from pollen in lake sediments suggests that the treeline was never more than 200 m higher than today (Lauer 1988).

Hence, as predicted from theory, treelines as a whole are self-stabilized vegetation boundaries (Slatyer and Noble 1992; Armand 1992) which do not follow climatic changes very rapidly. Their **current position** must thus be assumed to reflect an environmental integral over several hundreds of years – a point also to be considered when functional interpretations of treelines are attempted. There may be climatic phases during which treeline-trees operate either persistently below or close to their physiological limits, with the overall position of the treeline remaining almost unchanged. Seed production and seedling establishment does respond to such climatic alterations (e.g. Szeicz and MacDonald 1993; Holtmeier 1994; Cuevas 2000), but apparently this does not induce a big shift in the treeline. Certain thresholds of change must be exceeded or fundamental disturbances such as windbreaks or fire must come into play before treelines move significantly. Therefore, it was suggested that the altitudinal position of the treeline is not a particularly useful criterion for assessing effects of rapid climate change (Slatyer and Noble 1992; Noble 1993). However, growth responses of forest trees at or just below the alpine treeline, are sensitive indicators, as evidenced by a large body of dendro-ecological studies (e.g. Schweingruber 1987; Villalba et al. 1990; Nicolussi et al. 1995; Paulsen et al. 2000).

Palaeorecords also indicate that **human influence** on highlands including alpine treelines is a worldwide phenomenon both in the temperate zone (Eijgenraam and Anderson 1991; Bortenschlager 1993; Holtmeier 1994) and the tropics (Flenley 1979) and dates back several thousands of years. Most of today's high altitude grasslands at or below the climatic treeline are related to some sort of human impact. There is clear evidence that fire is the cause for depressions of treelines in the tropical African mountains (Hedberg 1964; Beck et al. 1986; Smith and Young 1994; Wesche 2000) and in many other tropical mountains (Deshmukh 1986), and temperate zone treelines have been burned as well, as indicated by charcoal layers in soils of the lower alpine zone which have been dated at between 700 and 3700 years old (Neuwinger 1970), and below the current treeline position at 4700 years before present (Markgraf 1969; references in Zoller 1987).

Attempts at a functional explanation of treelines

So far correlations between climate and treeline have been discussed, but not the mechanisms which could dictate a climate driven transition from forests to the alpine life zone. It was shown that on a global scale, climatic high altitude treelines occur at similar growing season temperatures (means from 5.5 to 7.5 °C) whereas season length varies between 2.5 and 12 months, and many other climatic constraints which have been suggested as contributing to alpine treeline formation show large regional variation. In order to dissolve this apparent discrepancy which limits explanatory attempts, I suggest **separating "modulative" (regional) from "fundamental" (global) drivers**. Most of the literature relates to modulative forces on treeline position (regional fine tuning). In the following I will try to explore what the possible global determinant (with locally less precise predictive potential) might be. Irrespective of scale, five groups of – partly interrelated – mechanisms may be distinguished:

1. The **stress** hypothesis: repeated damage by freezing, frost desiccation or phototoxic effects after frost impair tree growth.
2. The **disturbance** hypothesis: mechanical damage by wind, ice blasting, snow break and avalanches or herbivory and fungal pathogens (often associated with snow cover) may remove a similar amount, or more biomass or meristems than can be replaced by growth and development below certain temperatures (see 4).
3. The **reproduction** hypothesis: pollination, pollen tube growth, seed development, seed dispersal, germination and seedling establishment may be limited and prevent tree recruitment at higher altitudes.
4. The **carbon balance** hypothesis: either carbon uptake or the balance between uptake and loss are insufficient to support maintenance and minimum growth of trees.
5. The **growth** limitation hypothesis: synthetic processes which lead from sugars and amino acids to the complex plant body may not match the minimum rates required for growth and tissue renewal, independently of the supply of raw materials (e.g. photoassimilates).

Alone or in combination, type 4 and 5 mechanisms may lead to insufficient **tissue maturation** when the growing season is short, and may, in seasonal climates, sensitize plants to type 1 and 2 damages or prevent reproduction. Most of these explanations of the treeline emerged from observations in certain temperate zone mountains (Alps, Rocky Mountains). The question to be asked here is what controls alpine treelines worldwide between 70°N and 55°S lat, including all climatic zones, with and without seasons, and across such different **phylogenetic groups** as for instance Pinaceae, Podocarpaceae, Fagaceae, Rosaceae and Ericaceae (see Wardle 1993)? Which of the explanations developed for temperate zone conifers would lose power, and which would gain, if one adopts a global perspective?

Whatever the common, basic mechanism is that leads to the abandonment of massive tree-stems and tall, closed canopies at a certain altitude, it cannot be associated with **seasonality**, because treelines are formed in non-seasonal climates as well, and as has been shown above, at surprisingly similar temperature regimes. Furthermore, treelines establish in maritime areas with neglegible frost risk, they are also found in areas with hardly any wind, and in complete absence of ice blasting or snow damage. In the following I will discuss the above listed potential causes of treeline formation with a global perspective.

Historically, the growth limitation hypothesis was favored (e.g. Däniker 1923) but became overshadowed by the carbon balance hypothesis, in particular photosynthetic CO_2 fixation, once technical know-how attracted science to this field of research. In the 1960s and 1970s, the old idea of winter desiccation, first substantiated by data in the 1930s and early post-war period, came into favor, later supplemented by evidence that insufficient tissue maturation may be important. Frost per se, snow mold or rust (*Chrysomyxa*) and mechanical damage, were discussed controversial, but were rarely considered critical for established trees, other than as modulating the chances of regrowth in the "kampfzone" or "krummholz" belt (Däniker 1923; Schröter 1908/1926; Turner 1968; Holtmeier 1974; Frey 1977; Tranquillini 1979; Larcher 1985a; Oberhuber et al. 1999; see below; Fig. 7.9).

Treeline and climatic stress

Frost damage is a potential contributor to treeline formation. However, in climates with a thermal season, the dangerous period is not the coldest part of the year, when frost tolerance of treeline species generally exceeds environmental demand. Consequently, Brockmann-Jerosch (1919) found no correlation between annual absolute minima of temperature and treeline position in the Alps, and considering the global distribution of treelines it is obvious that no such correlation does exist at this largest scale either. For conifer taxa Sakai and

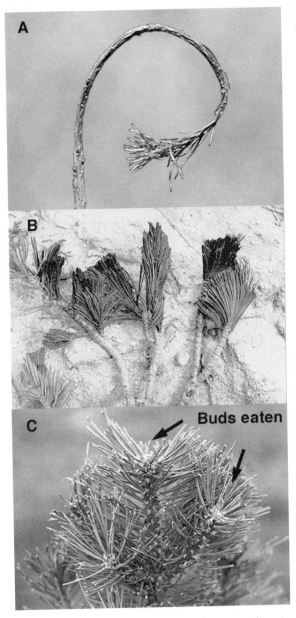

Fig. 7.9. Damage of young trees above the alpine treeline: *A*, early autumn frost killed non-hardened, immature leading shoots of *Larix decidua* (before needle fall), which desiccated during winter (photograph taken in spring, 2200 m central Alps); *B* snow mold damage (possibly *Herpotrichia* sp.) in lower pine branches under late lying snow (Rocky Mts, Utah); *C* all terminal buds of *Pinus mugo ssp. uncinata* browsed by snow hen (site as *A*)

Larcher (1987, p. 215) illustrate that the global variation of low temperature resistance in buds of treeline-trees ranges from −70 to −5 °C and parallels the latitudinal trends of minimum temperatures at treeline altitudes. At higher latitudes, critical situations occur during rapid freezing in periods of incomplete hardiness in autumn or in the later part of the winter and early spring when mature tissue may be damaged by temperatures between −6 to −10 °C (for references see Tranquillini 1979; Sakai and Larcher 1987; Gross 1989 and the recent work by Perkins and Adams 1995). Because of temperature inversion, high altitude trees are not necessarily exposed to lower temperatures than low altitude trees, and are not more sensitive (e.g. Sundblad and Andersson 1995). Tranquillini (1979) and Larcher (1985a) conclude that frost damage does not threaten survival of trees in the temperate zone treeline ecotone, but may contribute to distorted growth by partial injuries (cf Däniker 1923). At tropical treelines, freezing damage could theoretically occur during clear nights at any time of the year, but again, data by Sakai for New Guinea (in Sakai and Larcher 1987) and by Larcher (1975) and Goldstein et al. (1994, p. 142) for Venezuela indicate that it is rather unlikely that frost damage plays a decisive role in treeline formation. Since minimum temperatures occur during the night, it is important to note that radiative cooling to below air temperature under a clear sky is likely to be less in narrow-leaved tree crowns with good convective coupling to the atmosphere than in unscreened low stature vegetation with poor convective exchange (Grace 1988; Squeo et al. 1991; Germino and Smith 2000).

Winter desiccation is one of the most widely assumed causes for treeline formation. Damage of needles and branches can be caused by late winter water losses not replaced by the water supply because of frozen soil or stem bases (e.g. Michaelis 1934; Larcher 1963a, 1985a; Tranquillini 1979, 1982; Sowell et al. 1982; Sakai and Larcher 1987; Hadley and Smith 1990). Stem freezing may also induce xylem cavitation (Tyree and Sperry 1989) which can inhibit water flow during subsequent warmer periods. By mentioning "winter" it is clear that this cannot be a cause for treeline formation in general (Troll 1961; Dahl 1986), but perhaps this is a component which contributes to the "fine tuning" of treeline position in some temperate zone mountains. The problem with winter desiccation is that if damage occurs, this will only become visible much later, and a causal interpretation becomes problematic (Larcher 1963a, Wardle 1981b). So, most of what we know about winter desiccation or frost drought is inferred from asymmetric (i.e. sun-side) shoot browning observed much later. Yet the occurrence of critically low leaf water potentials in woody plants near temperate zone treelines in late winter is clearly documented (see above references). As already suggested by Kerner (1869, p. 41) summer predisposition has been proven to correlate with the degree of winter desiccation, e.g. by insufficient **maturation** of leaf cuticles or buds (Baig and Tranquillini 1980; Tranquillini and Platter 1983; Wardle 1981b). Direct damage to needle surfaces by winter conditions such as **cuticle abrasion** has also been suggested as contributing to excess moisture loss (Holtmeier 1974; Hadley and Smith 1983; see discussion by Grace 1989; Fig. 7.10).

However, the phenomenon of winter desiccation at alpine treelines does not seem to be common, even within the temperate zone. No indication or possibility of winter drought has been seen by Sakai (in Sakai and Larcher 1987) in *Pinus sylvestris* and *Pinus banksiana*, by Slatyer (1976) in *Eucalyptus pauciflora*, by Marchand and Chabot (1978) in *Abies balsamea* and *Picea mariana*, by McCracken et al (1985) in *Nothofagus solandri*, and by Grace (1990) again in *Pinus sylvestris*. Although young, leafless stems may be prone to desiccation as well, deciduous trees seem to be at less risk of being seriously affected (Barclay and Crawford 1982, Richards 1985, Tranquillini and Plank 1989) and perhaps dominate some northern treelines because of this (e.g. *Betula* and *Larix* species).

The critical question is whether water losses during late winter atmospheric conditions can exceed available **moisture stored in branches** and stems. For the Scottish treeline, Grace's (1990)

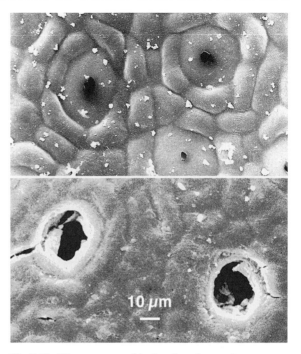

Fig. 7.10. Winter stress and leaf surface damage in *Eucalyptus pauciflora* at the treeline in the Snowy Mountains (Australia). The scanning electron micrographs (×800) are for current season (*top*) and over-wintered leaves (*bottom*). Note, at the treeline *E. pauciflora* sheds nearly all previous season (worn out) leaves as soon as the next generation of leaves expands in spring.

calculations indicate that this is physically impossible. Larcher (1963a) showed that needle desiccation strongly depends on the attached (at least periodically unfrozen) stem volume. Detached needles reached critical water deficits twice as fast as those on whole detached branches exposed to the late winter climate. Hadley and Smith (1990) showed that needles may lose 40% of their moisture in only 5–18 days at the Rocky Mountain treeline if no replacement from branches occurs. The bigger the tree, the less likely critical water deficits in needles will occur, explaining why all reports about winter desiccation damage are for relatively young trees, mostly individuals from the "kampfzone". Marchand (1991) went a step further by detaching whole trees, and could demonstrate by comparison with uncut controls that significant water supply from the root zone does occur in late winter. Branches under snow remain close to water saturation.

In summary, winter desiccation may be a problem for young trees above the treeline in some parts of the temperate zone (according to Tranquillini 1979, it sets the krummholz limit in the Alps), but cannot explain the worldwide alpine treeline phenomenon as such. Given that the length of the growing season at conifer treelines with similar seasonal mean temperatures varies between 3 months at subarctic to 7–8 months at subtropical latitudes (with some conifer species also found at equatorial treelines), season length is not a common determinant of treelines, even though upslope excursions of seedling populations are limited by maturation problems in some higher latitude mountains.

Carbon acquisition, carbon investment and growth

Forty years of leaf gas exchange studies in treeline trees have revealed no particular disadvantages compared with low altitudes, except for the reduction of the length of the active period in extratropical areas. Many studies have illustrated the relative insensitivity of **photosynthesis** in treeline trees to temperature in the range of prevailing daytime field temperature because of thermal acclimation and the broad shape of the photosynthetic temperature response (Pisek and Winkler 1958, 1959; Slatyer and Ferrar 1977; Häsler 1982; Goldstein et al. 1994; Rada et al. 1996). Substantial rates of photosynthesis have repeatedly been measured at 0 °C, and one fourth to one half of maximum rates are reported for +5 °C. Modest altitudinal reductions in photosynthetic capacity reported in the literature disappear in most cases when rates are expressed by unit leaf area rather than by dry mass, because a dry mass basis reflects the well-known altitudinal reduction of specific leaf area (greater fraction of cell wall material), rather than a reduction of metabolic activity of cells. In evergreens, reduced specific leaf area is balanced by increased leaf life span, hence a func-

tionally meaningful assessment of the leaf carbon balance would have to account for leaf duration. Similarly, much of the divergence seen in the literature with respect to altitudinal trends in respiration disappears when actual tissue temperatures during relevant periods are considered (for leaves only the night is relevant, because daytime losses are already accounted for by *net*-photosynthesis). Comparisons of rates of high versus low altitude tissues measured at equal temperatures are ecologically rather meaningless.

Allowing for these points, the photosynthetic capacity of treeline trees per unit leaf area is relatively high, thermal acclimation of photosynthesis optimizes seasonal carbon gain, water supply is less restrictive to gas exchange than at low altitude, recovery of photosynthesis after cold nights is surprisingly rapid and respiratory losses at habitat temperatures do not exceed those at low altitude because of lower temperatures during the winter and at night. Thus, the **annual leaf carbon uptake** is largely a function of season length (in extratropical mountains) and the doses of solar radiation (review for the temperate zone by Tranquillini 1979; tropical tree data by Goldstein et al. 1994). In this respect, treeline trees do not differ from alpine plants, whose main limitation to CO_2 uptake (within a given season length) is quantum supply and not the temperature during daylight hours (cf. Scott et al. 1970; Körner 1982; Körner and Larcher 1988; see Chap. 11). In extratropical mountains season length may only matter because (1) respiratory losses during the dormant season burden the annual carbon balance, and (2) because there is as a critical point at which the growth rate or rate of tissue renewal cannot be supported.

This is not the place for a detailed evaluation of **whole plant carbon relations** of treeline trees, but the following numbers on respiratory carbon losses, obtained from Tranquillini's (1979) synthesis illustrate some important aspects for the temperate zone. A 5 month dormancy under winter snow cover in the central Alps creates **respiratory costs** of about 7% of the seasonal carbon uptake in seedlings of *Pinus cembra*. Most recently Wieser (1997) demonstrated that respiratory losses during the 3 coldest months of the winter in *Pinus cembra* shoots can be covered by a single day's carbon gain in the growing season. Hence, the dormant (i.e. cold) period is a much smaller burden than might have been assumed from its duration. Despite some additional investments in freezing protection, the carbon balance rather improves in colder winters because of reduced respiratory losses. In areas with a long and warm autumn and spring, substantially higher losses may occur (Schulze et al. 1967). Compared with the tropics, carbon gains during the 15 to 25% longer daylight periods in the growing season at higher latitudes alone can equalize the dormant season respiratory costs. The year round maintenance of a stem in 7 m tall trees at the treeline was estimated to consume 17 (*Larix*) to 23% (*Pine*) of annual net photosynthetic carbon-gain, but, because of the cooler climate (and also because of lower specific rates), these are only 60–70% of the losses estimated for low altitude trees (Tranquillini 1979). Finally, all estimates of root respiration compiled by Tranquillini point at exceptionally small losses, resulting from very low root/leaf mass ratios and cold soils, but exports to ectomycorrhiza (which has been suggested to be very important for treeline trees, cf. Moser 1966, Wardle 1971) are unknown. Hence, treeline trees are not necessarily burdened by a greater respiratory load than trees at low altitude, rather the reverse seems to be the case.

Taken together, the above gas exchange data do not support the idea that insufficient photosynthetic activity during a given growing season or excess respiratory losses and a marginal annual carbon balance can conclusively explain the temperate zone treeline. Both the growing season moisture regime and the nutritional status of needles in treeline trees are rather more favorable than in trees at low altitude, an observation that holds for most high altitude vegetation (Tranquillini 1979; pp. 52, 80; Körner and Mayr 1981; Körner and Cochrane 1985; Körner 1989b; see Chaps. 9 and 10). Comparative studies by Benecke et al. 1981 in the Alps, and elevational transects in California (Mooney et al. 1964), Australia (Slatyer

1978) and New Zealand (Benecke and Havranek 1980) match the picture derived from Tranquillini's data. With daytime leaf temperatures between 3 and 17 °C (mostly 5–10 °C), tropical high altitude *Polylepis* exhibits rates of photosyntheses of ca. 7–8 μmol m^{-2} s^{-1} year round, which stand against night time respiratory losses in leaves of between 0.2–0.3 μmol m^{-2} s^{-1} (extrapolated to the prevailing 0 to 2 °C night-time temperatures; data from Goldstein et al. 1994). The gain to loss ratio is ca. 30 and thus, 2 to 3 times better than any reported for trees at lower altitudes, suggesting (at least for *Polylepis*) factors other than the leaf carbon balance to be limiting tree growth also at tropical treelines.

What remains to be considered is the overall investment of assimilates by trees compared with shrubs, grasses and herbs in the alpine zone. Unfortunately, **dry matter allocation** is extremely poorly documented for treeline trees, despite the fact that it may be more important than leaf photosynthetic rates as a determinant of the carbon balance (Körner 1991). This reflects the human fascination in electronic machinery compared with spades, but lacks any scientific rationale. Sporadic mentions of leaf, stem and (even less frequently) root dry matter of young (!) trees, suggest no particular altitudinal change of allocation patterns. The few numbers reported by Larcher (1963a), Oswald (1963) and Tranquillini (1979) for young *Pinus cembra* trees (but older than 20 years) near the treeline indicate leaf mass ratios (LMR) of between 13–25% (leaf mass versus total plant mass; when root data were missing, root mass was assumed to contribute 25% to total mass). Oswald's (1963) data for 20 to 52-year-old *P. cembra* in the uppermost forest (needle-stem-root fractions of 24–49–27%) and the kampfzone 200 m above treeline (25–50–25%) really do not indicate any trend across the ecotone. If one extrapolates the declining LMR with increasing tree age that can be extracted from Oswald's data, a value of 10% (independent of altitude) is approached between 100 and 150 years of age. However, for seedlings of uniform provenance Benecke (1972) showed that leaf:root mass ratios increase from 1 to 3 in *Picea abies* and from 1.5 to

2.5 in *Pinus mugo*, as one moves from optimum montane habitats to the treeline. The LMR numbers mentioned above for relatively young trees are much higher than the 3–4% commonly reported for mature lowland conifers, and match the mean of 21% known for herbaceous plants of both high and low altitude (no change of LMR with altitude; Körner 1994; Fig. 12.15). The above mentioned leaf:root ratios are even larger than in many perennial forbs (0.5 to 1, Körner 1994). New data for the Swiss treeline (Fig. 7.11) suggest no significant alteration in dry matter allocation across the treeline ecotone. At least for young trees at the treeline, the autotrophic:heterotrophic tissue ratio does not appear to be disadvantageous and is quite high.

What is the **significance of a large stem fraction** (>70%) of total biomass for the carbon balance in older treeline trees? When compared with patterns of dry matter allocation in low stature alpine plant communities, one finds that such high heterotrophic fractions are not uncommon. The above mentioned mean of 21% for LMR was cal-

culated for many species and sites, including some of the highest ranging species such as *Ranunculus glacialis* which operate successfully with 10% of dry matter in green leaf blades (of 10 weeks duration, but higher photosynthetic rates) and 90% in non-green tissue (see Chap. 11), similar to LMR in the deciduous *Larix* shown in Figure 7.11. However, these heterotrophic compounds are active and respire, whereas in trees, much of the stem is metabolically inactive and resembles a dry matter compartment comparable to the dead leaf coat in tropical giant rosettes (also about two thirds of their total plant mass according to Monasterio 1986).

On the basis of these sporadic observations on trees, and in comparison with some very successful alpine plant species – shrubs with massiv woody structures in particular – it seems that the possession of a woody stem alone is not necessarily critical for the carbon balance at treeline altitudes. But the formation of stems certainly adds to the lifetime burden of the carbon balance, the older a tree becomes. Whether this relates to carbon balance problems or not, it needs to be noted that the mean stem fraction decreases stepwise from the tree life form to shrubs and perennial herbs – the latter representing the dominant life form at highest altitudes (Körner 1994). In the "endless" season of a humid tropical treeline climate, the burden of a stem should be less important, and the treeline could be expected to advance to lower isotherms, if its position in the temperate zone were "depressed" by the carbon-cost of a stem – a trend not detectable from available data.

A conclusive picture of the carbon relations of treeline trees is not yet available, and will only be obtained by modeling the annual and long-term carbon balance, the greatest uncertainties of which are constraints to meristematic, i.e. sink activity, the fractionation of dead and live tissue, tissue specific respiration and the unknown belowground sinks. As will be illustrated below, to start from leaf gas exchange data is a rather hopeless approach because CO_2 assimilation is unlikely to be the bottleneck of growth at the treeline, even though correlations may be found (Scuderi et al.

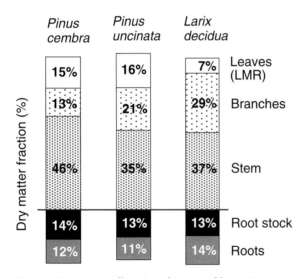

Fig. 7.11. Dry matter allocation of 23-year-old trees (n = 24 trees per species) sampled across the alpine treeline ecotone at Stillberg, Davos, Switzerland (2040 to 2180 m). The leaf mass ratio (LMR) is the amount of leaf dry matter in % of total tree dry matter (including roots). (Bernoulli and Körner 1999)

1993). Currently available data from growth analysis (Tranquillini 1979) are not in favor of the carbon limitation hypothesis. While the carbon balance may control distribution boundaries of trees along light or aridity gradients (Boysen-Jensen 1949), this does not seem to be the case at the alpine treeline (Tranquillini 1979, p. 80).

A hypothesis for treeline formation

Above, it has been illustrated that **a global perspective** eliminates possible candidates as treeline determinants, even though they may contribute to the determination of the treeline position in some parts of the temperate zone. Neither the stress hypothesis nor any hypothesis relating to seasonality or mechanical disturbance is able to explain the occurrence of treelines at similar mean growing season temperatures with otherwise widely diverging climates across the world. Tranquillini (1979, p. 80) stated: ". . . one can safely conclude that neither the rapid decrease in tree height, nor the total elimination of woody plants is primarily due to inadequate dry matter production" (sensu "carbon assimilation").

For a discussion of limitations by **reproduction**, I refer the reader to Körner (1998). It is not obvious why alpine plants, including many woody shrubs, should be able to establish seedlings, and trees should not (once viable seed is locally produced or brought in). Since diaspore dispersal from below the treeline and successful germination is observed so regularly (e.g. Hättenschwiler and Smith 1999 and the excellent analysis by Cuevas 2000), centennial recruitment waves (Klötzli 1991) should be possible even if treeline source trees did not produce viable seed. The question is, why do seedling populations above the treeline not develop a forest, but remain nested in the graminoid or shrubby ground cover, or form crippled scrub?

A **unifying theory** for the alpine treeline must account for discrepancies such as the two- to threefold (!) difference in the daily doses of quantum flux at treelines in the central Alps or subtropical Andes, compared with the treeline in New Guinea, which is almost permanently enveloped in clouds (Körner et al. 1983). It must further account for a change in season length from 2.5 months in arctic-alpine birch treelines to 12 months in equatorial tropics, and for regions with heavy snow pack and non at all. Under all these conditions we find treelines at seasonal mean air temperatures of between 5.5 and 7.5 °C. The thermal conditions at natural climatic treelines clearly need further investigation, but the consistency in available temperature data it is too obvious to be dismissed as a meaningless correlate.

On the basis of what was discussed above I **hypothesize** that there is a minimum mean temperature that permits sufficient production of new cells and the development and differentiation of functional tissue of higher plants which is unrelated to the carbon balance. In the following, I will present some evidence in favor of both this growth limitation hypothesis and its implications for life form-microclimate interactions. In simple words, I suggest that if a growing tree were a house under construction, the limitation of "growth" is not a question of availability of mortar and bricks, but depends on the workmen creating the walls. They may stop working if it is too cold, despite a lot of building material. Suppliers may even slow down or stop delivery (Hoch et al. 2002).

The published evidence in support of hypotheses 1 and 2 (with consequences for 3) for certain climatic regions is not questioned here, but in view of the global patterns illustrated above, and in view of the mechanisms discussed below, one is forced to consider these limitations as regional and "modulative", on top of more fundamental limitations to be explored. Removing these typical temperate zone modulative impacts would have little effect on treeline position within a given ecotone, but may allow a more gradual transition from dense forest to shrubby tree growth.

The concept of a mean **minimum temperature for tree growth** in which carbon sinks rather than carbon sources control production is only a special case of a more general rule. The reason why such growth limitations affect trees first, and shrubs and forbs only at much higher elevations, has to do with life form specific effects on micro-climate. Shoot apical meristems of trees can not benefit from radiant canopy warming during the day or stored warmth in the topsoil during the night, as subsoil leaf meristems of many alpine graminoids and rosette forbs or dwarf shrubs do (e.g. Körner and Cochrane 1983, Grace and Norton 1990), but experience convective cooling through tight atmospheric coupling. Thereby tissue expansion may become blocked periodically as will be discussed later and trees "lose" a substantial fraction of the season and most nights (even during otherwise warm periods) for structural growth.

In addition, trees – particularly when forming dense canopies – efficiently prevent **soil heat flux** and radiative warming of their own root zone. The tree life form evolved as a means for light competition under warm soil conditions, a point made earlier by Slatyer and Noble (1992). Closed tree canopies at the treeline create cold soils which impair root activity (e.g. Däniker 1923; Shanks 1956; Wardle 1968; Ballard 1972; Munn et al. 1978; Körner et al. 1986; Holtmeier and Broll 1992; Kessler and Hohnwald 1998; Figs. 7.12 and 7.13), a point so far not seriously considered in the treeline discussion. In Montana (2300 m) for instance, −50 cm soil temperatures were nearly 5 K lower under forest compared with adjacent grassland during the summer (Munn et al. 1978). As a consequence, snow cover often lasts longer in high altitude forests, compared with adjacent treeless terrain (Alps, Tranquillini 1979, p. 55; Fig. 7.14). As a rule, similarly oriented grassland soils within a few hundred meters above treeline, exhibit root zone temperatures above those found under closed forest near the treeline (Körner et al. 2003). Halloy and Mark (1996) report means of between 5.9 to 8.1 °C for the three warmest months for phyto-geographically comparable above treeline sites at 46, 26 and 0° latitude. As a consequence of ground shading, ground temperatures under closed tree canopies at treeline do not differ with slope exposure (unpubl. data from Mt. Patscherkofel, Innsbruck, Austria), which may explain why treeline altitudes in humid regions also do not differ with slope exposure, as revealed by a GIS analysis

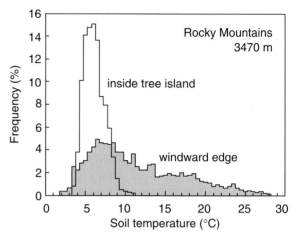

Fig. 7.12. Forest trees create a cold root zone. Soil temperatures within and at the windward edge of a stunted forest tree island of ca. 4 m diameter close to the alpine treeline on Niwot Ridge, Rocky Mountains at ca. 3470 m altitude (19 July-4 September). Note the narrow, i.e. strongly buffered and lower range of temperatures below the tree cover (centered at 6 °C) compared with the wide range found underneath the shrubby, low ground cover (highest frequency of 7–8 °C, median between 8 and 9 °C). (Holtmeier and Broll 1992)

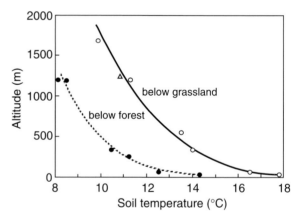

Fig. 7.13. The altitudinal variation of soil temperature under forests (*Nothofagus menziesii*) or adjacent grassland in South Island, New Zealand in midsummer (February 1981). Temperatures were measured at midday at 20 cm depth. (Körner et al. 1986; *triangle* Greer 1978)

for the whole Swiss Alps (Paulsen and Körner 2001) and the Mexican volcanoes (Beaman 1962).

In other words, I suggest that the tree life form is **limited** at treeline altitudes by possibilities for

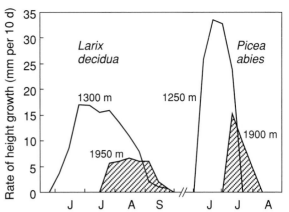

Fig. 7.15. Dynamics of height growth in *Larix decidua* and *Picea abies* at montane and treeline altitude field conditions. (After Oberarzbacher 1977, in Tranquillini 1979)

Fig. 7.14. Late laying snow under treeline trees often shortens the growing season as compared with adjacent open terrain. Sierra Nevada of California, Tioga Pass area (mid July 1983)

investment rather than by **production of assimilates.** The hypothesis centers around the assumption that the treeline is not caused by carbon shortage, but is created by sink inhibition as a result of low temperature. Dahl (1986), developing similar ideas, proposed that ATP supply (mitochondrial respiration) is the critical factor, but many other processes involved in tissue formation (e.g. protein synthesis) may be limited by low temperature. The fact that treelines occur at similar seasonal mean temperatures, with partial pressures of CO_2 (not the mixing ratio) varying from 94% of values occurring at sea level in subarctic-alpine treelines to only about 55% at the upper limit of tree growth in some subtropical mountains (a drop not fully balanced by enhanced diffusion, see Chaps. 3, 9 and 10), adds weight to the idea that carbon supply is not decisive.

Growth trends near treelines

No direct evidence for a threshold temperature for the production and differentiation of a critical mass of new cells in growing tree tissue is available. However, there is substantial information on growth responses to temperature in treeline trees which stands in obvious contrast to any attempt to stress carbon acquisition as an explanation for the abrupt limitation of tree growth. Again, I refer to the excellent compilation of data by Tranquillini (1979) which supports the growth hypothesis so obviously, that I was surprised not to find an explicit statement in this direction.

Tranquillini describes ten examples (two shown in Figs. 7.15 and 7.16) in which **altitudinal variations of tree growth** were studied from seedlings to mature trees in a number of different temperate mountain regions, and both in terms of short-term (daily) rates as well as in terms of cumulative growth. In all cases, dramatic reductions of growth rates (factors of 2 to 4) were found at altitudes close to the treeline, often across gradients of only 200–300 m of altitude (see also Ott 1978). Solar radiation does not significantly change over such gradients in the temperate zone, nor does the 1 to 2 K temperature difference exert a significant change in photosynthesis, as is well documented in the relevant gas exchange literature. The change

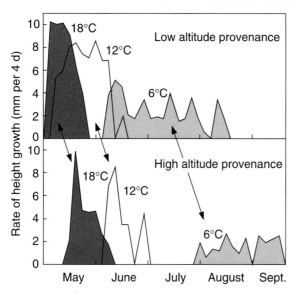

Fig. 7.16. Growth responses of high and low altitude provenances of *Picea abies* seedlings to temperature in a controlled environment. Note the dramatic reduction of maximum growth rate between 12 and 6 °C (factors 2 and 3.4). The trend is even more pronounced in the high altitude provenance, which may have to do with a genotypic growth restriction in high altitude ecotypes; see chapter 1). Source as in Figure 7.15

Fig. 7.17. Substantial radial tree growth (2–5 mm a^{-1}) occurs close to the treeline. The photograph shows a trunk cut in a dense *Pinus hartwegii* forest at 3850 m elevation in Mexico (for site see Fig. 7.1), which is only 150 m below the treeline (corresponding to a ca. 0.9 K higher mean air temperature as compared with treeline altitude)

in season length may be 2–3 weeks, nothing compared with the differences seen along global latitudinal gradients of treelines. In other words, there exists a tremendous discrepancy between altitudinal trends in growth and the potential for photosynthetic gas exchange (Fig. 7.17).

Referring to his own data on *Larix*, Tranquillini shows that no new roots are formed at 4 °C soil temperature. Turner and Streule (1983, in Schönenberger and Frey 1988) observed roots at the treeline with root windows, and found no root growth at soil temperatures below 3–5 °C, with 5% of maximum rates only reached once soil temperatures exceeded 6 °C (Fig. 7.18). Havranek (1972) found a linear correlation between root temperature and photosynthesis in *Pinus cembra* between 0 and 7 °C (Fig. 7.19), but not at higher temperatures. Cold nights may also result in photoinhibition of photosynthesis the next day (Germino and Smith 1999), perhaps because of

low-temperature inhibition of growth (end-product inhibition). In situ root zone temperatures in adult trees also exert immediate, almost linear effects on stomatal conductance and photosynthesis (once a low threshold of between 1 and 4 °C is exceeded) with no indication that this response is associated with cold soil induced needle water deficits (Day et al 1989, Körner et al. 1995). Similarly, Scott et al. (1987) found a linear correlation (r = 0.95) between cumulative **root zone temperature** and shoot elongation in subarctic *Pinus taeda* experiencing atmospheric temperature and moisture conditions quite favorable for photosynthesis, again indicating tight control of shoot activity by a below ground signal. Also working with *Pinus taeda*, Bilan (1967) demonstrated the existence of root growth threshold temperatures at mixed temperature regimes of between 10/1.7 and 10/4.4 °C (24 h means of between 6 and 7 °C). Hellmers et al. (1970) were able to block growth in potted *Picea engelmannii* seedlings, by exposing them to daytime temperatures of 15, 19, 23, 27 or 35 °C but a night temper-

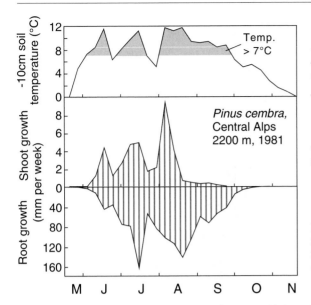

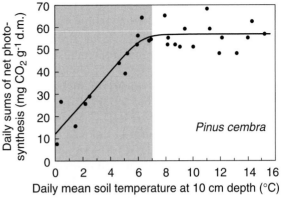

Fig. 7.19. Tight control of daily sums of photosynthesis in natural regrowth of *Pinus cembra* at the alpine treeline by root zone temperature in the 0–7 °C range (central Alps). (Havranek 1972)

Fig. 7.18. In situ temperature dependence of root growth in *Pinus cembra* in the Swiss central Alps at 2200 m. (After Turner and Streule 1983, in Schönenberger and Frey 1988; see also Turner et al. 1982)

ature of only 3 °C. At a 7 °C night temperature, significant growth (>10% of optimum) occurred only when daytime temperatures were between 19 and 23 °C; at colder or warmer daytime temperatures the 7 °C night treatment also reduced growth to negligible rates. A 15/23 compared with a 15/11 °C regime produced seedlings 28 times larger. Seedlings of *Betula pubescence* ssp *tortuosa* at the alpine treeline of northern Sweden showed no net growth or nutrient uptake at 5 °C, irrespective of substrate type or nutrient supply (Karlsson and Nordell 1996). Loris (1981) monitored electronically the radial thickness of *Pinus cembra* for 2 full years in the Alps. Radial cambial growth during the growing season ceased whenever temperatures fell below ca. 5 °C, a temperature permitting needles to assimilate at over 60% of the maximum for overcast and 25% of the maximum of bright weather conditions (Pisek and Winkler 1959). Grace (1989) and James et al. (1994) monitored shoot height extension of *Pinus sylvestris* at the treeline in Scotland and found a low threshold meristem temperature of between 5 and 6 °C in

native trees, and between 6.5 and 7.5 °C in potted seedlings. A ca. 5 °C limit for in situ root and shoot growth is reported for *Larix decidua* by Häsler et al. (1999). Low temperatures in one season also significantly affect growth in the next season. Mikola (1962), Roberts and Wareing (1975), Junttila (1986) and others have shown that the number of leaf primordia in the bud is affected by late-season temperature and determines next year's shoot elongation – **a developmental and not a resource driven control of growth.**

Altogether these experiments clearly document the significance of a critical temperature for growth under otherwise favorable conditions for photosynthesis.

The threshold for direct effects on growth and development appears to be higher than 3 °C and lower than 10 °C, possibly in the 5.5 to 7.5 °C range most commonly associated with treeline positions according to Figure 7.4.

Evidence for sink limitation

From tree-ring analysis it is known that rather minute changes in temperature seen in **tree rings** of treeline trees (e.g. Mikola 1962, Schweingruber et al. 1988, Grace and Norton 1990; Paulsen et al.

2000). There is no gas-exchange related basis to explain a doubling of ring width for a 2–3 K warmer season (or 300–500 m lower altitude), even if one accounts for some auto-correlation with sunshine hours.

A recent large-scale survey of mobile carbon pools (carbohydrates and lipids) in pines across the treeline ecotone revealed an altitudinal increase with a maximum at the tree limit in all three test regions (Fig. 17.20). The trends were not related to osmotics, but reflected contributions by the starch and lipid fractions and became even enhanced when using tissue volume rather than dry weight as a reference (increasing tissue density with elevation). Treeline trees completely defoliated before the current year flush had almost restocked C stores by the end of the season (Li et al. 2002). A step increase in CO_2 concentration using free air CO_2 enrichment at the Swiss treeline (Hättenschwiler et al. 2002) induced further increases in mobile C stores in both needles and branches. These data are in obvious conflict with the carbon limitation hypothesis. It is too early to comment growth responses, but any initial stimulations are likely to be transitory in nature, given the above data.

I want to close this chapter by again referring to Däniker (1923) who assumed, based on the knowledge at the beginning of this century, that the necessary warmth of the ground is missing in the shade of trees, and, intrigued by the rapidity of the reduction of tree vigor near the treeline, was convinced that a minimal "quantity of warmth" is required for what he called "life activity". It seems that there is little to add to this view 80 years after its publication, except that we do now have a lot more data to support it. Attempting a mechanistic understanding, Däniker studied tissue formation, cell size and cell wall properties by simple light microscopy, driven by the awareness that – what we would call developmental cell biology – bears a great explanatory potential for the alpine treeline phenomenon as a low temperature **boundary for tree growth per se**. The cellular processes involved open a wide field of research in which molecular physiology and plant ecology need to become partners (see Chap 13). The em-

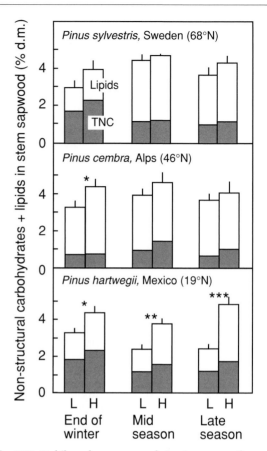

Fig. 7.20. Mobile carbon compounds in pines across the treeline ecotone in N Sweden (*P. sylvestris*, 68°N), the Alps (*P. cembra*, 47°N) and Mexico (*P. hartwegii*, 19°N). There is a significant site effect, but the elevation effect across sites is highly significant. (Hoch et al. 2002, Hoch and Körner 2003)

pirical evidence available today largely supports the view of the treeline as a thermal boundary for growth and formative processes.

Trees, with their elevated canopy, negatively influence their shoot and root zone temperature, and thus co-determine their own distributional limit at high altitudes. In contrast, the low stature of alpine plants creates warmer tissue temperatures (Chap 4), and thus allows them to grow at altitudes high above the treeline, at temperatures often higher than those experienced by treeline trees. Once more, this is the **result of** interactions between **plant morphology** and climate.

8 Climatic stress

Mountain tops, just like hot deserts are among those environments where life is dominated by climatic severity (Fig. 8.1). Climatic extremes play a key role in selection and evolution of alpine taxa. Whether those species and ecotypes which make their fortune in these "stressful" environments are "stressed" depends on whether stress is defined as a limitation of biomass production or of persistent presence and propagation (see Chap. 1). The two are not necessarily linked. In the first definition, alpine plants would rank as severely stressed, in the second, the answer is uncertain and may be "no" for plants at the alpine distributional center of some highly specialized species. However, independently of which concept one prefers, daily life in the alpine zone requires processes and structures which minimize climatic impact by avoiding or tolerating what is commonly considered stressful in a cold climate.

There are many **definitions of stress** in the literature but none is without shortcomings (see the discussion by Larcher 1987, Osmond et al.1987, Jones et al. 1989). Narrow definitions do not fit the variable nature of the many stresses plants are exposed to, and vague definitions do not help either. I suppose that the most important point is that stress is not seen as necessarily negative for an organism. To some extent stress is an essential component of successful life – and not only in plants. In Larcher's (1987) words "stress contains both destructive and constructive elements, it is a selective factor as well as a driving force for improved resistance and adaptive evolution". By no means should "stress" be used as a synonym for all sorts of constraints that limit growth to rates below the genetic potential, because in this sense all plants, except those growing in physiologically "optimized" culture conditions would be permanently "stressed" and "stress" would become synonymous with "normal" life (see the discussion on limitations in Chap. 1).

In this chapter, the discussion of stress will largely be restricted to thermal influences that cause losses of existing tissue (destructive stress) rather than the limitation of photosynthesis (see Chap. 11) and the production of new tissue. Thus, all gradual growth limitations falling under the category "suboptimal" will not be considered as "stressing", just as biotic interactions such as competition for light and soil resources are not included. Many of these gradual growth constraints are significant co-determinants of survival of severe climatic stress. In physically demanding environments fast and unlimited growth leads to fragile, intolerant plants, which underlines the ecological irrelevance of a concept in which stress is confused with limitation in general.

Low temperature stress and its survival by plants is one of the few fields in the alpine literature that has exhaustively been reviewed before (Larcher 1985b; Sakai and Larcher 1987; Beck 1994), allowing me to keep the re-assessment of this topic relatively short. Among other potentially important climatic stresses, exposure to extreme heat and ultra-violet radiation will be discussed. Mechanical stress has been touched on in Chapters 5, 6 and 7. Drought stress, rarely fatal in the alpine zone, will be dealt with in Chapter 9. Overall, these subjects will be treated largely at the phenomenological level, and the reader interested in cellular and molecular mechanisms will be referred to the relevant literature.

Fig. 8.1. A late spring freezing damage to leaves of *Rumex alpinus* at 2450 m elevation Swiss Alps (8 July 1998; see also color Plate 3d at the end of the book)

Survival of low temperature extremes

Freezing stress plays a decisive role in global plant distribution (e.g. Larcher and Bauer 1981; Sakai and Larcher 1987; Woodward 1987) and is the first environmental "filter" a species has to pass to become "alpine". Since this selective filter operates over very long time scales, it can be assumed that the natural vegetation of an area is adjusted to cope with local low temperature extremes. Survival at the species or population level does not necessarily imply that all individuals or all types of tissues remain unaffected. Hence, occasional injuries due to low temperatures are possible, which in turn alter the contribution of affected species to cover and biomass production (Körner and Larcher 1988). The likelihood of damage by subfreezing temperatures depends on the leeway between climatic extremes and tolerance limits during certain parts of the year.

As discussed in the previous chapter, periods with predictable cold weather (winter) are unlikely to be critical because frost hardening provides sufficient protection unless the "normal" environment is disturbed (e.g. by removal of snow in species dependent on snow protection, see Chap. 5). Some fully hardened alpine species have been shown to survive dipping in liquid nitrogen (Larcher 1980, Sakai and Larcher 1987). The situation is different when subfreezing temperatures occur unpredictably in the form of freeze-thawing cycles in spring or autumn or as freezing episodes during the growing season, and year-round during clear nights in the tropical-alpine zone. Under such conditions alpine plants may lose a substantial fraction of above-ground tissue, but are unlikely to be killed (Larcher 1985b, Sakai and Larcher 1987).

A typical **late freezing** situation in the alpine zone of the Alps (Tirol) was described by Kerner (1869) as follows:

"During the second half of last June we had a heavy snowfall which covered the alpine zone, including south-exposed slopes, which had already become snow-free up to 8000 feet. The snow persisted down to the treeline for 7 days. For all low stature plants, this exceptionally cold period passed safely, but taller perennial forbs and shrubs, which had already started to flush, and had partly protruded through the snow, suffered severely. Most young leaves were destroyed or desiccated during the following warm weather. The young leaves of (the evergreen) *Rhododendron ferrugineum* in the lower alpine belt, which had emerged shortly before, were killed outright. Even the new needles of the crippled spruces in the krummholz belt were frozen to death. However, higher up the slope, where *Rhododendron* had not yet flushed, the frost passed without any disadvantage for these shrubs".

These observations contain three important points: (1) life form (height above ground), (2) developmental stage (time) and (3) snow cover, all co-determine the risk. Annual absolute minima in meteorological records do not capture such detail.

While late freezing events reduce options for a productive season, **early freezing events** in autumn reduce gains as well. Leaf injury before natural senescence abruptly terminates photosynthesis and causes a loss of stored non-structural assimilates such as sugars, starch, lipids, proteins (which may be one third of leaf dry matter), and a loss of about half of the leaf mineral nutrients, most of which would have been recovered during normal senescence. In addition, the immature seed crop of late flowering species is lost.

Another example: during the night of 27/28 August 1995, a cold northwesterly front crossed the central Swiss Alps, bringing a heavy snow fall, followed by clear weather. Minimum 2 m air temperature during the first night dropped to −4.1 °C, followed by −5.4 °C 2 days later (much of the snow had melted by that time) and −4.3 °C after all snow had disappeared (Fig. 8.2). Ground temperatures possibly dropped 2–3 K lower due to radiative

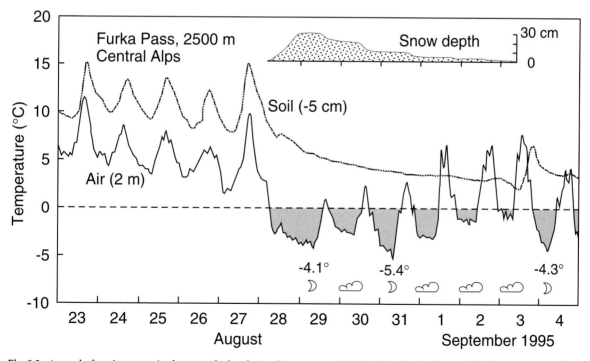

Fig. 8.2. An early freezing event in the central Alps during late summer 1995. Explanation in the text. Leaf temperatures of prostrate plants were possibly 2–3 K lower than the 2 m air temperature due to radiative cooling once snow had melted. (Unpubl. data)

cooling. The following damage was observed near the summit of the Furka Pass at 2500 m altitude (300 m above treeline), where a lush alpine flora exists, and where the above temperatures were recorded: all species of Asteraceae and Campanulaceae and some other late flowering species lost their entire diaspore crop. A large number of typical high alpine species such as *Oxyria digyna* lost a substantial fraction of their fully active leaves (see color Plate 3 at the end of the book). All forbs taller than 10–15 cm (the remaining snow depth on day 3 when air temperature reached its absolute minimum) and which were not bent to the ground, were decapitated by freezing (e.g. Apiaceae such as *Ligusticum mutellina*, the Asteraceae *Doronicum clusii*). While some species have already started to senesce in August, many remain active until late September or October (e.g. *Achillea* sp.), hence the losses in effective season length for these species were 2 to 5 weeks.

A **worldwide comparison of freezing tolerance** of alpine plants is rather difficult, because a great variety of "resistance" definitions, methods and pre-treatments have been applied, with hardly two data sets being readily comparable. For the current purpose I have selected data for critical temperatures ranging from first visible damage to 50% damage (Tables 8.1 to 8.3).

The data for freezing tolerance in active tissue in Tables 8.1 to 8.3 and Fig. 8.3 show a strong species and organ specificity, and the tolerance ranges in communities reflect the local temperature regimes. In low stature plants, freezing tolerance is not associated with growth form, leaf sclerophylly or plant family. Dwarf shrubs, cushion plants, herbaceous rosettes and sedges together cover a common range which is least negative in the temperate zone as exemplified for the Alps (−4 to −10 °C) where these limits also coincide with the low temperature limit of net photosynthesis (Pisek et al. 1967; Larcher and Wagner 1976). Flowers of early flowering species, such as *Saxifraga oppositifolia* may resist lower temperatures than leaves (−10 °C or less, data from Kainmüller, in Moser et al. 1977). Seedlings of alpine plants tolerate temperatures similarly low to leaves of

Table 8.1. Freezing tolerance in leaves of actively growing herbaceous alpine plants across the globe. (See also Table 8.2 and Fig. 8.3)

Species	Temperature (°C)
Alps (cool temperate)[a]	
Primula minima	−3
Senecio incanus	−4
Geum reptans	−4
Soldanella pusilla	−4
Oxyria digyna	−6
Ranunculus glacialis	−7
Central Asian Pamir (warm temperate)[b]	
Carex melanantha	−14.5
Dracocephalum discolor	−13.0
Potentilla pamiroalaica	−13.0
Saussurea pamirica	−12.0
Aster heterochaeta	−11.5
Sibbaldia tetrandra	−11.0
Ranunculus glacialis	−11.0
Primula pamirica	−10.0
Leontopodium ochroleucum	−10.0
Andes of northern Chile (subtropical)[c]	
Adesmia echinus	−14.2
Adesmia subterranea	−12.3
Calceolaria pinifolia	−20.0
Chaetanthera acerosa	−19.0
Gymnophyton spinosissimum	−12.0
Menonvillea cuneata	−16.3
Viola chrysantha	−20.0
Mt. Kenya (tropical)[d]	
Happlocarpha rueppellii	−13
Ranunculus oreophytus	−14
Senecio purtschelleri	−14
Carduus chamaecephalus	−14
Carduus chamaecephalus (4500 m)	−15

[a] first visible leaf damage in herbaceous species. According to authors, there is little difference in temperature between first visible and 100% damage, and thus 50% damage (compiled from various sources by Larcher and Wagner 1976).
[b] survival criterion not specified, selected data for plants treated by 12 h frosting at 4300–4800 m altitude (Tyurina 1957 cited by Sakai and Larcher 1987, p. 221).
[c] 50% reduction in photometric absorption of ethanol extracts of tetrazolium treated samples; middle of the dry summer in northern Chile (29°S, 3700 m altitude); standard deviation for 6–8 samples per species is ca. ±1 K (Squeo et al. 1996).
[d] 50% of maximum increase in conductivity after freezing treatment in submersed samples. (Beck 1994).

Table 8.2. Organ specific freezing tolerance in dehardened and fully hardened temperate zone alpine dwarf shrubs (data compiled by Larcher and Bauer 1981) and plants forming compact cushions (Larcher 1980; Kainmüller 1975) in the Alps. Numbers are temperatures (°C) at which 50% of the samples were damaged. Numbers in brackets are the maximum resistance in the fully hardened state in winter

Species	Type of organ			
	Leaf	Bud	Stem	Root
Dwarf shrubs				
Empetrum nigrum	−8 (−70)	–	– (−30)	– (−30)
Loiseleuria procumbens	−6 (−70)	– (−40)	−10 (60)	– (−30)
Vaccinium vitis-idea	−5 (−80)	– (−30)	−8 (−30)	– (−20)
Calluna vulgaris	−5 (−35)	– (−30)	−5 (−30)	– (−20)
Cushion forming plants				
Saxifraga oppositifolia	−10 (−196)	–	−19 (−196)	−25 (−196)
Silene acaulis[a]	−7 (−196)	–	−8 (−)	−11 (−196)
Carex firma	−7 (−70)	–	−6 (−)	−8 (−70)

[a] Junttila and Robberecht (1993) report −9 (−30) °C for subarctic-alpine *S. acaulis*.

Table 8.3. Frost tolerance in tropical-alpine plants in the Venezuelan Páramo at 4200 m altitude (°C, survival criterion as in Table 8.2). Note, the relatively tall shrubs of *Hinterhubera* and *Hypericum* reach these limits by supercooling, whereas the other low stature plants tolerate freezing. (Squeo et al. 1991)

Species	Type of organ		
	Leaf	Stem	Root
Hinterhubera lanuginosa	−12.3	−13.8	–
Hypericum laricifolium	−10.9	−11.2	–
Senecio formosus	−9.3	−7.9	−3.7
Castillea fissifolia	−14.8	−11.7	–
Arenaria jahnii	−18.8	−19.1	–
Azorella julianii	−10.6	−9.2	−4.0
Draba chionophila	−14.8[a]	−12.0	−14.0
Lucilia venezualensis	−14.3	−11.7	−9.8

[a] −14 °C for 4700 m, Azocar et al. (1988).

adult plants (Sakai and Larcher 1987, p. 223), but their establishment may be additionally restricted by ice heaving of soil (see Chap. 6).

Comparable tropical-alpine species in Venezuela and East Africa tolerate temperatures twice as low (−9 to −19 °C). Shrubs in the Hawaiian alpine zone (20°N) tolerate temperatures down to −11.8 and < −15 °C (Lipp et al. 1994). Numbers for the active season in continental warm-temperate or summer-dry subtropical mountains like the eastern Pamir and the Chilean Andes cover a similar range of −10 to −20 °C. The greater unpredictability of low temperature extremes in tropical and subtropical compared with temperate mountains apparently requires consistently higher frost tolerance.

The data for herbaceous plants shown in Fig. 8.3 permit a direct comparison of freezing tolerance of alpine and lowland species of one region, measured during the peak of their growing season by the same procedure. The full range from first injury to complete damage for the 33 lowland species is −2.2 to −6.5 °C (four species not tested at low enough temperatures). The range for the nine alpine grassland species from 2000 m higher elevation is −5.5 to −9.6 °C, with *Nardus* and *Carex* hardly affected at the lowest temperatures tested. The mean ca. 3 K difference is small given that the adiabatic lapse rate corresponds to an 11 K difference (see Chap. 3), and that radiative cooling during clear nights is more pronounced at high altitude. There is a tendency for graminoids to resist lower temperatures better than broad-leaved species. According to Larcher (1985b) active low altitude pasture grasses and alpine graminoids are both killed at around −7 °C. Growth and photosynthesis during the growing season apparently are incompatible with greater frost tolerance, and thresholds vary little with altitude.

A special case are giant rosettes in the tropical-alpine environment. Data for Andean, East-

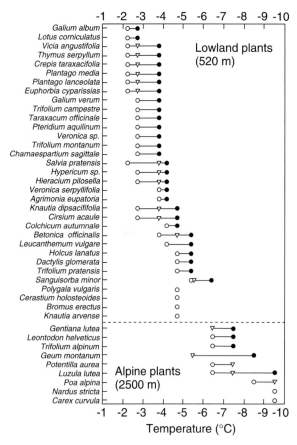

Fig. 8.3. A comparison of frost tolerance in alpine and lowland herbaceous plant species during peak season i.e. the second half of July at high altitude (2470 m, Furka Pass, Swiss Alps) and mid May at low altitude (550 m, calcareous grassland near Basel). Ranges are indicated as beginning (*open circles*), half (*triangle*), and full damage (*dark circle*). (Unpubl. data)

Table 8.4. Critical temperatures for frost survival (50% damage) in mature leaves of tropical giant rosette species of the Venezuelan Andes (G: Goldstein et al. 1985; S: various sources after Squeo et al. 1991), Mt. Kenya (Beck 1994) and Hawaii (Lipp et al. 1994)

Species	Temperature (°C)
Venezuela, 4200 m	
Espeletia schultzii	−11.2 (G), −12.0 (S)
E. moritziana	−11.3 (G), −10.6 (S)
E. spicata	−9.5 (G), −11.3 (S)
E. lutescens	−10.2 (G)
E. timotensis	−11.9 (S)
Mt. Kenya 4100–4200 m	
Senecio kenyodendron,	−5 to −14
S. keniensis	−5 to −10
Lobelia telekii	−14.5 to below −20
L. keniensis	−10 to below −20
Haleakala, Hawaii 2740 m	
Argyroxiphium sandwicense	−14.8

measurements at the same site in other years (Squeo et al. 1991). According to available temperature records, the Andean *Espeletia* species are unlikely to experience the extremes they are able to tolerate (−9 to −11 °C), whereas for the African species, Beck actually measured a temperature as low as −16 °C, clearly within the range which was found critical (−5 to −20 °C depending on weather-prehistory). Beck's data illustrate the need of repeated measurements in order to also account for such climatic extremes.

African and Pacific representatives of this spectacular plant type are listed in Table 8.4. The data from Beck, which includes repeated measurements over several periods from 3 different years illustrates a large temporal variation, even in such an equatorial environment (not presented in detail here). This variability possibly reflects the immediate weather-prehistory and/or phenological rhythms. In contrast, the data from Goldstein et al. presented here, differ little from various other

Avoidance and tolerance of low temperature extremes

The mechanisms by which alpine plants manage to cope with sub-zero temperatures fall into four categories. Plants may to a variable extent:

- Avoid exposure to low temperature extremes
- Avoid freezing of exposed tissue
- Tolerate freezing
- Substitute damaged tissue

Avoidance of exposure to low temperature extremes is achieved in three ways: (1) by phenology (in extratropical mountains), (2) by morphology (life form), and (3) by microhabitat preference. Alpine plants in seasonal regions show very distinct phenological rhythms reflecting the long-term adjustment to a periodically cold climate (e.g. Soeyrinki 1938; Sörensen 1941; Mooney and Billings 1961; Heide 1985; Pangtey et al. 1990).

By seasonal **phenology,** plants avoid exposure of sensitive tissue to intolerable temperatures. This requires temporal control of sprouting, flowering, seed maturation and senescence of summer green (deciduous) leaves. Annual plants have adopted the most extreme phenological mode of avoiding exposure to freezing temperatures. However, annuals are rare at alpine altitudes because of obvious difficulties of seedling establishment under the adverse climatic conditions (see Chaps. 2, 6 and 16). The mechanisms by which seasonal phenology is controlled are similar to those controlling seasonally variable freezing tolerance in persistent tissues, which will be discussed later.

Four phases of development (and hardiness) may be distinguished: (1) the dormant state in winter during which even warm spells or daytime radiative warming in snow-free vegetation will not activate plants (sensitive organs shed, maximum hardiness in persistent tissue). The low ratio of day/night duration and low night temperatures co-determine this quiescence. (2) In late spring the developmental blockade becomes loose, and opportunistic behavior (depending on actual temperatures) determines the annually variable onset of growth and the loss of full frost hardiness. (3) The third phase, the fully active period in summer with minimum resistance, is terminated in most alpine plants by photoperiod (critical day/night ratio), i.e. in contrast to spring activation, largely independently of the actual temperatures (examples provided by Prock and Körner 1996). This assures timely completion of the seasonal growth cycle and prevents loss of active tissue by freezing. (4) During the following fourth phase, resources are recovered from active above-ground tissue by natural senescence, and seed maturation is completed before the first critical frost events. Perennial tissues increase hardiness under the additional influence of decreasing temperatures (Pisek and Larcher 1954; Schwarz 1970; Sakai and Larcher 1987; Junttila and Robberecht 1993, and references therein).

Morphology, in particular plant size and the position of regeneration buds (e.g. Raunkiaer's system of life forms) is the best known and most obvious means of avoiding low temperature extremes (Larcher 2003). In regions with winter snow cover, tall plants are at greater risk than small plants. During snow free periods and in the tropics in general it is not clear whether low or tall plants are at greater risk with respect to freezing damage, because of the complex interaction of convection and radiative cooling. In mixed communities, low stature plants receiving some shelter from taller ones are likely to profit under radiative freezing. However, in open vegetation prostrate plants are at greater risk because radiative losses of heat are not sufficiently mitigated by convective heat transfer as result of poor coupling to the atmosphere (Grace 1988). This may explain the vertical stratification of freezing tolerance and the survival mechanisms observed by Squeo et al. (1991) in the Andes. Leaves closely attached to the ground were found to be 3 K more tolerant and, in contrast to leaves of "supercooling" tall plants, exhibited freezing tolerance (see below).

The great abundance of graminoids (often tussocks) and sessile herbaceous plants (often rosettes) in the alpine life zone of all latitudes is one of the most obvious examples of morphological avoidance of low temperature extremes. In most cases such plants have their apices, their basal leaf meristems (graminoids), their leaf primordia or premature leaves (dicotyledonous plants) and their premature reproductive organs several centimeters below the soil surface, protecting them from deleterious freezing temperatures (Fig. 8.4). As much as half of the total leaf mass may be below the ground because of the deep insertion level (Chap. 12) – a substantial structural and metabolic cost for this mode of avoidance.

Fig. 8.4. In most alpine plants, vegetative shoot apices and leaf meristems are buried several centimeters below the ground, and thus are not exposed to low and high temperature extremes. *Carex curvula* and *Ranunculus glacialis*, both from the Alps (2500–3000 m), and *Perezia* sp. from the northwestern Argentinan Andes (4250 m).

Seedlings and young plants often lack this safety, but with increasing plant age, contractile roots cause a progressive retreat of the meristematic zone below the ground (see also Chap. 6). Buried meristems have many other advantages. During the growing period, they profit from stored daytime warmth during the night, they avoid the zone of most severe surface heating during hours of intense direct solar radiation, and they are safe from grazers and mechanical damage.

Compact growth forms such as cushion plants profit from delayed night time cooling due to the heat capacity of either the moisture stored within the cushion or in the underlying soil or rock (unpubl. data by Larcher in Sakai and Larcher 1987). Prostrate dwarf shrubs are likely to profit in a similar way from close attachment to the ground or burial in the litter layer or top humus (e.g. *Salix herbacea* and other alpine *Salix* species).

A classic example of morphological protection from low temperature extremes is found in tropical giant rosettes, which close during the night and screen the sensitive apical meristems (Beck 1994). In addition, Beck reports that some of these (*Lobelia* sp.) excrete up to 2 l of fluid into the tight base of the rosette, submerging the core of the rosette in a thermal buffer. The protective nature of the dead leaf coat around stems of tropical giant rosettes is well established (Monasterio 1986).

Microhabitat selection is of obvious significance if one compares open versus sheltered sites, snowbeds versus ridge-tops etc. (see Chap. 4 and 5). The associated differentiation of frost tolerance was illustrated by Larcher 1980 (see Fig. 5.9) and Larcher and Siegwolf (1985).

In summary, the recipe for avoiding low temperature extremes is to restrict activity to safe periods and withdraw sensitive tissue during the cold season from exposed positions, to stay small and hide under snow or litter, or for taller, more resistant plants, to maintain all meristems below ground, as in grasses and most rosettes, and select topographically safe sites. Once the potential of these measures is exhausted the next step is to avoid tissue freezing.

Avoidance of freezing is possible in two ways, by osmotic adjustment, and by "supercooling". The first, and least effective mechanism is **freezing point depression** by accumulating solutes such as sugars. Although solute concentration, and thus osmotic pressure in alpine plants and treeline trees has been shown to undergo characteristic seasonal changes associated with the seasonal course of temperature (Blum 1926; Michaelis 1934; references in Sakai and Larcher 1987), the gain in tolerance usually does not exceed 2 K, because it needs 1 mole of solutes for 1.8 K freezing point depression, which, at the same time, causes the osmotic potential to drop by −2.2 MPa. Compared with lowland plants, alpine plants and treeline trees do not appear to exhibit consistently higher solute concentrations (Walter 1931; Turesson 1933; Mooney and Billings 1965; McCracken et al. 1985; Goldstein et al. 1985, Pantis et al. 1987; Chapin and Shaver 1989; Earnshaw et al. 1990), which suggests that this is not the main way of achieving low temperature resistance. However, it needs to be noted that the accumulation of sugars and other solutes at low temperatures may fulfill more than just osmotic functions (e.g. metabolization of other substances or provision of energy and direct cryoprotective effects on biomembranes, see below).

More effective (though risky, as will be shown) is a process called **supercooling** which, by avoiding nucleation, permits leaf and stem tissues to cool substantially below freezing point without freezing. This delay of solidification of water is favored by the compartmentalisation of plant water into cells and vessels, by cell wall impregnation with lignin, by the specific structure of water in solutions and by the absence of particles which initiate the crystallization process. In areas with regular, but not too low freezing temperatures (reliably never below −12 °C) such as in parts of the tropical-alpine life zone of the Andes, it appears to be an important mechanism for frost survival, as was documented for giant rosettes and taller shrubs (Goldstein et al. 1985; Squeo et al. 1991). In mature leaves, the supercooling effect is of limited duration (several hours) and is therefore called short-term supercooling.

When temperatures below supercooling capacity occur, supercooling becomes a fatal strategy, because freezing takes place abruptly, not permitting cellular water to feed intercellular ice formation (as will be discussed below), which would ensure gradual protoplast dehydration. In such environments, early freezing, rather than a delay of freezing ensures survival. Mucilage, frequently found in leaves of freezing tolerant alpine plants, was suggested to assure early nucleation and minimizes supercooling right from the beginning (see discussion in Lipp et al. 1994; mucilage micrographs in Körner et al. 1983). Thus, short-term supercooling is risky when temperatures below the nucleation point are possible – the reason why, in contrast to the Andean *Espeletia*, African giant rosettes, which experience lower temperatures, do not appear to employ supercooling, but are freezing tolerant, just as Andean plants from much higher altitudes are (Azocar et al. 1988).

In tissues with unexpanded or otherwise small cells such as in buds or seeds, but also in woody tissue, persistent **deep supercooling** to temperatures below −35 °C has been found. For instance, Becwar et al. (1981) reported xylem water of treeline trees to supercool to −40 °C. Explanations for this extraordinary physical phenomenon are not yet available – a difficult, but important field of future research (see the review by Sakai and Larcher 1987).

Where low temperature extremes exceed short-term supercooling capacity or equilibrium freezing-point depression, true **freezing tolerance** is required for plants to survive. I have noticed that this term often causes confusion, because it is mistaken as an indication of tolerance of whole tissue freezing, while in reality it is only water outside the protoplast (apoplast-water) which freezes – otherwise the tissue would be killed. No plant has been found to survive the formation of ice crystals within the cytoplasm. In the very short term, ice formation in close proximity to live cells by itself is a protecting process, because thermal energy is released at the liquid-solid transition, which can be measured upon nucleation as an abruptly increasing temperature (so called "exotherms").

Slight supercooling (which occurs even in a glass of water), leads to a first **exotherm** when apoplast freezing commences. A second exotherm (or sometimes several exotherms) may indicate nucleation in different tissue compartments which undergo supercooling to lower temperatures, or osmotic freezing point depression, but more commonly this indicates the freezing of the protoplast and cell death (some unexplained gaps between such a second exotherm and the actual injuring temperature are reported in the literature, e.g. Squeo et al. 1991). The frequently cited survival of submersion in liquid nitrogen is a very special situation in which super-hardened plants are frozen so quickly that amorphous ice is formed which apparently does not recrystallize in a damaging way during the thawing process.

Freezing tolerance means tolerating **dehydration** of the protoplast due to rapid transfer of water to a growing body of ice in the intercellular space. In mature leaves, including those of alpine plants, intercellular air spaces commonly account for 25–30% of the total leaf volume (Körner et al. 1989a). Ice may also form between the cell wall and the plasmolytic protoplast. In more compact tissue such as in buds, ice fills the niches resulting from progressive protoplast shrinkage. While intercellular ice is formed, leaves become darker in appearance – an impression enhanced during thawing when the mesophyll becomes infiltrated (Fig. 8.5). In this situation photosynthetic gas exchange is blocked. In freezing tolerant plants the infiltration water is re-absorbed by cells in 1–4 hours after thawing and leaves return to the original lighter coloration and normal photosynthetic activity.

Freezing tolerance requires plants to first control nucleation (see review by Andrews 1996) and second, to retain the plasma membrane in a fluid state so that water can be freely transferred from the interior of the cell as the extra-cellular freezing process progresses. Membrane fluidity is maintained by a characteristic phospholipid composition developing during cold hardening, with plants from cold climates generally exhibiting more fluid membranes (Sakai and Larcher 1987; Beck 1994). Because of the associated dehy-

Fig. 8.5. Intercellular ice formation withdraws water from cells and leads to filling of air spaces, often causing leaves to turn dark (enhanced during thaw). The photograph shows spotty infiltrations in Vaccinium myrtillus, recovering from freezing during the last night (mid May, near the treeline in the Alps).

dration of the protoplast, a third requirement for the survival of extra-cellular ice formation is the stabilization of biomembranes both by structural adaptation and by protective (adhesive) substances such as raffinose and trehalose (Sakai and Larcher 1987, p. 117; Larcher 1995, p. 362). These compounds assure the preservation of high order molecular structures during the distorting phase of extreme desiccation. The close link between freezing and dehydration tolerance finds expression in their synchronized and parallel seasonal courses (Pisek and Larcher 1954; Larcher 1963b) and the positive effect of drought pre-treatment on frost tolerance (Larcher 1985b; Sakai und Larcher 1987).

One aspect which has come up recently in a paper by Halloy and Gonzalez (1993) deserves further investigation: these authors observed higher survival of seedlings at the same low temperature but at reduced atmospheric pressure or at increased altitude. If there were such a link between pressure and freezing sensitivity it would, besides the ecological implications, matter where (in terms of ambient pressure) experimental cooling treatments of alpine plants are conducted. Halloy and Gonzalez (1993) suggested that rapid membrane repair in the critical temperature range may benefit from reduced oxygen partial pressure. On the other hand, hypoxia has been shown to reduce frost resistance in winter cereals as well as in roots and shoots of trees (references in Larcher 1985b, p. 283).

In cases where avoidance of low temperatures, avoidance of freezing, or freezing tolerance are insufficient, frost damage will occur and plants will lose part of their organs or even total above-ground biomass. In such situations survival depends on the buffering capacity of organ losses at the whole plant level. **Repair and replacement abilities** of a plant require morphological and developmental plasticity (Bell and Bliss 1979; Körner 1995). In seasonal climates responses will depend on the time of damage. Losses of young leaves and shoots due to late frost at the beginning of a 10 week alpine season leaves little time for replacement. The preformation of (embryonic) future leaf cohorts and sufficient storage reserves are crucial for a second flush. Species flowering early from preformed buds will rarely produce a successful second cohort of flowers. Late season losses of reproductive units in late flowering species are finite. The great risk of reproductive failure due to freezing damage (together with problems of seedling establishment) has led to abundant clonal propagation in alpine taxa.

Although alpine plants do not possess greater storage organs than lowland plants (Körner and Renhardt 1987), the root system is commonly larger and carbohydrate and lipid reserves are substantial (see Chap. 12). Diemer (1996) has shown that some high altitude specialists can tolerate repeated losses of much of their annual leaf crop by browsing with no obvious long-term harm, which is in line with the suggestion discussed later, that alpine plants under most situa-

tions are not carbon limited. Since graminoids and sessile rosettes have their vegetative apices at safe depth below the ground (Fig. 8.4), replacement of tissue losses will be a matter of activating these intact meristems, whereas plants with above-ground meristems may have to develop whole new shoots, the reason why above-ground meristems become increasingly rare at high altitudes.

Heat stress in alpine plants

The title of this section sounds like a contradiction. However, damaging heat stress in high mountain environments is a realistic possibility. As was illustrated in Chapter 4, prostrate plant growth causes a decoupling of the plant's environment from the ambient climate to such a degree that differences between tropical lowlands and glacier forefields may disappear periodically. What has evolved as the only way to escape the cold, namely the formation of a heat trapping morphology, can turn into a problem when solar radiation peaks, when top soils dry and winds calm. Maximum tissue temperatures may indeed surpass the heat tolerance threshold (Tables 8.5 to 8.7).

Gauslaa (1984) compiled published data for a gradient from the Sahara to the Norwegian mountains, and the surprising result is seen in Table 8.6. Unlike low temperatures, the maximum heat plants have to cope with varies little, despite the fact that the annual duration of exposure to such extremes is certainly different. The numbers once more document the enormous effect of plant canopy structure in cold environments. While air temperatures differ by 24 K between these habitats, mean maximum leaf temperatures are quite similar, in fact, almost identical in the Alps and Mauritania. Given these patterns, one would expect similar heat tolerance.

Indeed, mean heat tolerance is surprisingly similar among these and other habitats (Table 8.7). The difference between the North African desert and the Norwegian fellfield is only 1.4 K. However, in each of these habitats, species differences do exist, which are related to the specific microhabi-

Table 8.5. Examples of maximum leaf temperatures measured under full solar radiation in midsummer in adult plants, under arctic-alpine (Gauslaa 1984), temperate-alpine (Larcher and Wagner 1976; Cernusca 1976) and subtropical-alpine (Breckle 1973) conditions

Regions/Species	T_{leaf} (°C)	T_{air} (°C)	T_{leaf}-T_{air} (K)
Central Norway 900–1400 m			
Silene acaulis	45.5	21.0	24.5
Rubus chamaemorus	42.0	16.5	25.5
Saxifraga oppositifolia	39.9	17.0	22.9
Dryas octopetala	38.0	20.3	17.3
Artemisia norvegica	38.0	19.0	19.0
Loiseleuria procumbens	37.2	19.4	17.8
Central Alps 2000–2300			
Sempervivum montanum	54	22	32
Carex	47	16	31
Arctostaphylos uva-ursi	44	25	19
Loiseleuria procumbens	43	23	20
Primula minima	38	14	24
Saxifraga oppositifolia	35	15	20
Hindukush 4350 m			
Carex nivalis	40	12	
Primula macrophylla	38	13	

tat conditions. Table 8.8 shows the amplitude of heat tolerance for Norwegian alpine plants (43.5 to 53 °C). Larcher and Wagner (1976) reported values between 45.5 and 55 °C for the central Alps, with the highest number for the alpine facultative CAM-plant *Sempervivum montanum*. A maximum temperature of 60 °C is reported by Kainmüller (1975) for *Carex firma*, growing on steep, equator facing calcareous rocks at 2300 m altitude above Innsbruck.

Gauslaa tried to group his large data set from which the examples in the Tables were extracted, and he found similar ranges of heat resistance for plants from alpine heath and open lowland habitats on shallow soil (LD 50 ca. 51 °C), for rock and scree vegetation, dry alpine snowbed communities, mixed lowland forests (LD 50 ca. 48 °C), for wet alpine snowbed communities, mire- and spring-vegetation and lowland forest understory (LD 50 ca. 46 °C). Standard deviations in each group are between ± 2 to 3 K, hence are substantially greater than the differences between means in the above global comparison. Heat pre-

Table 8.6. A comparison between maximum leaf temperatures from the African desert to the Norwegian alpine zone, compiled by Gauslaa (1984 from his own data; Larcher and Wagner 1976; Lange 1959; Lange and Lange 1963)

Region	Number of plant spp.	T_{leaf} (°C)	T_{air} (°C)	T_{leaf}-T_{air} (K)
Norway (arctic-alpine)	10	36.8 ± 1.7	19.3 ± 0.7	17.4 ± 1.8
Alps (temperate-alpine)	6	43.5 ± 2.7	19.2 ± 1.9	24.3 ± 2.4
Spain (coastal-mediterranean)	19	40.4 ± 0.9	31.7 ± 0.7	10.6 ± 1.0
Mauretania (subtropical desert)	15	44.4 ± 1.3	43.7 ± 1.4	0.7 ± 1.2

Table 8.7. A global comparison of mean heat tolerance (50% survival) for plants from contrasting environments including alpine (A). (References in Gauslaa 1984 and original data from Larcher and Wagner 1976)

Region	number of species	Mean temperature	Authors
Mauretania	52	49.4	Lange (1959)
Spain	39	50.4	Lange and Lange (1963)
Puerto Rico	39	49.1	Biebl (1964)
Greenland	23	48.1	Biebl (1968)
Norway (A)	89	47.6	Kjelvik (1976)
Norway (A)	118	48.0	Gauslaa (1984)
Alps (A)	10	47.1[a]	Larcher and Wagner (1976)

[a] Means calculated from highest temperatures with no damage and temperatures at 100% damage. This may underestimate the true 50% survival threshold.

Table 8.8. The variability of heat tolerance (50% suvival) in a selected number of alpine species from Dovre Fjell, Central Norway. (Gauslaa 1984; data in brackets by Larcher and Wagner 1976 from the Alps)

Species	Temperature (°C)
Herbaceous species	
Oxyria digyna	43.5 (46.5)[a]
Arabis alpina	44.1
Ranunculus pygmaeus	46.2
Ranunculus glacialis	47.1 (46.5)
Ranunculus nivalis	47.6
Astragalus norwegicus	47.5
Gentiana nivalis	47.5
Draba alpina	48.2
Hieracium alpinum	50.5
Potentilla crantzii	51.7
Prostrate woody plants	
Salix herbacea	46.5
Salix reticulata	47.6
Cassiope hypnoides	47.5
Loiseleuria procumbens	51.8 (50.0)
Arctostaphylos uva-ursi	52.9 (52.0)

[a] Calculation as in Table 8.7.

treatment ("heat hardening") can cause an upward shift of the 50% survival threshold by about 2 K. The most extreme difference was found between completely dehardened plants and plants hardened by maintaining them for 24 hours at close to lethal temperatures. Under such conditions a shift from 45.1 to 50.1 °C in LD 50 was found in *Oxyria digyna* and from 46.6 to 48.7 °C in *Ranunculus glacialis*.

In summary, these data illustrate similar heat stress and similar heat tolerance in lowland and alpine environments, independently of latitude. The range of heat tolerance thresholds among species is rather narrow, the acclimation potential is comparatively small and so is the difference between peak temperatures found in the field and observed heat tolerance, particularly in the most prostrate alpine plants growing in isolation or with large spacing on shallow soil. The establishment of seedlings on such microsites is only possible in spring or autumn or in a very wet seasons (see discussion of topsoil temperatures in Chap.

4). The frequently observed persistence of patches of bare humic soil within alpine vegetation is likely to be related to heat stress. Gauslaa (1984) reports that he had measured surface temperatures in *Silene acaulis* of 45 °C for 2 h and found this part of the cushion with brown leaves the next day. Since the preceding days were rainy there was no doubt that extreme heat (and not desiccation) had damaged these leaves. In the central Alps, brown, leafless patches on the southern part of *Silene acaulis* cushions are quite common. Compact cushions or prostrate plants narrowly attached to dark otherwise non-vegetated ground are extremely vulnerable to overheating (Larcher 1977), and are thus restricted to high altitudes or sub-polar latitudes. These are the only regions were plants of this growth habit can survive. For all other alpine growth forms heat stress is a potentially critical factor during seedling establishment on equator facing slopes because the short season forces recruitment into the hottest part of the year.

Heat tolerance is a matter of membrane stability and parallels frost tolerance which leads to a counter-intuitive seasonal trend with peak heat resistance in winter-hardened tissue and minimum resistance in summer (e.g. Lange 1961; Schwarz 1970; Kainmüller 1975). For instance, Kainmüller's data show a heat resistance in *Saxifraga oppositifolia* of 55 °C in midwinter and 51 °C in midsummer. The seasonal amplitude in heat tolerance is much smaller than that in frost tolerance and commonly does not exceed 5–8 K.

These considerations illustrate a dilemma of alpine plant life. Upright growth does not capture enough warmth to facilitate growth above the treeline (see also Chap. 7). Prostrate growth, on the other hand, creates a warmer, favorable microclimate during the day, but bears the risk of periodic overheating. From an ecosystem point of view, the long-term stability of alpine slopes and their precious soils is best assured under such stress situations by the presence of a mixed vegetation where the various climatic risks are distributed among various morphotypes. In this respect high biodiversity is the best "stress management".

It is obvious that any damage of alpine vegetation that opens the ground and allows surface heat to accumulate (see Chap. 4) can lead to sustained bare spots, which may be eroded before vegetation has a chance to return – an important aspect for alpine land use and conservation (Körner 1980; Chapin and Körner 1995).

Ultraviolet radiation –
a stress factor?

In popular literature, UV-radiation is regularly mentioned as a key factor in alpine plant life, however, this is not based on sound scientific evidence. Perhaps this view emerged from mountaineers' frequent experience of erythema. While low temperature stress at high altitudes is documented and understood reasonably well, the morphogenetic and potentially stressful influence of UV-radiation is not. A century of interest in this field of experimental alpine ecology has seen many attempts at answering the question of whether alpine plants are dwarfed, develop small and thick leaves and colorful flowers because of enhanced UV-radiation. In the very early literature (cf. Schröter 1908/1926), this was assumed without real facts and the story has been perpetuated over the years. Today there is still little, if any, evidence in favor of this hypothesis. However, before entering this field in more detail, a few words on the meteorological aspects of UV-radiation.

"Ultraviolet" refers to the invisible, short wave part of the **solar spectrum** in the range of ca. 200 to 400 nm, of which the most dangerous wavelengths, shorter than 280 nm(UV-C), never get to the ground. UV-C is absorbed by oxygen splitting and by interactions with the resultant ozone in the earth's uppermost atmosphere at 25 to 30 km altitude. The longest wavelength segment of the UV spectrum, between ca. 320 and 400 nm(UV-A), is transmitted through the atmosphere in a similar way to visible solar radiation and exerts little extra impact on organisms. It is the 280–320 nm band (UV-B) and its variable atmospheric transmittance and damaging potential, which makes UV

ecologically and medically so significant. As has been mentioned in Chapter 3, the UV-B contribution to total solar radiation tends to increase with altitude, as do the frequency and intensity of extremes of total solar radiation, but long-term means are strongly dampened by cloudiness, which tends to increase with altitude as well.

Often, the consequences of reduced screening of the horizon at high elevation, the more intense total solar radiation, strong reflectance from snow or bright ground, and – in the case of sun burn – the longer outdoor exposure during mountain tours, are confused with UV-B intensity effects, but in reality are largely explained by the overall radiation dose. Researchers, like mountaineers prefer bright days, which leads to a biased impression of environmental conditions for alpine plants! **Altitudinal trends** in UV-B for clear days are likely to be steeper than those in means across all days. While, in relative terms, UV-B radiation may be even more intense under overcast conditions at high compared with low altitude, the long-term dose strongly depends on the relative frequency of such weather. But "clear day only" data also vary regionally. While pronounced increases are reported for the Alps (e.g. Blumthaler et al. 1993), the differences reported by Caldwell (1968) for the Rocky Mountains Front Range in Colorado are comparatively small.

Since absorbance of any component of solar radiation by the atmosphere depends on the path-length of solar rays through the atmosphere, there is a strong latitudinal decrease of UV intensity measured in the direction of incoming radiation. On a horizontal plain, the reduction is even greater. Also, the ozone layer is thinnest near the equator and thickest near the poles (Caldwell et al. 1980, 1989), which further enhances **latitudinal differences**. This statement refers to midsummer conditions and does not account for recent periodic depletions of the ozone shield, particularly in polar regions (the so called "ozone hole"). The combination of latitudinal and altitudinal trends yields maximum UV-exposure in tropical-alpine and minimum in arctic lowland plants. Tropical-alpine plants in the Andes are exposed to more than sevenfold the total daily effective UV-B dose experienced by plants in sub-arctic Alaska during the peak of the growing season (Caldwell et al. 1980; Robberecht et al. 1980).

As with freezing stress, an important question is whether, and how, UV-B radiation reaches potentially sensitive tissue. In other words, "**avoidance**" must be known before tolerance can be discussed. In leaves, avoidance of mesophyll exposure to UV-B radiation can be achieved by (1) reflectance at the surface, (2) absorption through pubescence, waxes, the cuticle and outer cell walls, (3) absorption by epidermal cells, and (4) by placing sensitive tissue such as meristems below ground.

Reflectance of UV-B radiation on leaf surfaces is reported to be low (less than 10%, see references in Caldwell 1968, DeLucia et al 1992), but may be enhanced slightly by pubescence just like visible short wave radiation (Gauslaa 1984). Caldwell compared UV-A reflectance of 13 montane (ca. 1800 m) and 18 alpine species (ca. 3750 m) and found no difference between the two groups, with reflectance ranging from 1.5 to 7%. I am unaware of reflectance data for UV-B in alpine plants, but it can safely be assumed that the majority of incident UV-B is likely to be absorbed and not reflected, hence, unlike some flowers, leaves are "UV-dark".

The first systematic analysis of leaf epidermal **UV-B transmittance**, including alpine plants, was conducted by Lautenschlager-Fleury (1955) in Basel. She discovered that isolated epiderms of a broad variety of herbaceous species transmit between 2 and 25% of incident UV-B, hence, according to her data, 75 to 98% is absorbed before reaching the mesophyll, and in most cases, it made little difference whether epiderms mounted on a special holder in a spectro-photometer were turgid or dried. Fused silica optical fiber studies in intact leaves of herbaceous plants conducted more recently by Day et al. (1992) revealed much higher transmittance of 18–41% in meadow plants near the treeline. The discrepancy between the two data sets is unlikely to be due to measurement "errors", but reflects the fundamental difference of how transmittance is defined. The situation is similar

to radiation measurements in snow (see Chap. 5) where back-scatter may more than double readings taken by a pointsensor within the absorbing matrix, compared with readings at a "dark plane" below a considered substrate layer. Point measurements (largely independent of acceptance angle) within the absorbing tissue reflect the light regime at this point, but do not represent "transmittance" of the overlaying stratum in the physical sense according to Lambert-Beer's law. However, this will not affect the comparability of data within one measurement technique.

Lautenschlager-Fleury's data show (1) consistently higher **absorbance by** upper versus lower **epiderms**, (2) that the absorption is largely associated with solutes rather than with cell wall properties, and (3) that both rapid short-term as well as long-term acclimation does occur, which is (4) associated with intensity of visible solar radiation rather than UV-B per se. In addition, Caldwell (1968), Caldwell et al. (1982) and Ziska et al. (1992) demonstrated (5) significant ecotypic differences in UV-B transmittance or mesophyll sensitivity. In Lautenschlager-Fleury's study, epiderms of alpine plants exhibited the lowest transmittance of all studied groups of plants (Table 8.9). While 1.5 to 8.4% of incident radiation was transmitted in the alpine group (Tab. 8.10), the range was 5.0 to 16% in fully sun exposed forbs from low altitudes. Smaller overall epidermal transmittance of only ca. 3% and no such altitudinal differences were found by Caldwell between the plains (1700 m) and alpine altitudes (3450–3750 m) which paralleled the negligible difference in UV-radiation data along this transect. Perhaps this is associated with the dry continental climate of these plains compared with the more humid and cloudy climate at the alpine site in the Rocky Mountains. Robberecht et al. (1980) report epidermal transmittances of UV-B below 2% in tropical alpine plants and values exceeding 5% at temperate and subarctic latitudes. In their morphological comparison, Day et al. (1992) showed that epiderms of herbaceous plants generally exhibit the highest transmittance, compared with almost zero transmittance in conifer needle epidermis and grasses

Table 8.9. A comparison of upper epidermal UV-transmittance in various groups of herbaceous plants. (Lautenschlager-Fleury 1955, number of species in brackets)

Plant group (habitat)	Transmittance (%)
Lowland plants from deep shade	17.8 (7)
Lowland plants from open fully sunlit habitats	9.3 (1)
Mediterranean plants from open sites	7.9 (3)
Alpine plants (Central Alps 1800–2500 m)	5.5 (14)

Table 8.10. Mean epidermal transmittance of UV-B in field grown leaves of alpine plants of the central Swiss Alps. (Lautenschlager-Fleury 1955)

Species and altitude	Epidermal transmittance (%)	
	Upper	Lower
Primula auricula 2100 m	1.5	2.0
Gentiana punctata 2300 m	2.5	8.5
Gentiana purpurea 1800 m	3.3	8.0
Daphne mezereum 1950 m	3.9	6.6
Gentiana verna 2100 m	4.5	3.7
Saxifraga aizoon 1950 and 2500 m	4.7	3.3
Primula viscosa 2200 m	5.1	7.5
Veronica bellidioides 2300 m	5.3	10.1
Erigeron uniflorus 2500 m	5.7	9.5
Anthyllis vulneraria 1950 and 2100 m	7.1	12.1
Cardamine resedifolia 2300 m	6.7	19.0
Campanula thyrsoides 2200 m	7.2	22.8
Lotus corniculatus 1900 m	7.8	12.3
Sempervivum tectorum 2100 m	8.4	12.3

or woody dicots, which hold an intermediate position. Epidermal transmittance thus appears to be correlated negatively with life expectancy (Day 1993) and/or positively with metabolic capacity of leaves.

In summary, leaf epiderms effectively protect the mesophyll from UV-B exposure and thus equalize temporal (season, weather), habitat (sun, shade) and large-scale altitudinal and latitudinal differences in UV-B exposure of mesophyll tissue of plants. UV-B absorbance is associated with active metabolism in the epidermis and shows high acclimative potential. There is no indication that the mesophyll of alpine plants is exposed to substantially higher UV-B levels than the mesophyll in comparable lowland plants of similar latitude.

Among the substances responsible for UV-B absorbance, **flavonoids** and related phenolic compounds are most important (e.g. Lautenschlager-Fleury 1955; Klein 1978, Robberecht and Caldwell 1983; Larson et al. 1990; Ziska et al 1992; Gonzalez et al. 1993) which, when extracted confirmed the trends seen in intact epiderms (Lautenschlager-Fleury 1955; Caldwell 1968). The amount of soluble flavonoids in leaves can be remarkable: Veit et al. (1996) found that up to 10% of the dry mass of leaves of the fern *Cryptogramma crispa* collected at 2050 m altitude in the Swiss Alps were flavonoids. These authors also confirmed Lautenschlager-Fleury's observation that flavonoid concentrations can rapidly be adjusted to prevailing UV-B regimes and follow the diurnal course of solar radiation (Fig. 8.6).

Among other protective agents, anthocyanids (in reddish leaves of some alpine plants during the cold season, cf. Caldwell 1968; see color Plate 3 at the end of the book), carotenes and waxes are

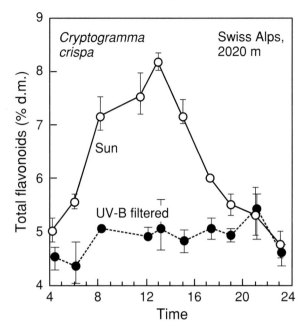

Fig. 8.6. Alpine plants have effective means of preventing "sun burn": large amounts of soluble flavonoids absorb UV-B. In the case of the fern *Cryptogramma crispa*, studied at 2050 m in the Swiss Alps (Ticino), the concentrations vary diurnally between 4.5% and almost 9% of leaf dry mass (*open symbols*; data for a clear day at the end of June). When screened with UV-B absorbing filters (*dark symbols*), the predawn base level is maintained throughout the day. The experiment also included a control with UV-transmissive filters under which the course of flavonoid concentration paralleled the unscreened curve (not shown). Removal of the UV screen after 10 months of treatment caused the curves to merge within hours. (Veit et al. 1996)

mentioned in the literature. It is unknown whether all these protective substances can sufficiently prevent the potentially **mutagenic effects** of UV-B radiation (Caldwell 1971) in high mountain floras (but see below). Since tropical mountains experience particularly high UV-B loads, Lee and Lowry (1980), and Flenley (1993) and others have suggested **evolutionary effects** as well as influences on altitudinal limits of tropical montane forest. Physiological, biophysical and genetic data in favor of such UV effects in alpine plants are not available.

It is often overlooked that **cell division** and a large fraction of **cell differentiation** in alpine plants happens below the ground (see the discussion on escape from freezing above). In the case of dwarf shrubs and cushion plants, bud scales envelop the developing leaf or shoot. In graminoids in particular, cell differentiation is largely completed when leaves emerge, and there is no difference in cell size of leaves between alpine and lowland plants (Körner et al. 1989a, b). Also, flowering buds are commonly preformed below ground. Reports about the repressive influence of UV-B radiation on mitosis are all from experiments with tissue cultures or bare root tips under direct UV-B radiation (Klein 1978). Exposure of pollen during pollination and of pollen tubes during stigma surface penetration may be the single most critical step for mutagenic effects (Flint and Caldwell 1983). These authors showed that anthers filter out over 98% of UV-B, and that UV reflectance by the corolla is largely in the UV-A range, which is not considered mutagenic. Embryogenesis takes place under many layers of cells, and if negative mutagenic influences were effective during this phase of life, one would expect selection for a thicker pericarp and integuments in alpine versus lowland plants – not an obvious feature, but worth being explored – irrespective of potentially effective chemical filters, as found in anthers.

This observational evidence does not rule out the possibility of indirect morphogenetic influences of UV-B radiation via signal transduction from mature tissue (hormones), but direct effects on developing tissue, specific to the alpine environment, appear unlikely. Finally, if such direct or indirect influences of UV-B radiation occurred, one would expect latitudinal gradients in alpine plant stature, which is not seen in nature. Many arctic and subantarctic plants in misty coastal ranges are extremely stunted and dwarfed, but receive very little UV-B compared with temperate lowland plants, not to mention equatorial plants. The big "global experiment" of alpine plant life at all latitudes and across different climatic regions once again assists in narrowing down the most likely common denominator for the explanation of alpine dwarfism: low temperature (see Chap. 14).

Having said this, what have we learned from **UV-B experiments**? Results of six experiments with alpine plants under UV-B absorbing window glass or special filters were published between 1908 and 1955 (see review by Klein 1978). All suggest UV-B is effective. Screening UV-B reduced some of the morphological and anatomical characteristics of typical alpine plants. However, missing controls or replication, confounding environmental influences of screens and the horticultural nature of some of these early studies make it difficult to interpret the observations. It was shown repeatedly in greenhouse experiments with crop plants that enhanced UV-B can reduce growth and alter plant morphology (Klein 1978; Tevini et al. 1983) and more recently, rather moderate experimental UV-B enrichment has been shown to affect plants in the field as well (Caldwell et al. 1995; Rozema et al. 1997), but whether the in situ removal of the natural "impact" of UV-B in the alpine zone has the opposite effect is uncertain.

The first and so far only quantitative approach in this direction is Caldwell's (1968) classic screening experiment on Niwot Ridge at 3750 m altitude in the Rocky Mountains. In a fully replicated design, herbaceous alpine vegetation was screened by UV-B absorbing or UV-B transmitting plastic film filters for two growing seasons. In short, most growth and developmental parameters studied in five species did not suggest any significant change. In one species (*Trifolium parryi*), screening of UV enhanced flowering, in some species it slightly stimulated leaf elongation but, for instance, in *Carex rupestris* percent plant cover was reduced during certain periods when UV was excluded. Seasonal biomass yield per plot was slightly stimulated in *Trifolium* and *Geum rossii*, unchanged in *Oreoxis alpina* and *Carex* and somewhat decreased in *Kobresia myosuroides*. If persistent, such minute responses may translate into a **biodiversity effect** in the long term, but in accordance with the above observational data, there is absolutely no indication that natural UV-B exerts destructive stress or imposes significant fitness constraints in the field. As small as these responses were, Caldwell detected pronounced alterations in leaf extract absorbancies (reduced under UV-exclusion), indicating that plants did indeed perceive the alteration in UV-exposure. When abruptly exposed to unfiltered late summer sun, no damage was observed in previously UV-protected plants (in contrast to the commonly observed damage in shade plants brought into sun), underlining the observations by Lautenschlager-Fleury (1955) and Veit et al. (1996) that visible solar radiation alone guarantees the maintenance of some base-level UV protection.

When arctic and temperate- or tropical-alpine congeneric species or conspecific ecotypes were exposed to a 15 h UV shock treatment, a surprising dichotomy of responses was observed (Caldwell et al. 1982). In high latitude ecotypes/species **photosynthesis** was reduced to 20–30% of that in unshocked plants, whereas low latitude ecotypes/species were depressed to only 50–78% of controls and these responses were independent of UV-B pre-treatment, except for one out of eight species/ecotypes (*Plantago lanceolata* collected in the Peruvian Andes). In contrast to these latitude specific responses at the chloroplast level, epidermal damage seen after this artificially high UV-B stress was not significantly different between the arctic and alpine group. Epidermal damage and photosynthetic damage are thus fundamentally different photo-biological reactions.

It appears that UV-B tolerance at the **chloroplast** level is also altitude specific. Lütz (1987) observed close associations of peroxisomes with chloroplasts and mitochondria and particularly high peroxidase activity in alpine *Ranunculus glacialis*. Accessory pigments are also part of this second "defense-line" (Wildi and Lütz 1996). Bergweiler (1987) in his very detailed analysis found a substantially greater abundance of carotenes in congeneric alpine versus lowland species. But the literature is not univocal in this field, with some researchers also reporting increased concentrations (Morales et al. 1982), others finding no elevational trends (Seybold and Egle 1940; Polle et al.

1992), and there are even reports of altitudinally decreased carotene contents (Todaria et al. 1980). Part of this inconsistency may have to do with the rather varied references used for expressing carotene content (fresh or dry mass, leaf area).

Experimental exposure of seedlings of alpine and lower altitude species of *Aquilegia* to continuously **enhanced UV-B** under controlled conditions did not affect photosynthetic capacity and stomatal conductance, but flavonoid contents increased, plants remained shorter and produced more leaves, and this UV enhancement effect was less pronounced in the high altitude species (Larson et al. 1990). Artificial UV enhancement also reduced plant height in 14 out of 33 species from an elevational cline in Hawaii, and again the high elevation provenances showed a smaller response (Sullivan et al. 1992). Trends were also seen in a comparison between congeneric alpine and lowland plants in the Alps (Rau and Hofmann 1996). Photosynthetic capacity studied in a subsample of these species was maintained in the high

elevation group, but reduced in low elevation provenances (Ziska et al. 1992), but it must be remembered, that these experimental plants did not grow under normal sunlight.

In conclusion, solar UV-B radiation is unlikely to be an important constraint for growth and development of alpine plants. UV-B radiation has been an evolutionary selective force which has led to genetic differentiation of species from regions of contrasting UV-B regimes (Caldwell et al. 1982). There is no indication that the stunted growth of alpine plants is a direct response to UV-B stress, despite the fact that experimentally enhanced UV-B can moderately shorten plants under greenhouse conditions. For wild plants, evolved and grown under natural sunlight, UV-B radiation does not appear to exert "stress". The only exeption may be the release from snowcover to full sunlight at the beginning of the season at higher latitudes, but even under these conditions adaptive mechanisms have been illustrated to be very effective (see Chap. 5).

9   Water relations

At high altitudes plant life is commonly less constrained by moisture shortage than at low altitudes, but periodically and regionally variable water shortage does occur, and contributes to the overall impact of the physical environment on mountain plants (Fig. 9.1). In this chapter I will first discuss the overall hydrological situation, the land area based water balance and soil moisture at alpine altitudes – aspects which have strong links with Chapters 4 and 6. With this background in place, I will then summarize the current knowledge on alpine plant water relations.

Ecosystem water balance

At a given point in the landscape (e.g. the rooting zone of a plant) the water balance during the growing season is determined by three major components:

- The net input of moisture (precipitation P plus drainage from uphill locations D_i, minus surface runoff D_r, and deep seepage D_s)
- The losses due to evaporative processes E (including plant transpiration E_p, and direct evaporation of moisture from the soil E_s, and of surface wetness intercepted by plant structures E_i)
- The change in soil moisture ΔR

leading to the general water balance equation:

$$P + D_i = E_s + E_p + E_i + D_r + D_s + \Delta R \qquad (9.1)$$

The sum of all vapor-fluxes is commonly termed **evapotranspiration**. On flat terrain, or when D_i and D_r are equal (slope driven drainage in and out of a considered patch equal), or when the considered area is large and includes a complete water shed (no D_i), and over long periods when plus and minus deviations in soil moisture cancel to zero, the equation becomes simplified to

$$P = E + D. \qquad (9.2)$$

In this case, knowing two elements allows calculation of the third. Precipitation, the most commonly available parameter of the three, tends to increase with altitude at lower elevations, but exhibits no regular pattern at higher elevations (either increasing further or showing reversed trends) as was discussed in chapter 3. E can be obtained either directly by micrometeorological measurements or calculations, or by weighing lysimeters (containers with vegetated soil monoliths), or indirectly, for larger areas (whole "catchments") and long periods by subtracting river flow (runoff) from precipitation. Locally, there is also the possibility of measuring D directly from water captured under weighing or non-weighing lysimeters (which force all runoff into D_s). These various methods have been employed for estimating the water balance of alpine terrain in temperate zone mountains (e.g. Baumgartner et al. 1983; Isard and Belding 1989; Körner et al. 1989c). Information from subtropical and tropical mountains in this respect is rather limited. The following examples include data from the Alps, Rocky Mts., Caucasus, New Zealand, southeast Australia and northwest Argentina.

Precipitation often – as exemplified in Figure 9.2 for the Alps – continues to increase with increasing altitude (see Chap. 3). However, even if precipitation did not increase with altitude, or

Fig. 9.1. The three major life strategies plants have evolved for coping with the variable availability of water are present in the alpine zone and may be found coexisting on a hand size patch of vegetation: (1) desiccation tolerant (poikilohydric) plants such as lichen and mosses, (2) higher plants, using stomatal control, leaf area dynamics and deep rooting, and (3) succulents, which store water. Here a section of alpine turf with *Cetraria* sp, *Carex curvula*, *Trifolium alpinum*, and *Sempervivum montanum* (*center front*). (Swiss Alps, 2500 m)

even slightly decrease, the E/P ratio would still decrease in such climates, because of the marked reduction of annual E.

On an annual basis, there is an obvious increase in moisture surplus with altitude. These direct determinations of the seasonal water balance yielded similar **altitudinal gradients of evapotranspiration** to those obtained in three completely independent approaches using catchment data or meteorological calculations (Fig. 9.3). The annual vapor loss at 14 grassland sites in the central Alps drops from nearly 700 mm at low altitude near Innsbruck (47°N) to 210–250 mm at the upper limit of the alpine grassland belt.

The linear elevational reduction of E in Figure 9.3 (see also Steinhäusser 1970) is fully explained by the altitudinal decline of season length. When the seasonal vapor losses are divided by the number of snow-free days (a range from 90 to 310 days) a seasonal daily mean of 2.3 mm d^{-1} (range 1.9 to 2.6 mm d^{-1}) for all sites is obtained, with no significant altitudinal trend, despite substantial climatic differences across this almost 2 km range in elevation. This phenomenon results from at least three facts: (1) the growing period at highest altitudes falls in the warmest part of the year, whereas long cool periods in spring and autumn are included in the snow free period at low altitude; (2) as illustrated in Chapter 4, low stature

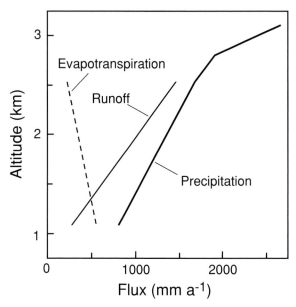

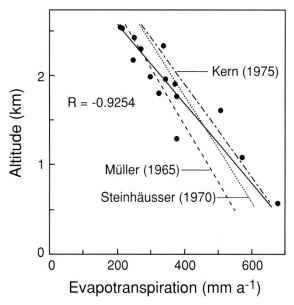

Fig. 9.2. The altitudinal variation of annual precipitation, evapotranspiration, and the sum of all drainage processes (runoff) for grassland areas in the eastern central Alps. (Data from lysimeter studies by Wieser et al. 1984)

Fig. 9.3. The altitudinal variation of annual evapotranspiration (data points) at 14 grassland sites between 580 and 2530 m in the eastern Alps (treeline at 2000 m, data from 1975–1978). The three labeled lines are estimates from three completely independent approaches. Steinhäusser (1970) used catchment data for whole river systems, Müller (1965) and Kern (1975) used meteorological calculations (mainly based on temperature gradients) for two different parts of the Alps. (Körner et al. 1989c)

alpine vegetation heats up substantially under direct solar radiation, which enhances driving forces for evaporation (Smith and Geller 1979; Cernusca and Seeber 1981); (3) periodic moisture shortage at low altitude causes stomata to down regulate plant transpiration, a phenomenon not seen at highest altitudes (Körner and Mayr 1981; see below). In contrast, uniform, well watered, potted horticultural plants exposed at different altitudes (closely coupled to respective atmospheric conditions) showed a consistent reduction of transpiration with increasing altitude (Whitfield 1932).

The figure of 2.3 mm d^{-1} represents a seasonal mean across all weather conditions. Typical **bright day vapor losses** are 3.7 and 4.5 mm d^{-1} for high and low altitude respectively (Körner et al. 1989c, Fig. 9.4). Nakhuzrishvili and Körner (1982) report 4 to 5 mm d^{-1} (extremes 7 mm d^{-1}) for pastures near the treeline in the central Caucasus. Numbers for bright periods at midsummer in Rocky Mountain alpine grassland range from 3 to 5 mm d^{-1},

depending on slope, with a mean for east and west slopes of 3.6 mm d^{-1} (Isard 1986, see the discussion in the next section). Almost exactly the same range (2.6 to 5.3 mm d^{-1}) was estimated by Geyger (1985) for various types of closed vegetation in the high-Andean semi-desert of subtropical northwest Argentina between 4100 and 4800 m altitude (Fig. 9.5).

For each of the above mentioned 14 sites in the Alps, the individual components of the water balance equation have been determined (with separate inputs for the snow-free and snow-covered period; Körner et al. 1989c). As an example, Figure 9.6 shows data for the most typical natural alpine grassland in the Alps, a low stature, sedge dominated mat (LAI of 2.3). Physiognomically similar vegetation types occur in the alpine zone all over

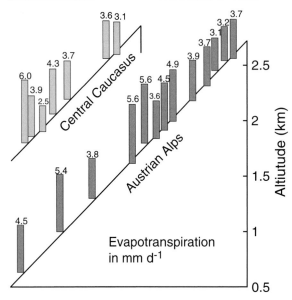

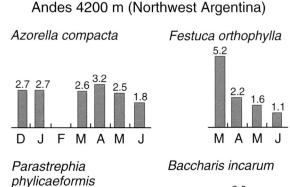

Fig. 9.4. Midsummer evapotranspiration in grassland on clear days in the central Alps and the central Caucasus. Data from periods of peak biomass development, i.e. early June at low altitude and mid-to late July at high altitude. Note that peak leaf area index along this gradient decreases from ca. 7 at low altitude to <2 at the highest site. (Compiled from various sources by Körner et al. 1989c)

Fig. 9.5. Seasonal variation of estimates of daily transpiration of high andean semi-desert vegetation at 4200 m altitude in the Sierra del Aguilar, northwest Argentina. Fluxes refer to unit ground area vapor losses of patches of closed vegetation calculated from leaf transpiration and LAI, and are given for the favorable part of the year. (Without evaporative contributions from soil under vegetation and areas without plant cover; Geyger 1985)

the world; for instance, in the central Caucasus at 2500 m, on Niwot Ridge in the Rocky Mountains at 3600 m, or in the Southern Alps of New Zealand at 1800 m. Figure 9.6 shows that most of the water leaves this ecosystem by **runoff** during snowmelt or by deep seepage in summer, with only a rather minor part of 7.8% of total precipitation lost via plant transpiration (all year E = 17%). In other words, the water discharge at this and possibly many other humid alpine sites, is largely liquid, driven by runoff and seepage rather than by evaporative fluxes. This is an important point for ecosystem stability, because it underlines the significance of mechanical properties of plants, dense root systems in particular, for protecting alpine slopes from surface erosion. In contrast, 70% of annual precipitation (a four times greater fraction) leaves temperate zone low altitude sites in the form of water vapor, of which roughly two thirds passes through plants.

Vapor loss from alpine vegetation was found to strongly depend on **vegetation type** (e.g. Pisek and Cartellieri 1941; Mooney et al. 1965). Lysimeter data for clear days in the Alps for the same altitude and exposure show a range from 2.1 mm d^{-1} in a dry lichen heath to 8.2 mm d^{-1} in a wet *Polytrichum*-moss carpet in a snowbed community (Körner 1977). Within graminoid vegetation numbers varied between 3.3 in low sedge turf to 5.1 mm d^{-1} in moist meadow stands. Bare, but humid soil averaged at 4.1 mm d^{-1}, hence may evaporate more then vegetated ground, a phenomenon observed repeatedly in other places as well. Isard's (1986) numbers for Niwot Ridge cover a similar range from 3 mm d^{-1} in moist shrub tundra to 5.5 mm d^{-1} in wet meadow. Across a fellfield knoll, Isard found pronounced south-north (but no east-west differences), but these differences were accompanied by changes in both soil and plant cover, hence do not allow the separation

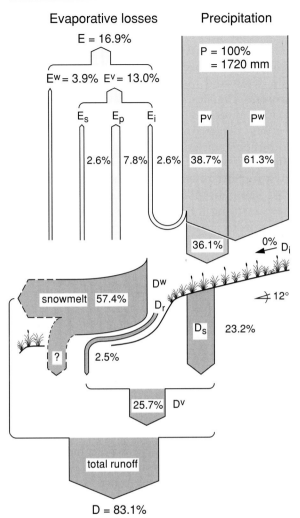

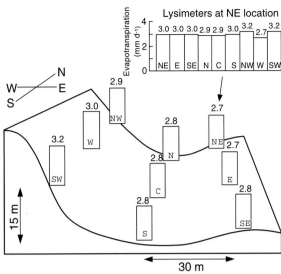

Fig. 9.7. The spatial variability of evapotranspiration across an exposure gradient in alpine grassland at 2430 m altitude in the Alps. Means for *in situ* lysimeter measurements are for 20 bright days, those for lysimeters intermittently translocated to one common location (NE) are for 9 days, in 1978 and 1979. (Körner *et al.* 1989c)

Fig. 9.6. The partioning of annual precipitation to various evaporative and runoff fractions in an alpine grassland at 2300 m altitude in the Alps (see the example on color Plate 2d at the end of the book). Superscript v indicates fluxes during the vegetation period and w indicates fluxes during the remaining part of the year (winter). For *symbols* see Eq. (9.1). (Körner et al. 1989c)

of **exposure effect** from vegetation effect. Figure 9.7 illustrates a situation where slope-only effects within a structured, but overall south oriented landscape, with very similar vegetation and soil (excluding a north slope) could be tested with mobile lysimeters. In this case, evapotranspiration neither differed between slopes (largely east versus west), nor did it differ significantly between monolith-origin, when all lysimeters were exposed at one spot. These data are in line with Isard's observation and suggest little mid-season slope effects on daily evapotranspiration as long as extreme north slopes are excluded, and vegetation is not too different.

A dependency of water budgets on vegetation type was also reported by Mark and Holdsworth (1979) who found maximum water yield from natural undisturbed snow tussock grassland in New Zealand (55–68% of precipitation as runoff and/or seepage water), which is more than is yielded from exotic pastures or even bare soil. A special feature of snow tussock is the ability of its long, thin leaves to intercept fog, which contributes to the higher seasonal water yield. Mark et al. (1980) showed that natural alpine vegetation has a positive influence on catchment value and soil water balance when compared with introduced pasture mixtures, a similar finding to earlier reports by Costin (1966) for the Australian Alps. However, traditional land use on well maintained alpine pastures may also reduce evapotranspiration as was shown for a short, sheep-grazed turf near the treeline in the central Caucasus, and for a cattle pasture in the Austrian Alps, both compared with adjacent abandoned pastures (Nakhuzrishvili and Körner 1982; Körner et al. 1989c).

Except for the study by Geyger (1985), no such experimental data seem to be available for alpine vegetation outside the temperate zone, but this deficit can possibly be filled by remote sensing (see Chap. 4). Also, digital terrain data in connection with an **energy balance** model have been successfully adopted to predict moisture distribution in alpine terrain (e.g. Blöschl et al. 1991). At high soil moisture (a common situation in the alpine zone, see below) evapotranspiration has been shown to linearly correlate with net radiation (Isard and Belding 1989). The high fraction of solar energy converted to latent (i.e. evaporative) heat also causes the **Bowen ratio** (the ratio of sensible/latent heat) of closed alpine vegetation in the humid temperate zone to be comparatively low, commonly below 0.5, and rarely above 1 (Cernusca 1977; Cernusca and Seeber 1981; Tappeiner and Cernusca 1996). In their grassland study, Cernusca and Seeber (1981) noted an increase in Bowen ratio in the upper part of their transect , but this had to do with a reduction of LAI from 7 to 2 and an absolute increase in heat convection.

In the subtropics and part of the tropics, precipitation does not continuously increase with altitude, as was discussed in Chapter 3. In some of these areas, precipitation drops so dramatically above the condensation zone that alpine semi-deserts are formed, as for instance in some parts of central Asia and the southern Andes (see examples above and below).

Soil moisture at high altitudes

Whether the common altitudinal reduction of the evaporation/precipitation ratio leads to improved plant water supply strongly depends on soil structure, the actual pattern of rainfall distribution during the growing season, and plant cover. The term ΔR in Eq. (9.1) encapsulates the capacity of soils to store moisture, which in turn depends on the depth and quality of the rooted profile. In seasonal climates, soils usually enter the growing season saturated and whether moisture pools become depleted depends on the initial size of these pools, seasonal refilling and the length of the season.

A number of authors have pointed to the fact that **snow distribution** and snowpack duration have a strong influence on soil moisture during the growing season (see Chap. 5). However, this is partly a coincidental, not straightforward causal relationship. Relief driven, moisture accumulation in summer is often found in the same places as snow accumulates in winter, and these sites are special in terms of soil conditions and microclimate, and thus bear a rather specific vegetation for several reasons. Snowbeds have repeatedly been shown to exhibit poor soil development, with high fractions of sand, and thus poor water holding capacity, despite high early season moisture (Isard 1986; Fig. 9.8). Neuwinger (1980) found that exposure to **wind** during the growing season had the highest predictive power for soil desiccation. Hence, it is the topographic complex which controls soil conditions, and snow pack is part of this complex but not its sole cause. The longer the growing season, the less important are influences of snowpack, and the more influential other relief derived factors become, which thus dominate in the tropics.

Which soil moisture matters? The top 2 cm of soil may desiccate periodically even in the wettest mountains of the world. The moisture in the largely inorganic material underground, perhaps 1 m below the surface, may not dry up even in semi-arid mountain climates. Soil moisture assessments in alpine terrain commonly cover the top 15 cm of the profile, the zone most relevant for biological activity including nutrient cycling. However, alpine plants almost always have a small fraction of deeper roots, not uncommonly reaching to 1 m **depth**. These roots secure some water supply even under conditions when surface soils desiccate. Consequently, available soil moisture data for the top 15 cm of soil profiles are relevant for topsoil root and microbial activity, but may not sufficiently explain plant water status – an important caveat for the following paragraphs.

Associated with the altitudinal decrease of the E/P ratio, soil moisture tends to be higher at high compared with low altitude, provided profiles are similar in depth and quality. How deep are alpine soils, and how much moisture can they store? As was stated in Chapter 6, undisturbed alpine soils are often characterized by richness in raw humus (top of profile) and in rock debris (deeper profile). The former commonly enhances moisture storage capacity, the latter diminishes it. Humus rich soil layers in the alpine zone may have pore volumes up to 70% (Franz 1979; Wolfsegger and Posch 1980), and, on the other hand, coarse debris may occupy as much as 90% of deeper soil layers, a volume unavailable for moisture storage.

Figure 9.8 shows maximum total available, and physiologically unavailable soil water contents for a series of **soil profiles** along a transect from the heart of the alpine grassland belt to its upper margins in the Alps. Roughly 50–60% of the profile can be filled by capillary water (which resists gravity) after snowmelt or saturating rainfall, which is typical for alpine soils (cf. Neuwinger 1980; Isard 1986). Top layers usually have higher (55–65%) and deeper layers have lower (35–45%) pore volumes, and approximately half of the maximum soil moisture (after drainage of macropores) may be utilized by plants – the remaining moisture is hardly accessible. In sandy soils the usable fraction is higher, but the overall storage is much lower, because of greater pore sizes, and thus greater drainage. Profiles 5, 6 and 7 from snowbeds or a site with eroded topsoil store much less moisture compared with well drained intact profiles under closed alpine grassland. The total amount of easily available moisture (soil water potential > −0.03 MPa) after snowmelt varies between 180 and 340 mm under grassland and 65 to 110 mm

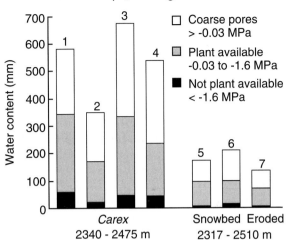

Soil profiles between 2340 - 2510 m, Central Alps, Grossglockner area

1+2 alpine pseudogley, 3 m apart, well drained
  3 as 1+2, steeper slope, solifluction front, alpine brown earth
  4 very steep, alpine brown earth
  5 E-exposed plateau, pseudogley
  6 pseudogley
  7 as 4, top 40 cm eroded

Fig. 9.8. Water storage in alpine soils for full (rooted) soil profiles varying in depth from 30 to 140 cm in the Grossglockner region of the Austrian central Alps. Each *bar* illustrates the water content in millimeter water column at saturation (*blanc*), and at soil water potentials of −0.03 MPa and −1.6 MPa, the latter conventionally assumed to represent moisture unavailable to plants. Note the higher moisture pool in intact, drained profiles under alpine grassland between 2340 and 2475 m a.s.l. (*left*) and the lower pools in snowbed soils and in an eroded profile between 2417 and 2510 m a.s.l. (*right*). The treeline in this area is around 2050 m altitude. (Wolfsegger and Posch 1980)

under the eroded surface or in high altitude snowbed soils.

Given these field data, the following theoretical considerations illustrate the **likelihood of soil moisture depletion** in the temperate alpine zone by (1) assuming a season length of 100 days with seasonal E = 230 mm (2.3 mm d^{-1}) and estimating required additional P, or (2) by estimating the maximum duration of moisture availability under continuously bright weather and a constant daily E of 4 mm. The effects of various combinations of soil depth, pore volume and debris content on overall moisture depletion under average (1) or bright (2) weather conditions are illustrated in Table 9.1.

This exercise shows that a profile of 1 m depth (with no rock debris and average physical soil properties) would not require additional rainfall during a standard 100 day alpine growing season ($2.3 \, mm \, d^{-1}$) to prevent turgor loss in plants. In fact, 20 mm of free moisture would be left by the end of the season. The moisture pool would suffice to support evapotranspiration for 2 rainless months of bright weather ($4 \, mm \, d^{-1}$). Alpine soils meeting these parameters may exist in places (e.g. profiles 1 and 3 in Fig. 9.8), but are rare. Over the same profile depth, but with 80% coarse substrate, potentially available soil moisture would support only 22 average or 13 bright days. In contrast, the moisture stored in a rather shallow profile of only 25 cm depth, but of high humus content and with less debris (a rather typical situation in many places) would meet the requirements of 41 average or 24 bright days. Thin crusts of highly organic soil on rocks require rainfalls at least every 3rd day (30 mm thickness) or every 11th day (100 mm thickness) in order to prevent plant desiccation.

These various **scenarios** are helpful in estimating the likelihood of physiologically effective moisture shortage in the alpine zone under given rainfall regimes. For instance, in all but the first example, more than 150 mm of precipitation are required during the growing period to prevent critical depletion of pools. Of course, such comparisons need to account for additional losses by runoff (D_r) during heavy storms, for interception losses (E_i) and temporal patterns of precipitation. However, scenarios 3 (1 m profile with 80% coarse substrate) and 4 (25 cm profile with high humus and less debris), which represent abundant situations, clearly illustrate that 50 to 60 mm of rainfall per month during the growing season would eliminate the possibility of moisture shortage under alpine climate conditions. Since precipitation tends to peak during the temperate zone summer and exceeds those amounts substantially in many places, and since soil reservoirs can buffer temporal irregularities of supply, the chances for the occurrence of direct drought effects on plants are small, and are restricted to soils with particularly constrained rooting volumes. This does not preclude periodic desiccation of surface layers, which may occur at any time.

It should be noted that the above scenarios did not consider the possibility of plants actively

Table 9.1. The importance of humus, debris and profile depth for the likelihood of soil dehydration during a temperate alpine growing season

No.	Depth (mm)	Debris (%)	Pores (%)	AW (%)	R	2.3/d (d)	+P (mm)	4/d (d)
1	1000	0	50	50	250	109	0	63
2	1000	80	50	50	50	22	180	13
3	500	30	50	50	88	38	142	22
4	250	10	60	70	95	41	162	24
5	100	0	70	60	42	18	188	11
6	30	0	70	60	13	5	217	3

AW, the fraction of easily available water of maximum soil water content at the beginning of the season (after snow melt); **R**, resultant total available soil moisture reserves after snowmelt; **2.3/d** or **4/d**, the maximum duration of soil moisture without additional precipitation assuming a daily evapotranspiration of 2.3 mm (seasonal mean including all weather conditions) or 4 mm (continuously bright days only); **+P**, the additional precipitation required to balance the evaporative demand of 230 mm per 100 day season.

reducing transpiration during periods of high evaporative demand or prolonged soil moisture depletion (see below) which would increase moisture duration. Available soil moisture data (e.g. Fig. 9.8) are in line with these predictions. During the Europe-wide centennial summer drought of 1976, soil moisture never fell below −0.3 MPa in the B-horizon (ca. −15 cm) and below −3 MPa in the A-horizon in a southwest-facing, windswept *Loiseleuria procumbens* heath on Mt. Patscherkofel near Innsbruck (data by H Guggenberger, in Larcher 1977; Körner et al. 1980). Figure 9.9 illustrates that the situation described here for the central Alps finds a parallel in the Australian alpine zone at 36°S lat, with a longer growing season. Midsummer moisture reserves near the tree-limit are massive throughout the summer and far from risk of becoming depleted where closed vegetation is found. Very moist soils are reported for the afro-alpine life zone (Beck et al. 1981).

The **subtropical, high Andean region** in northwest Argentina, Peru and Bolivia is known for its aridity and thin plant cover. However, visual impression may be rather misleading when the moisture supply of these plants, which manage to cope with this environment, is considered (Figs. 9.10 and 9.11). Figure 9.10 illustrates two moisture profiles sampled in the late part of the South Hemisphere summer at 4250 m altitude in the Argentinan Cumbres Calchaquies, where annual precipitation of ca. 300 mm is restricted to the 6-month summer. Starting a few centimeters below a dry layer of sand, the deep profiles of this semi-desert plateau, exhibited surprisingly uniform moisture, and plants extend their roots as deep as one can reach (>1 m). The dry surface layer (together with snowless, dry and cold winters) may limit plant life, but those plants established appear to have ample access to water. The same is true for the semiarid volcanic cinders of Mt. Teide on Tenerife, where 20–30 cm below a desiccated and insulating layer of volcanic debris, moist sand can be found even on south slopes (pers. observation). Similar observations are reported by Chapin and Bliss (1988) for pyro-

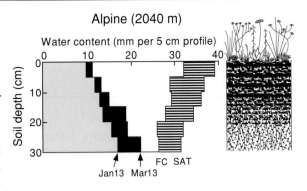

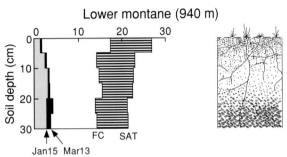

Fig. 9.9. Midsummer soil moisture profiles in the Mt. Kosciusko area of the Snowy Mountains in southeastern Australia (*arrows*). The lower mountain zone experiences a long summer drought and supports open *Eucalyptus* woodlands. The alpine site (on Mt. Perisher) is in the alpine dwarf shrub belt, a short distance from the uppermost tree individuals in this area. FC field capacity, SAT water content at saturation (laboratory determinations). The diagram illustrates a much better moisture supply at high altitude with no realistic possibility that moisture pools ever get depleted. (Körner and Cochrane 1985)

clastic soils on Mt. St. Helen. While top layers desiccated in summer, soil water potentials never dropped below −0.1 MPa 20 cm below the surface, and plants showed no indication of moisture shortage throughout the season). It is not soil moisture which restricts plant life on such almost bare slopes.

The data in Figures 9.10 and 9.11 illustrate an important fact, possibly true for many situations conventionally rated as "extremely dry": sparsely vegetated dry land surfaces are not necessarily "extremely" dry in a physiological sense for native species. Geyger (1985) summed up the annual

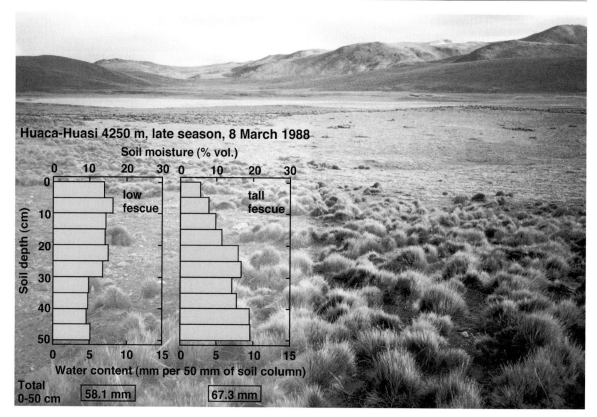

Fig. 9.10. Two 50 cm soil moistures profiles in the Cumbres Calchaquies of northwest Argentina (ca. 26°S) at 4250 m altitude, sampled in the late season (8 March 1988; Laguna Nostra visible in the rear part). The photograph illustrates the sample area with low (see also Fig. 16.14) and tall fescue grassland. With a leaf area index of <0.5 and the insulating sand surface, daily evapotranspiration is possibly not more than 1 mm. Since the depth of these sandy soils exceeds 1 m, the available moisture stored possibly exceeds demand for 2 months. Note the different moisture depletion profile below low and tall fescue (Ch. Körner and S. Halloy, unpubl. data)

transpiration per unit of ground area of patches of closed vegetation for her Andean sites between 4200 and 4800 m altitude, and arrived at numbers of 400 to 600 mm a^{-1}, exceeding mean annual precipitation of around 300 mm (see also Ruthsatz 1977). However, when accounting for actual percentage **ground cover by vegetation**, the total transpiratory losses per unit land area were reduced to 130–220 mm a^{-1}, leaving sufficient leeway for some bare soil evaporation and seepage. In other words, cover (or leaf area index averaged over the whole landscape) is such that sufficient moisture is retained and plants tap moisture from much larger areas than those actually occupied above the ground. **Plant spacing** in "arid" land is a most powerful means of controlling plant water relations, but mechanisms of density control are still poorly understood for low altitudes, and the question was not approached for such dry uplands. This control of water loss via plant density is a parallel phenomenon to increasing spacing of trees at treelines, which results in relatively warmer root temperatures (see Chap. 7), but the mechanisms controlling this type of spacing are not understood either. This brings us back to the topic of Chapter 1: no doubt the vegetation in the high

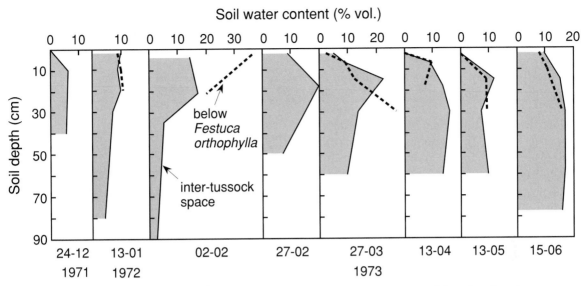

Fig. 9.11. The seasonal variation of soil moisture in a high Andean semi-desert in the Sierra del Aguilar of northwest Argentina (ca. 23°S). *Solid lines* are for inter-tussock space, *broken lines* show moisture below *Festuca orthophylla*. Note that soil moisture between January and June is never depleted below 10% volume (soil water potential remained above −0.1 MPa). Only moisture below 3% is unavailable to plants (water potential <−1.5 MPa) in these sandy soils. (Geyger 1985)

Andean zone (see Ruthsatz 1977) would look rather different were there ample water, but this is a different question from whether plants composing the existing natural vegetation are physiologically constrained (see the hierarchy of controls in plant water relations in the next section). Also, in polar "deserts" it was found that water stress for established plants was minor and soils remained effectively saturated throughout the growing season, despite negligible rainfall (Gold and Bliss 1995).

One of the driest alpine regions of the world is the **Pamir in central Asia.** At elevations between 3800 and 4800 m, alpine semi-desert, dry grassland, dwarf shrub and cushion plant assemblages can be found with genera such as *Eurotia, Artemisia, Potentilla, Astragalus* and *Kobresia* playing important roles (Sveshnikova 1973, Izmailova 1977; Agakhanyantz and Lopatin 1978). By all standards, low precipitation (ca. 300 mm per year), missing winter snow cover, extreme surface desiccation of soils and relative humidities of air down to less than 10%, force plants in the Pamir to cope with very little moisture, while at the same time they have to survive temperatures of −48 °C in winter. Thanks to the very detailed long-term study of plant water relations by Sveshnikova, we know that plants growing in this environment in reality experience little drought stress. While actual soil moisture deeper in the ground is unknown, plant data suggest that there is enough for those managing to live in this environment. Leaf water deficits measured during the growing season are small, osmotic potentials are high (> −2 MPa during the peak growing season), and both diurnal and within-season variations are similar to those reported for arctic or temperate lowland plants. Leaf transpiration rates peak between 12.00 and 16.00 h, with no pronounced reductions throughout the growing season, and rates decrease only in autumn. These observations once more emphasize that plant selection and adaptive community responses to potential moisture shortage operate via low ground cover (LAI and low leaf to root ratios (morphology), and result in conservation of soil moisture so that physiological limitations of existing tissue are unlikely to occur (see Chap. 1).

That shortage of moisture is unlikely to exert major direct physiological limitations on alpine plant life is a conclusion already drawn by Walter (1931) from his studies on Pikes Peak, Colorado (see also below). In the temperate zone, rooted **soil profiles** containing at least a net sum of 300 mm of fine substrate (about 80 mm of easily available capillary moisture at field capacity) seem crucial, however. Deeper profiles would be required in coarse or sandy substrates, but if soil temperatures are too low, moisture at such deep horizons may not be accessible for roots, as was suggested by Beck et al. (1981) for Mt. Kenya and Perez (1987) for the upper Venezuelan páramos. In continental, subtropical mountains with semi-arid climates, or whenever soils are shallower or too cold, periodic moisture shortage may become part of alpine plant life. However, as discussed above, a visual impression of aridity at alpine altitudes or climatic data alone (e.g. Leuschner and Schulte 1991) must not be taken as a self-evident indication of physiologically effective moisture shortage for the plant specialists which inhabit such areas. Using deep soil exploration combined with reduced aboveground structures (low LAI), plants may effectively avoid soil moisture depletion even in areas that look rather arid.

Soil coverage by rock **debris, gravel, cinders or coarse sand** positively contributes to moisture storage in the root zone of alpine plants (Schröter 1926, Pisek et al. 1935; see also the discussion by Perez 1987; Fig. 9.12), while at the same time such soil covers restrict the development of a closed, transpiring plant cover. As will be discussed below, it is the indirect effect of topsoil desiccation, the blockage of microbial activity and nutrient recycling – a phenomenon functionally different from direct drought stress – which often explains poor growth and development of plants in alpine areas with periodic drought. Under such conditions plants may not exhibit signs of water stress (turgor

Fig. 9.12. Soils topped by scree, gravel, cinders, sand and any other coarse substrate are commonly moist, but mechanically serverely limiting plants. *Top*: *Papaver rhaeticum* on dolomite scree, southern Alps. Bottom: cinders on Mt. Teide (Tenerife; same location as Fig. 6.2) with moist substrate visible at ca 20 cm depth

loss) but still suffer from surface drought through limited access to nutrients. Reduced growth may, in fact, contribute to a slower depletion of deep moisture sources. Obviously, there is a link between direct and indirect modes of drought stress (see also below).

Plant water relations – a brief review of principles

Before discussing plant water relations in alpine plants, four general points need to be clarified which relate to (1) plant life strategies, (2) types of responses to water shortage in plants, (3) methodological implications, and (4) the distinction between direct and indirect effects of low soil moisture. Knowledgable readers may wish to skip this section.

Plant life strategy and water relations

A trivial but important point to start with: in order to persist within a given environmental matrix, plants need to optimize growth and reproduction, and not just gas exchange (including water consumption). With this logic in mind, the optimization of CO_2 assimilation versus transpirational losses at leaf level, a topic which unfortunately dominates the relevant literature, becomes one aspect among many, and not necessarily one of high priority. **Control of growth** involves a multitude of costs and benefits of investments (see Chap. 12), with risks and trade-offs all subordinated a certain "life plan", which in turn encounters morphological (architectural), developmental (temporal) and mutualistic components. For instance, photoperiodic control of flowering and cessation of vegetative growth are often associated with reduced moisture demand. Seen in this wider perspective, plants cannot be expected to have evolved a behavior oriented to maintain or optimize a single process or a certain variable such as water potential, or a certain vapor loss/carbon gain ratio, as long as these relations remain within a tolerable range (Schulze and Chapin 1987).

Plant controls of water status

The second important point is that **stomata**, commonly the centerpiece of any discussion of plant water relations, are one out of several determinants of water loss, and they fulfill a short-term fine-tuning function, which, when the longer-term controls are perfectly tuned, should exert little additional constraints to plant gas exchange. Restrictions of stomatal diffusive conductance cause a reduction of returns of the underlying photosynthetic machinery in the leaf (amortization). Adaptive trends to any shortage of moisture could be expected to rather favor plants with less total photosynthetic machinery (e.g. less leaf area) which operates at "full power", rather than investment in photosynthetic over-capacity suffering from periodic "shut down" (with respect to available moisture). On the other hand, it makes no sense to develop leaf diffusivities which exceed demand for CO_2, the reason why plants tend to adjust stomatal conductance so that any further increase would hardly affect CO_2 uptake, but would create disproportionate water loss (Cowan and Farquhar 1977). The longer-term controls of plant water relations are indeed more complex, and involve adjustments of dry matter investments and phenology.

A useful **hierarchy of controls** from short term to long term is the following: (1) stomatal responses to evaporative demand, (2) stomatal responses to a root signal indicating periodic (diurnal) reductions of moisture availability in the root-soil interface, (3) negative feedbacks on stomata by leaf turgor by water potential falling below a critical value (4) reduction of leaf area per plant and/or increased root production, (5) reduction of ground cover per unit of land area (LAI), (6) replacement of species from less to more drought adapted ones. In nature, the significance of controls increases from 1 to 6. If 4 to 6 were perfectly adjusted, 1 to 3 would hardly come into action.

Since leaf water potential became measurable, views about its functional significance have changed. Low leaf water potential may not only result from insufficient uptake (low soil moisture, cold soil) or impaired transport (frozen stems, xylem cavitation), but may also result from high flux with high soil moisture, given that certain hydraulic resistances do always exist. Hence, above a critical level (e.g. turgor loss), water potential alone has no explanatory power – a reduction may equally well be associated with either high flux with high soil moisture or almost no flux with serious drought-stress. The significance of negative water potential feedbacks on stomata has commonly been overestimated. Unless turgor is about to be lost, reduced leaf water potentials exert little influence on stomata and are flux dependent, rather then vice versa.

Implications for methodology

From the above it is obvious that a full description of plant water relations requires four parameters: a measure of **flux** (transpiration or leaf conductance), a measure of **supply** (soil moisture, including deep horizons), a measure of atmospheric **demand** (vapor pressure deficit) and a measure of the resultant water **status** of the plant (e.g. water potential or relative water content). Selecting literature by completeness of data sets in these respects, would leave little to be discussed here for alpine plants. Hence, the following discussion will, by necessity, also lean on numerous data sets which contain only one or two of the above parameters, thus often precluding a conclusive interpretation.

Direct and indirect effects of soil moisture on plants

As was discussed in the soil moisture section, two modes of drought stress need to be distinguished. Commonly, effects of "moisture shortage" are believed to be fully accounted for by assuming that if topsoils dehydrate, plants have difficulties in maintaining turgor, and this IS "drought stress". However, the more frequent, and much more influential effect of topsoil dehydration is on nutrient availability rather than plant water status (which is commonly supported by a few deep roots). Most of what has been seen as moisture stress (particularly in alpine plants), in essence is **drought-enhanced nutrient shortage**.

Furthermore, moisture shortage in plants can also occur when moisture is still present in the soil, but remains unavailable or hardly available to roots because of a too low temperature (Tranquillini 1982; Larcher 2003; see Chap. 7). Hence, in a similar way to nutrients, one needs to distinguish between presence and availability of moisture. Moist but cold soils may be "physiologically dry" as was stated already by Schimper (1898).

Water relations of alpine plants

The literature in this field is substantial, and mainly includes information on leaf controls of water losses and shoot water status (for earlier assessments see Tranquillini 1963; Courtin and Mayo 1975; Körner and Mayr 1981). This Section will first explain static and dynamic elements of alpine plant water relations, will then address resistance to extreme drought, and close with a brief section on "specialists" such as succulents, cushion plants and giant rosettes.

Water loss from leaves – static features

When one considers averages over whole assemblages of plant species forming the low stature ground cover at high altitudes, leaf epiderms tend to exhibit higher **maximum** vapor **diffusive conductances** and higher **stomatal frequencies** than comparable assemblages from lower altitudes (for reviews see Körner and Mayr 1981; Körner et al. 1989a; Figs. 9.13 and 9.16). Pisek and Cartellieri (1934) and Berger-Landefeldt (1936; Fig. 9.19) noted the comparatively high rates of transpiration of many alpine plants (they used vapor loss of wet filter paper disks as a reference). Higher leaf conductances in high compared with low altitude plants were also reported by Rawat and Purohit (1991) in three out of four species in the central Himalayas. A comparative analysis of altitudinal transects from mountains with different altitudinal gradients in moisture supply (Alps, New Guinea, New Zealand, southeast Australia), revealed no relatedness of these trends with local moisture (Körner et al. 1983, 1986, 1989a; Körner and Cochrane 1985). However, the only one of these transects with a pronounced elevational reduction of solar radiation, the transect in New Guinea, showed a pronounced altitudinal reduction of stomata density and maximum leaf diffusive conductance, suggesting that light rather than moisture or partial pressure of CO_2 is responsible for these altitudinal trends in leaf characteristics.

Not only does the density of stomata and maximum leaf diffusive conductance tend to increase with altitude, there is also a trend for leaves to shift stomatal abundance from the lower to the upper **leaf side**, even in New Guinea, where overall stomatal density decreases with altitude. For instance *Ranunculus saruwagedicus* on top of

Mt. Wilhelm (4420 m) has 64 adaxial and 9 abaxial stomata per mm^2, similar to leaves of *Potentilla* at 4800 m on Anapurna in the Himalayas and *Cerastium kasbek* at 3750 m in the Caucasus (Körner et al. 1989a). Some species known to be hypostomatous at low altitude develop some stomata on the upper leaf side at high altitude (e.g. *Vaccinium myrtillus*). Plants with leaves which hug the ground or form dense cushions may even be epistomatic, i.e. have stomata restricted to the upper leaf side, as for instance *Saxifraga oppositifolia*, *Silene acaulis*, *Primula minima* in the Alps and *Saxifraga exarata* in the central Caucasus (see the review by Körner et al. 1989a). These trends have already been described in great detail for temperate zone mountains by Wagner (1892), Schwartz-Clements (1905), and Lohr (1919), and by Espinosa (1933) and Spinner (1936) for the tropical Andes. According to Spinner's comparisons, the frequency of species with leaves with only abaxial stomata decreases from 75% at low

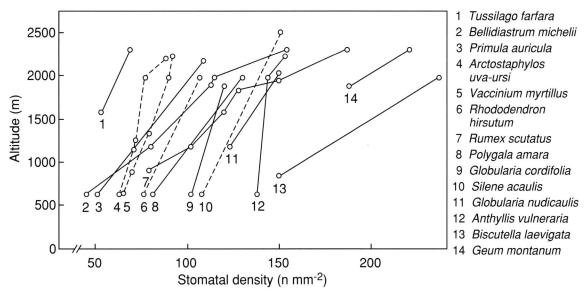

1 *Tussilago farfara*
2 *Bellidiastrum michelii*
3 *Primula auricula*
4 *Arctostaphylos uva-ursi*
5 *Vaccinium myrtillus*
6 *Rhododendron hirsutum*
7 *Rumex scutatus*
8 *Polygala amara*
9 *Globularia cordifolia*
10 *Silene acaulis*
11 *Globularia nudicaulis*
12 *Anthyllis vulneraria*
13 *Biscutella laevigata*
14 *Geum montanum*

Fig. 9.13. Intraspecific altitudinal variation in stomatal frequency of herbaceous plants (*solid lines*) and dwarf shrubs (including one cushion species; *dashed lines*) in the Alps. Each line is for one species sampled at contrasting altitudes. Since epidermal cell size does not change with altitude in a systematic manner similar trends are seen in stomatal index. The same trends were documented in other mountain areas. Only 1 out of 17 species tested in the Alps did not follow this pattern (not shown). (Körner et al. 1989a, where further details are provided)

altitude to 23% in the high Andes, where species with exclusively adaxial stomata represent 21–29% of the flora. These trends in leaf anatomy led Spinner to conclude that the high Andean flora is "apparaissent comme des mésophytes à envelope xéromorphe" (look like mesophytes in a xerophytic envelope). It should be noted that the trend of increasing stomatal frequency with increasing altitude may become reversed at extreme altitudes (e.g. Körner et al. 1986, 1989a).

A number of plant families are represented by alpine species with **furrowed leaves** and the abundance of such species is often increased at high altitude (e.g. species of Ericaceae, Asteraceae, Poaceae, Fig. 9.14). Inspection of stomatal distribution in such species often reveals stomata outside the furrow (e.g. *Tetramolopium macrum* and *Drapetes ericoides* in New Guinea or *Calluna vulgaris* in the Alps), and if bound to furrows, stomata often protrude from the epidermal surface (e.g. *Loiseleuria procumbens*, some *Festuca* species; additional examples for Andean alpine shrubs in Spinner 1936; cf. Körner et al. 1989a). Except for these genera, known to produce furrowed leaves also in their low altitude representatives, leaf folding, leaf rolling etc. are features not seen very frequently among alpine taxa, but some of the species which possess such leaves reach high abundance. These structures seem to protect the lower epidermis from occlusion by excess moisture (surface wetting) rather than from excess transpiration (see below).

It is often claimed that **leaf pubescence** is particularly striking at high altitudes, and that this has to do with water saving strategies of plants. This vision is derived from the presence of some very prominent representatives of highly pubescent species at high altitudes (e.g. the edelweiss and other *Leontopodium* species in Eurasia and similar Asteraceae in Japan and New Zealand, *Anaphalis* species in southeast Asian Mountains, the famous hairy clusters of *Saussurea* in the Himalayas, *Lupinus*, *Culcitium* and *Espeletia* species in the Andes and *Senecio* giant rosettes in Africa). However, a quantitative account of alpine floras reveals that high pubescence is in fact quite rare at both low and high altitudes. Though, Halloy

and Mark (1996) report an increasing frequency of species with at least some leaf pubescence with increasing elevation, the overall fraction of such species hardly exceeds 30% of all species present. Remarkably, the fraction increases with increasing climatic wetness. The majority of alpine species are not pubescent, and many species reaching extreme high altitudes have completely hairless leaves (e.g. *Ranunculus glacialis* in the Alps shown on Figs. 8.4 and 15.6, *Ranunculus sericophyllus* in New Zealand).

Leaf pubescence in general and in mountain plants in particular is unlikely to represent a means of diffusion control. Any enhancement of **leaf boundary layer** resistance reduces the relative effectiveness of stomatal control of transpiration (a "constant resistor" on top of a "variable resistor" dampens control), and the anatomical features of alpine leaves from around the world (discussed above) illustrate that plants rather tend to position stomata so that minimal aerodynamic restrictions are added to the diffusion path. Explanations for extreme pubescence of leaves in some species may have to do with screening of radiation, buffering of short term oscillations of ambient humidity, herbivory and pathogen defense or avoidance of surface wetting (see also the discussion by Larcher 1975; Baruch 1979; Körner et al. 1983; Goldstein et al. 1989; Halloy and Mark 1996). In tropical giant rosettes, pubescence of leaves (Fig. 9.15) adds to the screening efficiency during night-time nyctinastic rosette closure and certainly does protect stomata from wetness (and thus occlusion) in foggy weather, when films of water cover the woolly surface, while the epidermal surface underneath remains dry.

Taken together, these static features of leaves point to a rather unproblematic water regime. Since one of the most universal trends in alpine leaf anatomy is increased leaf thickness (increased length of the leaf-internal diffusion path; Chap. 11), the greater stomatal density and maximum leaf diffusive conductance may be associated with the maintenance of a low resistance to CO_2 transfer to chloroplasts. Since molecular gas diffusivity increases with altitude (with decreasing total atmospheric pressure, Gale 1972), the

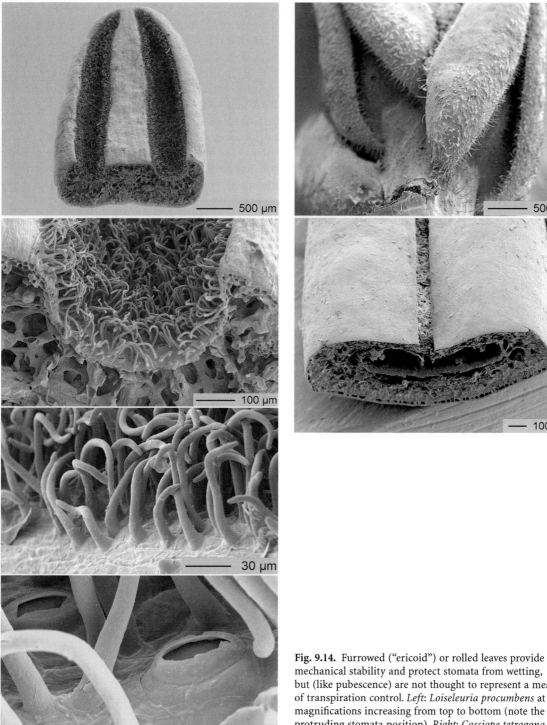

Fig. 9.14. Furrowed ("ericoid") or rolled leaves provide mechanical stability and protect stomata from wetting, but (like pubescence) are not thought to represent a means of transpiration control. *Left*: *Loiseleuria procumbens* at magnifications increasing from top to bottom (note the protruding stomata position). *Right*: *Cassiope tetragona* (*top*) and *Empetrum nigrum* (*bottom*)

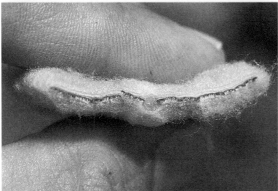

Fig. 9.15. Leaf pubescence in *Espeletia* species from 3600 to 4100 m altitude in Venezuela and Ecuador. The thick felt of hairs enhances reflection of strong radiation, prevents leaf surfaces from wetting during foggy periods and – in conjunction with night-time closure of rosettes – contributes to insulating the apical region against radiative freezing

area (leaf area per leaf mass) tends to be lower at high altitude, total leaf area per total plant mass (leaf area ratio) is slightly reduced at high altitude. In contrast, the total length of fine roots produced per unit of leaf area is almost five times as large in alpine compared with lowland forbs (Körner and Renhardt 1987; Fig. 12.18). Whatever the cause of this trend, it indicates a better rather than worse exploration of the substrate by thin and active roots. This may be seen as an adjustment to lower soil temperatures or a compensation for reduced mycorrhization at high altitudes, or may simply reflect slower root turnover (see Chap. 12).

Stomatal behavior and leaf water potential in alpine plants

As a rule of thumb, stomatal restrictions of vapor loss are reduced at high altitude. This is not only reflected in the static features as discussed above, but also in **diurnal and seasonal stomatal dynamics** (Körner and Larcher 1988). At low altitude, leaf diffusive conductance (stomatal opening) tends to peak during morning hours and then declines. Alpine plants often exhibit maximum leaf conductance during the warmest part of the day, i.e. midday and early afternoon, and take longer to reach full opening in the morning, which is possibly related to low morning temperature (Fig. 9.16).

Prostrate dwarf shrubs on wind-exposed south slopes in the Alps have been shown to undergo little diurnal variation in leaf conductance throughout the main part of the growing season (Körner 1976). *Loiseleuria procumbens*, producing a peculiar microclimate in its dense canopies, in essence opens its stomata in late spring and closes them in autumn (not even reducing conductance by more than 50% during nights in summer). Miller et al. (1978) review the earliest porometer studies in the Rocky Mountains for the two alpine species *Bistorta bistortoides* and *Caltha leptosepala*, and found no consistent changes in conductance during the season, with conductances generally staying high, and they conclude that

potential for high leaf transpiration at high altitudes is enhanced further.

At the whole plant level, important "static" characteristics related to water relations are leaf mass ratio (the ratio of leaf mass versus total plant mass) or – more relevant – fine root length/leaf area ratio. Both are invariable in the short term, but may reflect long-term adjustments. Leaf mass ratio in herbaceous alpine plants does not differ from that in low altitude plants and amounts to ca. 20 ± 2% for communities studied in the Alps, the northern Scandes, and the northwest Argentinan and Ecuadorian Andes (Körner and Renhardt 1987; Körner 1994; Fig. 12.15). Since specific leaf

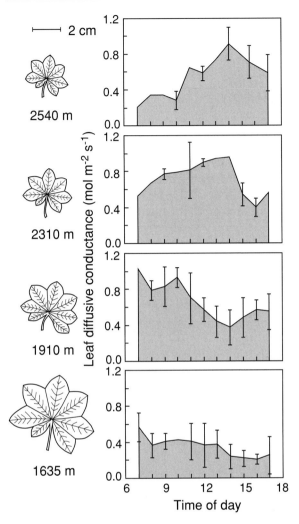

Fig. 9.16. Diurnal trends in leaf diffusive conductance in the genus *Alchemilla* studied at various altitudes in the Alps. Note the increase in maximum conductance and the shift of the period of maximum conductance from morning to early afternoon as altitude increases. (Körner and Mayr 1981)

recent observation by Enquist and Ebersole (1994) and Bowman et al. (1995) that regular addition of water to alpine vegetation on Niwot Ridge had no effect on leaf conductance or water potential of plants. Johnson and Caldwell (1975, 1976) tested alpine species under controlled conditions and concluded that stomata of *Geum rosii* and particularly *Carex aquatilis* responded to vpd before water potentials could drop to critical levels. Field data for evergreen dwarf shrubs in the Alps by Körner (1976) and for *Mimulus* in the Sierra Nevada of California (Field et al. 1982) led to the same conclusion. Also, stomata of afro-alpine *Lobelia* were found to be unresponsive to the water potentials observed in the field, but sensitive to vpd (Schulze et al. 1985). To separate effects of vpd (or rate of transpiration) from those of water potential, synchronous measurements of water potential and leaf conductance are necessary, while vpd changes under otherwise constant and high soil moisture. Figure 9.17 illustrates such a situation where, by means of a fan, naturally dry above-canopy air was forced into the normally sheltered and humid leaf boundary of *Loiseleuria procumbens*. The impact of dry air induced an immediate recovery (!) of shoot water potential as a consequence of fast stomatal closure. It is obvious that the above mentioned "unresponsiveness" of the stomata in this species is a consequence of its high aerodynamic canopy resistance, which does not usually permit its stomata to experience high vpd. Once the canopy climate is disturbed, stomata are in fact rather sensitive. Similar responses have been found in a small herbaceous rosette, *Primula minima* (Körner 1980). Without protection by its dense canopy *Loiseleuria* has also been shown to dehydrate quickly in the dormant state in late winter (Larcher 1957).

As was explained earlier, **leaf water potentials** alone do not permit conclusions about soil moisture shortage, because the same leaf water potential may result from true shortage and closed stomata or fully open stomata (high fluxes) at otherwise high soil moisture. In the latter case, a lowered water potential reflects hydraulic resis-

water does not limit production except in very dry years or over short periods in particularly dry places.

When diurnal restrictions of stomatal conductances were observed in the field, these were in most cases found to relate to evaporative demand (**vapor pressure deficit**, vpd) rather than leaf water potential – an observation consistent with the

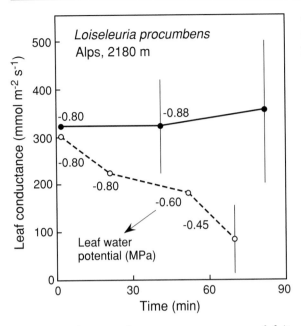

Fig. 9.17. *In situ* stomatal responses to vapor pressure deficit (or its consequence, the rate of transpiration) in alpine *Loiseleuria procumbens* growing in very humid soil. The humid canopy climate was disturbed by a jet of dry air created by a fan at time zero. Note that ventilated leaves (*lower line*), show a recovery (increase) in leaf water potential, while stomatal conductance is reduced. Conductances were converted from total to projected leaf area by a factor of 3.1; *bars* are for SD for 2–3 shoots. (Körner 1980)

tances rather than low soil moisture. The few available data sets for water potentials combined with leaf conductance suggest that stomata of alpine plants (similar to most other fully sunlit plants) are relatively insensitive to changes in leaf water potential unless species specific values of −1.2 to −2.0 MPa are experienced.

Pollock (1979), for instance, saw little evidence of stomatal closure due to low leaf water potential in alpine *Chionochloa* in New Zealand, but noted that most of the variation in water potential (>−2 MPa) was explained by changes of vpd at otherwise relatively high and stable leaf conductance. On Niwot Ridge, stomata of *Bistorta bistortoides* hardly responded to water potentials less negative than −1.6 MPa (which was the full range experienced by this species in the field) and the slight reduction in conductance was possibly confounded by vpd effects. *Caltha leptosepala* did respond, but only below −1.2 MPa (Ehleringer and Miller 1975). In the Medicin Bow Mountains of Wyoming Oberbauer and Billings (1981) noted that *Trifolium parryi* and *Potentilla diversifolia* did not respond to leaf water potential unless a threshold of −1.7 MPa was surpassed. These authors studied water relations of alpine plants along a topographic gradient during an extremely dry summer (only ca. 130 mm of rainfall between the end of June and 1st September, most of which fell in the second half of August). Under these unusual conditions they were able to demonstrate that shallow rooted species on exposed ridges can experience severe drought stress even at high altitudes (3300 m), while deep rooted ones like *Trifolium parryi* were affected little, irrespective of location. During Oberbauer and Billing's observations topsoil moisture (0–15 cm) on a leeward slope dropped below the permanent wilting point for 35 days (46 days on a ridge), but the moisture of deeper horizons was unknown. Topsoil desiccation is also a frequent phenomenon in south facing alpine grassland in the Alps, but a large series of diurnal courses of leaf conductance in *Carex curvula* and *Festuca halleri* for bright midsummer days over three consecutive years revealed very little variation in leaf conductance (in fact none at all in *Festuca*), while leaf water potentials regularly reached −1.8 MPa at noon, as a result of intense transpiration (Körner et al. 1980). This indicates (1) that stomata of these graminoids are insensitive to such reductions in water potential, and (2) that deep soils always retained sufficient plant available moisture to support unrestricted midday or early afternoon transpiration despite visible topsoil desiccation.

In a few other cases, water potentials or leaf water content alone were reported and the ranges commonly support the view that drought stress is uncommon in closed, low stature alpine vegetation (e.g. Bliss 1964). Water potentials as low as −4 MPa for sites where the majority of species exhibited extremes higher than −2 MPa and which receive

July–August rainfalls of ca. 250 mm (e.g. Miller et al. 1978) either reflect very poor soil conditions on ridges, shallow snowbed substrates, or exceptional weather situations (Oberbauer and Billings 1981), but are not typical for alpine conditions in general. Since measuring short-leaved herbaceous material with a pressure chamber is a rather delicate job, in particular when one considers the rather inappropriate equipment available when this method first came into use, some of these extreme numbers may also reflect mechanical xylem damage during pressurization. Data for various altitudes in the Alps were compiled by Körner and Mayr (1981) and indicate that most common bright day minima at alpine altitudes are around −1.5 MPa, with the highest minima observed in cushion plants (never below −0.9 MPa, see also Körner and DeMoraes 1979) and the lowest in graminoid tussocks (−1.9 MPa).

Classic functional types of plants known for reduced water consumption or increased water use efficiency, such as plants with the **C4-pathway** of photosynthesis or plants exhibiting **crassulacean acid metabolism (CAM)** are found at alpine altitudes as well, but at greatly reduced abundances. Tieszen et al. (1979, Kilimanjaro) and Pyankov and Mokronosov (1993, central Asia) by using stable carbon isotopes, and Ruthsatz and Hofmann (1984) by inspecting leaf anatomy, documented a dramatic shift from C4 to C3 species as altitude increases (discussed in more detail in Chap. 11). Besides the well-known difference in thermal and moisture preferences between these plant groups, Ruthsatz and Hofmann suggested considering the elevational reduction of grazing pressure as an additional factor. C4 leaves tend to contain less protein, reducing their attractiveness to grazers as compared with C3 leaves, which consequently are eliminated from lowland grasslands. Moisture availability alone appears to be an insufficient explanation because Ruthsatz and Hofmann found that C4 plants do particularly well at lower elevation when they have ground water access.

Stable carbon isotope discrimination during leaf photosynthesis (see Chap. 11) can be used to estimate long term integrated stomatal, relative to carboxylation restrictions of CO_2 uptake. Drought induced stomatal closure reduces the overall discrimination of CO_2 containing the heavy ^{13}C isotope, and $\delta^{13}C$ values (as compared with a limestone standard) become less negative.

Plants from cold environments, alpine ones in particular, exhibit less negative $\delta^{13}C$ (a mean increase of 1.2‰ per 1000 m of elevation; Körner et al. 1991). However, this is unrelated to moisture stress (Arroyo et al. 1990; Körner et al. 1991; see Chap. 11). For instance, $\delta^{13}C$ for forb and grass species from similar elevations of 4100–4700 m are, in the sequence of decreasing precipitation, −25.9 for New Guinea, −25.0 for the Venezuelan Andes, −24.3 for Mt. Kenya, and −26.3‰ for the northwest Argentinean Andes (Körner et al. 1991). Hence, the last site with the lowest precipitation (300 mm a^{-1}) exhibits similar $\delta^{13}C$ than the first and moistest site (>3000 mm a^{-1}).

Within a given elevation, one could still expect local differences in moisture availability to exert an additional effect. This was tested along a wetness gradient in the Andean semi-desert of northwest Argentina at 4250 m altitude (Fig. 9.10). As Figure 9.18 illustrates, there is no significant reduction of discrimination between the swampy ground at the shore of a pond and the "dry" outcrops of the rocky semi-desert. These numbers support the conclusions derived from the soil moisture studies discussed above, that plants growing in this alpine "semi-desert" experience little physiologically effective water shortage.

In summary, these more recent findings on alpine plant water relations from various parts of the world are in line with observations published during the earlier part of the century by Senn (1922), Walter (1931), Pisek et al. (1935), Berger-Landefeldt (1936), and Schenk and Härtel (1937). What Senn concluded from his many comparisons of leaf transpiration of alpine versus lowland herbaceous plants may hold for most situations: "Irgend einen Einfluss einer Erschwerung der Wasseraufnahme konnte ich bei Alpenpflanzen bisher nicht konstatieren" (thus far I have been

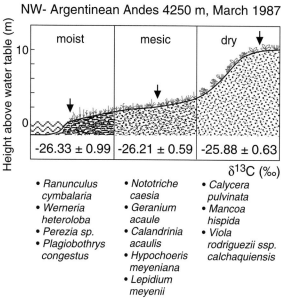

NW- Argentinean Andes 4250 m, March 1987

moist	mesic	dry
-26.33 ± 0.99	-26.21 ± 0.59	-25.88 ± 0.63

Height above water table (m)

$\delta^{13}C$ (‰)

- *Ranunculus cymbalaria*
- *Werneria heteroloba*
- *Perezia sp.*
- *Plagiobothrys congestus*

- *Nototriche caesia*
- *Geranium acaule*
- *Calandrinia acaulis*
- *Hypochoeris meyeniana*
- *Lepidium meyenii*

- *Calycera pulvinata*
- *Mancoa hispida*
- *Viola rodriguezii ssp. calchaquiensis*

Fig. 9.18. Carbon isotope discrimination ($\delta^{13}C$) in herbaceous plants (rosettes) along a moisture gradient from the shore line of "Laguna Nostra" to semi-desert slopes (see Fig. 9.10) at Huaca Huasi in the Cumbres Calchaquies of northwest Argentina, 26°S, 4250 m altitude, precipitation 300 mm a^{-1}. Note that there is no significant change across this transect, suggesting that moisture is equally available to plants. (unpubl. data)

unable to detect any constraints to water acquisition in alpine plants). It should be noted that Senn worked in the relatively drier central part of the Swiss Alps, and he referred to water consumption and not to possible indirect impacts of topsoil desiccation on mineral nutrient supply, which are likely to be important at times in all alpine regions. The data for the Pamir by Sveshnikova (1973) discussed above, illustrate that Senn's conclusion holds even for rather dry high altitude situations.

Osmotic potentials in alpine plants

Adequate moisture supply should find its expression in a low osmotic pressure (less negative osmotic potential) in cells. **Cell sap concentrations** and osmotic potentials were studied in alpine plants long before pressure chambers and leaf psychrometers became available. With no exceptions, the data point at rather low to moderate solute concentrations during the growing season. The most complete assessment by the cryoscopic method was conducted by Pisek et al. (1935, Fig. 9.19) and revealed a range from −0.7 to −2.1 MPa across a wide spectrum of plant communities and exposures. The least negative numbers were from *Rumex scutatus* and *Oxyria digyna* and the negative end is occupied by ericaceous dwarf shrubs, with the majority of other species falling between −1.0 and −1.7 MPa. Breckle (1973) reports a range of −0.8 to −2.2 MPa for the summer-dry state in the Hindukush of Afghanistan (4000–4350 m altitude) with the majority of species falling between −1.1 and −2.1 MPa. Similar numbers were found by Sveshnikova (1973) in the Pamir. For 2850 m altitude in central Japan Nakano and Ishida (1994) report osmotic potentials for turgid dwarf shrubs between −1.4 and −1.8 MPa. High andean plants collected in the above mentioned northwest Argentinan semi-desert highlands at 4050 m altitude cover a range from −0.7 to −1.3 (with some even less negative values for alpine aquatic plants not considered here; Gonzalez 1985). These few examples may suffice to underline the comparatively low demand in solutes needed for alpine plants (including those from semi-arid climates) to maintain turgor during the active season.

Pisek et al. (1935) also studied the **diurnal and** – in evergreen species – the **seasonal variation** in osmotic potential in the alpine zone. Under bright weather, the diurnal amplitude was commonly around 0.2 MPa and reached a maximum of 0.6 MPa in some alpine grassland species. Pisek et al. (1935) noted that plants from scree slopes, which showed highest (i.e. least negative) osmotic potentials, also exhibited the smallest diurnal variations, which they consider as a further indication of the good water supply in such habitats. Figure 9.20 shows their classical example for massive solute accumulation in evergreens in winter, reaching the lowest potential of −3.8 MPa

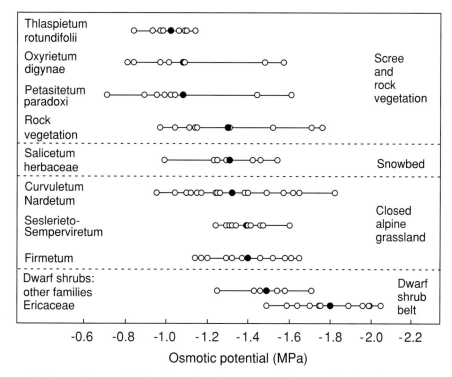

Fig. 9.19. Spectra of leaf osmotic potentials in alpine plants from various habitats in the Alps (*black points* are the means). Note the highest potentials (lowest cell sap concentrations) in plants from open fellfields and scree slopes with poor cover. (Pisek et al. 1935)

in the wind-edge species *Loiseleuria procumbens*. Part of this depression of osmotic potential is due to dehydration because of missing snow cover. When rehydrated, the potential reached only −2.75 MPa. The active enrichment with solutes in the fall is explained as a cryoprotective measure.

Turesson (1933) investigated the possibility of **genotypic** osmotic potentials in plants from contrasting altitudes in Scandinavia, by growing high and low altitude ecotypes under common low altitude climates. In most cases, high elevation provenances had significantly higher osmotic potentials (i.e. lower osmotic pressure) than low altitude provenances (as identified by plasmolysis). Under desiccation stress, the high altitude provenances wilted earlier, and took longer to recover than the low altitude provenances, from which he con-

cluded that alpine ecotypes are not better adapted to desiccation stress as some earlier authors had suggested.

Desiccation stress

From the above, it is clear that tissue desiccation in higher plants must be a rather rare phenomenon at high altitudes, and if it occurs, is restricted to plants growing on extremely exposed sites on shallow substrates (Neuner et al. 1999) or to situations where frozen soil blocks the water supply in the absence of snow cover (largely confined to the temperate zone). The possibility of winter desiccation in treeline trees has been discussed in Chap. 7. In principle, similar conclusions apply to **evergreen alpine dwarf shrubs**, although the

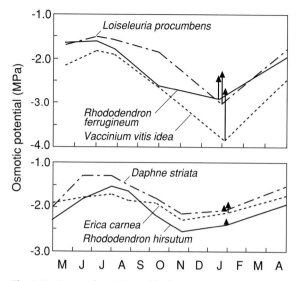

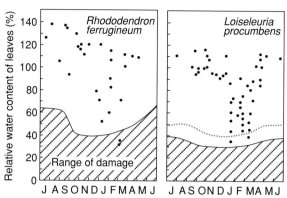

Fig. 9.20. Seasonal variation of leaf osmotic potential in alpine dwarf shrubs. *Upper diagram*: plants with little snow cover in winter. *Lower diagram*: plants under deep snow. *Arrows* indicate leaf water potentials after rehydration. The *length of the arrow* illustrates the "passive" contribution to osmotic potential by dehydration in winter. (Pisek et al. 1935)

Fig. 9.21. The seasonal variation of leaf water content and the critical water content for the occurrence of desiccation damage (*hatched area*) in two evergreen dwarf shrub species in the Alps. Data for 1950–2000 m (*Rhododendron*) and 2150 m (*Loiseleuria*) on Mt. Patscherkofel (Innsbruck, Austria) pooled from various years and authors and compiled by Larcher (1972, 1977). The critical water content represents the boundary at which 10–15% of the leaves are damaged. The additional *dotted line* for *Loiseleuria* indicates the 5% damage limit. The four points falling within the critical range are exclusively from very exposed microsites with no snow cover (e.g. *Loiseleuria* stretching over a rock)

phenomenon is even less likely to be abundant, given the small stature of these shrubs and their more likely coverage by snow **during winter**. According to Larcher's summary (1972), evergreen dwarf shrubs in the Alps can lose between 50 and 60% (% d.w.) of their turgid water content before damage occurs, and such dehydrations are restricted to the period between February and April and to plant parts stretching over rocks for extensive periods without snow protection while soils remain frozen (Fig. 9.21). In species regularly experiencing such situations, as for instance in *Loiseleuria procumbens*, water loss during this period is strongly inhibited by endogenously controlled stomatal closure, and excessively warm periods cannot break this dormancy, which is essential for survival (Körner 1976; Tranquillini 1982). Vapor loss is unlikely to be predominantly cuticular (as is often assumed), but largely confined to the weakest points of the leaf surface, the closed stomatal pore (leakage) and the thin,

hardly cutinized outer cell walls immediately at the pore. Under such extreme conditions, deeply furrowed leaves may be advantageous. Evergreen species restricted to sites with guaranteed snow cover, such as *Rhododendron ferrugineum* desiccate within only 3 days when snow is removed, because they do not possess such a strict control over stomata (Larcher and Siegwolf 1985).

Active summer-green alpine species have hardly been studied in this respect. But two observations can be reported, which may be typical for short stature **alpine grassland** as it occurs in many alpine regions of the world. First, such turfs may exhibit extreme tolerance to desiccation. Small 10 × 10 cm blocks of 5 cm thick soil swards with *Carex curvula* and *Primula minima*, collected from 2300 m in the central Alps, were allowed to desiccate in a growth chamber at 6 °C for 44 days. By the end of this period, swards appeared completely dry, and soil moisture was reduced to 4% of oven dry weight. Leaves of *Carex* had turned grayish-green, were visibly distorted (wilted) and leathery. All mature rosette leaves of *Primula* were desiccated and brown. Within 8 days after rewatering, 82 of 84 *Primula* rosettes had resprouted, and *Carex* leaves were fully turgid and green as if nothing had happened (Körner et al. 1980).

Second, and following from the above, the common brownish overall appearance of such

Fig. 9.22. Dead, often distorted leaf ends are a typical late season phenomenon in temperate alpine grasslands around the world and do not signal water shortage, but reflect normal recovery of nutrients before winter. (This example is *Carex curvula*, Alps; similar patterns are seen in high altitude *Chionochloa* grasslands in New Zealand or *Kobresia* grasslands in central Asia and in the Rocky Mountains)

alpine turfs, particularly in the later part of the season in temperate zone climates, is not related to desiccation due to drought. This phenomenon is due to normal leaf turnover, which in graminoids is associated with terminal **leaf dieback.** The rather rigid senesced leaf ends are not shed and decompose slowly, hence the brownish appearance (see color Plate 2d at the end of the book; Fig. 9.22). This phenomenon is observed irrespective of microsite moisture. These dead structures represent an effective windbreak and improve the canopy microclimate (Körner 1980). Studies in the Alps have shown that this late season browning can be drastically delayed if N fertilizer is applied (unpubl. data by C. Heid and C. Körner), indicating a link with nutrient recovery.

There is a large variation in individual alpine plants' transpirational water loss, covering a range from 60% to less than 10% of the simultaneous evaporation from wet blotting paper when stomata are fully open (Fig. 9.23). Closure of stomata can reduce transpiration by 20- (forbs) to 200-fold (succulents, some evergreens). However, most of the time such restrictions do not occur and are not necessary in the alpine life zone. This and the previous sections have illustrated that alpine plants are mostly not constrained by water shortage. Desiccation stress in higher plants in the alpine life zone is a very rare phenomenon, restricted to special microhabitats.

Water relations of special plant types

Alpine succulents, potentially employing the water saving crassulacean acid metabolism (CAM), are known from all climatic zones. They inhabit rock terraces and sun exposed locations on a shallow substrate (see color Plate 3a at the end of the book). Examples of important genera are *Sempervivum* and *Sedum* in the holarctic zone and in the tropical and subtropical Andes the genus *Echeveria* and various genera of Cactaceae

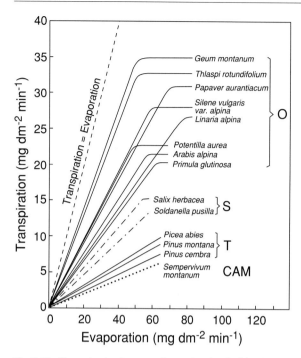

Fig. 9.23. Transpiration/evaporation ratios (moist blotting paper as a physical reference) illustrate the intensity of leaf water loss per unit leaf area at a given ambient evaporative demand. Note, those species exhibiting highest relative rates of vapor loss decouple from atmospheric forcing by stomatal responses (*plateauing lines*). *O* Open alpine grassland or scree vegetation; *S* snowbed vegetation; *T* evergreen treeline trees. (Simplified after Berger-Landefeldt 1936)

(e.g. *Tephrocactus, Trichocerreus*) and Orchidaceae (e.g. the genus with the shortest name, *Aa*). Succulents may either not perform any CAM, exhibit facultative CAM, (not operative during moist periods), or perform full CAM (always engaged). All three possibilities are also found in the alpine zone.

The widespread and species rich genus *Sedum* has never been found to engage CAM in the alpine zone (for details on alpine CAM plants, see Chap. 11). For *Sempervivum*, facultative CAM had been reported for the Alps, but carbon isotope data indicate that the CAM contribution to total carbon uptake is relatively small. Hence, normal C3 photosynthesis is the dominating mode of CO_2 assimilation in these succulents. *Echeveria* sp. from

4000 m altitude in Mexico performs "mild" CAM with $\delta^{13}C$ around −16 to −17‰. Medina and Delgado (1976) noted that *E. columbiana* employs full CAM during dry periods, but normal C3 photosynthesis during rainy periods in the Venezuelan Páramos, a region with a positive annual water balance. A single observation in high Andean cacti suggests persistent full CAM engagement with $\delta^{13}C$ of −12‰, which means almost no water loss during the day (unpubl. data from NW Argentina). In summary, the limited evidence on alpine succulents suggests that some of these specialists make no or little use of the water saving CAM mechanism, and are mainly water storers. Succulence, in these cases, appears to guarantee survival of short-term extremes on "hot spots", but does not significantly contribute to night-time CO_2 fixation.

Cushion plants, in their most compact growth form (color Plate 1 at the end of the book), not only store large detritus-bound nutrient pools and host intense microbial life in an otherwise often bare landscape, but also store large amounts of moisture. Rauh (1939, p. 474) allowed a cushion of *Silene acaulis ssp. excapa*, collected at 2900 m in the Alps, to dehydrate in his office for 1 month without reaching lethal desiccation. In his classical monograph, Rauh demonstrated that the cushion type of growth form (of which he distinguishes several sub-types) results from genotypic bud dominance and branching rules and does not represent a phenotype response related to environmental impact, including water shortage. Since cushion plants are commonly deep rooted, storage of moisture beneath the leaf canopy is more likely a byproduct of litter and raw-humus accumulation (and very beneficial to nutrient cycling) rather than being of special adaptive advantage for plant water status as such. As was shown by Körner and De Moraes (1979), moisture loss per unit cushion surface area is small, because of the intrinsically low leaf area index, and water potentials hardly drop below −0.6 MPa, even during bright midsummer weather. This parallels findings of Sveshnikova (1973) in the semi-arid Pamir, where cushion plants exhibited no signs of tissue water shortage.

Cushion shaped plant species are most abundant in cool or humid climates, and absent in the lowlands of the temperate and boreal zone and the tropics. Rauh (1939) reviewed the earlier literature, according to which approximately half of all cushion plant species occur in the southern high Andes and in Patagonia, 16% in central Asia and the Himalayas, 14% in New Zealand and the Sub-Antarctic Islands, and 12% in the European mountains. He noted their minor contribution to the Arctic (3%), Africa (3%) and North America (ca. 2%). Given this distribution, the cushion morphology cannot be seen as a prime adaptation to drought (see also Chaps. 2, 4 and 11). Inherently cushion shaped thorn-scrub (not just shaped by grazing) as it is found in wind-swept high altitudes in North Africa, the Mediterranean and continental semi-deserts of western and central Asia (cf. Breckle 1973; Hager and Breckle 1985), also seem to benefit more from litter trapping and obvious mechanical advantages than from specific benefits related to water relations (Körner 1993).

The third and final "special case" are **giant rosettes**, which are found in the afro-alpine and equatorial Andean flora at altitudes between 3600 and 4600 m (see Chap. 2). Smaller versions of this life form can be found in many other mountains of the world (Smith 1994). These often tree-like plants have to maintain water supply through a thin, up to 4 m long, stem which may freeze during the tropical-alpine night.

When night temperatures are above zero, both *Dendrosenecio* species found at 4200 m altitude on Mt. Kenya showed little, if any reduction in leaf conductance during the day (Schulze et al. 1985). The caulescent *Lobelia* shows either no stomatal reductions or (on very dry days) an almost linear response to vapor pressure deficit. However, after a $-5\,°C$ night frost, followed by a relatively humid day, the sessile *Lobelia* showed almost constant high leaf conductance, whereas the tall *Dendrosenecio brassica* had slightly reduced leaf conductance by 10 a.m., perhaps as a consequence

of cold or still frozen xylem. Dead leaves coating the stems of giant rosettes efficiently prevent night-time freezing of the xylem (Monasterio 1986) under most situations, but fire may remove that screen.

Night-time cooling of almost bare soils around giant rosettes in the high Andes is supposed to restrict moisture availability and make rosettes periodically dependent on moisture stored in the pith of their stem. For the Andean *Espeletia* Meinzer et al. (1985) and Goldstein et al. (1985) documented a pronounced increase of the pith volume/leaf area ratio with increasing altitude, and suggested an adaptive improvement of water relations. According to Baruch (1979), *Espeletias*, just like their afro-alpine convergent species, neither exhibit particular stomatal restrictions nor do they experience critically low water potentials. The lowest value he found was $-1.5\,MPa$ during the dry season. However, the work by Perez (1987) suggests that under exceptionally adverse soil conditions, drought stress seems to occur. According to Perez, the likelihood of moisture stress increases with the altitudinal decline of fine substrate, but decreases with the age and size of giant rosettes. Apart from such critical soil situations, available data suggest that moisture supply does not commonly exert a significant direct limit on tropical-alpine giant rosettes – according to Beck et al. (1982) there may even be a waterlogging problem at the base of slopes on Mt. Kenya. Giant *Senecio* and *Lobelia* on Kilimanjaro (Tanzania) are restricted to moist places (gullies, riversides, moorlands).

In summary, observations in all three of these special life forms are in line with the overall picture derived from studies of other life forms that water relations play a less decisive role for alpine plants than in comparable plants growing at low altitudes. Disregarding some very special situations, which were discussed above, comparative studies along altitudinal gradients suggest that plant water relations are not critical for the functioning and survival of alpine plants. Extreme water shortage, as it occurs in some alpine areas, selects for specific life forms and types of plant communities in a similar way to low elevations. A result of this selection is that even under very dry alpine conditions, physiological signs of drought stress are almost non-existent. However, the following chapter will illustrate that nutrient cycling and plant nutrition are likely to be affected by periodic topsoil moisture shortage also at high altitudes.

10 Mineral nutrition

The distribution and productivity of plant species in alpine regions is clearly affected by variation in soil fertility. This variation often occurs at spatial scales of one to a few meters and can be quite extreme, as illustrated by Figure 10.1. In regions with traditional alpine cattle farming such as the Alps, a multitude of unintended "fertilizer experiments" demonstrate how local nutrient addition (redistribution) can influence alpine vegetation (Fig. 10.2).

However, it is questionable whether a shortage of mineral nutrients at alpine altitudes can explain the small size and slow growth of undisturbed alpine plants and the structure of alpine plant communities in general. The growth of **individual plants** can almost always be stimulated by nutrient addition, and alpine plants are no exception. However, in a closed community, the expansion of one individual (or one plant species) causes another one to become restricted. Therefore, plant **communities** (assemblages of certain plant species) are never nutrient limited, because the addition of nutrients in the long term creates new communities, replacing the old ones. Hence, "nutrient limitation" is a question of definition (Chap. 1). If one views an ecosystem as an assemblage of organisms, each of which having its specific niche and function, all nested in, and depending on a specific environmental matrix, a concept of limitation largely drawn from agronomy makes no sense. The current composition of an ecosystem reflects all these potential constraints to the individuals, and removing some of these constraints would lead to the loss of this structure, including several or all of the previously "limited" species.

What people commonly have in mind when discussing nutrient limitation is the **productivity** of a site, irrespective of community structure. In this more agronomic sense, removal of limitation automatically means facilitating growth of other, more nutrient demanding, more vigorous species (Chapin et al. 1986). The need for a clear distinction between influences on genetic structure and dry mass production applies to any change in resource supply, but is particularly critical when the roles of nutrients in low productive natural systems are discussed.

Fertilizing alpine vegetation in the Rocky Mountains and in the Alps (see below) has been found to stimulate growth, and nitrogen addition was most effective. Hence, nitrogen may be considered most limiting to alpine plant productivity. While supplies of all other soil nutrients ultimately depend on substrate weathering, nitrogen supply depends predominantly on biological activity – both through dinitrogen fixation and recycling. Consequently, low alpine soil temperatures have repeatedly been cited as being the major constraint for the supply of plant available nitrogen (e.g. Bliss 1971; Billings 1974). This chapter will therefore **focus on nitrogen** as the key nutrient, but phosphate and other nutrients will be considered as well (see the synthesis by Bowman and Seastedt 2001). The first two sections provide a short account of mineral nutrients in soils and an assessment of plant nutrient concentrations. Then I will discuss nutrient cycling, nutrient budgets, nitrogen fixation and alpine mycorrhiza. Finally, responses of plants to nutrient addition will be summarized.

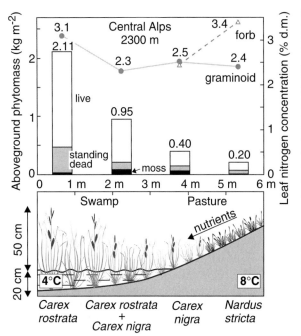

Fig. 10.1. Nutrients matter! Steep nutrient gradients lead to dramatic differences in plant biomass and species composition even at very high altitudes (here 300 m above the treeline in the Alps, Wattens/Tirol). A gulley with no outlet traps leached nutrients from surrounding slopes (see color Plate 2 at the end of the book). Rainfall in this area is high (>1200 mm/year with a peak during the growing season), hence upslope plants never experience any moisture shortage. Note that leaf nutrient concentrations vary little, and high leaf nitrogen concentrations occur at both ends of the transect. (Unpubl. data)

Fig. 10.2. Regular fertilizer addition (here cattle manure) can convert low stature alpine heath into a luxurious "herbage crop". Though total aboveground leaf dry matter may not be different, plant and leaf size can differ dramatically. *Rumex alpinus* leaves shown here reach a size of 25 cm × 40 cm. The massive rhizomes produced by these plants were formerly fed to pigs. Photograph taken at 2100 m elevation near the climatic treeline at Furka Pass, Swiss Alps

Soil nutrients

The concentration of total nitrogen in alpine soils is correlated with soil organic matter concentration (SOM). The more humus, the greater the total nitrogen pool per unit land area. However, the **C/N ratio** commonly becomes lower the less organic matter is contained in soils (Table 10.1). Poorly developed soils at extreme high elevation with only sparse vegetation may exhibit C/N ratios of as low as 5 to 8, in contrast to ratios of more than 40 in acid raw humus soils under ericaceous shrub. The major root horizon (i.e. the top 10 cm of the profile) in closed alpine grassland commonly shows C/N ratios between 10 and 20, similar to what is observed in low altitude grasslands and deciduous forests. The very detailed analysis by Rehder (1970) for the Alps revealed that low C/N ratios of 8 to 11 are found at moderately productive sites (with *Poa*, *Sesleria*, *Nardus*) and higher values of 12 to 14 are typical for less productive sites (*Carex curvula*, *Carex firma*). Hence, annual biomass production is negatively correlated with total N concentration in the soil (Fig. 10.3), just as it tends to decline with SOM concentration. For the top 20 cm of the profile, Rehder calculated total **organic N pools** of 4.4 to 8.8 t N ha^{-1} for these alpine grasslands (approximately 0.7 to 1 kg N m^{-2} for full profile depth). Again, these numbers are very similar to those for low altitude ecosystems (a range of 0.75 to 1 kg N m^{-2}, with an extreme of 2 for

Table 10.1. Nitrogen concentrations and C/N ratios in alpine soils

Sites/soils	Depth	%C	%N	C/N	pH
High altitude raw soils with sparse vegetation					
4 alpine sites with *Oxyria digyna* (USA, 3500 m)	–	0.3	–	5.4	–
Open tussock flats (NW Argentina 4250 m)	0–15 cm	0.7	0.1	6.4	–
	20–45 cm	0.3	0.07	4.6	–
Closed alpine grasslands					
2 types of grassland (NW Caucasus 2750 m)	0–5 cm	7.8	0.8	10	4.0
	10–15 cm	2.0	0.25	8	3.8
3 types of grassland (Rocky Mts., 3500 m)	0–10 cm	14.5	1.2	12	5.1
8 types of grassland (Swiss Alps, 2600 m)	0–5 cm	9.5	0.5	19	–
	5–10 cm	–	0.3		–
Alpine sedge mat (Tirolian Alps, 2550 m)	0–10 cm	–	1.0	–	3.8
(Hohe Tauern Alps, 2300 m)	0–5 cm	10	0.6	17	5.6
	10–15 cm	1	0.036	28	4–5
6 types of grassland (Bavarian Alps 2000 m)	0–5 cm	17	1.5	11	5–7
Alpine dwarf shrub heath					
Pioneer heath (Tirolian Alps 2050 m) humus layer	–	6	0.2	30	3.8
Loiseleuria heath (Tirolian Alps 2000 m) organic layer	5–20 cm	42	1.0	42	2.6
humus layer	20–25 cm	8	0.3	27	3.3
weathering layer	25–50 cm	4	0.3	12	4.2
Vaccinium heath (Tirolian Alps 2000 m) organic layer	5–25 cm	45	1.6	28	2.6
humus layer	20–30 cm	29	0.9	32	3.3
weathering layer	30–65 cm	5	0.3	17	4.5
Subtropical and tropical alpine vegetation					
Mt. Wilhem (Papua New Guinea, >3450 m)	tundra	12	0.5	23	5.0
	tussock	15	0.9	17	6.1
	bog	19	1.3	15	5.7
Mt. Kenya (sites with giant rosettes, 4150 m)	slope	6	0.4	15	5.2
	valley	14	1.0	13	6.1
Páramo el Banco (Venezuela 3800 m)	slope	5	0.3	17	4.1
	valley	10	1.6	6	4.9
Andean semi-desert (NW Argentina 4800 m)	nival	1.3	0.1	13	5.1
	tussock	1.2	0.1	12	5.2

Rounded means (standard deviations were omitted and soil horizons were simplified for the sake of clearness). References from top to bottom: Mooney and Billings (1961); Ch. Körner and S. Halloy (unpubl.); Rabotnov (1987); Fisk and Schmidt (1995); Galland (1982); Holzmann and Haselwandter (1988); Danneberg et al. (1980) and Posch (1980); Rehder (1976a); Neuwinger (1972); Larcher (1977); Wade and McVean (1969, cited by Hope et al. 1976); Rehder (1994); Barnola and Montilla (1997); Ruthsatz (1977).

coniferous forests, compiled by Killham, 1994). Galland (1982) reports generally much higher C/N ratios for similar grasslands in the Swiss National Park (mostly between 14 and 30). Nitrogen to Phosphate ratios in the soils studied by Galland varied between 5 and 16 (mean of 9 ± 3, n = 8).

These organic pools of nitrogen and other nutrients tied up with SOM are not directly avail-

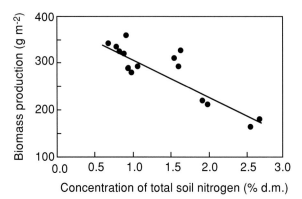

Fig. 10.3. Net seasonal aboveground biomass accumulation and concentrations of total N in alpine soils correlate negatively because total soil N is a function of soil organic matter concentration, which in turn, is associated with soil acidity. (Rehder 1970)

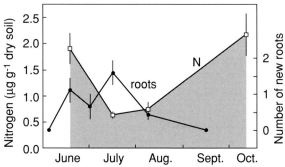

Fig. 10.4. The seasonal variation of total mineral nitrogen in soils and the rate of root initiation (presence of new root tips in excavated individuals of *Bistorta bistortoides*) in alpine *Kobresia* grassland of the Rocky Mountains in Colorado. (Jaeger and Monson 1992)

able to plants and must be "mineralized", and made soluble, to become accessible. However, nutrient **concentrations in aequous extracts** of soils also do not provide a direct measure of true nutrient availability but reflect a momentary picture of free nutrients, and concentrations may be very low due to instantaneous uptake by plants and microbes (midseason $<10\,\mu g\,N\,g^{-1}$ substrate, i.e. 10^4 to 10^5 smaller than those for total N; Rehder 1970; Jaeger and Monson 1992). Since it is the nutrient addition rate to the root, rather than the nutrient concentration in the soil solution which determines plant growth (Ingestad 1981), extractions of soils at a point in time will always be inconclusive with respect to the actual supply situation. In a well coupled (natural) plant-soil system, very little free soil nutrients may be found, even though plant demand may be met. The more actively plants grow, the less likely it is to find freely **available nutrients** in the soil, with soil microbes tying up mineralized N during the remaining periods. The highest free nutrient concentrations are observed at snow melt in spring and in late fall (Rehder 1976b; Lipson et al. 1999; Jaeger et al. 1999; Fig. 10.4).

Commonly, **ammonium** nitrogen dominates the soil solution in organic and acid alpine soils. However, as Rehder (1970) has shown for calcare-

ous soils with a pH of around 6, this is not due to greater microbial ammonium release, but results from faster **nitrate** resorption by plants. In ion exchange resin bags exposed in sedge dominated alpine grassland of pH 4, Arnone (1997 and pers. comm.) captured 1.3 to 4 times more $NO_3^- $ N than NH_4^+ N. Atkin (1996) and Michelsen et al. (1996) concluded that NO_3^- is an important N source also in the arctic, and evidence is accumulating that soluble organic nitrogen is an important source of N in cold soils as well (Chapin et al. 1993; Michelsen et al. 1996; Eviner and Chapin 1997; Lipson and Monson 1998; Lipson et al. 1999; Miller and Bowman 2002). The latter authors also showed that $\delta^{15}N$ in the field mirrors distinct preferences for either NH_4^+ (high $\delta^{15}N$) or NO_3^- (very negative $\delta^{15}N$) among alpine species (pot experiments).

The nutrient status of alpine plants

Except for extreme situations, the concentration of mineral nutrients in plant tissues is not directly related to nutrient supply, but largely depends on the way plants use the nutrients which they absorb. Nutrient concentrations may be low due to severe limitations of supply, but they may also be low because of dilution through fast growth (as a result of accumulation of carbon compounds). Tissue nutrients may reach very high concentrations because of excess supply or because they concentrate as a result of retarded growth. Tissue nutrient analysis can tell us how plants invest mineral nutrients and whether plant tissues have a potential for low or high metabolic activity (e.g. maximum rate of photosynthesis), because nutrient (N) concentration, protein concentration and physiological activity are commonly closely correlated.

Concentrations are rated as high or low by comparison and general experience. For instance, if someone finds midseason leaf nitrogen concentrations in deciduous or herbaceous leaves of only 1.5% of dry mass, this would be considered very low (permitting only relatively low rates of photosynthesis). The same concentration would be considered normal or high in evergreen leaves. In herbaceous plants, 3 to 4% N would be considered a high concentration. Were the concentration around 2.5% one could safely conclude that the investigated leaf operates in a normal range (is neither depleted nor short in N). Conclusions about growth limitations at the whole plant level are not possible. A plant may just have stopped producing new leaves (e.g. because of exhausted available soil N) and thereby is maintaining an "adequate" nutrient level in existing leaves. Thus, "nutrient status" in the current context means the supply status of cells in an existing tissue, in comparison with published knowledge or some reference sites. In a comparative approach, tissue nutrient concentrations can at least tell us whether or not tissues are seriously deficient in certain nutrients.

Results of a global screening project of peak season mineral **nutrient concentrations in tissues** from alpine plants revealed that mineral nutrient concentrations, most consistently those of nitrogen (but also phosphate), increase with increasing elevation (Körner 1989b; Fig. 10.5). Leaves of plants from the highest elevations may contain as much as 4–5% N. This trend is observed across climatic zones, plant life form, a multitude of species, and is not restricted to leaves, but includes

roots and non-woody stems as well. The trend is enhanced when N per unit leaf area is considered, because leaves become thicker with increasing altitude. Detailed studies in single species over narrow altitudinal gradients (Körner and Cochrane 1985, Woodward 1986, Morecroft et al. 1992a) underline the universality of these N-patterns.

However, such elevational comparisons are only valid as long as comparable sets of species within a certain climatic range are considered (see the discussion in Chap. 1). It holds within forbs, grasses, shrubs and trees, but means for **whole communities** composed of species of these life forms may not fit this pattern if the relative abundance of these life forms changes with altitude.

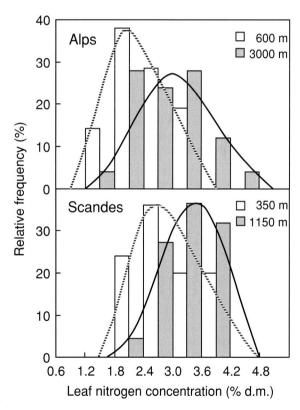

Fig. 10.5. Nitrogen concentrations (% dry matter) in leaves of comparable plant species from low and high altitude in the Alps and the northern Scandes. Alps: 2.87 ± 0.73 *versus* 2.40 ± 0.62 (n = 25/21 species, p = 0.02); Scandes: 3.18 ± 0.51 *versus* 2.80 ± 0.58 (n = 22/25, p = 0.02). (Körner 1989b)

Since many sedges, tussock grasses and sclerophyllous dwarf shrubs produce tissues with little N and/or much C, their greater abundance at high altitude may reduce the nitrogen concentration in a mixed biomass sample. A typical example is illustrated by Fig. 10.1, where the community mean at the right end of the profile is dominated by a low N tussock grass (*Nardus stricta*), but the small forb fraction of the biomass is very rich in N. In other words, the increase of % N with altitude in coherent sets of species does not mean that the overall N concentration in the whole stand biomass must follow the same trend. Furthermore, if soil chemistry, moisture supply or cloudiness show radical changes along an elevational gradient (as in some tropical mountains with arid tops), low/high altitude comparisons, in the strict sense, with respect to tissue N become meaningless as well. This is why the above mentioned survey centered on transects with only moderate changes in soil moisture (e.g. parts of the Alps, New Zealand, the Scandes and New Guinea).

This data set revealed a remarkable **latitudinal trend** as well (Fig. 10.6). Alpine plants (in essence forbs) of polar mountains exhibit much higher leaf N concentrations than alpine plants in equatorial mountains, but are proportionally thinner (Körner et al. 1989a; Fig. 11.3). Consequently, N per unit leaf area changes little with latitude. This global "natural experiment" hints at some possible causes for the observed patterns in N concentration in alpine plants. Tropical-alpine species are not restricted by a short season and invest/accumulate more C per unit leaf area (in fact, they contain on average 2.4 times as much carbon per area as arctic-alpine forbs). Accounting for day-length differences, their potential carbon gain per year (at equal photosynthetic capacity) would be ca. 2.9 times bigger than in the arctic (Körner 1989b). This comparison suggests that the product of leaf % N and effective season length is maintained fairly constant across latitudes. It appears that carbon rather than nitrogen per leaf area varies in response to season length under such marginal life conditions. Greater carbon (and thus diluted N) concentration in leaves of tropical-alpine plants is

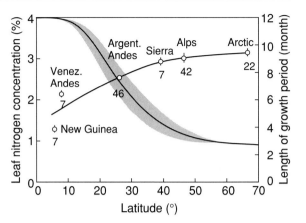

Fig. 10.6. The latitudinal variation of leaf nitrogen concentration in herbaceous plant species close to the regional upper limits of higher plant distribution. Numbers indicate the number of species sampled in each region. Note the inverse relationship with length of season (*shaded*). (Körner 1989b)

associated with longer leaf life (C amortization; see Chap. 12), greater mechanical rigidity and reduces attractiveness for herbivores.

On a whole plant basis, a surprisingly similar fraction of nitrogen is partitioned to leaves and special storage organs in high and low altitude forbs (data for the Alps, Fig. 10.7). Alpine forbs, however, invest less N in stems and more in fine roots, largely a consequence of the overall patterns of dry matter allocation (see Chap. 12).

Whatever limits mountain plants in migrating to even higher elevations, it does not seem to be a general limitation of metabolic capacity per unit leaf area by low levels of leaf nitrogen. There are no indications of any unusual mineral nutrient shortages in tissues of alpine plants. Concentrations in alpine forbs compare well with those in plants from fertile agricultural land at low altitudes. This may be related to the fact that in cold, infertile habitats plants tend to adjust growth to shorter growing seasons by "luxurious consumption" of nitrogen (Chapin et al. 1986; Körner 1989b). Direct growth limitation due to low temperatures, particularly during nights, may further restrict nutrient "dilution" (Körner and Pelaez-

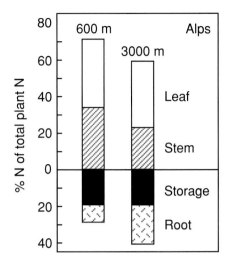

Fig. 10.7. Average patterns of nitrogen partitioning in lowland (n = 21) and alpine (n = 25) forb species. The stem compartment includes all non-leaf aboveground plant parts. (Körner 1989b)

Menendez Riedl 1989; Morecroft and Woodward 1996). A test with *Betula* seedlings by Weih and Karlsson (2001) suggests the involvement of active, acclimative responses to low temperatures. These conclusions do not preclude nutrient limitation of biomass production at the ecosystem level, and also do not exclude the possibility that nutrient addition may stimulate growth. These data rather suggest that alpine plants grow such that relatively high mineral nutrient levels are maintained in tissues.

Nutrient cycling and nutrient budgets

Mineralization of organic matter, atmospheric deposition, nutrient release from the mineral soil, and biological fixation of molecular nitrogen are the four major pathways through which plants obtain their nutrient supply. Given the remote position of most alpine habitats with respect to natural and anthropogenic atmospheric sources of nutrients, and because temperatures are low, all four sources may be expected to be comparatively

weak and co-limit alpine plant growth. Severe limitations of overall **microbial activity** have indeed been documented for alpine soils (e.g. Rehder 1970; Schinner 1982; Holzmann and Haselwandter 1988), and low temperature limitation of nutrient uptake is also a well-known phenomenon (Chapin 1978; Karlsson and Nordell 1996). However, Brooks et al. (1996) noted that soil activities may be higher than one might expect, given the low soil temperatures. Morecroft et al. (1992b) found either no change or even an absolute increase of mineralizer activity with altitude across two transects of 600 and 800 m difference in altitude in Scotland. However, other environmental factors, not linked *a priori* to altitude may overshadow the consequences of cooler climate alone, in particular if elevational differences are small (e.g. local effects of precipitation and runoff, type of vegetation, geology and soil acidity, slope activity, land use).

A classic technique of assessing rates of **mineralization** in soils is to exclude the major consumer, the higher plant. By using so-called buried bags or incubation tubes with intact soil cores, it is possible to approximate net rates of microbial release of nutrients from soil organic matter. Ground-area based rates, which are required to compare sites and relate responses to biomass are, however, rare.

For the **temperate zone**, data sets by Rehder (1970) and Fisk and Schmidt (1995) indicate that between 0.1 and $5\,g\,N\,m^{-2}$ are released annually in alpine grassland, with means of around $1\,g\,m^{-2}\,a^{-1}$ for intact soil cores found by Fisk and Schmidt on Niwot Ridge appearing to be the most robust estimates. These authors also noted that seasonal variability of microbial activity was greater than landscape variability across three very differently exposed sites. They also suggested that microbes themselves had a very high and variable N demand, which strongly co-determined N availability to plants. On average, total microbial N pools were roughly ten times as big as the monthly rates of mineralization on Niwot Ridge.

Very little information is available for the **tropics and subtropics**. Ruthsatz (1977) used the buried bag technique at eight sites between 2600 and 4800 m altitude in the northwest Argentinan Andes, a rather dry area with semi-desert vegetation along most of the transect. In the middle part of the transect (3700–3800 m) she found a pronounced seasonal variation of free soil nitrate, with a maximum of 1 g nitrate-N m^{-2} during the dormant period, but very small pools (<0.1 g nitrate-N m^{-2}) during the growing period. Between 3700 m and 4800 m, where the total above-ground biomass (shrubs and grasses) averages somewhere between 2 and 4 t ha^{-1}, annual mineralization was extremely small (no data), so that Ruthsatz concluded that plants must have access to other N sources, presumably microbial N$_2$-fixation.

In **arctic-alpine** fellfields, Jonasson et al. (1995) found soil microbes to contain 3.5%, 7% and 35% of all soil C, N and P respectively. When they added sugar, microbial C increased, but microbial N and P pools remained unaffected. When they added NPK, **microbial biomass** C was not affected, but the microbial N and P pools increased by a factor of two (N) or more (P), further confirming the sink strength of alpine soil microbes for nutrients (cf. Lipson and Monson 1998). By adding fungicide these authors showed that the majority of microbial biomass was not fungal. N limitation of alpine soil microbes was also demonstrated by a less intrusive supply of carbon to microbes via long-term CO$_2$-enrichment of vegetation, which had no effect on microbial biomass or respiration (despite a proven C surplus in the system) unless extra mineral NPK was provided as fertilizer (Niklaus and Körner 1996, Körner et al. 1997). Hence, the common assumption for low altitude soils that microbes are primarily C limited, does not seem to hold for alpine situations. **Plant-microbe competition** for N and P in closed alpine vegetation can apparently be severe, and microbes appear to be co-limited by both C and nutrients such as N and P in such undisturbed, late successional high altitude systems.

N and P immobilization are not necessarily coupled, and their strong dependency on **soil type** may override any altitudinal effects. Jonasson et al. (1993) compared mineralization in subarctic alpine fellfields with lowland heath in northern Sweden and found higher rates of N mineralization but close to zero P mineralization in alpine soils, whereas the reverse was true on peaty heath soils at low altitude. By using soil transplants they demonstrated that these differences were soil-specific and were unrelated to the 4–5 K temperature difference between sites. Such soil-specific responses may also explain the higher N mineralization observed by Morecroft et al. (1992b) in the alpine zone of Scotland.

A **seasonal release** of only 1 g N m^{-2} (see Fisk and Schmidt, above) is clearly not enough to support the annual substitution of total biomass nitrogen even in very sparse alpine vegetation. Although data on land-area-based N in biomass are rare for alpine vegetation, and great uncertainty exists about the below-ground living fraction, a mean of about 10 g N m^{-2} for grasslands appears plausible (estimates of 6 to 20 g N m^{-2} reported by Rehder 1976a; Evans 1980; Smeets 1980; Schäppi and Körner 1997; Arnone 1997) of which 60–70% is below ground. Since the seasonal net biomass production commonly does not exceed 200 g dry matter m^{-2}, a seasonal demand for "new" N of 2–3 g m^{-2} per season would still exceed the estimated seasonal release by mineralization rates by a factor of two to three. The seasonal incorporation of N corresponds to roughly 15% of the peak season biomass N pool in these grasslands (a substantial fraction of the biomass is perennial).

The "missing" N fraction is obtained through plant internal re-allocation, through atmospheric deposition largely in meltwater, but also via N fixation. **Resorption** of N from old senescing tissue and allocation to new tissue, either directly via leaf replacement or via intermediate storage of N are well documented phenomena for arctic-alpine (Skre 1985; Jonasson 1989; Karlsson 1994), temperate alpine (Jaeger and Monson 1992; Schäppi and Körner 1997) and subtropical/tropical alpine plants (Sundriyal and Joshi 1992; Beck 1994). According to Schäppi and Körner, **recovery** of N from green leaves during senescence is 68% in *Carex curvula* and 49% in *Leontodon helveticus* in the Alps, and phosphate recovery from senescing leaves was reported by Smeets (1980) to reach 80% in this sedge species. Jaeger and Monson (1992) report that 60% of the total nitrogen found in new shoots of *Bistorta bistortoides* was re-allocated N from belowground biomass stores. Hence, a minimum of 50% plant internal recycling of N and other mineral nutrients appears to be a conservative estimate.

Linked to investment and recovery dynamics, nutrient concentrations (particularly in leaves) vary dramatically with season (Fig. 10.8). All sorts of functional attributes of these **seasonal changes** have been discussed in the literature, although they largely reflect two independent processes, not necessarily related to the nutrient supply status of active cells: (1) the accumulation of carbon due to tissue maturation in the first part of the season – in essence a dilution process, which has little to do with nutrition per se, and (2) nutrient recovery during tissue senescence in the late season. While nutrient content per cell may not vary in the first phase (dry matter is just an inadequate reference) it does in the second phase. As Fig. 10.8 illustrates, a stable intermediate phase may not exist within

the short alpine summer of temperate and sub-polar climates.

As in other nutrient poor environments, alpine plants, particularly those on bare soils and in open vegetation in windswept habitats have developed remarkable structural adjustments in order to close their nutrient cycle with minimal risk of losses. The **cushion** life form and dense rosettes represent a compromise between plant internal recycling and "open" recycling via litter and soil. As was discussed in Chapter 6, cushion plants must be seen primarily as litter- and thus, nutrient-traps, which at the same time create a very favorable micro-environment for decomposers and adventitious roots. Another remarkable piece of nutrient recycling machinery was described by Beck (1994) for giant afro-alpine *Senecios*, where stem-born adventitious roots explore the attached decomposing leaf coat for nutrients.

The mineral nutrient cycle may also be enhanced through **herbivory** occurring in all alpine environments (see end of Chap. 15). Mammals graze alpine vegetation up to the highest peaks (Halloy 1991; Diemer 1992) and as much as half of the annual nitrogen investment in alpine grassland phytomass (bio- and necromass) can be removed by ruminant grazers (Sundriyal and Joshi 1992; Fig. 10.12). In natural alpine grassland in the Alps (300 m above the treeline), the exclusion of moderate cattle grazing (which historically replaced part of natural grazing by ibex and chamoix at these elevations) had quite unexpected effects. After 6 years, total peak season phanerogam biomass in the fenced plot was 16% lower than in the continuously grazed land outside the fenced area (standing dead plant mass and litter, reflecting cumulative effects of several years, was reduced by −34%). Forb representation in the fenced communities was significantly reduced by the time of sampling (S. Schneiter and Ch. Körner, unpubl.). Thus, extensive grazing had a positive effect on both biomass production and species richness and is responsible for a substantial fraction of nutrient cycling in this system (Fig. 10.9). Less obvious are invertebrate grazers, which can affect plant communities more than larger

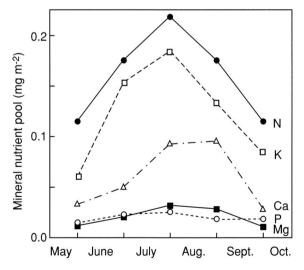

Fig. 10.8. The seasonal variation of nutrient pools in phanerogam biomass of sedge-dominated alpine grassland at 2300 m altitude in the central Alps. (Smeets 1980)

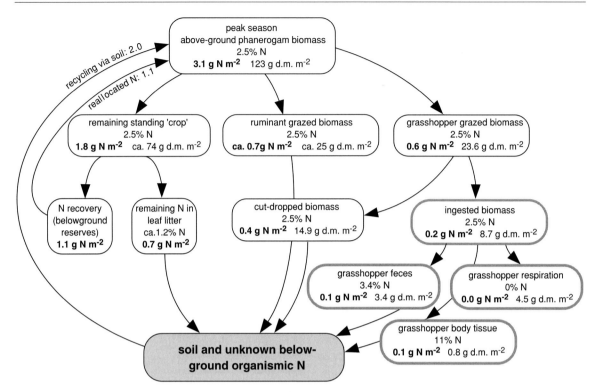

Fig. 10.9. The role of grasshopper and ruminant grazing in the aboveground nitrogen cycle in an alpine sedge-dominated grassland. "*Cut-dropped*" means that grasshoppers harvested leaves, but dropped them uneaten on the ground. Numbers for ruminant grazing are only rough estimates (in this case for short cattle visits in midsummer which substitute former ibex and chamoix grazing), and recycling of this biomass is patchy (dung, urine) with a mean spatial recurrence of 20–50 years. From data by Blumer and Diemer (1996), Schäppi and Körner (1997) and S. Schneiter and Ch. Körner (unpubl.). See also Fig. 10.16

animals. Blumer and Diemer (1996) found that alpine grasshoppers may recycle between 20 to 30% of above-ground biomass (Fig. 10.9). Since this interrupts late season nitrogen recovery within plants, at least half of the nitrogen contained in grazed biomass is diverted into the microbial nutrient cycle (for recent US literature see Dearing 2001).

Complete **nutrient budgets** for alpine vegetation are rare in the literature, particularly because so little is known about below ground processes. From ecological theory, late successional, stable alpine vegetation should have a balanced nutrient budget with small net gains and losses and nearly constant overall pools. This is in essence what was

found for alpine dwarf shrub communities during the International Biological Programme (Larcher 1977). As an example, Fig. 10.10 illustrates the nutrient fluxes of three key nutrients in a *Loiseleuria* heath. These evergreen shrubs annually recycle 40% of their above-ground P, and 80% of their aboveground K. The major flux of K is via canopy leaching, whereas litter represents the main flux for P. The pools in the litter compartment are remarkable. Of the total aboveground live and dead plant material, 48% of P, 54% of Ca, 52% of Mg, and 80% of Fe are found in litter (only 22% of K) at peak season. This is a typical situation for many alpine ecosystems. **Litter pools and litter decomposition** in some alpine tussock grass-

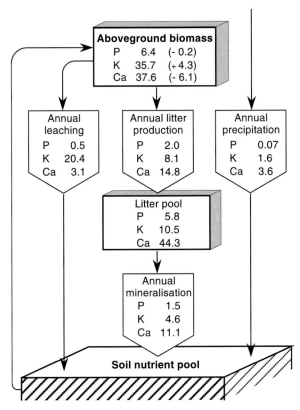

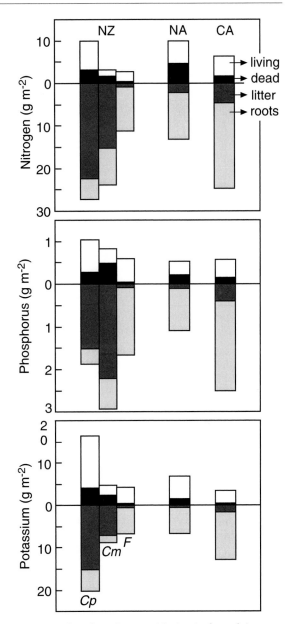

Fig. 10.10. The mineral nutrient budget of an alpine dwarf shrub canopy at 2000 m near Innsbruck, Austria (kg ha^{-1} a^{-1}). For other nutrients see the original paper by Larcher (1977)

lands may be even larger, in particular, when belowground dead plant material is included (see the example for New Zealand in Fig. 10.11).

As a rule, more mineral nutrients are contained in belowground biomass than in aboveground biomass (Fig. 10.11), but very little is known about the mean residence time of nutrients in roots. Sundriyal and Joshi (1992) mention **turnover times** for roots of Himalayan alpine grassland of 2.2 years for P, 1.7 years for N and 1.5 years for K. Their nutrient flux model (N fluxes shown in Fig. 10.12), indicates a total seasonal influx of 15 g N m^{-2} into phytomass, of which one third ends up in necromass (litter and standing dead material) during the growing period.

Fig. 10.11. Mineral nutrient partitioning in three alpine snow tussock communities in New Zealand (*NZ*) and alpine pasture land in the Alps (*NA* northern Alps, *CA* central Alps). Note the massive nutrient pool in litter in the two *Chionochloa* (*C. pallens, C. macra*) snow tussock communities compared with the *Festuca novae-zealandiae* (*F*) community in New Zealand. (Evans 1980; Smeets 1980; Rehder 1976a)

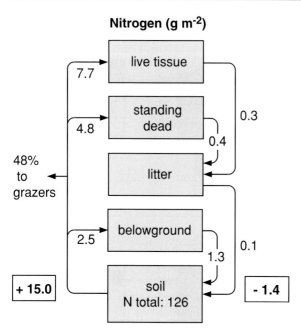

Nitrogen (g m⁻²)

live tissue

7.7

standing dead

4.8

0.3

0.4

48% to grazers

litter

belowground

2.5

0.1

1.3

+ 15.0

soil
N total: 126

- 1.4

Fig. 10.12. Accumulative seasonal fluxes of nitrogen (g m⁻²) into and out of various phytomass compartments in Himalayan alpine grassland. (For P and K see the original paper by Sundriyal and Joshi 1992)

Alpine **P dynamics** have been studied much less than N dynamics and are most often discussed in relation to mycorrhization (see below). In contrast to N fertilization, P fertilization appears to have a much weaker (if any) effect on plant growth in alpine grassland. Only a small group of species (mainly non-mycorrhizal ones) were responsive to P fertilization in the trials of Theodose and Bowman (1997, see also the later section on this topic). No significant effects were seen in late successional alpine grassland in the Alps (J. Bolliger and Ch. Körner, unpublished). Perhaps this is so because alpine plants have evolutionarily adjusted their growth and efficiency of P acquisition to the long-term supply status, as studies in arctic and alpine ecotypes of *Carex aquatilis* by Chapin and Oechel (1983) suggest. They found that roots from cold, phosphate-poor soils had higher rates of uptake under standardized conditions than ecotypes from warmer, more fertile sites. They also concluded that the local availabilty

of P had a more pronounced selective effect on P-uptake characteristics than temperature.

Nitrogen fixation

Both symbiotic and free bacterial **di-nitrogen fixation** have been documented for alpine ecosystems (for literature see Holzmann and Haselwandter 1988). With increasing altitude, the relative contribution of symbiotic (largely legume) di-nitrogen fixation decreases, whereas that of cyanobacteria increases. This is simply because legumes are less successful at high altitudes than other phanerogam families, and are altogether absent in the highest life zone above the snowline. However, the overall seasonal contribution of both these N sources is small compared with N derived from mineralization.

Where **legumes** are present in the alpine zone, their roots are always nodulated (pers. obs. in various parts of the world). Whether bacteria in nodules are active can be estimated by the acetylene reduction assay or, with less certainty, by stable ^{15}N isotope techniques. *Rhizobia* do not discriminate against this heavy form of N, whereas decomposers in the soil do, and thereby alter the isotopic composition of litter-derived N (both positively and negatively, depending on the fraction considered). Table 10.2 illustrates that legumes from high altitudes tend to have $\delta^{15}N$ values close to the atmospheric value of zero, whereas non-legumes most commonly have negative $\delta^{15}N$, hence, are ^{15}N depleted (but there are noteworthy exceptions, to be discussed later).

Evidence from acetylene reduction assays by Wojciechowski and Heimbrook (1984; see also Johnson and Rumbaugh 1986) confirmed symbiotic N_2-fixation in *Trifolium dasyphyllum* at 3650 m on Niwot Ridge, and Bowman et al. (1996), by combining field and laboratory studies, concluded that *Trifolium* species from the same site meet 70 to 100% of their N requirements by symbiotic N fixation (60–90% for nine species according to Jacot et al. 2000a). Where legumes are present, N fixation per unit land area was esti-

Table 10.2. Stable nitrogen isotope discrimination in legumes and non-legume species from cold high altitude or subarctic habitats (‰ $\delta^{15}N \pm SD$)

Site	Legume sp.	$\delta^{15}N$	Mean $\delta^{15}N$ for non-legumes (n)
Iztaccihuatl, Mexico 4000 m	*Lupinus* sp.	0.69	-4.55 ± 1.43 (8)
Niwot Ridge, Colorado 3600 m	*Trifolium nanum*	0.05	-1.59 ± 0.72 (2)
	T. dasyphyllum	−0.05	
Furka Pass, Swiss Alps 2470 m	*T. alpinum*	−1.02	−4.00 – (9)
	T. pallescens	−1.31	
	T. badium	−1.72	
	T. pratense ssp. *nivale*	−1.34	
N slope Alaska (foothills)	*Lupinus arcticus*	0.00	-4.90 ± 1.60 (2)
Abisko N Sweden (450–1150 m)	*Astragalus alpinus*	−1.50	-3.90 ± 1.70 (9)
	A. frigidus	−2.00	

Sources from top to bottom sites: Ch. Körner (unpubl.), Bowman et al. (1996), Ch. Körner (unpubl.), Nadelhoffer et al. (1996), Michelsen et al. (1996) who mention that the two *Astragali* were nodulated and had much higher leaf N than all the non-legumes, hence were N-fixing despite the negative $\delta^{15}N$. Cyperaceae, Ericaceae, hemiparasitic Scrophulariaceae and non-legume symbiotic N-fixers such as *Dryas* and *Alnus* were excluded from the means for non-legumes (see text).

mated to range between 0.1 and 0.8 g N m^{-2} a^{-1} by these authors. Using the same technique with a suite of potted species in high alpine sedge mats in the Alps, Holzmann and Haselwandter (1988) measured a 100 fold **nitrogenase activity** in legumes (*Trifolium badium*), compared with non-legumes. Such legume pots fixed an equivalent of 0.8 g N m^{-2} a^{-1} – exactly what Bowman et al. (1996) estimated (a maximum of only 0.008 g N m^{-2} a^{-1} was found in the non-legume pots). Irrespective of species, the nitrogenase activity in the pots was enhanced when plants were incubated at 22 °C in the laboratory, hence low soil temperature seems to inhibit activity in the field. For grassland in the Alps Jacot et al. (2000b) report symbiotic N contributions decreased from 16% at montane sites to 9% at low alpine sites (a consequence of reduced legume abundance, not activity). It should be added that there are also other than legume N$_2$ fixing symbiotic associations in alpine vegetation, the actinomycorrhizal ones by *Dryas* and alpine *Alnus* shrubs being well-known examples.

The fact that certain groups of species had to be excluded from the comparison in Table 10.2 in order to make it meaningful, shows that a multi-

tude of other potential sources for variability in $\delta^{15}N$ exist; some are known, but others are yet to be explored. The most important for alpine ecology appear to be the extreme $\delta^{15}N$ signatures of **Ericae** ($\delta^{15}N$ as low as −9‰) and **Cyperaceae** ($\delta^{15}N$ as high as +3.5‰). Michelsen et al. (1996) and Nadelhoffer et al. (1996) explain these differences by different N sources (soil horizons, melt-water, species of N compounds) and type of mycorrhiza. For example, discrimination against ^{15}N during mineralization leads to higher $\delta^{15}N$ values in older, more recalcitrant humus fractions (cf. Nadelhoffer et al. 1996). Perhaps, sedges have access to these otherwise very stable forms of organic N. Ericaceae, on the other hand, seem to have access to extremely ^{15}N depleted N sources, most likely from newly mineralized organic N, perhaps made available through the ericoid mycorrhiza. The high $\delta^{15}N$ in sedges can not be explained by the absence of mycorrhiza, since, for example, *Carex curvula* ($\delta^{15}N$ 0.0 $\pm$ 1.23‰; unpublished data) clearly has endomycorrhiza, whereas *Carex vaginata* ($\delta^{15}N$ −0.50 $\pm$ 0.4‰) studied by Michelsen et al. (1996) was non-mycorrhizal, but both species show similar and comparatively high

$\delta^{15}N$ (for mycorrhiza, see below). It also seems possible that these sedges profit from N_2 fixation at their root surface. Cyanobacteria in the rhizoplane were indeed documented by Nosko et al. (1994) for 10 arctic graminoids (bacterial species isolated were *Clostridium, Desulfovibrio, Klebsiella, Azospirillum*). The separation of free living versus symbiotic bacteria appears artificial in view of such microbe-plant associations, but it still awaits testing as to whether this is a significant direct access route for plants to N.

There also seems to be a distinct association between **age and stability of soils** and $\delta^{15}N$ in plants (Fig. 10.13). Plants growing on very old and deep soil profiles with low pH and high humus accumulation were found to have less negative $\delta^{15}N$ (in the example shown in Fig. 10.13, around −2‰, disregarding legumes and sedges) compared with young, highly disturbed and inorganic substrates (around −7‰; in this case a glacier forefield, released from ice only few decades ago). An explanation of these patterns is difficult. Glacier forefield plants could be expected to live in essence on soluble N in meltwater (atmospheric deposition), whereas late successional vegetation strongly depends on N cycling (accumulation of ^{15}N in the soil). Whatever the reason for such differences, it is obvious that N sources differ a lot in these otherwise similar groups of alpine plant species growing in close proximity at the same elevation.

Because of the rarity of legumes in alpine vegetation and over long periods, di-nitrogen fixation by free cyanobacteria seems to be more important, at least at moderate alpine altitudes. Substantial activities of **free living di-nitrogen fixers** have been documented for wet meadow sites in the Rocky Mountains by Wojciechowski and Heimbrook (1984). However, the significance of N derived from cyanobacteria seems to lie more in the long-term accretion of N compounds in developing soils, rather than in the immediate provision of plant available N. This is because estimated seasonal inputs are very low. The 5 mg N m^{-2} input per season reported by Wojciechowski and Heimbrook (1984) for Niwot Ridge would require 200

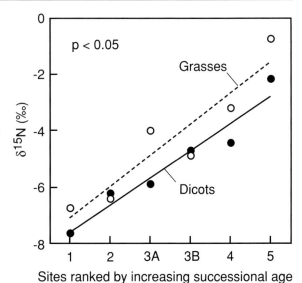

Fig. 10.13. A profile of $\delta^{15}N$ in grasses and non-legume dicots along a series of sites differing in soil age and soil stabilty, from a largely inorganic, highly disturbed substrate in a glacier forefield, **1**, to very old, acid SOM-rich soil, **5**. Sites are seen in Fig. 4.3, with the glacier forefield site in the center and the sites **3A** and **3B** to the left and site **2** to the *right* (sites **4** and **5** not visible, but at the same elevation). (Unpubl.)

years to support one season's phanerogam flush. Very low rates of annual N_2 fixation by free living bacteria were also reported for arctic tundra in interior Alaska by Chapin et al. (1991). Waughman et al. (1981) conclude that less than 5% of the estimated annual plant uptake in the tundra is obtained through biological nitrogen fixation, which includes cyanobacteria activity in associations with **lichens** (e.g. the genera *Peltigera* and *Stereocaulon*) and with mosses, often forming cryptogam crusts (Fig. 10.14). Haselwandter et al. (1983) found no evidence for the presence of cyanobacterial activity near the upper limit of higher plant life in the Alps (3000 m).

Because nitrogen fixation per season is so small, nitrogen supply to alpine plants largely rests on recycling. However, an additional source of soluble N compounds is precipitation, particularly in **meltwater**. Relatively high concentrations of both ammonium and nitrate in meltwater

Fig. 10.14. Cryptogam crusts (3–5 mm thick) formed by algae, tiny and leafy liverworth (here *Nardia breideri, Kiaeria falcata, Anthelia juratzkana*; id. P. Geissler, Genevé) or mosses such as *Polytrichum sexangulare*, lichens and perhaps also cyanobacteria, cover large areas of young and sparsely vegetated alpine soils, particularly in wet and cold places (near Muttenhorn glacier, Fig. 4.2, Swiss Alps 2500 m). The *arrow* indicates a spot with the crust broken off. The $\delta^{15}N$ values for the microscopic bryophytes and the remaining fraction of the organic crust were identical, and surprisingly negative: −6.6‰. This means that bryophytes use the same N source as the pooled other organisms, and that this source is not immediately derived from bacterial dinitrogen fixation but underwent massive ^{15}N discrimination (the N-concentration was 1.2% of dry matter in both components, but mineral dust could not be eliminated completely). (Unpubl.)

were reported by Haselwandter et al. (1983) and Bowman (1992). According to Bowman the annual input through snowmelt on Niwot Ridge varied between 0.1 and 0.6 g N m⁻², depending on snowpack. Present natural sources of soluble atmospheric nitrogen are increasingly supplemented by anthropogenic ones, causing annual rates of soluble **N deposition** to reach 0.5 to 1.4 g N m⁻² in the central Alps (Psenner and Nickus 1986, Graber et al. 1996; see Chap. 17). It is unclear how this is already affecting, or will affect alpine vegetation. One consequence could be a shift from predominantly N to more P-limited situations. The annual soluble **P deposition** by precipitation in the central Alps was found to be only 7 mg m⁻² (Psenner and Nickus 1986), approximately 30 times less the annual requirement for biomass production (assuming a 10:1 tissue ratio of N:P). Hence, even when accounting for such deposition scenarios, P provision remains relatively more dependent on soil processes than N provision.

Mycorrhiza

All known types of mycorrhiza occur in alpine vegetation: ectomycorrhiza (e.g. *Salix, Dryas, Polygonum, Kobresia* sp.), ericoid mycorrhiza (Ericaceae), vesicular-arbuscular (VA) mycorrhiza (most forbs, grasses and some sedges) and even orchid mycorrhiza. Non-mycorrhizal plant species can also be found (Gardes and Dahlberg 1996). A further special category are so-called dark-septate hyphae associations (many species, including *Carex*) which have also been proven to be mutualistic (cf Haselwandter 1987).

Mycorrhization generally declines with increasing altitude, but **VA mycorrhiza** and dark-septate hyphae are found even in the highest rock and scree habitats, though at strongly reduced abundance (Haselwandter 1979; Read and Haselwandter 1981; but see the exception found by Väre et al. 1997; Fig. 10.15). Completely isolated plants above 3000 m in the Alps were largely free of mycorrhiza, but dark-septate hyphal fungi could be seen (data for *Ranunculus glacialis, Cerastium uniflorum* and *Poa laxa*; Read and Haselwandter 1981). An analysis of fine root length and biomass in plants from this site (Körner and Renhardt 1987) revealed a significantly more extensive root system relative to low altitude congeneric species. This may be viewed as compensation for reduced mycorrhization (see Chap. 12).

The ubiquitous presence of **dark-septate hyphae**, even in plants free of "conventional" mycorrhiza suggests some functional significance of these endophytes. By comparing sterile and inoculated individuals of *Carex firma*, Haselwandter

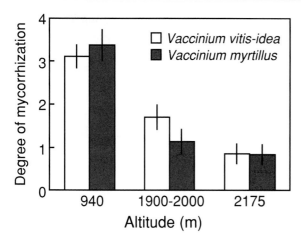

Fig. 10.15. The altitudinal decline of ericoid mycorrhization in *Vaccinium myrtillus* and *Vaccinium vitis-dea* in the Alps. Values are mean glucosamine concentrations of mycorrhizal roots in $\mu g\,mg^{-1}$ root dry matter. The treeline is at ca. 2000 m in this area (Haselwandter 1979)

and Read (1982) demonstrated that growth and phosphorus uptake increased when these hyphae were present, suggesting a symbiotic function. Hence, it is uncertain whether plants at uppermost altitudes become independent of fungi. Given the low plant cover, slow growth, short growing season and the abundance of nutrients in meltwater at such sites, Haselwandter et al. (1983) concluded that demand for symbiotic associations under such life conditions may be absent or weak.

Formation of mycorrhiza is not only species and altitude dependent but is also influenced strongly by local soil conditions. Lesica and Antibus (1986) compared high alpine calcareous and siliceous sites in Wyoming and Montana with less than 60% plant cover and found that all but two out of 32 phanerogam species were mycorrhizal (including two species of *Carex*). Species of *Draba* (Brassicaceae) and *Astragalus* (Fabaceae) were non-mycorrhizal. According to these authors, mycorrhization was more intense on calcareous soils, which may be due to lower availability of phosphate. **Soil specific variation** of mycorrhization was also observed by Barnola and Montilla (1997) who found less mycorrhizal infection (exclusively VA) in mixed root samples from

poorly drained and fertile places, compared with well drained and less fertile slope sites at 3800 m altitude in the Venezuelan Andes. These authors also report that the degree of mycorrhization was highest in the upper 5 cm of the profile with only half as many infections in roots at 10–30 cm depth. They found one *Carex* species to be without mycorrhiza, while another Cyperaceae species (*Eleocharis*) formed mycorrhiza.

Taken together, these observations indicate that mycorrhiza is a common element of alpine plant life, but that it is more prominent in lower alpine altitudes and in infertile soils, becoming rare only in isolated plants on high mountain peaks, where soils are largely inorganic. There is no rule about non-mycorrhizal plant families but, as at low altitude, non-mycorrhizal species often belong to Cyperaceae and Brassicaceae, with no obvious drawback for such species.

While the studies reported above provide a relatively good picture of the presence of mycorrhiza in alpine vegetation, much less is known about **mycorrhizal function**. What we know is correlative. For instance, Mullen and Schmidt (1993) found a temporal relationship between the development of arbuscules in the snowbed species *Ranunculus adoneus* and subsequent plant phosphorus accumulation (Fig. 10.16). The evidence provided above about nitrogen isotope abundance also hints at a specific function of mycorrhiza in N acquisition (Michelsen et al. 1996). Furthermore, the pot experiments of Haselwandter and Read (1982) with alpine *Carex* species clearly showed that phosphate supply can be improved by the presence of endophytic dark-septate hyphae. When alpine plants received elevated CO_2 for four consecutive seasons they did not grow more, but non-structural carbohydrates in tissues were significantly increased, suggesting a carbon surplus (Schäppi and Körner 1997). However, the degree of mycorrhization in the dominant species, *Carex curvula*, was not significantly altered (R. Schertler, pers. comm.; see Körner et al. 1997). Hence, the degree of mycorrhization does not appear to depend on the carbon supply status of the host, at least not in the case of this late successional sedge.

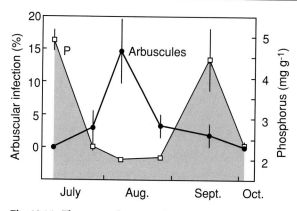

Fig. 10.16. The seasonal course of mycorrhizal arbuscule development and shoot concentrations of phosphorus in the Rocky Mountains snowbed species *Ranunculus adoneus*. (Mullen and Schmidt 1993)

Responses of vegetation to variable nutrient supply

What happens to alpine plants when nutrient availability is increased? This may occur naturally by meltwater flows from late snowfields or by animal dung. Effects of nutrient addition are not always as dramatic as illustrated by Figures 10.1 and 10.2. In the short term, responses range from zero to large stimulation of growth, and consequences of prolonged **fertilization** may still occur several decades later (Hegg et al. 1992, who revisited the site of the first scientific fertilizer trial in alpine grassland by Lüdi 1936). The message from Lüdi's classic experiment in *Nardus* grassland at 2000 m altitude (near the climatic treeline in the front ranges of the Alps in Switzerland), is that the regular annual addition of NPK fertilizer (approximately 40 kg N ha^{-1} applied as ammonium sulfate) completely changed the community structure and resulted in a seven-fold increase in biomass (useable yield without stubble) reaching 350 g dry matter m^{-2}. The gain was largely due to a stimulation of a few grass species. A number of species like *Vaccinium vitis idea* and *Gentiana kochiana* disappeared. Fertilization of Lüdi's plots was repeated between 1946 and 1950 and leaf nutrient

concentrations (N and P) were still enhanced 30 years later (Hegg et al. 1992).

Even most barren sites, such as a glacier forefield with very sparse pioneer vegetation (center of Fig. 4.3), show a massive fertilizer response. Within two seasons, the addition of 100 kg N m^{-2} per season created lush *Poa alpina* dominated "meadows", green patches which could be spotted from a great distance (Fig. 10.17).

From unpublished nitrogen fertilizer trials at higher altitudes in the Medicine Bow Mountains (Wyoming), Scott and Billings (1964) concluded that nitrogen availability was not limiting plant growth. Similarly Diemer (1992) found no effects of fertilizer addition on population dynamics in *Ranunculus glacialis* at 2600 m in the Alps. Morecroft and Woodward (1996) report that Scottish alpine *Alchemilla* showed no growth stimulation when 100 kg ha^{-1} ammonium nitrate N was added. These observations are surprising, even if one accounts for rapid losses of fertilizer by nutrient leaching in these rather wet environments. When provided with the equivalent of 40 kg N ha^{-1} a^{-1} as liquid fertilizer in small doses over the first half of the alpine growing season, a doubling of aboveground biomass production occurred in a sedge community in the Swiss central Alps at 2500 m by year 2, and persisted into year 4 of treatment (no change in below-ground biomass; Körner et al. 1997). Such rates of wet N deposition are currently observed in many parts of lowland Europe.

A pronounced stimulation of biomass production of a similar sedge community in the Austrian central Alps (2300 m; treeline at 2000 m) was observed by Haunold et al. (1980; 100 kg N ha^{-1} as ammonium N). Their ^{15}N trial resulted in a unique data set on the **uptake dynamics of N fertilizer**. ^{15}N was applied once, early in the season of year 1. Results showed that (1) significant amounts of ^{15}N in plants are found only during the year of treatment (13–20% of total N applied recovered by plants). (2) Uptake of ^{15}N dropped to 5% in the following growing season, and to 0.5% in the 3rd year. (3) Hardly any labeled N was found in soil horizons below 30 cm depth (1%), and 60–70% (year 1) of the total label were located in the top

Fig. 10.17. The visible effect of fertilizer addition on a barren glacier forefield after 2 years of treatment (plot size 1 m²; site in the center of Fig. 4.3; Heer and Körner 2002)

1–3 cm of the dense root felt (predominantly in organic form). By year 2, 50% of all label was immobilized in this layer. More than 70% of all applied N remained in the soil until year 3. (4) Only 1% of the total label was extractable (soluble) from the top 3 cm 49 days after application, and the fraction was reduced to 0.2–0.5% in the following year. (5) Concentrations of free soil nitrate remained negligible. (6) Parallel lysimeter studies confirmed close to zero leaching of applied N. This is a most convincing demonstration of the potential N binding capacity of alpine soils and may in part explain some of the contradictory results mentioned above. In 2002 nearly half of all ^{15}N applied in 1974 was still present in the top soil (M Gerzabek, G Haberhauer, E Haunold, pers. commun.).

Further explanations for different results of fertilizer experiments emerged from various trials of Bowman and co-workers on Niwot Ridge. These authors showed that **responses to fertilizer are species, life form and habitat (soil) specific,** and may take more than a year to become apparent (Bowman et al. 1993, 1995; Bowman 1994; Bowman and Conant 1994; Theodose and Bowman 1997). Increased nitrogen supply, alone (their dry site) or in combination with P (wet site), but not P alone, stimulated growth (160 kg N ha^{-1}, supplied liquid in bi-weekly portions). Just as in Lüdi's experiment, grasses responded more than forbs, and sedges (*Kobresia*) not at all. Among forbs, only the non-mycorrhizal ones (species of *Thlaspi, Draba, Saxifraga, Sedum*) showed a positve response to P addition. Also, since graminoids were more abundant on the well drained ridge they studied, the ridge community as a whole responded more in relative terms (+70 to +100% depending on parameter) than did a wet meadow on flat terrain. In no case was the community response the result of physiological changes at the individual shoot or the leaf level (e.g. shoot mass or rates of leaf CO$_2$-exchange). Also, dwarf willows, supplied with 250 kg N ha^{-1} slow release fertilizer pellets, responded more on the ridge than in flat terrain, but the majority of responses tested were smaller than +20%, similar to what a 5 year fertilizer trial in alpine ericaceous dwarf shrubs yielded (Körner 1984). Seastedt and

Vaccaro (2001) repeated Bowman's fertilizer trials across a snow pack gradient in the same area. Although statistical power was very low, there was a trend to enhanced N effects (100 to 200 kg N ha^{-1} a^{-1} over 4 years) on biomass in sites with prolonged snow duration, but the gain was surprisingly small given the agronomic fertilizer dose. P addition alone (10 to 20 kg P ha^{-1} a^{-1}) had similar effects, and N and P combined produced a doubled biomass, also when snow duration was increased by a snow fence, but not in a natural snow bed site (zero response). N addition, but clearly not P addition, led to a reduced species richness, although both N and P fertilization caused similar biomass increments. It seems, therefore that N addition can have negative effects on some species, irrespective of the communities' biomass response (see Chap. 17).

A recent fertilizer trial in arctic-alpine vegetation in north Sweden (U Molau, J Alatalo and M Weih, pers. comm.) showed significant stimulations of grasses by year 2, and of deciduous shrubs by year 3, with evergreens becoming outcompeted, again suggesting a loss of diversity.

In nature, available nutrients are often not uniformly spread over the rooted soil matrix, but have a patchy distribution (e.g. bird droppings, dead grasshoppers). Are alpine plants able to "sense" such patches and respond rapidly enough before the patch disappears through seepage or absorption by microbes? Because I found no answer to this important question in the literature, I present results of a small pilot study conducted in 1997 in Fig. 10.18. This widespread late successional alpine sedge responded rapidly and explored the small pockets of nutrient enriched sand with six times more new fine roots than were found in sand-only patches. This extremely slow growing species is not at all slow when "food" is provided! As was mentioned above, *Carex curvula* responded by almost doubling its aboveground biomass when the whole stand was fertilized, but root biomass remained unaffected when nutrients where, so to speak, everywhere. Responses to nutrient hot spots in an otherwise nutrient poor matrix are obviously quite different.

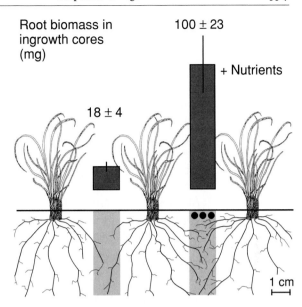

Fig. 10.18. Alpine plants forage for mineral nutrients. Nutrient rich "patches", installed in stands of the late successional alpine sedge *Carex curvula*, are rapidly explored by roots within one 10-week season (15 mm cores of 45 mm depth filled with acid sand and topped with slow release full fertilizer pellets at the onset of the preceding winter). The new root mass contained in the nutrient rich patches (six times the amount found in controls) corresponds to 550 g m^{-2}. Swiss central Alps, 2500 m. (Unpubl.)

By applying small amounts of ^{15}N labeled fertilizer to alpine communities and studying the **species specific capture** of ^{15}N, Theodose et al. (1996) demonstrated that tissue N concentration (and ^{15}N incorporation) was not an important predictor of the relative abundance of a species. In fact, ^{15}N uptake rate was negatively correlated with percent cover of a species. Their results suggest that high resource uptake rates were not indicative of competitive ability, but may instead be a mechanism by which rare alpine species are able to coexist with competitive dominants.

Supplies of nutrients in most of these fertilizer experiments were massive, and not likely to occur in nature, but they illustrate a potential maximum range of responses to doses of N which, in the long run, most likely exceed the immobilization capacity of soil microbial communities. Similarly pronounced species specificity of nutrient responses, a strong response of grasses in particular, were reported earlier for low altitude arctic tundra (Shaver and Chapin 1986).

Responses of individual alpine plants to fertilization under controled conditions in greenhouses or growth chambers revealed more pronounced responses in early **successional**, faster growing species than in late successional, slower growing species (Scott and Billings 1964; Chambers et al. 1987; Atkin and Cummins 1994). All six Australian alpine species tested in sand cultures by Atkin and Collier (1992) with six concentrations of a complete fertilizer solution grew more, the higher the concentration, but shoot nitrogen concentrations did not change (except for two species at the highest concentration). In other words, growth occurred such that levels of tissue nitrogen (protein) remained constant.

Overall, these and the community responses described above suggest that alpine plants, in most cases, exhibit growth responses to fertilizer in a similar way to wild lowland plants. Similarly, Chapin (1987) concluded that N and/or P addition promoted plant growth in every arctic community thus far examined. In general, grasses are the most responsive group to fertilization, followed by forbs, deciduous shrubs and evergreen shrubs.

A fertilizer-induced stimulation of growth of alpine plants by no means indicates improved **fitness**. Even species not immediately outcompeted by others after fertilization may be affected very negatively (Körner 1984; Fig. 10.19). For example, fertilized alpine *Vaccinium myrtillus* was found to sprout 10 days earlier than non-fertilized plants (increased late frost risk), and entered winter with green, immature buds (increased early frost risk). Fertilized *Loiseleuria procumbens*, the leaves and internodes of which massively increased in size, was progressively killed by snow mold between the third and the fifth winter after the treatment began. Hence, the concept of nutrient limitation needs to account for fitness in a broader sense, including **stress resistance** which allows persistence in alpine environments.

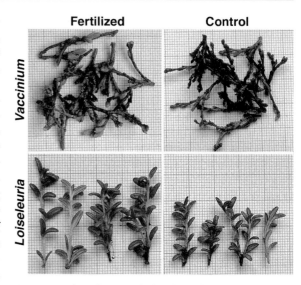

Fig. 10.19. The influence of a low dose of NPK fertilizer on alpine dwarf shrubs after 5 years of treatment. In *Vaccinium myrtillus*, phenology (25 May, 1982) is affected (note the earlier leaf appearance in fertilized shoots, *left*), the slow growing wind-edge species *Loiseleuria procumbens* produces leaves almost twice as big as normal, but becomes very sensitive to fungal infections. (Mt. Patscherkofel, 2000 m, Innsbruck, Austria)

While nitrogen and phosphate can play a key role in determining alpine plant growth, alpine plant distribution is strongly influenced by the abundance of **calcium**. A vast literature exists on how species abundance and distribution are influenced by the presence or absence of calcareous or dolomite rocks. Species are often grouped into "**calcicole**" (found on Ca-rich substrate) and "**calcifuge**" (not found on Ca-rich substrate), with the latter also containing more potassium in their tissue than the former (e.g. Passama et al. 1975; Tosca and Labroue 1981). However, for most species, site preferences are not that strict and for many "calcicole" species very little Ca, perhaps from dust deposits from distant Ca-rich substrates, may be sufficient (e.g. *Dryas octopetala*). Even classic calcifuge (*Rhododendron ferrugineum*) and calcicole (*Rhododendron hirsutum*) species in the Alps may occur next to each other on calcareous rock overlaid by acid rendzina soils.

Under culture conditions, NPK fertilization often eliminates the substrate specificity. Yet, substantial ecotypic differentiation exists and even subspecies occurring strictly on a particular substrate type have been discribed (e.g. Erschbamer 1990,1996). The addition of Ca to acid alpine grassland can also increase growth and lead to massive species replacement (Lüdi 1936; Hegg et al. 1992).

In **summary**, in this chapter I have attempted to demonstrate that alpine plants commonly have tissues with equally high (or even higher) nutrient concentrations than similar plant types from low altitudes. When provided with additional nutrients (more nitrogen in particular) over several years, most alpine species grow more, but peak season tissue concentrations often change little. However, for comparing tissue nutrient (N) status, dry matter is a problematic reference because the presence of more or less carbon compounds does not necessarily indicate that cells have less or more nutrients for their metabolic demand. Similar nitrogen, but varying carbon investments per unit leaf area across the globe's alpine regions, suggest that growth controls leaf nitrogen levels or vice versa, which ensures similarly high metabolic activities per leaf area over a wide range of nitro-gen supply situations, including conditions of very poor supply (Körner 1989b). On average, it is the amount of carbon rather than the amount of N that varies per unit leaf area with altitude and latitude. Biological nitrogen fixation contributes very little to the annual nitrogen demand of alpine vegetation, and plants depend heavily on recycling of nutrients. However, N fixation contributes significantly to the long-term buildup of the ecosystem's N capital. Stable N isotope studies indicate that alpine plants use a variety of different soil N sources, but mechanisms are hardly understood. In seasonal climates of the temperate and subpolar zone, meltwater is an (increasingly) important source of soluble mineral nutrients. Key components of an understanding of mineral nutrition of alpine plants are internal nutrient cycling, turnover times of live and dead tissues (roots in particular), and the largely unknown roles that free soil microbes and mycorrhiza play in plant nutrient supply in cold climates (Jonasson et al. 1995). N addition reduces species richness not only via a general increase of plant–plant competition, but via species-specific (below-ground) effects not yet understood.

11   Uptake and loss of carbon

Plant dry matter production is the net result of uptake and loss of CO_2, the two processes considered in this chapter. The photosynthetic uptake of CO_2 by a plant depends on the specific activity of its leaves and the total amount (mass or area) of leaves present per total plant mass. Leaf photosynthesis depends on three processes: (1) the diffusion of CO_2 from the free atmosphere to the site of carboxylation in chloroplasts, (2) the biochemical reduction of CO_2 and formation of sugar through photosynthesis, and (3) the removal of carbon assimilates from the site of synthesis. Each of these depends again on a number of internal and external determinants. The diffusion and the biochemical fixation of CO_2 are usually well coordinated so that the plant's investment in photosynthetic machinery per unit leaf area is efficiently utilized and diffusional constraints are kept small. The third process includes both transport and demand problems. If there is no demand for carbon assimilates (no active sinks), neither transport nor photosynthetic limitations matter because end products will soon inhibit photosynthesis.

The demand for carbon assimilates in the plant body is controlled by three main factors: (1) the developmental stage of a plant (i.e. its readiness to grow, which varies during the year), (2) the availability and activity of structural, reproductive, storage or metabolic (respiratory) sinks for C compounds, and (3) the availability of other essential resources (mineral nutrients, water) and environmental conditions (e.g. temperature) which co-determine sink activity (i.e. the biochemical processing of primary assimilates into more complex components or process energy.

The loss of CO_2 through various respiratory processes (for maintenance, growth and nutrient uptake) depends on the age, quality and activity of tissues and the major external driver is temperature. Just as with CO_2 uptake, it is not only the tissue specific rate of respiration, but also the amount of various types of respiring tissues per total plant mass that determines the overall loss.

Thus, neither the rate of CO_2 uptake per unit leaf area, nor the respiratory losses of a specific tissue per se are useful for predicting plant growth. Growth involves a multitude of additional determinants, most importantly the fractionation of assimilates into autotrophic and heterotrophic structures (a plant's strategy of dry matter investment), which will be discussed in Chapter 12. This list of elementary controls of a plant's CO_2 exchange also illustrates that a multitude of factors could become critical for the net carbon uptake of whole plants, and the biochemical control of leaf photosynthesis is one of these, in many cases not the critical one. It is important to keep this wider view in mind, while leaf photosynthesis of alpine plants is first considered in isolation.

Photosynthetic capacity of alpine plants

For more than a century it has been known that leaves of alpine plants are not inferior to their lowland relatives as far as **photosynthetic capacity** (A_{cap}) is concerned. A_{cap} is defined here in the sense of Larcher (1969) as the highest rate of photosynthesis (A) that can be measured under optimum temperature, light and moisture conditions and normal ambient CO_2 concentration in fully mature (non-senescent) leaves. During his work in the Alps near Chamonix and in the Pyrenees, Bonnier (1890b, 1895) was possibly the first who noted with surprise how much CO_2 alpine plants can capture, compared with low altitude plants. The leaf-anatomical basis for this – a greater mesophyll thickness – was documented for a large number of species by Wagner in 1892. The first detailed analyses of in situ alpine leaf photosynthesis were those by Henrici (1918, 1921) of Basel and by Cartellieri (1940) of Innsbruck. It took another 20 years until their results were confirmed by modern infrared gas analysis (for reviews see Pisek 1960; Billings and Mooney 1968; Friend and Woodward 1990).

In situ photosynthetic capacity in perennial species of alpine altitudes varies (in $\mu mol\, m^{-2}\, s^{-1}$) between 3 and 30, and reflects leaf morphotype, leaf longevity and microhabitat conditions (Fig. 11.1). Most herbaceous plants have rates between 12 and 22 (mean of 16), dwarf shrubs range from 4 to 14, tropical giant rosettes from 3 to 8, dwarf rosettes of the important arctic-alpine genus *Saxifraga* from 3 to 5. Maximum rates in CAM plants are below 3 (but see the specific section). A_{cap} of lichens is commonly below 1. Annual plants, which are rare at alpine altitudes, and confined to warm microhabitats, exhibit similar A_{cap} as perennial forbs (12 to 17 $\mu mol\, m^{-2}\, s^{-1}$, when measured in potted plants transferred to a low altitude growth chamber; Reynolds 1984).

The data set for the Alps by Körner and Diemer (1987) was collected at exactly the same place where

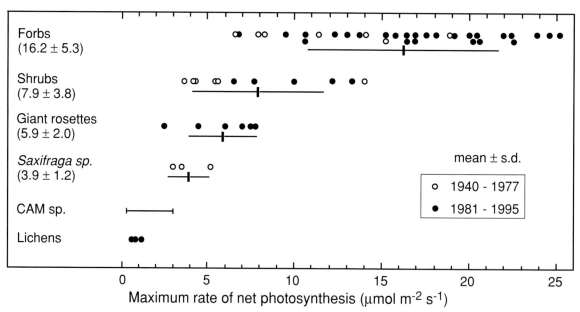

Fig. 11.1. In situ photosynthetic capacity in alpine plant species as measured under local pressure conditions at altitudes between 2000 and 4300 m (with a few exceptions, between 2600 and 3600 m). Authors (sorted from top to bottom of data): Cartellieri (1940), Billings et al. (1966), Moser et al. (1977), Chapin and Oechel (1983), Abdaladze (1987), Körner and Diemer (1987), Rawat and Purohit (1991), Terashima et al. (1993), Bowman et al. (1995), Larcher (1977, data from four dwarf shrub species measured in field grown shoots in the laboratory at 600 m were adjusted by −10% to account for a mix of short-term adjustment and pressure difference), Bowman and Conant (1994), Nakano and Ishida (1994), Schulze et al. (1985), Goldstein et al. (1989), Moser et al. (1977), Billings et al. (1966), Wagner and Larcher (1981, SLA assumed to be 1), Sonesson (1986), Larcher and Vareschi (1988)

Cartellieri (1940) worked in the 1930s with his titration machine and the two water bottles connected by a rubber tube, which had to be lifted alternately to create a known gas flow through the leaf cuvette. For *Ranunculus glacialis* and *Geum reptans*, Cartellieri reported 20.8 and 14 µmol m^{-2}s^{-1} for A$_{cap}$, which compares surprisingly well with the 19.1 and 13.7 µmol m^{-2}s^{-1} measured in the same species in the mid 1980s with electronic mass flow controllers and infrared gas analysis.

A comparison of photosynthesis in alpine plants with that in lowland plants needs to account for the difference of partial pressure of CO_2. This is also crucial for the on-site calibration of gas analysers (hence a lot of doubtful data in the literature). For the lower 5 km of the atmosphere, total **atmospheric pressure,** and with it the partial pressures of all atmospheric gases decreases by approximately 10% for every 1000 m of altitude (for details see Jones 1992). For example, at 2600 m where most of the data from the Alps were collected, the pressure is ca. 21% less than at 600 m altitude, the low altitude reference location. This must not be confused with the mixing ratio of atmospheric gases, which does not significantly change with altitude in the 5 km range relevant here (for CO_2 see Zumbrunn et al. 1983; Körner and Diemer 1987; Matson and Harriss 1988; deviations of a few ppm are observed at the vegetated planetary boundary and locally where fossil CO_2 is emitted). A suite of (dimensionless) units are in use for describing CO_2/"air" mixing ratios, but for the practical purposes relevant here, they are equally accurate and numerically identical (ppm, µl l^{-1}, µbar bar^{-1}, Pa MPa^{-1}, µmol mol^{-1}). In 1997, the mixing ratio of CO_2 in air was ca. 360 ppm, which corresponds to a partial pressure of 360 µbar (36 Pa) at sea level, 340 µbar at 600 m and 270 µbar at 2600 m altitude (which, by coincidence, corresponds to the pre-industrial partial pressure of CO_2 at sea level).

A negative effect of reduced **partial pressure of CO_2** on photosynthesis at high altitudes has often been postulated (e.g. Decker 1959; Billings et al. 1961; Mooney et al. 1966; Halloy 1981). However, decreasing partial pressure of CO_2 does not occur in isolation but is accompanied by other pressure related changes, and its direct influence on plants may partially be diminished or enhanced by three factors: (1) the **oxygen partial pressure** decreases as well, hence **photorespiration**, i.e. the oxygenase activity of the CO_2-binding enzyme "Rubisco" becomes reduced, and (2) "thinner" air allows CO_2 molecules to diffuse faster through stomata and the intercellular spaces in the leaf (see discussion in Körner and Diemer 1987; Körner et al. 1991); (3) air temperature drops and so does leaf temperature for much of the time, which counteracts (2) and enhances (1). The two diminishing factors (1 and 2) cannot balance the effect of declining CO_2 partial pressure, because they are not as influential on photosynthesis as CO_2-partial pressure is in these ranges of concentration.

When stomata are open and photosynthesis reaches a maximum, constraints by **gas diffusion** limit photosynthesis by only one fifth compared with four fifth of the total CO_2 uptake resistance contributed by intracellular – largely biochemical – processes (Körner et al. 1979), thus enhanced gas diffusion can never fully compensate the pressure effect on photosynthesis (as was suggested by Gale 1972). If extrapolated to the extremes, this assumption would suggest a maintenance of photosynthetic rates while pressure approaches zero. On the other hand, low temperatures suppress photorespiration more than carboxylation and the overall outcome of all these interactions depends on the actual microclimate and is difficult to predict (Terashima et al. 1995).

Terashima et al. (1995) modeled the interaction of these pressure/altitude related influences on photosynthesis and – by adopting various assumptions – arrived at the conclusion that the reduction of photosynthesis by reduced pressure is at most 23% for 3000 m, compared with the theoretical ca. 29% reduction by the actual drop of partial pressure of CO_2. Hence, the sum of the counterbalancing processes accounts for a ca. 6% reduction of the negative pressure effect under assumed warm canopy conditions, and ca. 10% at rather cool temperatures. As was shown in Chapter 4, temperatures are not always lower in short-statured alpine plants than in low altitude plants, particularly not when high radiation permits maximal photosynthesis.

In summary, the partial pressure of CO_2 at the surface of mesophyll cells will be progressively reduced with increasing altitude and alpine plants do have to cope with a (though not pressure proportional) diminished availability of CO_2, just as animals and humans have to cope with reduced oxygen pressure on high mountains. At the upper limit of higher plant life (see Chap. 2), the ambient CO_2 pressure is only half that at sea level. Whether and how this might affect actual leaf photosynthesis will be discussed below.

It is worth noting here, that **animals** can commonly compensate for low oxygen partial pressure to some extent by enhanced ventilation, but this is not true in all cases. For example, bird embryos inside **eggshells** (an analog to the porous leaf epidermis) depend on molecular diffusion just as leaves do. Interestingly, shells have been found to become more porous as the altitude of nesting habitats increases, which was explained as a diffusional adaptation to reduced oxygen partial pressure (Rahn et al. 1977; Rahn and Paganelli 1982). If this is a causal rather than coincidental link to pressure, there would be no reason for such adjustment, if enhanced molecular diffusion did fully compensate the drop in partial pressure of O_2.

A screening for **photosynthetic responses to CO_2** of over 20 comparable lowland and alpine herbaceous species revealed that alpine species have a similar mean A_{cap} when measured at their local CO_2 partial pressure. When measured in situ at equal partial pressure (in gas controlled leaf chambers, see color Plate 4 at the end of the book), alpine plants were almost always superior (mean A_{cap} +20% at the 2600 m experimental site; Körner and Diemer 1987, Körner and Pelaez-Menendez Riedl 1989; Fig. 11.2), a trend already noted by Henrici (1918) from comparing rates among her alpine and lowland transplants. Little or no short-term acclimation of photosynthesis to altered CO_2 partial pressure had been observed in manipulative experiments (Mooney et al. 1966; Körner and Diemer 1994). Decker (1959) grew clones of sub-

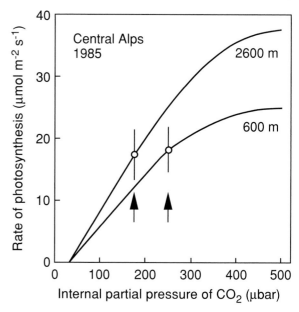

Fig. 11.2. Summary of responses of the rate of photosynthesis to intercellular partial pressure of CO_2 in groups of alpine (2600 m) and lowland (600 m) herbaceous plant species (see color Plate 4e,f at the end of the book). Arrows point at rates of photosynthersis (and SE) at the respective local intercellular partial pressures. (From data in Körner and Diemer 1987; Körner and Pelaez-Menendez Riedl 1989)

alpine and lowland *Mimulus* in a low altitude greenhouse and did not detect origin specific CO_2-responses.

Equal A at lower CO_2-pressure or higher A at equal CO_2-pressure in alpine as compared with lowland plants indicates that their specific capacity to fix CO_2 must be higher than in lowland species. What explains this higher photosynthetic **efficiency of CO_2 utilization** (ECU) per unit leaf area? The primary reason is the construction of the leaf mesophyll. On average alpine forbs produce one additional layer of palisade cells (2–3 as compared with 1–2 at low altitude) and thus, have thicker leaves (Fig. 11.3, Wagner 1892; Körner et al. 1989a). This would already suffice to facilitate higher A_{cap} if cells were equally well equipped with photosynthetic enzymes. However, as was shown in Chapter 10, alpine plants also tend to have

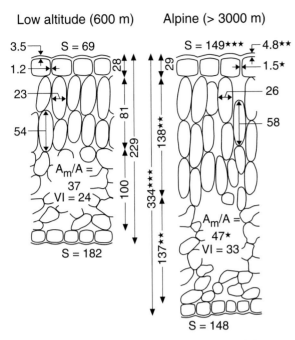

Fig. 11.3. Summary of quantitative leaf anatomy of alpine and lowland taxa of herbaceous plants in the Alps. *S* Stomatal density (n mm^{-2}; see also Chap. 9), *VI* intercellular air volume (% of total leaf volume), *Am/A* ratio of mesophyll cell surface area exposed to intercellular spaces/projected leaf area (m^2 m^{-2}), all other dimensions in μm. Means for 18 to 23 species from 580–700 m altitude and 24 to 27 species from 2650–3250 (mostly 3000) m altitude. *Asterisks* in the high altitude diagram indicate statistical significance of altitudinal differences; *no asterisk* means not significant. (Körner et al. 1989a)

higher concentrations of N per unit dry matter than lowland plants. Leaf N is effectively proportional to leaf protein of which at least half is involved in photosynthesis. The combination of greater **mesophyll thickness** and higher tissue N concentration, causes the amount of protein per unit leaf area to increase with altitude even more than per unit dry mass (see also Chap. 10).

While greater leaf thickness plus greater leaf protein concentration are plausible explanations of higher rates of photosynthesis at an equal partial pressure of CO_2, these differences must not be seen as a direct response to the elevational reduction in CO_2 pressure. There are other poten-

tial reasons for such changes in leaf characteristics. Surprised by the similar rates of assimilation of his high and low altitude *Mimulus* clones exposed to elevated CO_2, Decker (1959) concluded that he had missed the "right" climate in his greenhouse for inducing altitude specific responses. Low temperature alone tends to induce greater leaf thickness and to increase leaf N concentration, independently of altitude (Körner et al. 1989a; Morecroft and Woodward 1996; Pyankov et al. 1999; see Chaps. 10 and 12), and the slightly higher frequency of hours with very high solar radiation in some mountains, plus reduced mutual shading in canopies both favor greater leaf thickness (see discussions in Körner et al. 1983, 1991). The higher leaf diffusive conductances (and stomatal densities) at high altitudes (Chap. 9), which parallel trends in photosynthesis (but see below), may also have to do with greater leaf thickness, because there is a greater amount of photosynthesizing tissue to be supplied.

Does **stomatal conductance** increase with the altitudinal increase in photosynthetic machinery per unit leaf area so that the ratio between intercellular versus ambient CO_2 pressure remains constant? Gas exchange analysis says no (Körner and Diemer 1987). As can be calculated from parallel measurements of leaf transpiration and CO_2 assimilation under fully controled atmospheric conditions in a special leaf cuvette (see color Plate 4 at the end of the book), the CO_2 partial pressure at the cell surface drops more with altitude than the external partial pressure. This means that the balance of the contribution to total uptake resistance of CO_2 is shifted towards greater diffusional limitation. The biochemical limitation is reduced, hence the mesophyll itself becomes a more efficient CO_2 sink, its demand for CO_2 exceeds supply rates and the **intercellular partial pressure** drops faster with altitude than the ambient one. The greater efficiency of CO_2 uptake (ECU) finds expression in a steeper initial slope of the plot of photosynthesis versus intercellular CO_2 partial pressure (such curves are obtained by supplying leaves stepwise with different CO_2-concentrations; Fig. 11.2).

The observations summarized above are from the central Alps and may represent a very local phenomenon. Since such measurements are very laborious and require the operation of high-tech equipment in remote places, it is difficult to obtain sufficient replication on a global scale for testing whether these patterns represent a globally valid principle. Fortunately, a very elegant indirect method exists for a large scale extrapolation – the use of natural **carbon isotope** discrimination.

Confirmed theory tells us that CO_2 containing the heavier ^{13}C isotope (ca. 1% of all CO_2 in the atmosphere) diffuses slightly slower than $^{12}CO_2$, hence arrives with a 4.4‰ decreased probability at the wet surface of cells. Rubisco has been shown to discriminate further against $^{13}CO_2$ (27‰ lower chance of binding). Re-fixation of respired CO_2 within the leaf also contributes to the effect. As a consequence, plant tissue contains less ^{13}C than is found in the atmosphere and this is conventionally expressed as a sample's $\delta^{13}C$ value, which is always negative. Compared with a limestone standard, the atmosphere is also ^{13}C depleted, and its $\delta^{13}C$ is ca. −8‰. Since diffusion discriminates less than carboxylation, the actual $\delta^{13}C$ value of a tissue reflects the balance between these two major discriminating processes. $\delta^{13}C$ of photo-assimilates in a leaf with no epidermis (no diffusional barrier) would reflect the pure biochemical effect, and thus would be very "negative" (ca. −35‰). A leaf with only diffusional discrimination and no biochemical discrimination (as is the case in plants using the C4-pathway) would have much less negative values (ca. −12‰, i.e. only ca. 4‰ less than in free air). Equations have been developed by Farquhar and Richards (1984) which permit the calculation of the degree of intercellular depletion of CO_2 partial pressure from tissue $\delta^{13}C$ which has been determined in a mass spectrometer.

Applying this concept to alpine plants from many parts of the world did indeed confirm a general validity of the above mentioned gas exchange data in the Alps. Plants from high altitudes discriminate less against ^{13}C as long as comparisons exclude moisture gradients (Fig. 11.4).

Despite increasing stomatal conductance with altitude, stomata appear to become more limiting in relative terms as compared with carboxylation (thereby also their relative influence on overall ^{13}C discrimination increases), because ECU increases even more. On average, $\delta^{13}C$ increases (becomes less negative) by 1.2‰ per 1000 m of altitude. Thus, high elevation plants do fix carbon more efficiently per unit leaf area than lowland plants throughout the world's humid mountain ranges.

It is essential for such studies (though often confused in the literature) that no **drought** gradient is associated with altitude, because drought induced reductions of stomatal conductance would create a similar effect. While it was explicitly stated that the data presented in Fig. 11.4 are for regions without significant moisture shortage (if there was one, it was at low altitude, which would rather reverse the trend), the "alpine drought" argument has been stressed as a possible explanation for the trend in Fig. 11.4 (Terashima et al. 1995; see chapter 9 for alpine plant water relations), but there is no substance to support this argument (see also Chap. 9).

Another alternative explanation is **plant stature**. Friend and Woodward (1990) have plau-

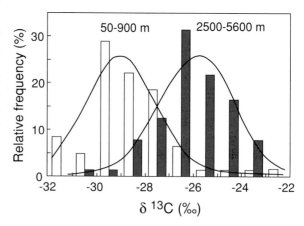

Fig. 11.4. The global frequency distribution of $\delta^{13}C$ in two altitudinally distinct groups of plant species of diverse life forms. *Left* Samples from altitudes between 50 to 900 m, mostly lower than 300 m; *right*, all samples collected at altitudes above 2500 m (up to 5600 m). (Körner et al. 1988)

sibly argued that the altitudinal increase in $\delta^{13}C$ could result from increasing aerodynamic resistance to gas diffusion (adding to stomatal resistance) as plant canopies become shorter and more compact. Data available to date do not support this hypothesis either. Upright, aerodynamically well coupled high altitude species of shrubs, tall forbs, and treeline trees all match the overall pattern (Körner et al. 1988). A comparison of protruding, tall versus prostrate, extremely compact species revealed no difference in $\delta^{13}C$ (Table 11.1).

One could also turn the argument around and assume that, because of greater aerodynamic resistance in low stature plant canopies, CO_2 diffusion from the soil to the free atmosphere should be slower. As a consequence, heavily ^{13}C depleted respiratory CO_2 could contribute more to photosynthesis than in vegetation with less dense leaf canopies at low elevation (more negative $\delta^{13}C$ in such compact life forms). Surprisingly, this effect seems to be very small. In leaves from the core of the most compact plant cushions found in the Alps, $\delta^{13}C$ was only −0.76‰ (±1‰, n = 5 species) lower than in adjacent fully exposed leaves of these individuals. This suggests that soil derived CO_2 contributes less than a mean of 5% to CO_2 assimilation of this special life form (unpublished data). In less compact life forms the contribution will be negligible, and hence $\delta^{13}C$ will be unaffected.

The most plausible explanation of the global altitudinal trend in $\delta^{13}C$ is the combined influence of enhanced efficiency of CO_2 fixation and increased CO_2 transport barriers somewhere inside the leaf at high elevation, as underpinned by Fig. 11.2, the work of Hikosaka et al. (2002) on Kinabalu (Borneo) and Kogami et al. (2001) on Mt. Fuji (Japan). From a comparison of altitudinal and latitudinal transects of tissue $\delta^{13}C$ by Körner et al. (1991) and from growth chamber experiments with *Nardus*, *Vaccinium* and *Alchemilla* (Friend et al. 1989; Morecroft and Woodward 1990, 1996), it became apparent that effects of low temperature play an important role. Radiation has no influence, since data for New Guinea, which were included in this survey, did not differ from other regions, despite the fact that solar radiation drops to one third of lowland values at alpine altitudes in this region. This data set was later complemented and confirmed by soil carbon analysis by Bird et al. (1994). Humus $\delta^{13}C$ represents a long-term and spatial integral of ^{13}C discrimination by plants. As illustrated by Fig. 11.5, soil and plant data for C3-vegetation in New Guinea show remarkably similar trends, adding weight to the suggestion that the worldwide elevational increase in $\delta^{13}C$ is unrelated to the plant's radiation regime.

There is substantial scatter in the plant data shown in Figures 11.4 and 11.5. This is very typical and reflects strong species-specific variations in $\delta^{13}C$. Körner et al. (1991) concluded that a range of 4‰ is to be expected within a single community of C3 plant species. In fact, the range among the 18

Table 11.1. A comparison of $\delta^{13}C$ in alpine plants from 2500 m altitude in Alps, selected by plant stature. Note that there are no differences between prostrate and taller species (R. Siegwolf and Ch. Körner, unpubl.)	Leaf canopy height above ground (mm)	Plant life form	$\delta^{13}C$ (‰)
	(a) 1–5	flat cushions	−26.38 ± 1.10 (n = 8)
	(b) 5–20	small rosettes, creepers	−26.00 ± 1.30 (n = 5)
	(c) 150–500	tall forbs, shrubs	−26.66 ± 0.88 (n = 5)

(a) *Androsace alpina*, *Loiseleuria procumbens*, *Minuartia sedoides*, *Salix serpyllifolia* (flat), *Saxifraga bryoides*, *S. muscoides*, *S. oppositifolia*, *Silene acaulis* (flat); (b) *Cerastium uniflorum*, *Linaria alpina*, *Salix serpyllifolia* (protruding), *Saxifraga paniculata*, *Silene acaulis* (protruding); (c) *Adenostyles leucophylla*, *Geum reptans* (tall individual), *Gnaphalium norwegicum*, *Rumex alpinus*, *Salix glaucosericea*.

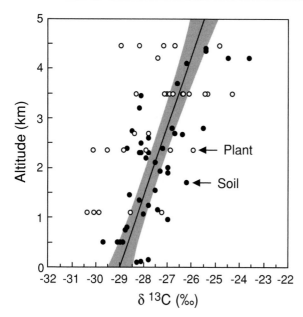

Fig. 11.5. The effect of altitude on soil humus and plant carbon isotope composition ($\delta^{13}C$) in C3 vegetation of New Guinea. The regression line plus 95% confidence limits (shaded) are for the plant data only. (Plant data by Körner et al. 1988; soil data by Bird et al. 1994)

new species analyzed for Table 11.1 was −24.87 to −28.68‰, i.e. again ca. 4‰, which indicates that only broad sampling or the use of soil data can produce habitat- or community-specific signals. Unreplicated data from three species of *Ranunculus* found growing next to each other at medium altitude, but belonging to altitudinally distinct floras indicated that there is also a strong ecotypic component in $\delta^{13}C$, further suggesting that direct pressure effects are not so important (Körner et al. 1991).

There is little evidence that (in addition to a thicker mesophyll) the biochemical capacity to fix CO_2 is enhanced at high altitude. While **Rubisco** in leaves of the tropical-alpine giant rosette species *Espeletia schultzii* was found to be more active in populations from 4200 m altitude compared with those from 3100 or 3550 m altitude in Venezuela (Castrillo 1995; in line with Baruch's 1979 observation for A_{cap} but not with those by Rada et al.

1998), no significant differences in either concentration or activity of Rubisco were found between alpine and lowland grassland species in the Alps by Sage et al. (1997). All other published data for Rubisco are from transplants or greenhouse grown plant material (see genotypic effects below).

Leaf chlorophyll concentrations tend to be lower or similar, but never higher in alpine versus comparable lowland plants (e.g. Henrici 1918; Seybold and Egle 1940; Mooney and Billings 1961; Todaria et al. 1980; Voznesenskaya 1996), despite numbers of chloroplasts per cell tend to increase with altitude (Miroslavov and Kravkina 1991). Bergweiler (1987) isolated thylacoids from chloroplasts in a series of alpine plants and found twice as many reaction centers for photosystem II. Chlorophyll a to b ratios are commonly higher in alpine than lowland plants (Seybold and Egle 1940, Bergweiler 1987). According to Bergweiler, a/b ratios in alpine forbs range from 4.2–5.3 as compared with 3.6–4.1 in lowland forbs, and carotenoids increase with altitude. All these characteristics are typical for plants from high radiation habitats, despite the fact that more frequent cloudiness often dampens the overall radiation doses in high mountains (see Chap. 3).

Which of the above discussed phenomena are **phenotypic**, which **genotypic**? While Bonnier (1895, later extended by Combes 1910) concluded from his classical **transplant** experiments that all morphological traits (including leaf structure) are fully acclimative, i.e. phenotypic, a series of later experiments illustrated that some physiological and anatomical differences between alpine and lowland leaves are retained when plants are grown under common conditions, hence, are ecotypic **traits** (review by Hiesey and Milner 1965). Henrici (1918) found that alpine forbs when grown together with con-generic lowland forbs had higher rates of photosynthesis. Similarly, Mooney and Billings (1961) reported higher A_{cap} for alpine compared with arctic provenances of *Oxyria digyna*. Alpine provenances of *Trifolium repens* retained higher A_{cap} than their low altitude relatives in Mächler and Nösberger's (1977) growth cabinet studies, and Woodward (1986) and Morecroft and Woodward (1996) compared Scottish altitudinal provenances of *Vaccinium myrtillus* and *Alchemilla alpina* and found some of the trends observed in the field were retained under controlled conditions. Higher A_{cap} also occurred in high versus low altitude populations of Californian *Achillea lanulosa* by Gurevitch (1992), a species whose provenances stood at the beginning of experimental testing of genotypic differentiation among altitudinal races (Clausen et al. 1948). However, there are also few examples where results of transplantation experiments did not match with these trends and showed either no (Decker 1959) or even reverse (Shibata et al. 1975) effects. In the latter case, high altitude provenances may have suffered from excessively high growth temperatures at low altitude.

No conclusive picture of genotypic traits at **chloroplast** level has emerged so far. Tieszen and Bonde (1967) reported lower chlorophyll concentrations in alpine versus arctic *Deschampsia caespitosa* and *Trisetum spicatum* grown in a common greenhouse, matching the observations by Mooney and Billings (1961) in *Oxyria* and by Mooney and Johnson for *Thalictrum*, but they also noted exceptions. Chabot et al. (1972) found no difference in Rubisco activity between arctic and alpine populations of *Oxyria digyna* as long as both grew in a cold room. Under warm conditions, alpine but not arctic provenances quickly lost activity, the reasons for which are not clear. Oulton et al. (1979) also found no significant differences in Rubisco activity in *Taraxacum officinale* genotypes from contrasting altitudes, but May and Villarreal (1974) had observed higher Hill reaction rates in the highest provenaces of this species as long as experimental temperatures were 25 °C or lower. At 35 °C the trend was sharply reversed. Pandey et al. (1984) collected Himalayan alpine *Selinum vaginatum* and grew plants in pots at 3600 and 350 m altitude. Rubisco activity was higher in the high altitude group, because, when exposed to the hot lowland climate, part of the Rubisco activity was replaced by PEP-carboxylase activity.

The fundamental problem with most of these **provenance and transplant tests** is the use of one common growth condition for plants taken from environments which differ. Most commonly these growth conditions are closer to those which lowland plants are adapted to, a treatment providing rather unequal (and not as assumed equal) opportunities for populations with contrasting environmental preferences. Hardly any of these controlled environment studies have used a fully reciprocal treatment design or at least growth temperatures in-between the test plant's original thermal environments. Which provenance will benefit and which will suffer, may largely depend on the selected common growth condition. With respect to temperature, the dilemma was avoided only in the classical alpine/arctic tests by Mooney and Billings (1961) and those by Tieszen and Bonde (1967), by comparing provenances from contrasting altitude but from similar thermal backgrounds. But in their case, the different radiation regimes of arctic and temperate-alpine latitudes interfered with the altitudinal comparison.

Though this problem cannot be resolved, the above examples for intact leaves suggest that ecotypic differentiation toward higher A_{cap} in alpine compared with low altitude populations does exist. It seems to be primarily associated with differences in leaf anatomy, and not with contrasting biochemistry. Leaf thickness, for example, remains higher in alpine *Ranunculus glacialis* than in lower altitude *Ranunculus nemorosus* irrespective of common growth temperatures, but in eight out of nine species, lower growth temperatures also created thicker leaves in both low and high altitude species, illustrating that there is also an important phenotypic anatomical component which can influence A_{cap} (Körner et al. 1989a). The inconsistency of biochemical data may have to do with a reference problem. It is difficult to separate trends in certain components or processes from responses of the reference itself (dry weight, fresh weight, area, chlorophyll etc.). This problem has first been noticed and discussed by Bonnier (1890b) and later by Billings and Mooney (1968), but has received insufficient attention.

Photosynthetic responses to the environment

Photosynthetic "capacity" refers to the most favorable life conditions. What are these, and how does photosynthesis in alpine plants respond to less favorable conditions? What are the main determinants of actual net photosynthesis in the field? In essence, the key points have been known since Henrici (1918) published her results. As confirmed many times since (e.g. Cartellieri 1940; Scott and Billings 1964; Moser et al. 1977; Körner and Diemer 1987), the major limitation of alpine plant photosynthesis is the photosynthetically active **quantum flux density** (QFD, 400–700 nm). In reality, temperature is relatively unimportant (in the sense of "non-limiting") during the growing season. The simple explanation is a combination of physiological thermal acclimation, and the fact that leaf temperatures are commonly not cold when the sun is out (see Chap. 4). Hence, thermal limitation of photosynthesis is largely restricted to low QFD situations, during which carbon gain is already restricted by light.

Thermal acclimation of photosynthesis in alpine plants can be characterized by five points.

- The temperature optimum of photosynthesis is adjusted to prevailing leaf temperatures during periods of high QFD, which permits maximal carbon gain.
- The photosynthetic temperature response curve is very wide, so that photosynthesis operates at 95% of maximum rates over a range of ca. 8 K.
- The temperature optimum of photosynthesis shifts with QFD so that the optimum is at low temperatures when QFD is low and at high temperatures when QFD is high (Fig. 11.6), as is known for non-alpine plants as well.
- Re-adjustments of the temperature optimum to prevailing temperatures is relatively fast (1 to a few days).
- Cold nights with subfreezing temperatures down to −3 to −6 °C do not affect A_{cap}.

As can be seen from Fig. 11.6, the **temperature optimum** of A in *Carex curvula* is almost identical

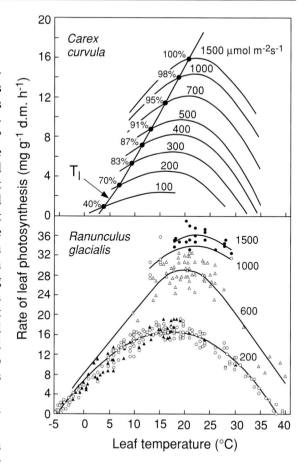

Fig. 11.6. The temperature response of leaf photosynthesis in alpine plants at different QFD. **A** For *Carex curvula* at 2300 m altitude in the central Alps. The line crossing these responses is a micro-meteorological correlation. It indicates the actual combinations of QFD and leaf temperature occurring during the 3 months growing season. At QFD >75% of the mean maximum, the line matches the temperature optimum of photosynthesis, but actual leaf temperatures are much lower than the optimum ones for photosynthesis when QFD is low. For example at 100 μmol m⁻² s⁻¹ QFD (5% of maximum), CO_2 fixation actually found in the field will only be 40% of that which would have occurred if the leaf temperature were 14 °C (the optimum temperature at this QFD) instead of 3.5 °C. (Körner 1982). **B** As above but for *Ranunculus glacialis* from 2700–3100 m altitude (triangles) and plants of the same source, but grown in pots at treeline (2000 m; circles), both measured in a low altitude (600 m) laboratory (Moser et al. 1977)

to the mean leaf temperature measured when QFD reaches $1000\mu molm^{-2}s^{-1}$. Thermal limitation of CO_2-fixation is really limited to low QFD periods, as indicated by the percentages in the diagram. An optimum temperature of 22 °C for A at saturating QFD is quite typical for alpine plants, and optima as high as 27 °C have been measured in situ (Körner and Diemer 1987). Cabrera et al. (1998) also report a 22 °C optimum for *Acaena* in 4200 m in the Venezuelan Andes. Unfortunately, it is a persistent belief that alpine heliophytes have much lower temperature optima than comparable low altitude plants (Todaria 1988). Temperature optima as low as 12–16 °C are found in shade plants or in treeline conifers (Pisek et al. 1973) and tall forbs (Cabrera et al. 1998), which are better coupled to the cool air. Lösch (1994) also reports a relatively low optimum temperature for photosynthesis of only 17 °C for seedlings of the giant rosette *Echium wildbretii* from the alpine zone of Tenerife, perhaps reflecting major vegetative activities during cooler periods in this high altitude semi-arid desert. In sunny habitats the temperature optima of A in alpine plants are not very different (ca. 2–4 K lower) to comparable plants at low altitude in mesic regions of the temperate zone. This holds for plants growing in the field under bright weather conditions. Under prolonged cool and overcast weather, plants adjust their photosynthetic temperature response within 1 or a few days, as many growth cabinet studies have illustrated (see reviews by Billings and Mooney 1968; Pisek et al. 1973).

Thermal acclimation of A to low temperatures primarily results from changes in the photosynthetic apparatus itself and is associated with **electron transport** in the **thylacoid membrane,** photosystem II in particular (Mawson and Cummins 1989). Acclimation studies by these authors with the arctic-alpine *Saxifraga cernua* revealed that the optimum temperature for whole chain electron transport shifted from 10 to 25 °C when plants were grown in 20 instead of 10 °C. Measured at 10 °C, electron transport rates by thylacoids which had developed under 10 °C was 4.2 times larger than rates at 20 °C grown plants. These

adjustments do not appear to require major structural changes in the protein-chlorophyll complexes but seem to result from what these authors call "interactive changes between thylacoid compartments", which can happen fast.

The **minimum temperature** for positive net photosynthesis in alpine plants commonly coincides with the temperature at which leaves freeze or start to show damage. During the growing season this is commonly between −2 and −6 °C (Fig. 11.7). Ice formation in the intercellular space is thought to impair gas diffusion and dehydrate the protoplast so that net photosynthesis ceases (see Chap. 8). When temperatures rise during the morning following a night with freezing temperatures intercellular ice melts and intercellular spaces infiltrate with water. Combined with stomatal closure, as is illustrated in Fig. 11.8, photo-synthesis will recover only once this water is re-absorbed, and gas diffusion is re-established, which may take several hours.

Alpine plants reach at least 20, often 30% of A_{cap} at 0 °C, if QFD is saturating, which is the case under less than 10% of full midday sunlight intensity at such temperatures. It was shown by Henrici (1921) that night temperatures as low as −6 to −8 °C may have no negative aftereffect on photosynthesis during the following morning, which is similar to what Blagowestschenski (1935) reported for alpine plants at 3800 to 4700 m altitude in the Pamir Mountains. Snow tussock in New Zealand (*Chionochloa* sp.) was also found to exhibit full photosynthetic capacity shortly after switching from a −4 °C pre-treatment to warm conditions (Mark 1975). The low temperature limit of photosynthesis in lichens is even lower, −11 °C, representing a conservative estimate for many species (e.g. *Stereocaulon alpinum* and *Cladonia alcicornis*), but *Peltigera subcanina*, for instance, had no apparent CO_2 fixation below −2.3 °C (Lange and Metzner 1965). In summary, low temperatures have surprisingly little effect on alpine plant photosynthesis.

High temperature limits of photosynthesis in alpine plants commonly vary between 38 and 47 °C (Fig. 11.7), temperatures which may occur

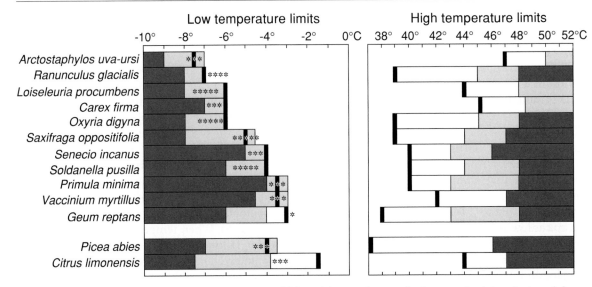

Fig. 11.7. The minimum and maximum temperatures at which positive net photosynthesis occurs in alpine plants and, for comparison, in a treeline and a Mediterranean tree species. *Black bars* indicate the endpoints of photosynthesis. *Light shaded* areas to the *left* and *right* mark leaf temperatures at which 50% damage occurs, *dark* areas mark 100% damage. *Star-stripes* indicate the temperature range at which intercellular freezing commences. (Larcher and Wagner 1976, Pisek et al. 1973)

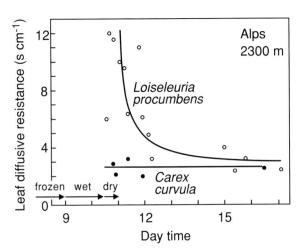

Fig. 11.8. The in situ response of stomata after a midsummer night frost of <−2.4 °C in *Loiseleuria procumbens* (like other Ericaceae, showing pronounced infiltration of intercellular spaces, see Fig. 8.5) and *Carex curvula* (not infiltrating) at 2300 m altitude in the eastern Alps. Note stomata open as infiltration resolves. (Körner 1977)

locally at ground level during dry periods with strong radiation (Chap. 4). Larcher and Wagner (1983) measured leaf temperatures in *Sempervivum montanum* at noon up to 52 °C, just short of heat damage, and causing respiratory losses to exceed CO_2 uptake. Although such extreme situations are very rare, 40 °C may be reached in this species on any clear day during the growing season (7% of all days). Seedlings on surface-dry, sun exposed ground will regularly experience such high temperatures, constraining their carbon balance. High temperature limits of photosynthesis in seedlings of alpine species on Tenerife (Canary Islands) are between 35 (*Echium* sp.) and 45 to 48 °C (*Erysimum* sp.; Lösch 1994).

Light requirements for A_{cap} are comparatively high in alpine plants, and range from 500 $\mu mol\,m^{-2}\,s^{-1}$ in "shade" adapted alpine species such as *Oxyria digyna* to >2000 $\mu mol\,m^{-2}\,s^{-1}$, with a mean of 1200 $\mu mol\,m^{-2}\,s^{-1}$ for various species (95% of saturating QFD; Körner and Diemer 1987). This high light requirement explains why A was found by many authors to closely follow QFD during the day (e.g. Cartellieri 1940; Billings et al. 1966; Fig. 11.12). Some alpine species do not saturate even at full midday sun and continue to increase A up to $3000\,\mu mol\,m^{-2}\,s^{-1}$ (e.g. *Ranunculus glacialis, Ligusticum mutellina*), conditions which may occur when direct midday sun is combined with diffuse radiation from thin, bright clouds (Körner and Diemer 1987; Terashima et al. 1993; see chapter 3). The ability to utilize such extremely high QFD (corresponding to the solar constant) may assist alpine plants to avoid photoinhibition and photo-damage. In addition, alpine plants exhibit greater protection by accessory pigments and **antioxidants**, which may facilitate the maintenance of high photosynthetic activities at high irradiation (Wildi and Lütz 1996; Streb et al. 1998; Fig. 11.9).

The light climate a leaf experiences strongly depends in its shoot morphology and the **leaf angle**, which are thus co-determinants of leaf photosynthesis (Körner 1982; Germino and Smith 2000). Across all variants of shoot morphology, four principal modes of leaf display may be distinguished in alpine plants: steep cylinders (e.g. tussock grasses), horizontal leaves (e.g. flat, disk-type rosettes), dome-shaped rosettes which combine all possible leaf inclination angles in a regular order in one shoot (e.g. giant rosettes, many cushion plants), and last, optically irregular structures with a mix of leaf angles of varying, but commonly not random inclination (e.g. dwarf shrubs). While all four of these **morphotypes** are well represented in the tropics, the abundance of flat structures declines with latitude. These leaf angles interact with the diurnal, seasonal and latitudinal variation of solar angles, but become irrelevant under diffuse radiation, particularly under dense cloud cover and in fog.

In the case of *Carex*, a simple geometric model for 47°N lat showed that the rates of photosynthesis on clear days in naturally positioned leaves would not differ from those for strictly vertical leaves. However, the photosynthetic gain would be very different (less) in horizontal leaves (Körner 1982). At medium to high latitudes, steeper leaf inclination causes **light interception** during the

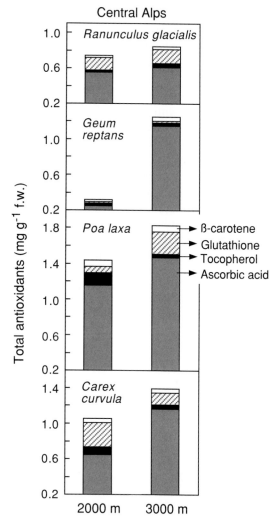

Fig. 11.9. Contributions of ascorbic acid, tocopherol, glutathione and δ-carotene to the total amount of antioxidants in alpine plant species from a site just below the treeline (2000 m) and at 3000 m in the central Alps. Note the massive overall increase in antioxidants with altitude, despite a very species specific composition of the antioxidant pool. (Wildi and Lütz 1996)

day to vary less, as long as leaves are sufficiently widely spaced. In contrast, leaves of rosette plants, both flat and dome-shaped ones, experience rather variable diurnal light interception, with the most extreme variation reported for tropical giant rosettes, in which some leaves are fully shaded while others receive maximum QFD. As a consequence, each leaf exhibits its own very specific diurnal course of photosynthesis, as was documented by Schulze et al. (1985) and Goldstein et al. (1989).

Light compensation points of photosynthesis in alpine plants are similar to lowland heliophytes (when measured in the field) and show the generally known decrease with decreasing temperature (Fig. 11.10). Since temperatures are much lower at high than at low altitude when radiation approaches zero, the light driven cessation of positive net photosynthesis will commonly occur at lower QFD than at low altitude (<20 μmol photons $m^{-2} s^{-1}$).

Direct effects on photosynthesis by climatic drivers other than temperature and QFD are probably small in most alpine regions. As was discussed

in Chapter 9, physiologically effective drought stress is rare in alpine plants but **moisture shortage** can become effective indirectly via a periodic interruption of topsoil nutrient cycling and thus plant nutrition. In line with this assumption, Enquist and Ebersole (1994) found no direct midseason effects of water addition on photosynthesis or relevant water relations parameters in *Bistorta vivipara* at 3800 m altitude in the Rocky Mountains. Since leaf nutrient concentrations have little explanatory value for plant nutrient availability, and since high A was associated with low water potential (and high flux) in their study, as it should (as long as turgor is maintained) they could not understand, why their watering treatment prolonged photosynthetic activity at the end of the season. From current knowledge, effects on nutrient availability are the most likely explanation for delayed senescence (see the discussion of fertilizer trials with *Vaccinium* in Chap. 10). Bowman et al. (1995) also observed no effects of water addition on photosynthesis in a series of Rocky Mountains species in both wet and drained locations.

Negative effects of reduced radiation on leaf photosynthesis can be lowered or even balanced by the positive effects of reduced radiative forcing on topsoil moisture, which in turn stimulates nutrient cycling and nutrient availability. A five year **shading treatment** in alpine grassland caused no reduction in plant biomass, but induced a pronounced softening of plant tissue, increased leafiness and ground cover, favored broad-leaved grasses (in terms of leaf area, not mass), reduced lichens, and halved the amount of peak season plant litter (Fig. 11.11), responses which were difficult to predict. Hence, moderate shading is not necessarily as negative for alpine plant growth as might be concluded from Figures 11.12 and 11.13, which consider direct effects of radiation on photosynthesis only, but do not account for such indirect effects on soil nutrients (see also below, effects of air humidity). Similarly, Michelsen et al. (1996b) observed no negative effect of long term shading on biomass of arctic-alpine plant communities.

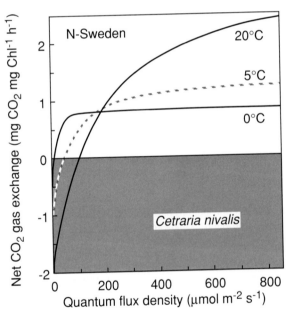

Fig. 11.10. The interactive effect of temperature and QFD on photosynthesis exemplified by data for the arctic-alpine lichen *Cetraria nivalis* in north Sweden. Because of the variable contribution of the fungus partner to dry mass, rates are expressed per unit of chlorophyll. Note, the increasing light demand for saturation of A as temperatures increase, which is complementary to Fig. 11.6. (Kappen et al. 1995)

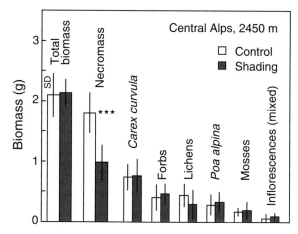

Fig. 11.11. The influence of long-term canopy shading on plant biomass in a low stature (>6 cm), sedge dominated alpine grassland (Furka Pass, 2450 m, Swiss central Alps). An area of 10 m² of homogenous alpine heath was screened from direct solar radiation so that canopies received −22% (overcast) to −44% (clear sky) of unscreened radiation while air circulation (wind, humidity, air temperature) remained unchanged (peak season harvest after 5 years of treatment in 1995; data for n = 8 microplots of 1 dm²). The predicted negative effect on photosynthesis (and thus growth) appears to have been balanced by positive effects on soil conditions (surface moisture), causing biomass to remain unaffected, but necromass to be reduced through faster decomposition. The shaded vegetation was taller and looked a lot more lush (larger SLA and LAI rather than more biomass). (Unpubl.)

The possibility that more frequent **leaf wetting** by dew, rain or fog at high altitudes can affect photosynthesis was examined by Brewer and Smith (1993). Under heavy dew they found photosynthesis of central Rocky Mountain plants to be reduced by 20%. **Leaf pubescence** in high altitude plants can prevent surface wetting (Brewer and Smith 1997). In tropical *Espeletia* it is quite obvious that pubescence keeps the gas diffusion path open while plants are soaked with atmospheric moisture (see Chap. 9).

Atmospheric humidity may indirectly influence photosynthesis during clear and warm spells, because stomata have been shown to be sensitive to vapor pressure deficit (or its consequence, transpirational flux) both in the temperate-alpine (Johnson and Caldwell 1975; Körner 1976, 1980)

and the tropical-alpine zone (Schulze et al. 1985; Goldstein et al. 1989), as was discussed in chapter 9. However, at wide stomatal apertures, changes in stomatal conductance have little influence on photosynthesis, and more substantial effects require vapor pressure deficits to increase above 15 mbar, which does not happen very often at alpine altitudes. In laboratory experiments, Lösch (1994) found no influence of vpd on photosynthesis of *Erysimum* species from the alpine zone of the Canary Islands and the Cape Verde Islands, and very small effects in other typical high altitude species. Leaf pubescence (Goldstein et al. 1989) or dense leaf canopies with high aerodynamic resistance (Grabherr and Cernusca 1977; Körner 1976) buffer many alpine plants against such influences (see Chap. 9). It is interesting that Young and Smith (1983) observed very beneficial effects of mid-summer cloudiness on photosynthesis in *Arnica latifolia*, an understory forb close to the treeline in the Medicine Bow Mountains of Wyoming. These plants may have profited from reduced vapor pressure deficit or reduced photoinhibition (cf. Germino and Smith 1999, 2000). Hence, under certain climatic conditions, clouds can have favorable or neutral effects (Fig. 11.11).

In summary, the photosynthetic responses discussed above illustrate that (during a given length of growing season) CO₂ uptake in alpine plants is mainly limited by QFD and much less by suboptimal temperatures, which is similar to what is known from low altitudes. The pronounced **heliophytic character** of leaves in alpine plants include genotypic characteristics. Under conditions where strong radiation affects topsoil moisture and thereby plant nutrient availability, plants may, however, profit from moderate shade or relief effects which reduce insolation, despite the overall dominance of the QFD dependency of photosynthesis at leaf level. The indirect (negative) effects of strong insolation on top soil processes possibly explain why lush alpine vegetation is often found on north or east exposed slopes. There seem to exist **trade-offs** between the photosynthetic demand for light energy and radiative forcing of evaporation, which can affect the nutrient cycle.

Daily carbon gain of leaves

The photosynthetic responses discussed above translate into characteristic diurnal courses of CO_2 uptake, of which one from the temperate zone for a typical cloudy day with rather variable light conditions is illustrated in Figure 11.12. The close correlation with QFD is so obvious because, under such weather conditions, QFD varies largely in the sensitive part of the light response curve. On clear days and in canopies with a low leaf area index (LAI 2 or less) and steep leaf inclination there is little change in A for most of the day, because leaves are light saturated from about 08:00 to 16:00 hours (midsummer, 47°N lat).

A rough estimate for the **total daytime net CO_2-uptake** of such sedges is typically 120 mg CO_2 per g dry weight under such conditions. Cartellieri (1940) estimated 150 mg g^{-1} for very active forbs. The mean for overcast days, as indicated by the dashed line in Fig. 11.12, is half this amount. In simple terms, CO_2 uptake on overcast days follows a triangular shape, compared with a trapezoid shape on clear days (Körner 1989c).

The daily carbon gain of *Carex curvula* is typical for many alpine species, and illustrates that

it takes about 14 clear or 28 overcast days for such average alpine leaves to absorb as much C as is contained in their structures (conversion factor 0.46 g C per g dry matter), numbers very similar to those which Cartellieri arrived at. Accounting for night-time respiratory losses (ca. 6–8% of daytime gains) and metabolic plus investment costs of leaf construction (about 0.6 g C per g leaf dry matter instead of 0.46; Chapin 1989) yields periods of 20 to 40 days of clear and overcast weather for **amortization** (see next section). Cartellieri (1940) plotted the daily leaf carbon gain against daily radiation sums for various measurement periods for *Ranunculus glacialis* and *Doronicum clusii* (two species with high QFD requirements) and arrived at an almost linear relationship for most of the data range.

The seasonal carbon gain of leaves

Given the facts discussed above, modeling the **combined effect of temperature and QFD** is all that is needed to predict leaf photosynthesis in alpine plants in most cases. Utilizing the frequency distribution of QFD and leaf temperatures

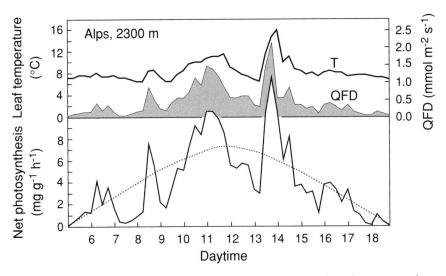

Fig. 11.12. The diurnal course of leaf photosynthesis (A) in alpine plants under cloudy midsummer weather exemplified by data for *Carex curvula* (Hohe Tauern, Alps, 2300 m). Note the close correlation of A with quantum flux density (QFD). The *dashed line* indicates the almost triangular mean course of A for such conditions, compared with a plateau or trapezoid shape for clear days with twice the total carbon gain. (Körner 1989c)

during an alpine season in the Alps it was possible to estimate the relative influence of QFD and temperature on the seasonal carbon gain of *Carex curvula* leaves (Fig. 11.13). Apart from the overwhelming influence of the length of the snow free period, it is obvious that clouds and mutual shading are far more important for seasonal CO_2 assimilation than temperature. Compared with a theoretical maximum carbon gain for a clear sky season and unlimiting temperatures (not accounting for any other potential constraints), the realized carbon gain is reduced by 40% due to insufficient leaf illumination and only by 7% due to suboptimal temperatures, as explained by the physiology of leaf gas exchange shown in Fig. 11.6.

During the 1045 daylight hours of the active growing season of about 12 weeks (23% of the year), *Carex* leaves in this canopy absorbed about 7.8 g CO_2 per g active leaf dry weight (ca. 4.4 g C per g leaf C), 53% of the potential maximum capture, and 19% less than if no mutual shading in the canopy had occurred during the whole sea-son (LAI 2.3). The *Carex* model revealed that 172 hours or only 17% of all daylight hours (QFD > 1500 μmol m^{-2} s^{-1}) contributed 34% of the seasonal yield, compared with 26% of the yield produced during 607 hours with QFD < 750 μmol m^{-2} s^{-1}. Clearly, short periods with strong radiation were more important than the predominant periods with weak light.

Utilizing a similar approach, but also accounting for the dependency of A on leaf age, Diemer and Körner (1996) arrived at a seasonal C-fixation of about 5.4 g C per g leaf C for *Ranunculus glacialis* at 2600 m altitude in the Alps (Fig. 11.14). The full season leaf carbon balance for this species, which also accounts for night-time respiratory losses and for construction costs of leaves, yields a gain of 4.6 g C per g of C investment in leaves. Taken together, these two studies for alpine plants of sunny habitats and with very active leaves suggest that leaves fix 4 to 5 five times as much carbon as they contain. However, carbon gain ratios may be as low as 1.2 g g^{-1} in plants growing in the shade of rocks, as was estimated

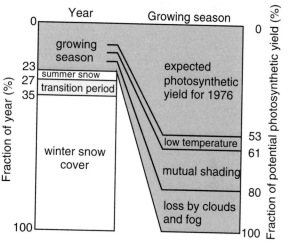

Fig. 11.13. The limitation of CO_2 uptake in *Carex curvula* by light and temperature at its natural alpine habitat. The theoretical maximum seasonal photosynthetic yield under assumed optimum conditions of a fully illuminated leaf is taken as a 100% reference. The photosynthetic yield modeled for the 1976 season (using polynomials from Fig. 11.6 and climate data) accounts, in a stepwise approach, for the less than optimal conditions in the real world. (Körner 1982)

for individuals of *Geum reptans* at the same site.

In the course of a season it took leaves of *Ranunculus glacialis* at 2600 m in the Alps 37 days from the time of emergence at snowmelt until leaf carbon investments were amortized (Fig. 11.14). For the above mentioned shade plants, 68 days

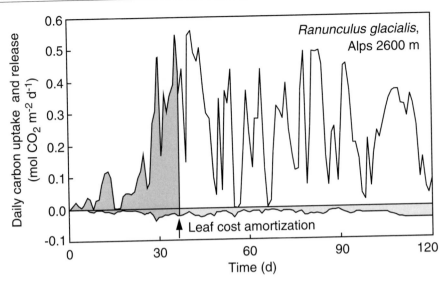

Fig. 11.14. The seasonal course of the leaf CO_2 balance (per unit leaf area) of *Ranunculus glacialis* at Mt. Glungezer, 2600 m, Tirol, Austria. The *shaded* part of the curve illustrates the period until the leaf has amortized its construction costs in terms of carbon. Oscillations of CO_2 assimilation largely reflect the seasonal course of cloudiness and short periods of snow cover. (Diemer and Körner 1996)

were required, about half the leaf life span of ca. 120 days. Comparative amortization periods for lowland forbs in meadows are around 20 days, which reflects the more favorable climate during early season leaf development and the lower construction costs per unit leaf area in lowland plants. The total amount of carbon lost by leaves from sunny habitats due to night-time respiration were 6–8% of carbon uptake at high and 4 to 10% at low altitude (see below). No consistent difference between the net carbon gain per unit leaf area during the life span of leaves in three alpine and three lowland species occurred, and this was equally due to the leaves' photosynthetic yield as to their respiratory losses (Diemer and Körner 1996). Evidence from a sensitivity analysis of this model revealed that the primary determinant of a leaf's carbon balance is its longevity, which does not differ significantly between alpine and lowland herbaceous species (see Chap. 13).

Since a highly positive leaf carbon balance is necessary for the maintenance of root growth, root respiration and reproduction, these data illustrate why alpine plant life is impossible when mean season length is less than a month, and why 6–7 weeks are the commonly observed limits in extreme snowbed habitats (see Chap. 5). One way of effectively utilizing such short season habitats is to reduce the carbon investment per unit leaf area (increase the protein concentration), exactly what has been observed at local as well as at the global scale (Chaps. 5 and 10). However, with seasons longer than that, alpine plants can acquire considerable more carbon than would be essential for survival. In fact, it appears that they can buffer considerable carbon losses due to herbivores with no negative consequences for vegetative vigor or reproduction even at highest elevations (Diemer 1996). The common 10 to 12 week season for assimilation in high alpine vegetation of the temperate zone is sufficient to produce substantial carbon surpluses, and even greater yields might be expected in subtropical or tropical mountains were plants also invest more C per unit leaf area and per unit of leaf nitrogen (Chap. 10) and produce more than one leaf cohort per year (Chap. 12).

C4 and CAM photosynthesis at high altitudes

The significance of plants using other than the C3-pathway of photosynthesis is low in the truly alpine life zone. Commonly, the C4 syndrome which involves a special leaf anatomy and PEP-carboxylase for CO_2 fixation, has an upper limit below or near the treeline. Tieszen et al. (1979) found no C4 grass above 3050 m on Mt. Kenya (Fig. 11.15) and Earnshaw et al. (1990) report the highest ranging **C4 species** for 3280 m in New Guinea (*Miscanthus floridulus*). Ruthsatz and Hofmann (1984), Geyger (1985) and Pyankov et al. (1992) describe C4 species for the dry subtropical mountains of northwest Argentina and central Asia at elevations up to 4200 m, and C4 annuals may be found as ruderals around 3000 m (i.e. below the treeline) in the Rocky Mountains of Utah and Colorado (e.g. *Muehlenbergia montana*, Brown 1977; Sage and Sage 2002). Altitudinal trends similar to those described by Tieszen et al. (1979) were reported by Cabido et al. (1997) for a transect in central Argentina (sites between 350 and 2100 m). Bowman and Turner (1993), who also reviewed more recent literature on the occurrence of C4 plants in cool climates, found clear genotypic differences in thermal sensitivity of gas exchange of altitudinally distinct populations of the C4 grass genus *Bouteloua*. As was mentioned earlier, the CO_2 binding enzyme PEP-carboxylase, which is characteristic for the C4 pathway, but occurs in C3 plants as well, was found to decrease with altitude relative to Rubisco in the C3 species *Selinum vaginatum* in the Himalayas (Pandey et al. 1984), hence there also seems to exist an acclimative intraspecific trend against PEP-carboxylase activity at high altitudes.

In contrast to C4 species, the succulent **CAM plants** (utilizing "crassulacean acid metabolism"), are much more abundant in the alpine zone, but are restricted to specific microhabitats where C3 species are excluded by drought or heat stress. CAM plants can fix CO_2 during the night by PEP-carboxylase, store it as malic acid in the vacuole (hence the succulence), and then engage the normal C3 pathway during the day for final reduction, while stomata can remain closed, and thus save water.

Some specialists such as species of the genera *Sempervivum* (Alps) and *Echeveria* (Andes and Mexico) have ranges which exceed the treeline substantially. *Sempervivum montanum* is found up to 3250 m altitude in the Alps, which is 1000 m above treeline (Larcher and Wagner 1983). A most important group of alpine CAM species are members of the Cactaceae family, which are represented by many species in the high subtropical and tropical Andes at altitudes of up to 4300 m, perhaps even more (e.g. *Tephrocactus* sp., pers. obs. in northwest Argentina, see also Ruthsatz 1977). While cacti seem to perform full CAM (see below and Chap. 9), *Echeveria columbiana* exhibits, in addition to pronounced night-time CO_2 fixation (up to $4.5 \, \mu\mathrm{mol \, m^{-2} \, s^{-1}}$; Medina and Delgado 1976), a small daytime fixation (maximum $0.3\text{–}1 \, \mu\mathrm{mol \, m^{-2} \, s^{-1}}$). In *Sempervivum montanum* CAM is facultative and becomes less engaged when the environment gets cooler

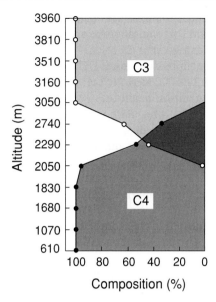

Fig. 11.15. The percentage of grass species which are C4 or C3, along an altitudinal transect on Mt. Kenya. (Tieszen et al. 1979)

(Wagner and Larcher 1981). At daytime temperatures < 10 °C (and high moisture) *S. montanum* behaves like a normal C3 plant. Still well supplied with water, but experiencing daytime temperatures of 15–25 °C weak CAM activity was observed, and the majority of CO_2 was still taken up during the day. Full CAM only occurred when plants were droughted and experienced very high temperatures during the day (50 °C is tolerated). This behavior allows plants to utilize both dry (and hot) midsummer conditions as well as moist (and cool) periods. The integral seasonal influence of CAM varies between sites, but seems to be significant in Tirol, and almost zero in the samples from the Swiss Alps (see Table 11.2 and Chap. 9).

As was discussed earlier in this chapter, stable carbon isotope discrimination of plant tissue, or its relative difference against a standard, $\delta^{13}C$, permit us to assess the long-term significance of CAM for alpine plant C acquisition. Values for $\delta^{13}C$ between −12 and −14‰ can be considered full CAM, values between −24 and −29‰ suggest no CAM engagement. Within a random sample of various high alpine succulents, the spectrum ranges from −12‰ in *Tephrocactus* to −28‰ in arctic-alpine *Rhodiola rosea*, a succulent which obviously is not engaging CAM (Table 11.2). Numbers for *Sempervivum* range from −18 to −24.3‰, indicating moderate to small CAM engagement over the whole growing season, in line with what Wagner and Larcher (1981) characterized as "weak CAM". Interestingly, the genus *Sedum*, with species in almost all alpine floras, covers a range from −23.2 to −27.1‰ (mostly around 24‰) suggesting that CAM plays a negligible if any role in the truly alpine representatives of this genus. These findings are in line with data largely (though not specified) from lower altitudes (500 m to the treeline) collected by Osmond et al. (1975; Table 11.2). These authors also tested a number of other species with succulent leaves such as seven species of *Saxifraga* (−25.5 ± 1.2‰), *Pinguicula alpina* and *Linaria alpina* (both −27.7‰), all non-CAM plants. Overall, these numbers illustrate comparatively

low CAM activity in temperate zone alpine succulents.

Tissue respiration of alpine plants

Though equally important, much less is known about respiration (R) than about photosynthesis of alpine plants. While alpine photosynthesis in principle depends on all components of the climate, in reality it is really the light regime and leaf age which matter (see above), and response functions are straight forward. The object is obvious and defined: a leaf. The most adequate reference unit is also obvious: because light is a vector, it is the light intercepting area that needs to be known. Nitrogen content per unit leaf area allows prediction of photosynthetic capacity with reasonable confidence, and with stable carbon isotopes we have an integrating tool. Unfortunately, the situation with respiration is much more complicated. First, there are many different tissues in a plant, all experiencing different (commonly unknown) micro-environments and respiring at different rates. Second, responses to the main external driver, temperature, are variable and often show rapid acclimation. Third, there are no good and easily defined tissue quality criteria by which "standard rates" can be predicted, and there is no sound rationale for any of the many possible reference units (though people mostly use dry mass). Finally, there is not one type of respiration (see below).

Oxidative release of CO_2 may be the result of mitochondrial respiration for (1) maintenance, (2) growth or (3) energy supply for nutrient uptake in roots. Only **maintenance respiration** will be discussed in this chapter. **Growth respiration** is commonly calculated from known energetic costs for certain plant compounds following the concept by Penning de Vries (see Chapin 1989 for arctic tundra plants), although these costs have never been evaluated for alpine life conditions. Also, the **respiratory costs of nutrient uptake** can not be measured directly, but may be estimated from laboratory data, tracer experiments and a suite of soil

Table 11.2. The contribution of crassulacean acid metabolism (CAM) to carbon gain in alpine succulents as indicated by $\delta^{13}C$ (original data, except for[a], which is from Osmond et al. 1975). Temperate zone samples are from peak season (flowering)

Species	Site	Altitude (m)	$\delta^{13}C$ (‰)
Subtropics			
Tephrocactus aff. bolivianus[b] (flowers and spines Nov. 97)	Cumbr. del Aconquijo (northwest Argentina)	4000 m	−12.2
Echeveria sp. (young leaves, March 96)	Iztaccihuatl (Mexico)	4000 m	−15.4
Echeveria sp. (dead old leaves, March 96)	As above	4000 m	−18.4
Sedum sp. (March 96)	As above	4000 m	−25.3
Temperate zone			
Sempervivum montanum	Patscherkofel (Tirolian Alps)	2200 m	−18.6[a]
S. montanum (dry site)	Furka Pass (Swiss Alps)	2480 m	−22.4
S. montanum (mesic site)	As above	2480 m	−24.3
Sedum alpestre	Avers (Swiss Alps)	2600 m	−23.7
S. alpestre	Saas Fe (Swiss Alps)	2700 m	−25.7
Sedum atratum	Grusin (Swiss Alps)	2600 m	−23.3
S. atratum	Aosta (Italy Alps)	2950 m	−24.5
Sedum lanceolatum	Niwot Ridge (Rocky Mts. USA)	3500 m	−27.2
Rhodiola rosea	Saas Fe (Swiss Alps)	2400 m	−28.0
Mean for 8 *Sedum* species from lower altitude, dry habitats	Eastern Alps (Austria)	unknown	−24.4 ± 1.4[a]
Arctic-alpine			
Rhodiola rosea	Kirkevagge (north Sweden)	800 m	−27.0

[b] collected by J. Gonzalez, Tucuman.

biological investigations. These costs include both the immediate demand for ion exchange in the root, and the indirect expenditures associated with mycorrhization and rhizosphere organisms, which are often treated as a separate carbon sink (Lambers and Van der Werf 1989). These are important carbon costs, particularly for slow growing plants of naturally infertile environments.

Photorespiration, i.e. carbon losses through oxygenase activity of Rubisco during photosynthesis are already covered by apparent net photosynthesis. However, photorespiration is an ecologically important process for the maintenance of the integrity of the photosyn-thetic machinery in a high light environment (Woolhouse 1986), particularly in cold climates (Öquist and Martin 1986). From studies with *Chenopodi-um bonus-henricus* (Heber et al. 1996) and *Ranunculus glacialis* (Streb et al. 1998) it became clear that photorespiration is essential for protection against photo-inhibition and the maintenance of photosynthetic electron transport under high radiation in highly productive plants. Slow growing plants such as *Soldanella alpina* and *Homogyne alpina*, in turn, rely on antioxidants (Streb et al. 1998).

Maintenance respiration or **"dark respiration"** of mature tissue (R), the main topic of this section, largely depends on two factors: (1) the specific activity of tissues and (2) temperature. Specific tissue activity in turn depends on a large number of determinants such as organ type (leaf, stem, storage organ, coarse root, fine root, flower), carbohydrate supply status, tissue age, plant developmental stage, daytime, climatic prehistory and stress situations. Though technically simple, respiration measurements are biologically tricky, because almost any experimental interference with the plant may induce changes. Also, concurrent leaf photosynthesis influences mitochondrial respiration. Simply darkening a leaf cuvette during the day, as was common practice for respiration measurements in the early days, is an inadequate approach. Leaves darkened during the day may respire twice as much as they would at equal temperature during the night. Furthermore, "dark" respiration of leaves in the light period is partially substituted by direct energy supply through photosynthesis, hence is reduced (Atkin et al. 1997). Therefore it is inappropriate to add respiration rates, measured during dark periods, to apparent net photosynthesis in order to construct something like a "gross photosynthesis". Given all these difficulties it is understandable why information on wild plant respiration, alpine plants included, is often inconsistent. Only a small fraction of the published data have been measured in situ and, in the case of leaves, at periods when "dark" respiration affects the carbon balance, which is during the night.

It is a century-old debate whether alpine plants respire less or more than lowland plants, or whether plants from cold climates (including polar latitudes) lose less or more respiratory CO_2 than those from warm climates. Does a *Betula* leaf in Greenland lose less CO_2 in the dark than the leaf of a tropical forest tree in Java? This is the question Stocker (1935) asked, and his data said no!

Note that Stocker asked the question in a specific way: he said "IN" Greenland (and it might have been better to refer to roots, because it is never dark in Greenland when leaves are out). Unfortunately, these two letters were forgotten during most of the following years of experimentation, which led to a vast confusion. The majority of data have been obtained at constant laboratory temperatures at which plants **genetically adapted** to life in low temperatures were often found to respire more than plants native to warmer climates (e.g. Pisek and Winkler 1958; Björkman and Holmgren 1961; Mooney and Billings 1961;

Mooney et al. 1964; Semichatova 1965; Klikoff 1968; McNulty et al. 1988). Others have misinterpreted these findings by the stereotype "alpine plants have higher respiration", which lacks a sense of realism. The only valid comparison in an ecological context is the comparison of R at actual thermal site conditions, rarely reported in the literature. Under these conditions, alpine plants commonly do not exhibit higher, and even sometimes lower R, as will be discussed below.

Table 11.3 illustrates respiration rates of field grown leaf and root tissue measured in high and low altitude laboratories immediately after harvest. On a dry weight basis, leaves respire about twice as much as roots. The extremely high rates per unit dry weight in *Oxyria digyna* are the result of a very high specific leaf area and are not seen when respiration is expressed per unit leaf area, highlighting the reference problem. At a common measurement temperature of 10 °C, alpine species respire more than lowland species. Since nighttime temperatures differ by ca. 10 K between sites, it is interesting to note that the alpine species tend to respire at similar or lower rates per unit dry mass at 10 °C as the lowland species at 20 °C.

The second problem associated with responses of R to temperature is short-term **acclimation** with time. The concept of Q_{10} and the messages derived from conventional instantaneous R-temperature response curves are doubtful, if not misleading when trying to understand R in the field. Such curves are commonly measured in plants precultivated at ONE temperature and exposed to varying temperatures within a short experimental period of <1 h. Under these conditions a mean Q_{10} of 2.3 for alpine as well as lowland species was found (Fig. 11.16), suggesting more than doubling of R for a 10 K increase in temperature (commonly measured in the range of 10 to 20 °C).

Such numbers are problematic for two reasons. First, the Q_{10} obtained in such experiments depends on the base temperature selected: the closer the lower reference is to zero respiration (e.g. −10 °C), the more Q_{10} tends toward infinity, so there is leeway for subjectivity. Second, and biologically important: a doubling of night-time temperature

Table 11.3. Rates of leaf and root respiration in mature lowland (L) and alpine (A) forbs measured at peak season (leaves during the night) with an oxygen electrode. Fresh samples were analysed within 2h after harvest, starting with low temperatures. Alpine site Mt. Glungezer, 2600 m Tirol/Austria (night-time temperatures ca. 5 °C), low altitude site, 600 m Botanical Garden Innsbruck (night-time temperature ca. 15 °C; U. Renhardt and Ch. Körner, 1988, unpubl.)

Plant species (A alpine, L lowland)	A/L	Rate of tissue respiration (mg CO_2 $g^{-1}h^{-1}$/mg $m^{-2}h^{-1}$)				
		10 °C	(n)	20 °C	(n)	Q_{10}
Leaves						
Oxyria digyna	A	7.2/218	(1)	12.2/352	(8)	1.7
Ranunculus glacialis	A	3.9/247[a]	(3)	6.6/374	(5)	1.7
Geum reptans	A	–		3.1/199	(4)	–
Poa alpina ssp. vivipara	A	4.1/249	(1)	4.4/279	(4)	1.1
Oxyria digyna (transpl.)	L	5.8/179	(2)	11.9/373	(5)	2.1
Ranunculus acris	L	3.1/149	(3)	4.9/280	(5)	1.6
Geum rivale	L	2.3/116	(3)	4.1/219	(5)	1.8
Poa annua	L	4.8/136	(2)	5.1/141	(3)	1.1
Fine roots						
Oxyria digyna	A	–	–	3.3	(1)	–
Ranunculus glacialis	A	1.3	(1)	1.6	(1)	1.2
Geum reptans	A	2.1	(1)	2.5	(1)	1.2
Poa alpina ssp. vivipara	A	3.4	(1)	5.7	(1)	1.7
Cerastium uniflorum	A	–	–	2.5	(1)	–
Coarse roots						
Oxyria digyna	A	1.7	(1)	3.1	(1)	1.8
Geum reptans	A	1.2	(1)	–	–	–

[a] in situ IRGA measurements: 206 mg $m^{-2}h^{-1}$ (M. Diemer pers. comm.)

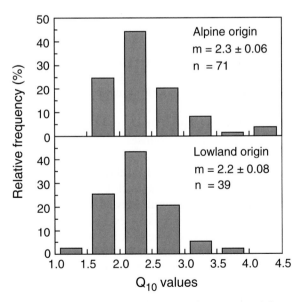

Fig. 11.16. The frequency distribution of Q_{10} (10–20 °C) for alpine and lowland plant species as measured in the laboratory during short term variations of temperature. (Larigauderie and Körner 1995)

rarely causes R to double in the field, because over periods of one to several days most plants re-adjust their metabolism so that the temperature effect may become zero (full acclimation) or diminished (Fig. 11.17). Given that the most significant components of respiratory loss are those by belowground organs (which contribute 50 to 80% to total biomass and respire 24h) and, by aboveground organs (green parts during the night only, with temperatures commonly <5 °C), short-term responses of R to large changes in temperature are relevant only for a small part of the season and a small fraction of non-photosynthesizing tissue immediately beneath and above the soil surface.

When measuring the **long-term thermal acclimation of respiration** (LTR_{10} in contrast to instantaneous Q_{10}) of potted alpine and lowland plants grown and measured at a 10K different temperature, a broad spectrum of responses was obtained, with some species showing no ($LTR_{10} = Q_{10}$) and some full adjustment, as exemplified in Fig. 11.17. Data by Skre (1985) for leaves of five

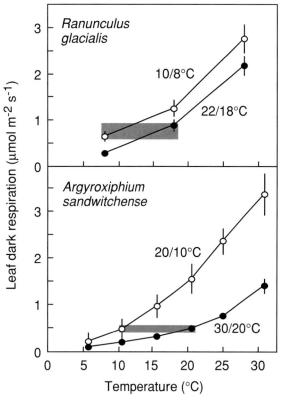

Fig. 11.17. The instantaneous temperature response of dark respiration in leaves acclimated to different temperatures. *Top*: *Ranunculus glacialis* acclimated to 8 and 18 °C growth temperatures before measurements were conducted at both these and a higher temperature (Arnone and Körner 1997). *Bottom*: *Argyroxiphium sandwitchense*, as above, but for plants acclimated to a 10/20 and 20/30 °C night/day temperature regime (Goldstein et al. 1996). Note, when measured at the growth temperature, no significant difference in R was detected in either case, indicating full acclimation (*grey bars*).

arctic alpine species also range from full to no acclimation for a change of 6–8 K in growth temperature. In contrast, Atkin et al. (1996a,b) tested leaves of six species of *Poa* from contrasting altitudes but grown at a common temperature of 20 °C and found no difference per dry weight (full acclimation) and a slight, not significant trend for higher rates per unit leaf nitrogen. Also, a lowland/alpine comparison in Australia by Atkin and Day (1990), including a wider spectrum of life forms, did not reveal any ecotypic differences

of leaf R when plants were grown and measured at 25 °C.

Remarkably, the three species with the least ability to adjust to a new, higher temperature in the study by Larigauderie and Körner (1995) were among the highest growing species of Europe, with *Saxifraga biflora* holding the record of 4450 m. This species exhibited an LTR_{10} of 5.5 and no acclimation, which means that these plants persistently lose 5.5 times as much CO_2 at 20 °C as they do at 10 °C. It is obvious that such plants will rapidly attain a negative carbon balance and die at higher temperatures and are therefore restricted to uppermost alpine altitudes. It would be interesting to check the species lists of lowland botanical gardens with an alpine section for alpine species missing in their collections and what the natural elevational range of these species is. *Ranunculus glacialis* is missing in all temperate zone outdoor collections (in Kew Gardens it grows on a cool bench), because it dies within a few weeks – despite leaves seeming to be able to acclimate (11.17), but the massive root system of this species may not. Dahl (1951) was possibly the first who suggested a high temperature, low altitude limit for alpine plants, because of overshooting respiration. However, *in situ*, the rate of leaf respiration in alpine forbs was found unrelated to the actual leaf carbohydrate charging, hence does not seem to be C-limited (McCutchan and Monson 2001).

Larigauderie and Körner (1995) illustrated a great spectrum of responses of thermal acclimation in both alpine and lowland plant species and the patterns in absolute R and degree of acclimation largely reflect genus specific traits. Thus, predictions at the community level cannot be made based on one or a few species. Origin of species also had no effect on the absolute rate of respiration when plants were grown and measured at 10 °C. However, when grown and measured at 20 °C, alpine provenances had a significantly higher mean leaf R, i.e., they were unable to proportionally downregulate R.

Recently, Collier (1996) conducted a similar screening experiment with 35 **arctic, subarctic and temperate taxa**. In this case, regardless of temperature regime (10 or 20 °C), leaf R measured at growth temperature was independent of geographic origin (Table 11.4), which is similar to what Chapin and Oechel (1983) described for *Carex aquatilis*. In Collier's survey, R of potted and fully acclimated plants was not significantly different among three provenance groups, hence showing no trend of inherently higher R in arctic populations. Given the two to threefold range of R within his provenance groups, this study underlined most impressively the potential pitfalls of studies with only one or a few species, when general principles are to be assessed.

As Collier and others have noted, a problem with such comparisons again is the reference. Results may differ when expressed per unit dry matter, fresh weight or leaf area because the relationship among these references also changes with climate, and there is no good advice as to the best reference. It seems, for a plant carbon balance consideration, a dry mass basis is a good compromise. When rates are to be related to function (e.g. carbon assimilation in leaves and nutrient assimilation in roots), area (leaves) and length (roots) might be more adequate.

In **summary**, a number of alpine plant species have been found to respire more than lowland plants, IF grown and measured at the same high temperature (no full acclimation). However, since the temperature in their habitat is NOT the same and commonly lower, their rates of in situ respiration are NOT necessarily higher, but rather may be lower in view of the lower soil and night temperatures – perhaps too low to achieve growth rates comparable to lowland plants (Körner and Larcher 1988; Friend and Woodward 1990; see Chaps. 7 and 12). Thermal adjustment to low temperatures – genotypic or not – allows alpine plants to achieve metabolic rates at these lower temperatures which low altitude plants would reach only at significantly higher temperatures. The explanation for this seems to be a combination of a greater number of **mitochondria** per cell, as suggested by Miroslavov et al. (1991) and Miroslavov and Kravkina (1991), and higher oxidative activity per mitochondrion (Klikoff 1968). Miroslavov and his coworkers also observed that the number of mitochondria per cell increases more with altitude than the number of chloroplasts, perhaps explaining the higher R/A ratio frequently observed in alpine plants (Mooney et al. 1964), when both R and A are studied at the same high temperature (which is not a good model for the natural climate).

Since perennial alpine plants live for much of their life at temperatures at or below 0 °C, it is surprising how little we know about respiration at such temperatures. The following two examples illustrate that respiratory carbon losses can be substantial during **cold periods**. Stewart and Bannister (1974) studied *Vaccinium myrtillus* shoots (thermally acclimated to their winter environment) in alpine Scotland at 2 °C and higher temperatures, from which one can deduce a rate of respiration of 0.6 mg CO_2 g^{-1} d.m. for 0 °C. A similar range of 0.3–0.6 mg CO_2 g^{-1} d.m. was measured during cold summer nights for *Carex curvula* leaves in the Alps (Körner 1989c). In view of the fact that respiration of 0.5 mg CO_2 g^{-1} d.m. could theoretically consume the total dry mass of such shoots in 140 days, such low rates of CO_2 release are ecologically very important, particularly in environments with a long dormant season.

Current evidence does not suggest that the respiratory carbon loss and the carbon balance of leaves of alpine and lowland herbaceous plants is any different (Diemer and Körner 1996). While leaf gas exchange can be studied in the field, it is impossible to study the whole carbon balance of a plant in situ because the root system is inaccessible, except through massive disturbance. Whether

Table 11.4. A comparative survey of arctic, subarctic and temperate zone leaf respiration in plants grown and measured at two temperatures (Collier 1996)

	Rate of leaf respiration in nmol CO_2 g^{-1} d.m. s^{-1}		
Origin of species:	Temperate	Subarctic	Arctic
Number of species:	7	14	12
R at 10 °C	34.7 + 8.9	35.3 + 9.8	35.6 + 8.0
R at 20 °C	31.2 + 9.2	28.4 + 8.6	29.2 + 9.7

the losses of carbon through respiration in whole plants differ also cannot be assessed by measurements at artificially high temperatures. Given the variability and species specificity of responses that has been seen so far, a broad screening for responses is required, which experimentally follows the concept of Atkin and coworkers (looking at the "productivity" of respiration rather than absolute numbers only), but adopts realistic growth conditions, soil temperature and nutrition in particular. Together with field data on actual plant responses and microclimate, this will allow modeling whole plant carbon losses during an alpine year. Even though genotypic effects can be marked, the effect of the environmental regime is greater (Billings and Mooney 1968), and hence deserves greater attention.

Ecosystem carbon balance

CO_2 uptake through photosynthesis and losses through respiration determine the balance of carbon flows in plants and whole communities. As long as plants grow, there is a net carbon flux into biomass, and when plants or parts of them die the flux is reversed. Together with soil microbe and animal respiration the net balance of all of these fluxes determines whether an ecosystem accretes or loses carbon. In late successional, quasi-stable systems and over large areas and long time scales these fluxes are commonly balanced. Over short periods, such as an alpine summer in the temperate or arctic zone, these fluxes are not balanced and correlate with the productivity of a site.

The net ecosystem CO_2 uptake (NEC), can be measured by exposing whole plots of intact vegetation to a transparent gas-exchange chamber. Aerodynamic methods, which assess CO_2 fluxes without enclosures, by using profile data, are an alternative approach, but the patchiness of alpine vegetation and gusty winds make it difficult to apply such methods at high altitude. During daylight hours such flux measurements reflect a mix of assimilatory and dissimilatory processes, from

which canopy photosynthesis itself is hard to separate.

Peak season CO_2 uptake of alpine vegetation has been measured on a few occasions only (Table 11.5). Per unit land area, NEC for good weather conditions appears to roughly follow a $30:20:10 \,\mu mol \, m^{-2} \, s^{-1}$ scheme from lowland crops to lush montane meadows and alpine vegetation. In the case of the alpine sedge community, NEC is similar to the A_{cap} of the dominating species, which means the whole system assimilates as much per unit ground area as leaves are known to assimilate per unit leaf area. For the *Loiseleuria* "cushion" shrub, which covered the ground fully, the intact canopy fixed twice as much CO_2 per ground area as fully illuminated isolated leaves fixed per unit leaf area. Aerodynamic exchange resistance and leaf shading in the canopy reduce CO_2 absorption per unit area within the canopy (Grabherr and Cernusca 1977).

The altitudinal reduction of maximum NEC for natural vegetation appears to correlate with the reduction in LAI. NECs in *Rhododendron* and low density prairie communities appear to match the alpine ones, as LAIs do, thus supporting the view that differences in A_{cap} contribute little to these trends. The higher numbers for prairie sites include communities with C4 species. The high end rates in some lowland crops may have to do with more uniform leaf quality and the inclusion of legume species such as alfalfa and soybean. According to Tappeiner and Cernusca (1996), light compensation points of NEC for ungrazed vegetation near the treeline range from 70 to $130 \,\mu mol \, m^{-2} \, s^{-1}$. This means that these closed communities achieve no net carbon gain when solar radiation is reduced to ca. 5% of the maximum (100 versus $2000 \,\mu mol \, m^{-2} \, s^{-1}$).

Mean rates of **night-time CO_2 evolution** (ecosystem CO_2 release, ER) of these systems are in the order of $1.5–2 \,\mu mol \, m^{-2} \, s^{-1}$ at peak season, i.e. about one sixth to one fifth of peak daytime NEC (references in Table 11.5). These efflux rates include above- and belowground plant respiration and CO_2 release from the substrate, largely through microbial activity. On a whole day basis,

Table 11.5. Peak season maximum rates of net ecosystem CO_2 uptake (NEC) in low stature vegetation from alpine and lower altitudes ($\mu mol\, m^{-2}\, s^{-1}$)

Type of community	Altitude	LAI	NEC$_{max}$	Ref.
Alpine zone (Alps)				
Carex community	2470 m	1.8	9.5	1
Loiseleuria monoculture	2000 m	3.2	10.8[a]	2
Subalpine zone (Caucasus)				
Rhododendron community	2200 m	3.8	13	3
Heracleum dom. community	2200 m	6.2	24	3
Upper montane zone (Caucasus)				
Hordeum dom. community	1850 m	5.4	19	3
Deschampsia dom. communirty	1750 m	5.8	22	3
Short and long grass Prairie (US)				
Low density short and tall grass	ca 2000 m	1.5	6–15	3
High density tall grass	ca 1500 m	ca 3	13–30	3
Lowland crops (US)	<500 m	4–5	22–45	3

References: *1* Diemer (1994), *2* Grabherr and Cernusca (1977), *3* Tappeiner and Cernusca (1996, including references for the two lowland data sets).
[a] This figure was calculated from an actual value of 12.7 obtained in intact monoliths of vegetation brought into a laboratory at 600 m altitude, at which the partial pressure of CO_2 was at least 15% higher than in the field.

the ER/NEC ratio will be bigger, largely because of lower than maximum NEC for much of the day. Given the lower soil and air temperatures, and the lower aboveground biomass of alpine compared with lowland ecosystems, these rates of night-time CO_2 release are quite remarkable and not much lower than in late successional lowland sites.

According to model calculations by Huber (1976) for the total **annual respiratory carbon losses** from living biomass in two alpine *Vaccinium* communities in Tirol (converted to dry matter units, 850 and 1640 g m^{-2} a^{-1}), half of the total respiratory CO_2 flux from plant tissue was released from belowground organs, and half from above-ground organs (despite 2.5 times greater below- than aboveground biomass). The ratios of carbon uptake and losses accounted for by plant respiration were 1.3 and 1.4 for the two communities. Because the ratio for the whole ecosystem should be close to 1 in such stable vegetation, the difference indicates that approximately one quarter of the annual carbon gain was recycled via microbial respiration in this ecosystem (biomass respiration

was derived from laboratory studies with isolated organs).

To date, the most exhaustive analysis of in situ **CO_2 evolution** from alpine vegetation and soil is from sites between 1600 m and 2500 m altitude in the Austrian Alps by Cernusca and Decker (1989a,b). Their full year studies in a *Carex* dominated community at 2300 m revealed a very close correlation of soil CO_2 release and **soil temperature** at 2 cm depth, contrasting with the commonly observed best fit with 10 cm depth temperature at low altitudes (a lag of 3–5 h was observed in the diurnal course when the temperature at 5 cm depth was used). **Soil moisture** exceeding 45% volume was found to depress CO_2 evolution by 30% (20 °C) to 60% (0 °C). Between 5 and 15 °C, the Q_{10} of soil CO_2 release during the main part of the growing season varied between 3.3 and 3.7 (rates at 10 °C of 1.9 to 3.2 $\mu mol\, CO_2\, m^{-2}\, s^{-1}$), indicating that temperature was much more influential on belowground compared with aboveground respiration (Fig. 11.18). Total ecosystem CO_2 release was composed of the following **partial fluxes**: 35% from substrate, 13% from roots, 11% from ground

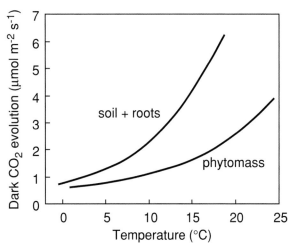
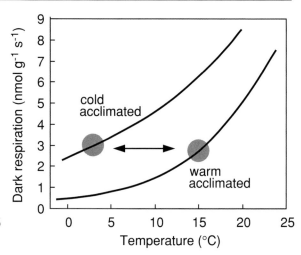

Fig. 11.18. The temperature dependence of midsummer CO_2 release in the dark from belowground (soil and roots) and aboveground (bio- and necromass) carbon sources in an alpine sedge community at 2300 m in the Austrian Alps. Note the greater temperature sensitivity of soil CO_2 evolution which possibly reflects the smaller in situ variation in soil temperature, compared with canopy temperature. (Cernusca and Decker 1989a)

Fig. 11.19. The immediate influence of preceding weather on the temperature response of mid season plant canopy respiration in an alpine sedge community at 2300 m in the Alps. The *upper curve* was measured after very cold weather with snow falls, the *lower curve* after very warm weather. Note the similar respiration at prevailing canopy temperatures indicated by the *shaded* area. (Cernusca and Decker 1989a)

litter, 13% from standing dead material (which is moist during much of the time) and only 28% was contributed from aboveground live plant parts.

The daily totals of CO_2 release from the soil surface decreased almost linearly from a peak of 360 mmol $m^{-2} d^{-1}$ (4 µmol CO_2 $m^{-2} s^{-1}$) early in the **season** (beginning of July; mean soil temperature 13 °C) to 40 mmol $m^{-2} d^{-1}$ (ca. 0.5 µmol CO_2 $m^{-2} s^{-1}$) at the end of the season in late October when soil temperature reached 0 °C. Peak season in situ plant canopy respiration measured at the same temperature (soil excluded) was more than twice as high on days preceded by very cold weather with snow fall and temperatures below 5 °C, compared with days preceded by very warm weather (Fig. 11.19), illustrating strong and fast thermal acclimation also at the plant community level. The overshooting in soil respiration seen after freezing periods by Brooks et al. (1997), may in part also have to do with such acclimatory processes. Cernusca and Decker (1989b) found no significant **altitudinal difference** of in situ respiratory losses of CO_2 per

g of phytomass across their gradient (a mean of 3 µmol CO_2 $g^{-1} s^{-1}$ at a field temperature of 10 °C).

A full season **carbon balance** for an alpine sedge community based on field measurements is illustrated in Fig. 11.20 for the Swiss Alps. CO_2 fluxes from various dates throughout the season combined with a model which links these gas flux data with actual microclimate data yielded a seasonal net CO_2 balance of 16 mol CO_2 m^{-2} season^{-1}, which would require a mean 0.7 µmol $m^{-2} s^{-1}$ efflux during the 270 day dormant season to become completely recycled. Since mobile carbon compounds will be utilized for spring growth (perhaps 10–15% of the carbon balance) respiratory carbon losses required to achieve a zero balance will be proportionally smaller. Otherwise, any annual net gain would accrete to the soil C pool, which is 8 to 10 kg m^{-2} in this old successional ecosystem. As was discussed in chapter 5, soils rarely freeze under snow in the Alps. In fact, there is evidence from various sites in this region that **soils under snow** remain at 0 ± 0.7 °C between −1 and −10 cm

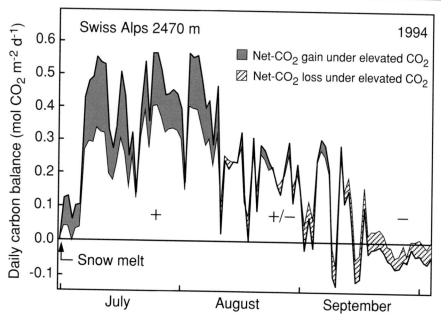

Fig. 11.20. The seasonal CO_2 balance of an alpine grassland (per unit ground area) in the Swiss Alps at 2470 m altitude at current ambient CO_2 and at elevated CO_2 (vegetation as shown in Fig. 11.13). The net CO_2 gain of $16 \, mol \, m^{-2} \, season^{-1}$ at ambient CO_2 corresponds to the sum of winter and springtime respiratory losses so that the annual balance approximates zero. Potential effects of the global rise of atmospheric CO_2 will be discussed in Chapter 17. (Diemer and Körner 1998)

depth for the whole winter (original data for 1996/97; the same was observed in the eastern Alps by Cernusca and Decker 1989a). This permits substantial microbial and root respiration in the order of $0.6 \, \mu mol \, m^{-2} \, s^{-1}$, as suggested by early and late winter measurements by M. Diemer (pers. comm.), sufficient to arrive at a zero annual carbon balance.

According to Brooks et al. (1997) the situation in the Rocky Mountains is different (more continental), and soils on Niwot Ridge freeze regularly, with temperatures falling as low as $-14\,°C$ at sites with poor **snow cover**, and still reach $-7\,°C$ under deep snow. This explains the overall low respiratory losses in winter reported by these authors. Mean rates during the period with snow cover vary with site from $0.03 \, \mu mol \, m^{-2} \, s^{-1}$ to $0.3 \, \mu mol \, m^{-2} \, s^{-1}$. Only at subalpine elevations did mean rates reach $0.6 \, \mu mol \, m^{-2} \, s^{-1}$ (references in Brooks et al. 1997). A comparison of their input and output estimates does not balance, hence there are some unac-

counted C sinks. These authors found no measurable soil CO_2 efflux at soil temperatures at or below $-5\,°C$, but a lot of variation at the prevailing temperatures of between $-2\,°C$ and $0\,°C$.

A comparison of modeled gains and losses of CO_2 for a sedge community at 2300 m in the Austrian Alps ended up with independent estimates of a $36.5 \, mol \, m^{-2} \, a^{-1}$ uptake versus a $37.6 \, mol \, m^{-2} \, a^{-1}$ loss, i.e. a zero C balance within the given uncertainty of such calculations (Körner 1989c). A mass balance approach (repeated harvests of all phytomass compartments over several years) in alpine dwarf shrub communities revealed no measurable net change over the years (Larcher 1977), again suggesting a balanced carbon budget in the long term, as is to be expected for such **stable natural ecosystems**. The combination of a warm winter with a miserable summer, or a winter with poor snow cover and frozen soils followed by a warm summer would probably lead to strong deviations from a balanced C budget.

The data presented in **this chapter** indicate that the slow growth and small stature of alpine plants, as compared with lowland plants is not associated with low rates of uptake and loss of carbon per unit of tissue. Photosynthetic leaf gains do not appear inferior, and there is no evidence that respiratory losses are enhanced in alpine plants. Apart from the overwhelming effect of the duration of the active season outside the tropics, it appears that the way in which alpine plants invest their assimilates and control their development plays an important role in understanding the overall appearance of alpine vegetation. The primary processes of carbon metabolism do not provide the answer.

12 Carbon investments

It is important, though trivial, that low temperatures affect **different life processes** in different ways. The life process most limited will most strongly influence growth and development. In the last century, when researchers watched submerged aquatic leaves for air bubbles in order to check whether the plant was photosynthesizing, a handful of ice added to such a water bath had illuminated the fundamental **dilemma between supply and demand** of carbon. By using these methods, Kraus (1869, p. 523) studied the effect of light and temperature on starch and sugar formation in plants in Würzburg and hit the point remarkably well:

"The low temperature limit of photosynthesis appears to be lower than that in any other plant process. Such a design may not only be useful, but appears essential. This assures that secondary processes such as cell division, cell enlargement, respiration etc. never fall short of substrate. It would be useless and even very ineffective if plants could grow and respire at 2 °C but were unable to produce the material required for these processes" (free translation from the more baroque German of those days).

Kraus's view is a good preface for this chapter, because it questions conventional beliefs focused at "photosynthesis first". Do we know which is affected more by low temperature: synthesis or investment of assimilates? This and the following chapter will provide evidence that in alpine altitudes, the latter seems more likely than the first. In the previous chapter it has been shown that rates of photosynthesis per unit leaf area are unlikely to limit growth of alpine plants to greater extent than in comparable lowland plants. It was also shown that the lifetime carbon balance of the unit leaf area is not necessarily lower in alpine plants, hence it is impossible to explain their slower growth on the basis of leaf gas exchange (Körner and Larcher 1988). The picture could change at a whole plant level, for instance, if the leaf area produced per unit of leaf mass or leaf mass per total plant mass was smaller at alpine altitudes and if the overall respiratory burden increased due to such allometric changes. The build up of mobile pools of carbon compounds (non-structural carbohydrates and lipids) could also affect the overall carbon balance. These are the topics considered in this chapter.

Non-structural carbohydrates

The abundance of **mobile compounds in plant tissues** may reflect two situations, with gradual transitions inbetween: (1) a demand for growth and adequate functioning of the plant (osmotic requirements, adjustments against stresses, short- and long-term reserves, transport), or (2) overshooting supplies, because of an imbalance between synthesis and demand (neutral or negative effects; "accumulation" after Chapin et al. 1990). The latter possibility is rarely considered but is one among many possible reasons why certain sugars, starch, fructans and lipids may be found in high concentrations in cold climates. An objective approach needs to consider both possibilities, and there are several reasons why a carbon overabundance problem may occur under certain alpine life conditions.

Research dating back to the beginning of the century showed that alpine, like arctic plants, are not short in total non-structural carbohydrates (TNC) as rated by tissue concentrations (e.g. Wilson 1954; Mooney and Billings 1965; Bell and Bliss 1979; Skre 1985; exellent review by Wyka 1999). However, an inspection of the literature, which largely considers single species, revealed a lot of variation and uncertainty about the justification of a generalization. A spot check on 38 truly alpine non-grass species at peak season at 2500 m in the Alps produced the pattern shown in Fig. 12.1. TNC concentrations in leaves range from 8 to 28% of dry weight, with a great **majority of species containing 15 to 20% TNC**. While in most species TNC is dominated by soluble carbohydrates (60 to 90% of TNC), eight species show the opposite pattern, including all five legumes in which starch dominates TNC. No other relatedness of these data with any other potential functional or taxonomic grouping is apparent. Also, micro-habitat had no significant influence by that time of the season. Obviously, species are different, which in some cases is related to their specific phenological rhythms (see below).

Whether these alpine TNC data are different from those in comparable plant species from low **altitude** was tested by collecting a similar data set from an infertile semi-natural lowland pasture. As illustrated by Fig. 12.2, the difference is small and not statistically significant, with slightly lower TNC concentrations at low altitude within a similar overall range. No significant altitudinal or latitudinal differences in mobile carbohydrates within selected species or genera were found by Mooney and Billings (1965, low/high temperate alpine), Chapin and Shaver (1989, temperate/arctic), Gonzalez (1991, low/high tropical alpine), Russell (1948, arctic/subtropical alpine), and Hnatiuk(1978, tropical alpine/subantarctic). If there was a trend, carbohydrates appear more rather than less abundant at the cool end of such gradients. However, in one respect there is a clear cold/warm difference: the lower the temperature, the less abundant becomes **starch** – except in legumes, but legumes become increasingly rare as

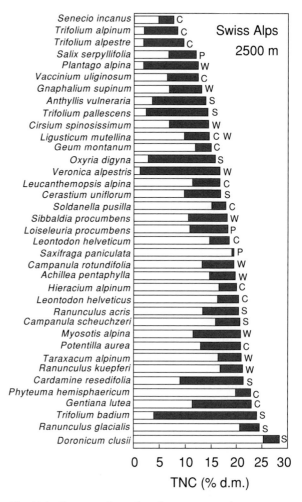

Fig. 12.1. Concentrations of total non-structural carbohydrates (TNC, % of total dry matter for a pooled sample from at least five individuals per species) in leaves of 37 alpine non-graminoid species sampled at peak growing season (27 July, 17:00–19:00 h) at 2500 m in the Swiss Central Alps (Furka Pass). Analysis for sugars (*light bars*) and starch (*dark bars*) followed the procedure described in Schäppi and Körner (1997). *Letters* indicate different sampling habitats. *P* exposed peak, *S* saddle, *C* flat *Carex* heath, *W* wet snowbed community. (Unpubl.)

altitude increases, and are absent at highest elevations. Enzymes involved in starch synthesis were found to be much more sensitive to low temperature than for instance enzymes of sucrose and fructose metabolism (Pollock and Lloyd 1987).

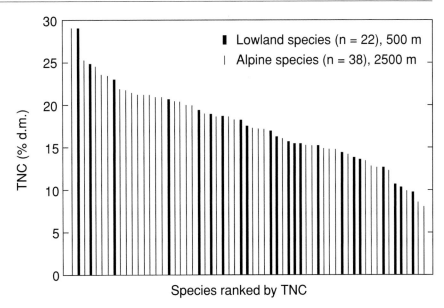

Fig. 12.2. A comparison of TNC in alpine and lowland herbaceous species. Alpine data and analysis as **Fig. 12.1.** The lowland samples are from a semi-natural calcareous grassland at 500 m altitude near Basel and mostly include species from the same plant families or even genera which were sampled at high altitude, all sampled at biomass peak in early June, again in late afternoon. (Unpubl.)

Given the pronounced **seasonality** in the alpine zone of extra-tropical mountains, mobile carbohydrate pools may be expected to undergo substantial seasonal variation, perhaps contrasting with the situation in the alpine tropics. The handful of studies on seasonal variation of TNC concentration in the **temperate alpine zone** (e.g. Mooney and Billings 1960; Bell and Bliss 1979; Skre 1985; Shibata and Nishida 1993; Wyka 1999) revealed three messages:

- There is never a critical depletion of TNC-pools. In the words of Mooney and Billings (1960): "all our results indicate that adequate carbohydrate reserves are present even at the lowest point in the seasonal cycle".
- TNC pools reach a minimum at the time of or shortly after snow melt; maxima in leaves can occur at any time between peak season (at peak flowering; Fonda and Bliss 1966; Wyka 1999) and late season. In belowground organs, TNC maxima commonly occur at the end of the season.
- Major refilling of diminished pools is fast (3–4 weeks) and often parallels the phase of maximum growth (Fig. 12.3, see also Fig. 12.6).

Ulmer (1937), who studied seasonal variation in sugars in conjunction with frost resistance in a series of alpine species in the Alps, found very little variation throughout the year, even none at all in some species. TNC concentrations are also fairly robust in the face of local variations in life conditions, both in summer (Fig. 12.1) and in winter, as was illustrated by Bell and Bliss (1979). These authors manipulated winter snowpack and compared transplants and found some, but not very pronounced, and consistent consequences for TNC (Fig. 12.4). TNC also differed surprisingly little across a steep snow melt gradient in early summer (Fig. 12.5). No significant variation in TNC concentration was observed in alpine grassland during the core 10 weeks of the growing season during 3 consecutive years in the Swiss Alps (Schäppi and Körner 1997).

A comparatively pronounced seasonal variation in TNC concentration was observed by Gonzalez (1991) in the **subtropical**-alpine zone of northwest Argentina (data from 3000, 4200 and 4600 m). TNC concentrations in leaves of the rosette forb *Calandrinia acaulis* varied seasonally between 2 and 15% in leaves and 3 and 47% in roots, with minima in late summer, and maxima

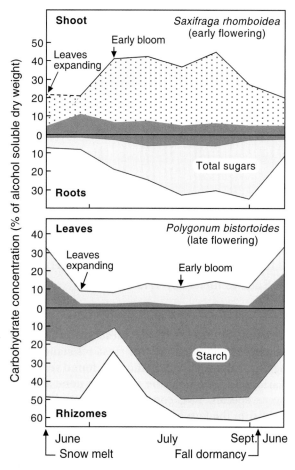

Fig. 12.3. The annual carbohydrate cycle in *Saxifraga rhomboidea* and *Polygonum bistortoides* at 3300 m in the Medicine Bow Mountains of Wyoming (% of alcohol-insoluble dry weight). The difference between the lines for TNC and total sugars represents starch concentration. (Mooney and Billings 1960)

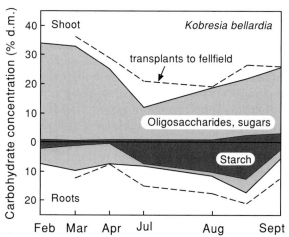

Fig. 12.4. The annual course of carbohydrates in *Kobresia bellardii* at 3600 m altitude, Trail Ridge, Rocky Mountains National Park, Colorado (% of alcohol insoluble dry weight). *Solid lines* for a site with moderate to low snow cover (22% of all days in February and March snow free) at which *Kobresia* dominates naturally, *dashed lines* for transplants exposed to a fellfield climate with poor snow cover (38% of all days snow-free in February and March). (Bell and Bliss 1979)

achieved within 2 weeks at the rather immediate onset of the dry alpine winter in March, opposite to the trends found in the temperate zone. Roughly the same pattern (though more variable and with much lower absolute concentrations) was exhibited by the alpine fern *Woodsia montevidensis*.

The only long-term data for the **tropics** are those for tussock grasses of the alpine zone between 3500 and 4000 m on Mt. Wilhelm, New

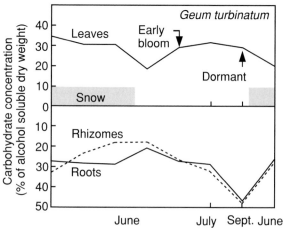

Fig. 12.5. The variation of carbohydrates in *Geum turbinatum* from the same location as Fig. 12.3 (same units), but for plants sampled across a snow melt gradient on June 20, plus samples from July 16 (flowering peak) and September 8 (dormant). (Mooney and Billings 1960)

Guinea measured by Hnatiuk (1978). Hnatiuk found no "seasonal" trends, but enormous variation from tussock to tussock, reflecting the decoupling of phenorhythmics from climatic drivers in this ever humid equatorial climate. However, he detected a clear diurnal variation in the concentration of sugars from a mean of around 12% (range 6 to 20) in the morning, to a peak of 25% (with much less variation) in the second half of the day. Starch levels showed the same diurnal trend, but concentrations were extremely low, mostly around 3%. According to Beck (1994), leaves of afro-alpine plants (*Senecio* sp., *Lobelia* sp., *Alchemilla argyrophylla*) contain up to 38% of their dry weight as sucrose, which represents 80–90% of TNC. The very low starch content also found in these tropical alpine species increased at the expense of sucrose when plants were grown in a warm greenhouse, in line with observations by Rada et al. (1985) who found that warm pre-treatment of *Polylepis* shoots reduces the soluble fraction of TNC in leaves by 2 to 3% within a few hours. Similar diurnal trends in soluble carbohydrates associated with temperature were described by Azocar et al. (1988) for high Andean *Draba*. New (unpublished) data from the Ecuadorian Andes show a mean leaf TNC concentration of only 9% (late afternoon), and confirm the relatively higher sugar concentration, compared with starch (again with the exception of legumes; Table 12.1). Soft leaved species like *Ranunculus* sp. and *Geranium* sp. show the highest concentrations (21 and 13% sugar; 3% starch), the rigid and pubescent, extreme high altitude *Culcitium* species the lowest (1–3% sugar, 2–3% starch), illustrating that volume density (cell wall thickness and sclerenchyma) strongly influences such means. The greater cloudiness in the Ecuadorian mountains may also add to the lower means.

Turnover times of mobile carbon pools in alpine plants are largely unknown (for turnover of whole organs such as leaves and roots, see Chap. 13). The only data I know are from an unintentional labeling experiment in the course of a CO_2 enrichment study in the Swiss Alps (Körner et al. 1997). After almost four seasons of exposure to

Table 12.1. Non-structural carbohydrates (% d.m.; means and range for n species) in leaves of alpine plants of the equatorial tropics (Ecuador; Guagua Pichincha 4450 m, Cayambe 4500 m, Páramo La Virgen 4000 m; unpublished data, for method see Schäppi and Körner 1997)

Group of plant species[a]	n	Sugars	Starch
Herbaceous non-legume dicots	12	6.8 (1–21)	3.4 (1–5)
Graminoids	2	8.8 (8–9)	1.5 (0–3)
Legumes	2	2.6 (1–4)	4.7 (2–8)

[a] genera included *Calandrinia, Cerastium, Culcitium, Erigeron, Geranium, Hypochoeris, Lachemilla, Niphogeton, Ranunculus, Valeriana, Werneria,* (herbaceous); *Agrostis, Luzula* (graminoid); *Astragalus, Lupinus* (legumes); samples collected by M. Diemer.

twice normal ambient CO_2 concentrations, the dominant species of the alpine grassland studied, *Carex curvula*, carried a strong $\delta^{13}C$ isotope signal of -37.0 ± 1.4‰, because the tank CO_2 had a $\delta^{13}C$ of -30 to -40‰ compared with -8‰ in normal air (the air mixture experienced by CO_2 enriched plants had a $\delta^{13}C$ of around -24‰). After termination of the treatment during the second half of the fourth season (10 August 1995) the fate of tissue ^{13}C was followed over two more seasons (Fig. 12.6). By mid September, most of the aboveground parts of *Carex* leaves are dead. The $\delta^{13}C$ for untreated *Carex* is -25.5 ± 0.8‰ at this site (see also Chap. 11).

Leaves emerging immediately after snow melt contain only about one quarter of "old" isotope label, hence three quarters of them are constructed from carbon assimilated during the last month of the previous season and current assimilation. In contrast, the inflorescence contains ca. 80% of carbon dating from before mid August of the previous season. By the end of the 1996 season, the mean for all current season leaf tissue was back to control levels, which means that the current leaf crop is largely a product of current season as-similation. There was no difference in isotopic signature between the basal regrowth tissue of established (2nd year) leaves and the tips of the next generation leaves, right from the beginning of the season (basal leaf meristems remain

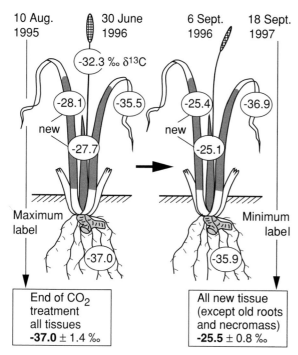

Fig. 12.6. Tissue carbon turnover in alpine *Carex curvula* as indicated by the disappearance of the stable carbon 13 isotope label in the course of 2 years after labeling with ^{13}C depleted CO_2 had been terminated. Note the fast removal of the signal in new leaves, indicating that mobile carbon pools older than 1 year are hardly utilized in leaf growth. The signal in new tissue at the base of old leaves and in tips of a new generation of leaves is similar, hence both appear to be supplied from the same carbon source. (Unpubl.)

active for three seasons). Thus, both meristems have the same C source. Some thick roots and attached dead leaf parts still carried the unchanged ca. −37‰ signal by the end of the second season after the treatment was terminated (September 1997). Although some recalcitrant TNC fractions may remain stored in belowground organs, these observations suggest turnover times of mobile carbon of less than 2 years. In fact, it seems that most of the active pool is replaced after less than 1 year and carbohydrate stores are rather mobile in the cold (Wallace and Harrison 1978).

Overall, the mean midsummer TNC concentrations seen during these individual plant studies in the temperate zone and the means of the few data

from New Guinea, the Andes and equatorial Africa fall within the upper half of the range shown in Fig. 12.1, suggesting (1) that TNC concentrations are indeed relatively high at high altitudes, and (2) a latitudinal difference in TNC concentration at these high altitudes does not seem to exist. However, much more data, particularly for subtropical and tropical mountains are needed before a conclusive picture will emerge. In the temperate zone, respiratory consumption in winter and during emergence at snow melt are the primary sinks for TNC. The warmer the winter, the thicker the snow cover, the more TNC becomes depleted (Stewart and Bannister 1974; Bell and Bliss 1979). When storage **rhizomes** are present, their TNC content shows most pronounced seasonal variation and is closely linked to the vigor of spring growth (Klimeš et al. 1993). Such rhizomes can store enormous amounts of **TNC per unit land** area: Mooney and Billings (1960) estimated an equivalent of $5 \, t \, ha^{-1}$ for patches of *Geum turbinatum*, which corresponds to the grain yield of a good wheat crop. Closed alpine grassland with $600 \, g$ biomass m^{-2} (above- plus belowground) with only 15% TNC would accumulate ca. $1 \, t$ of TNC ha^{-1}.

Reasons for, and the potential **functional significance** of, the relatively high concentrations of TNC in alpine plants are not obvious. Independent of function, high concentrations of TNC suggest either good supply or limited demand by structural or metabolic sinks. The most commonly discussed functional attribute of high sugar concentrations is their association with **frost resistance**. However, the logic is not straightforward as was discussed in Chapter 8, because concentrating osmotically active compounds is an inefficient way to avoid freezing, since 1 mole of solutes (osmotic pressure of ca 2.25 MPa) depresses freezing point by only 1.8 K. It is also not clear why alpine legumes should not require similar sugar concentrations for this purpose.

Since the concentration of soluble carbohydrates has been found to correlate with frost resistance in alpine plants, other membrane stabilizing functions of sugars are likely to be more important (Sakai and Larcher 1987, p. 114ff; Beck 1994). However, it remains difficult to separate causal from coincidental effects. Kaurin (1985) noted that a high carbohydrate concentration does not necessarily result in high frost resistance in *Poa alpina*, and Ulmer (1937), as mentioned above, found pronounced variations in frost resistance with comparatively small or no variation in sugars.

These field observations for alpine plants fit **general trends for plants growing in the cold at low altitudes** or being experimentally exposed to low temperatures. Given the significance of biological variation, more general patterns can only be obtained by considering large groups of species from geographically diverse origins. Such a survey – though not in alpine species, but very relevant for their understanding – was undertaken by Chatterton et al. (1989) who have screened 185 grass species from across the world (including a few montane species) and found remarkable patterns in TNC concentrations and their temperature responses. Grasses (in addition to Asteraceae) are special because outside the tropics they represent a family with the most consistent occurrence of **fructans**, carbohydrates whose solubility varies with their degree of polymerization, and which are stored in the vacuole, i.e. a place where they cannot directly interact with the bulk of cellular membranes. Chatterton et al. grew their plants at a 10/5 °C and 25/15 °C day/night growth regime (the first quite typical for midsummer near the treeline), and grouped them into **cool season** (C3, 128 species) **and warm season** (C4, 57 species) **species**. The following results were obtained for concentrations of leaf TNC or its components glucose, fructose, sucrose, starch and fructans:

- Concentrations of all TNC fractions were higher at low temperatures, irrespective of species group.
- Cool season species had higher concentrations of most TNC compounds, irrespective of growth temperature.
- Warm season grass species have no fructan, but almost all cool season grass species do.

- Of all compounds, fructans show the most dramatic increase with decreasing temperature.

Thus, TNC, and its component's concentrations in grasses (and possibly other groups of plant species) exhibit **phenotypic** (environmental) and **genotypic** (heritable) components, and are negatively correlated with temperature. The mean TNC concentration in grass leaves of the cool season group was 31% of TNC-free leaf dry weight at 10/5 °C compared with 10.7% at 25/15 °C (Table 12.2). Fructans increased from a mean of 1.2% at high temperature to a mean of 11.5% or one third of TNC, at low temperature. Although fructans do not universally contribute to TNC, their presence in grasses provides an elegant tool for understanding low temperature effects on TNC. Chatterton et al. plotted their cool season species data for the various TNC fractions against TNC concentrations and found a remarkable pattern: while all other compounds tested increased almost linearly with TNC concentration, with the regression intercept near zero TNC, the intercept for the highly linear regression with fructan concentration had a threshold of 15% TNC, which means fructans are the last component to become enhanced when TNC goes up (Fig. 12.7).

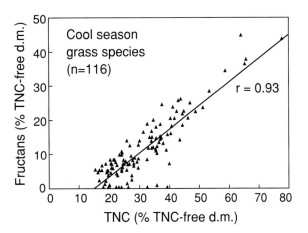

Fig. 12.7. The concentrations in fructan versus total non structural carbohydrates (TNC) per unit of TNC-free leaf dry mass in 116 grass species from cool climates grown at a 10/5 °C day/night growth regime. (Chatterton et al. 1989)

Table 12.2. Mean non-structural carbohydrate concentrations in 128 grass species from cool climates grown at 2 day/night temperature regimes (% TNC-free leaf dry mass; Chatterton et al. 1989)

TNC-components	25/15 °C	10/5 °C
Fructose	1.4	2.4
Glucose	1.8	2.9
Sucrose	2.3	5.8
Starch	4.1	8.6
Fructans	1.2	11.5
TNC	10.7	31.2

Table 12.3. Ranges of mobile carbon compounds and some other constituents determining agronomic fodder quality in eight seminatural or natural grassland communities between 1100 and 2500 m altitude in the Alps (woody elements, such as dwarf shrubs excluded; analysed by standard agronomic methods). Note, these are bulk biomass samples cut at 2–3 cm above the ground, also reflecting elevational changes in species composition (all numbers in % dry matter and for mature canopies harvested between end of June and mid September). (Spatz et al. 1989)

Type of compound	1100–1900 m	2300–2500 m
Mono- and disaccharides	5.0–10.0	2.1–7.3
Starch	0.3–0.9	0.3–1.2
Fructans	0.7–3.4	0.3–2.8
Raw protein	8.8–18.0	6.7–12.7
Lignin	2.0–6.0	3.2–8.1

Chatterton et al. (1989) concluded, that fructan accumulation in grasses in cold climates is not an alternative to starch formation (a substitute) but is an additional means of **disposing of carbohydrates in a harmless way** in the vacuole, in order "to allow photosynthesis to continue at cool temperatures when other storage pools are saturated". The very pronounced response of fructans to low temperature is a well-known phenomenon (e.g. Farrar 1988), but their role in alpine grasses has not been explored. In *Poa alpina* leaves at 2500 m altitude in the Alps, fructans contribute only 4% to a mid season TNC pool of 30%, less than half as much as was found in the Asteraceae *Leontodon helveticus* (B. Schäppi and Ch. Körner, unpubl.)

Since mobile carbon fractions also co-determine the **quality of high altitude fodder** for grazing ruminants, bulk biomass samples have been analyzed repeatedly by agronomists. Table 12.3 shows the ranges found in eight mountain hay fields and pastures in the Alps, all receiving either little (organic) or no fertilizer. Most components appear equal or less abundant at high altitude, but this reflects the changed composition of the vegetation, and is not related to physiological changes within defined organs of comparable species. The elevational trends are largely driven by the increasing abundance of sedges or hard-leaved tussock grasses at high altitude. Hence, just as with nitrogen (protein) concentration, one needs to distinguish between species-, genus- or morphotype-specific trends, and those in whole community harvests.

Experimental manipulations which affect the plant's carbon balance may shed further light on the reasons for high TNC in the alpine zone. In a 3 year very low dose fertilizer trial in alpine grassland (Schäppi and Körner 1997), which still doubled seasonal biomass accumulation, TNC concentration remained completely unaffected. In *Oxytropis*, very fertile test conditions did reduce TNC as a result of enhanced structural growth (Wyka 2000). When, in addition, plants were exposed to a doubled ambient CO_2 concentration over the same 3 year period, TNC concentrations increased in the majority of plant species, irrespective of the fertilizer treatment, but there was absolutely no CO_2 effect on growth (see Fig. 17.6).

In **summary**, all these observations suggests that high TNC concentrations in alpine altitudes are unrelated to biomass production or nutrient availability, but are associated with low temperature per se. The fact that stimulation of photosynthesis by elevated CO_2 further enhanced leaf TNC but not growth suggests that there is a problem with either carbon dissipation or investment.

Assimilate transport includes the critical step of **phloem loading**, which is known to be temperature sensitive. Gamalei et al. (1994) demonstrated that the endoplasmatic reticulum, which facilitates plasmodesmatal (i.e. symplastic) transport, collapses below 10 °C in warm adapted species, and as a consequence starch accumulates in the mesophyll of such species. However, cold climate species commonly bypass the constraints of symplastic phloem loading by adopting the apoplast pathway, which is much less temperature sensitive, and is comparatively faster (Van Bel and Gamalei 1992), and which leads to less TNC accumulation under otherwise equal conditions (Körner et al. 1995). Hence, a more likely explanation than a transport problem for the unresponsiveness of growth to CO_2 enrichment, and the overall high TNC levels in alpine plant tissues is a limitation of structural growth rates.

A great number of experimental studies in individual species of grass and non-grass species of low altitudes have added to the evidence that low temperatures enhance TNC, and several authors have attributed this phenomenon to the fact that growth is more sensitive to low temperature than is photosynthesis (Neales and Incoll 1968; Farrar 1988; Baxter et al. 1995). This aspect will be further discussed in chapter 13.

Lipids and energy content

In terms of carbon accumulation, a second important group of non-structural carbon compounds are lipids, the abundance of which has been found to increase with altitude. A survey of lipid concentrations along an elevational gradient on Mt. Olympus in Greece revealed a doubling from Mediterranean sea level to 2500 m altitude (Fig. 12.8). Although part of this change may be explained by decreasing contributions of lipid-depleted stem tissue as altitude increases, the trend is in line with observations by Headley and Bliss (1964; New Hampshire), Hadley and Rosen (1974; New Zealand), Tschager et al. (1982; Alps) and Höner and Breckle (1986; Greece), for temperate and Mediterranean alpine sites. Lipid concentrations tend to be very high in alpine plants.

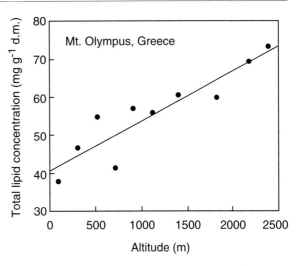

Fig. 12.8. Total lipid concentration in above-ground herbaceous biomass from an altitudinal gradient on Mt. Olympus in Greece. (Pantis et al. 1987)

Since the energy content of lipids is among the highest of all plant constituents, it is not surprising that linear correlations between total lipid content and leaf energy content have been found (Tschager et al. 1982; Pantis et al. 1987). In alpine leaves, lipid concentrations may reach 10% of dry weight, energy contents may exceed $20 kJ g^{-1}$ dry weight, and there is ultra-structural evidence for a very active lipid physiology in alpine leaves (Lütz 1987). Accordingly, the energy content of plant tissue tends to increase with altitude independently of latitude (Baruch 1982, Pipp and Larcher 1987). Lipid and energy contents are lower in belowground organs, but due to their greater total biomass contribution, Tschager et al. (1982) estimated that half of the total lipid pool is below the ground in evergreen dwarf shrubs and even more in deciduous dwarf shrubs. Common mid season leaf lipid concentrations in alpine Ericaceae are 5–8% d.w. (Table 12.4).

It was repeatedly observed that lipid concentrations in mature alpine plant tissue undergo comparatively little seasonal variation (Fig. 12.9). Headley and Bliss (1964) found almost no change with season, but they made the important obser-

Table 12.4. Total lipid concentrations in alpine Ericaceae in summer. Means in % dry weight. (Tschager et al. 1982)

Species	Leaves	Stems	
		A	B
Evergreen shrubs			
Calluna vulgaris	3.8	3.7	2.3
Empetrum hermaphroditum	7.3	4.2	4.8
Vaccinium vitis idea	6.0	2.6	1.9
Arctostaphylos uva ursi	8.1	4.5	3.1
Loiseleuria procumbens	9.0	8.0	4.8
Deciduous shrubs			
Vaccinium myrtillus	4.9	1.5	1.2
Vaccinium gaultherioides	5.5	2.7	2.0

A above ground stem, *B* below ground stem.

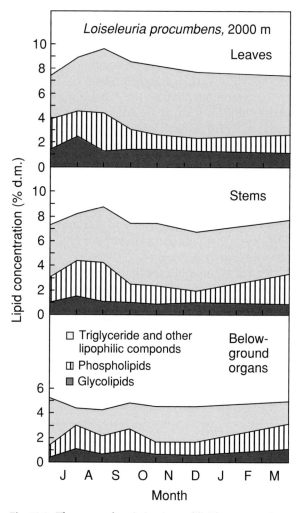

Fig. 12.9. The seasonal variation in total lipid concentration in *Loiseleuria procumbens* at 2000 m altitude in the Tirolian Alps. (Tschager et al. 1982)

vation that lipids accumulate more in old leaves. Accordingly, several authors reported little temporal or spatial change of energy content once tissues were matured (e.g. Headley and Bliss 1964; Bliss 1966; Brzoska 1973; Fig. 12.10). When early season and very late season data are included for herbaceous alpine vegetation, a seasonal reduction of energy content has been observed (Grabherr et al. 1980) which may have resulted from a "dilution" of energy content first by cellulose accumulation (leaf maturation) and then by recovery of leaf protein and/or lipids during senescence.

According to the analysis by Tschager et al. (1982) in *Loiseleuria*, two thirds of the total lipid fraction were found to be unsaturated **fatty acids**, largely with two and three double bonds, and only one third was found to be saturated. These relationships varied little among organs. The overall accumulation of lipids per unit land area estimated by these authors is remarkable: 2.4 t ha⁻¹ in biomass and 1.3 t ha⁻¹ in the litter layer, i.e. nearly 4 t ha⁻¹. Annual lipid turnover was estimated to be in the order of 1 t ha⁻¹.

In **summary**, a similar picture emerges for lipids as for non-structural carbohydrates in alpine plants. High concentrations, large pools per land area, and relatively little seasonal variation in concentrations. It is tempting to assume that these

high energy compounds are important reserves. However, why are these "reserves" accumulating in old evergreen leaves? Why are they abandoned with leaf shedding in evergreens and accumulate in the litter layer (as was observed by Tschager et al.) if they are so precious? These authors mention the possibility that lipid accumulation may have to do with growth limitation by mineral nutrient shortage, and they refer to studies in algae, in which lipids were found to accumulate

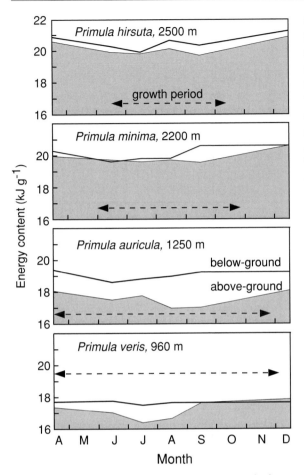

Fig. 12.10. The altitudinal and seasonal variation in leaf energy content in different *Primula* species in the Tirolian Alps. (Zachhuber and Larcher 1978)

when nutrients were short, an indication of carbon oversupply. The massive occurrence of lipids in litter really seems like a most wasteful management of carbon compounds by these alpine plants – not what one would expect were carbon a limiting resource.

Carbon costs of leaves and roots

In terms of a plant's carbon economy, it matters how much carbon is invested per unit of photo-synthesizing leaf area or per unit of length of fine roots exploring the soil. Growth analysis has shown that the leaf area per leaf mass (specific leaf area, SLA) and the root length per root mass (specific root length, SRL) are two key drivers of the relative growth rate. Slow growing species commonly found in habitats with low resource supply have been shown to be characterized by low SLA and high SRL (Ryser 1996; Atkin et al. 1996a). The following text considers investments in terms of plant dry matter, but these can be approximately converted to carbon by multiplying by 0.45.

Within the same morphotype (forb, grass, sedge, deciduous or evergreen shrub etc.) alpine plants commonly produce less leaf area per unit leaf dry mass, hence their **specific leaf area** is lower than in plants from low altitudes. This trend has been documented for whole groups of species or for individual species adapted to contrasting altitudes, both in the temperate zone and the tropics (e.g. Baruch 1979; Woodward 1983; Körner et al. 1983, 1986, 1989a; Pyankov et al. 1999). For example, in the Alps mean **SLA** drops from 2.3 to 1.9 dm^2 g^{-1}, i.e. −15% between 600 and 3000 m (Fig. 12.11). The difference is larger (−30%) when only congeneric species are considered. However, an aspect often overlooked is that most dicotyledonous species from alpine altitudes have sessile leaves or leaves with very short petioles. On average, they invest only 6% instead of 15% of total leaf mass in petioles, hence two thirds of the 15% greater mean investment of mass per area (1/SLA) are returned by savings in petiole construction (Körner et al. 1989a).

Explanations for reduced SLA at high altitude are greater leaf thickness and thicker cell walls, particularly in the epidermis (Körner and Diemer 1987; Körner et al. 1989a, Fig. 11.3). Experiments in which plants were grown under contrasting temperatures showed that these structural features are partly **genotypic** and partly **phenotypic**. A strong genotypic component was documented for grasses by Woodward (1979a, b), and Atkin et al. (1996). In their growth chamber experiments, the alpine species always had a lower SLA than the lowland species, irrespective of growth tempera-

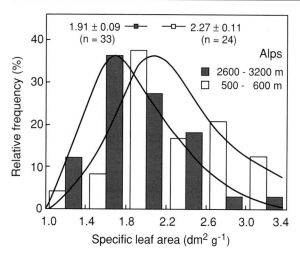

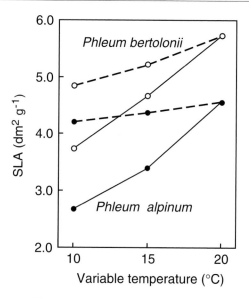

Fig. 12.11. Specific leaf area in herbaceous plant species from low (500–600 m) and high altitude (2600–3200 m) in the Alps. (Data pooled from Körner et al. 1989a and Diemer et al. 1992)

Fig. 12.12. The temperature response of specific leaf area in an alpine and a lowland grass species (*Phleum alpinum* and *Phleum bertolonii*). *Solid line* for plants in which belowground and aboveground organs experience the same temperature; *dashed lines* for plants in which only the belowground (including the apex) temperature was varied, while the aboveground temperature was maintained at 20 °C. (Woodward 1979b)

ture, but according to Woodward's study the relative effect of growth temperature was twice that of origin (Fig. 12.12).

No conclusive evidence exists for dicots (cf. Körner et al. 1989a). When grown at 8 and 18 °C five out of nine species reduced SLA at low temperature (only one significantly), two were indifferent and in three SLA was higher, but leaf maturation status may have been unequal. In an alpine/lowland **transplant** experiment, two out of four species decreased and two increased SLA, perhaps reflecting differential sensitivity to the drier growth conditions at low altitude. The only consistent and highly significant effect of growth of alpine species at low altitude was a reduction of the thickness of the outer epidermal cell wall by 15 to 45%. A cross continental transplant experiment with *Ranunculus* sp., *Geum* sp. and other dicots by Prock and Körner (1996) confirmed the stronger phenotypic and weaker genotypic nature of SLA. Temperate alpine species transplanted to the cooler arctic-alpine climate further reduced SLA, whereas arctic-alpines transplanted to a temperate-alpine climate slightly increased SLA. A similar trend was observed in low altitude latitu-

dinal transplants. Transplant experiments with *Alchemilla alpina* in Scotland also revealed predominant phenotype responses of SLA (Morecroft and Woodward 1996).

A **global comparison of** altitudinal differences in **SLA** indicates that trends are similar in forbs, shrubs and trees and along altitudinal gradients in the arctic, temperate, and tropical zones, and even appear to be relatively robust when faced with local variations in radiation and humidity. SLA in sclerophyllous species, for instance, was found to exhibit a strong elevational reduction in both New Zealand (Körner et al. 1986) and New Guinea (Körner et al. 1983), despite the fact that the latter gradient included an altitudinal reduction of radiation to about one third of the lowland value, whereas radiation does not change with altitude in New Zealand. Hence the common driver of changes in SLA with altitude appears to be reduced temperature.

However, in addition to temperature, there is also the influence of **season length**. SLA increases from the tropic-alpine to arctic alpine (see Chap. 10), and in the temperate zone plants found at late **snow melt** sites (typical snowbed species) have higher SLA than plants from early snow melt sites (Kudo 1996). Kudo showed that along snow-melt gradients, SLA correlates negatively with leaf life span. The shorter the leaf life span, the greater the SLA (the less carbon invested per unit leaf area; see also Chap. 5).

Investigating **potential mechanisms** of the reduction in SLA at lower temperatures, Woodward (1979b) treated potted grasses with either uniform temperatures of between 10 and 20 °C, or maintained aboveground temperatures at 20 °C, while thermostating the belowground part (including the apex) at between 10 and 20 °C. The results, shown in Fig. 12.12, indicate that the control of SLA is largely dominated by aboveground temperatures. An increase in leaf thickness at low temperature was also observed in this experiment and was associated with larger cells. For the eastern (rather cold and dry) Pamir Pyankov et al. (1999) showed that such mesophyll attributes strongly depend on leaf type, with dorsiventral leaves (most species) exhibiting fewer but larger cells and isolateral leaves having more but smaller cells. Hence, comparisons of plant assemblages from different elevations need to account for the representation of certain leaf types. What actually induces greater thickness of cell walls at low temperature is unknown (see Chap. 13). Overabundance of C assimilates could be one explanation.

While alpine leaves on average are more expensive (or luxurious) in terms of carbon per unit of light absorbing area, alpine **roots** are cheaper per unit of soil exploring length. While it is obvious that projected area is the most relevant reference for a flat, light capturing structure, the selection of a functionally meaningful reference for roots is more difficult. Accounting for the fact that thick old roots largely serve as means of transport, anchorage and storage, whereas young and thin roots are the primary structures for absorption of water and nutrients, length is considered a better reference than surface (which would give older roots more weight). Length of roots is also a more plausible bases for describing the degree of exploration of a unit of substrate volume.

On average, alpine forbs in the Alps produce 62 m of fine roots per g of dry root matter (**SRL**), compared with 41 mg^{-1} in comparable low altitude forbs (+51%) and the majority of roots in alpine forbs are 0.2 to 0.3 mm wide, compared with the 0.3 to 0.4 mm diameter most common in low altitude forbs (Körner and Renhardt 1987). Billings et al. (1978) report a SRL of 9–40 mg^{-1} in arctic tundra graminoids. Bliss (1956) reported that alpine plants produce finer and deeper roots than arctic plants. A comparison of alpine versus lowland grass species is yet to be done. Visual impression suggests very high SRL in alpine grasses as well.

Again, reasons for these altitudinal differences are speculative. The more efficient investment of carbon for soil exploration may have to do with the reduced mycorrhization at high altitude (see Chap. 10). Alpine plants may depend to a greater extent on direct resource acquisition than low altitude species. Ryser and Lambers (1995) found that grasses produce thinner roots (higher SRL) when grown with a low nutrient supply. A second reason may be an adaptation to the pulsed supply of dissolved nutrients at snow melt, which requires a very dense root felt to be efficiently utilized. A third reason may be an adaptation to low temperature, which slows liquid diffusion and reduces viscosity, and thus requires denser substrate occupation by roots. It is unknown whether these root characteristics are genotypic or phenotypic.

An important aspect in the interpretation of reduced SLA and increased SRL in alpine plants is **interspecific variation**. The data discussed before for the Alps are means for 20 or more species, covering a very wide range of values (between 1 and 3 dm^2 g^{-1} in the case of SLA). It must not be overlooked that pairs of alpine and lowland species could have been selected which showed opposite trends to community means. Even congeneric pairs of species bear a lot of randomness. In *Ranunculus* sp. for example, low altitude numbers vary from 1.53 to 3.10 and the high altitude representative shows $1.55\,dm^2\,g^{-1}$, whereas low altitude *Geum rivale* and *Geum urbanum* with SLAs of 1.90 and 2.15, compared with the alpine pair *Geum reptans* and *Geum montanum* with SLAs of 1.42 and 1.43 fit the mean pattern nicely. Even greater interspecific variation occurs in SRL, making comparisons among two or a few species doubtful. Given the fact that alpine species growing side by side at highest altitudes exhibit opposing specific leaf and root costs suggest great caution with simplistic functional interpretations. *Ranunculus glacialis* has fine roots of more than 1 mm diameter and SLA of 1.55; its neighbor *Cerastium uniflorum* has roots of less than 0.2 mm diameter and an SLA of 3.32. Both appear to do equally well in terms of biomass production and persistence at their 3150 m site in the Tirolian Alps. A more systematic survey for SLA and SLR in grasses and sedges is urgently needed.

Whole plant carbon allocation

How much do alpine plants invest in their autotrophic and heterotrophic components? As in an economy, the fraction of primary productive parts versus those which support these, and transport, utilize, invest or save their products, strongly determines the budget, which in the case of plants may materialize as growth. The major plant components that will be considered here are leaves, stems, special storage organs and fine roots (Fig. 12.13). Only forbs will be discussed, because no whole plant studies exist for other groups of plants. Since reproduction is the main reason for producing a stem in alpine forbs, stems stand for reproductive effort here, and include flowers. All available data are for mature plants studied at mid season at maximum aboveground biomass.

The result of a number of studies across the globe revealed a big surprise: the most important biomass ratio, the **leaf mass ratio**, LMR, i.e. the percent of green leaf dry matter within total plant

Fig. 12.13. Dry matter fractionation in alpine plants into flowers and stems, leaves, special storage organs and fine roots can influence whole plant carbon gain. Here, an excavated individual of *Primula glutinosa* at 3000 m elevation in the Ötztal Alps, Tirol

dry matter is a most conservative trait with community means between 19 and 25%; mostly around 21%. A small sample of species from an arctic-alpine site in northern Sweden had a community mean LMR of only 18%, but in Spitzbergen at 79°N lat, the mean was again 21% (Prock and Körner 1996). Data sets include more than 10, mostly more than 20 species per **field** site from low as well as high altitude in the arctic and the temperate zone, and from high altitude only in the tropics and subtropics (Körner and Renhardt 1987; Körner 1994; Prock and Körner 1996; Figs. 12.14, 12.15). With 21.6% of total life dry matter in leaves the famous Hawaiian *Argyroxiphium sandwichense* fits nicely into these community means (Rundel and Witter 1994).

As with SLA and SRL in the previous section, interspecific variation is enormous, with LMRs ranging from 8 to 48% (extremes represented by *Oxyria digyna* versus *Arabis alpina*), hence generalizations are again dangerous. When testing LMR against photosynthetic capacity or SLA, no correlations emerge among the species studied in the Alps. LMR in *Oxyria* is driven by the enormous

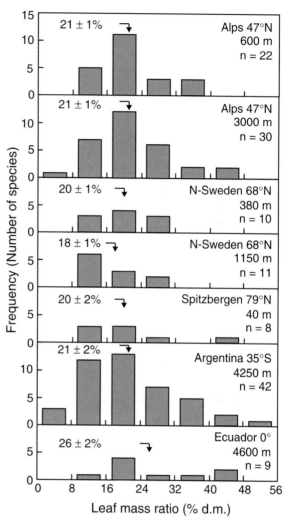

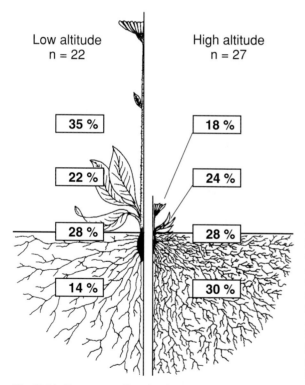

Fig. 12.14. Dry matter allocation in perennial herbaceous plant species from low and high altitudes in the Alps (stem plus flower, leaves, special storage organs plus thick roots, fine roots). (Körner and Renhardt 1987)

Fig. 12.15. Frequency distribution of leaf mass ratios in perennial herbaceous plant species excavated at various high and low altitude sites. Data for the Alps are from the Ötztal-Alps in Tirol and from low altitude meadows near Innsbruck (47°N lat). Data for Sweden are from Mount Slåttatjåkka near Abisko and Abisko itself (68°N lat) and those from Spitzbergen from fellfields near Ny Ålesund (79°N lat). Argentinean samples were collected in the Cumbres Calchaquies Mts. near Tucuman (26°S lat) and those from Ecuador near the upper limit of plant growth on Mt. Cayambe (0°). *Arrows* indicate means ± SE. (Körner and Renhardt 1987; Körner 1994; Prock and Körner 1996)

belowground structures, accumulated over many years of growth, whereas *Arabis* abandons shoots after 2 or 3 years with no persistent belowground structures, and with a large reproductive effort. As was shown in the survey by Körner and Renhardt (1987), there is a continuum of different LMRs and the above extremes are not outliers. Part of the variation may be associated with successional status of certain species. Chambers et al. (1987), suggested that a late successional status is associated with small shoot (leaf) mass and greater below ground mass.

The important **message** of these various field studies is that there is neither a significant altitudinal, nor a significant latitudinal difference in leaf mass ratio of perennial herbaceous plants. Very little is known about LMR under controlled conditions. **Growth chamber** studies with high and low altitude grasses by Woodward (1979a; *Sesleria, Dactylis, Phleum*), Atkin et al. (1996; *Poa*), and A. Larigauderie and Ch. Körner (unpubl.; *Poa*), uniformly showed no significant effect of origin on LMR. Temperature differences from 10 to 20 °C had either no effect (Woodward) or increased LMR by 21% (*Poa alpina*) or by 38% (*Poa pratensis*; A. Larigauderie and Ch. Körner, unpubl.). Overall, LMR appears to be a rather robust and invariable trait as long as similar morphotypes, closely related species or large enough groups of species are compared.

In **dwarf shrubs,** and possibly also in **tussock grasses** (both life forms which do not reach the highest alpine altitudes), the leaf mass ratio is smaller (around 10%), which results from the greater fraction of persistent belowground structures. Adult treeline trees (like trees in general) invest possibly less than 5% of their biomass in green structures (Körner 1994), but young trees are more like forbs in this respect (see Chap. 7).

It is tempting to assume that the **upper limit of certain life forms** has to do with an inherent limit in adjusting the autotrophic/heterotrophic ratio towards a smaller heterotrophic fraction. When massive stems become "unaffordable", one could imagine a shrub form with less heterotrophic investments becoming dominant, which is in turn replaced by herbaceous morphotypes which operate with even higher LMR. These aspects have been discussed in Chapter 7. Frost resistance cannot explain this altitudinal sequence of plant types, because woody species are by no means less resistant than forbs (see Chap. 8). However, the need for a more prostrate stature in order to capture more thermal energy, and mechanical constraints, are perhaps more plausible explanations for the reduction of supportive heterotrophic structures as altitude increases.

The second biomass fraction which shows no systematic change with altitude is that of **special storage organs** such as tubers, rhizomes and thick roots (Fig. 12.14). This appears to contradict widespread belief that alpine plants, particularly those in regions with a short growing season, invest more in such structures. It is important to note that this comparison is for forbs only. As was discussed in Körner and Renhardt (1987), there are no obvious reasons why forbs at low altitude in a seasonal climate should invest less in storage organs than alpine forbs, because lowland forbs are faced with much heavier competition in spring, and "sprouting, spreading and flowering first", is of even greater selective advantage than at high altitude, where spacing is often wide. However, some of the species sampled in Argentina and Ecuador have massive belowground stems or tap roots. In subtropical Argentina, five (very abundant) of the 61 herbaceous species examined had more than 50% of their dry matter in such structures (Fig. 12.16). This may be associated with three factors. First, severe frosts in almost snowless winters create enormous mechanical tensions during freeze-thaw cycles, perhaps better resisted by such structures. Second and not obvious by looking at the dry weight data only, these belowground structures are very voluminous and contain a lot of water, perhaps adaptive for periodic drought in this semi-arid altiplano (see Chap. 9). Third, such structures may assure long individual life, given the massive natural grazing pressure by Guanacos.

Fig. 12.16. While alpine plants commonly do not allocate more biomass to special belowground storage organs (tubers, rhizomes, thick roots) than comparable low altitude plants, these specimens from 4250 m elevation in northwest Argentina show massive investments in tap roots (more than 50% of total plant mass). The overall herbaceous flora of this area includes many more species without such structures, hence these are species specific characteristics adding to the variability seen in Fig. 12.15. From *left to right*: *Perezia* sp., *Hypochoeris meyeniana*, *Lepidium meyenii*, *Nototriche caesia*, *Plagiobothrys congestus*

The major change in plant carbon investment with altitude occurs in the **stem and fine root fraction.** Alpine forbs invest much less in stems (stem mass ratio, SMR) and much more in fine roots (RMR, Figs. 12.14 and 12.17), also causing the overall above-/belowground ratio to change in favor of below ground structures. The reduction of stem investments is the most obvious and well-known morphological change with altitude, and its genotypic nature was already acknowledged in the last century (Krasǎn 1882; Kerner 1898) and documented in many garden or transplant experiments (recently for instance by Bauert 1993, for *Polygonum viviparum*). Since roots are much more costly in terms of respiratory losses than green, and thus self-supporting herbaceous stems, this net shift from stem to root biomass potentially

causes a greater burden to the plant carbon balance. However, whether this will be so depends on the actual root zone temperature and the respiratory response to it, so far not assessed for an alpine versus lowland annual cycle. If greater root mass were associated with reduced mycorrhization as was discussed above, the net difference for the plant carbon balance would be less than indicated by dry weight ratios alone.

Constant biomass fractions of leaves and storage organs combined with reduced stem and enhanced fine root fractions at high altitude, cause the above versus total plant mass ratio in forbs to decrease from a mean of 57 ± 3% at low altitude to 42 ± 3% at very high altitudes (**shoot/root ratios** of 1.39 versus 0.75; Körner and Renhardt 1987). Above-/belowground biomass ratios for

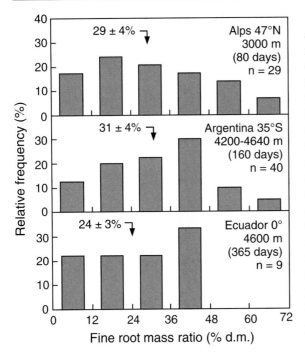

Fig. 12.17. The frequency distribution of root mass ratio (RMR) for temperate (Alps, 3000 m), subtropical (northwest Argentina, 4250 m) and tropical (Ecuador, 4600 m) herbaceous plant communities of the alpine life zone (only fine roots <1 mm; % of total plant dry matter). *Arrows* indicate means ± SD. (Körner and Renhardt 1987, and original data)

Rocky Mts. alpine forbs from 3000 m altitude vary between 0.5 and 1.0 (n = 18, Scott and Billings 1964), and a mean of 0.9 (n = 11) is reported by Nakhutsrishvili and Gamtsemlidze (1984) for alpine forbs from 3500 m altitude in the central Caucasus. Higher mean above-/belowground ratios of 1.5 were found in the 42 species examined at 4250 m in Argentina (despite the examples shown in Fig. 12.16; species as in Fig. 12.16 and 12.17). Decreased shoot/root ratios were shown to be inherent in cold adapted populations of the arctic-alpine *Carex aquatilis* (Chapin and Chapin 1981). Biomass ratios for whole communities (per unit land area), which also reflect species abundance, will be discussed in Chapter 14.

Describing biomass allocation by above/total, above-/belowground or shoot/root ratios is often not very meaningful, because the relative position of the soil surface is of doubtful functional relevance (Körner 1994). Part of the leaf (in tussock grasses and sedges) and stem fraction (in dwarf shrubs and cushion plants) are embedded in the litter or topsoil layer (see Fig. 8.4), and it makes no sense to amalgamate such buried structures with active roots. Because of these characteristics of grasses, sedges and dwarf shrubs (which occur at both high and low altitudes) and because of the often greater contribution of these life forms to alpine, compared with lowland flora, alpine plants are supposed to be characterized by massive belowground investment. This is clearly not true, or the difference is not as pronounced as believed, if comparisons are made within rather than across **morphotypes**. Although not yet systematically explored, it may be assumed that sedges, tussock grasses and dwarf shrubs from humid habitats at low altitudes exhibit allocation patterns not very different from their alpine counterparts. Hence, **an important distinction** needs to be made between altitudinal changes within certain morphotypes and changing abundances of certain morphotypes with altitude (for whole community biomass ratios see Chap. 15).

Mass ratios are one way of looking at carbon investment. In terms of organ-specific activities, other ratio descriptions are likely to be more meaningful. Two of these ratios will be addressed here: the **leaf area ratio** (LAR) and the **root length/leaf area ratio** (RLA). LAR is the product of LMR and SLA. Since LMR is rather unresponsive to altitude and SLA declines, by default the mean leaf area produced per total plant mass, LAR, declines with altitude as well. However, a ratio of one very active fraction (leaves) versus the bulk of remaining tissue, partly active, partly rather inactive, dilutes the functional significance of LAR when long-lived plants, with substantial perennial structures, are considered. In contrast, RLA disregards stems and storage organs (including all thick and old roots) and relates leaf area to fine root length, yielding a tighter functional relationship.

Such data are available only for an altitudinal gradient in the Alps. Resulting from a doubling of fine root mass ratio (RMR), a ca. 50% increase in specific root length (SRL) and a reduction in LAR (driven by SLA), RLA shows the most dramatic altitudinal increase of all allometric relationships (Fig. 12.18). Alpine herbaceous plants produce four to five times as many fine roots per unit leaf area as low altitude forbs. The change in RLA is so massive that potential errors in root excavations cannot affect it significantly.

In **summary**, alpine plants tend to accumulate equal or higher amounts of non-structural carbohydrates per unit of tissue mass, show greater concentrations of lipids, produce thicker leaves with thicker cell walls, develop less leaf area per leaf mass and higher root length per root mass, similar leaf mass ratios and, as a consequence of lower SLA and equal LMR, lower leaf area ratio (LAR) than comparable lowland species. These are all trends for means of whole assemblages of species.

The functional significance of these alterations can only be judged when the tissue specific metabolic rates under field conditions are known. For instance, the smaller mean LAR could be meaningless if the total respiratory losses and mycorrhizal investments of the root system were lower in alpine than in lowland plants. The high abundance of mobile C compounds suggests that other than carbon balance driven controls may be responsible for the observed patterns of overall dry matter allocation. The simple fact that plants growing next to each other, close to the upper limits of higher plant life in the Alps, exhibit strategies of dry matter allocation (LMR) at the opposite extreme of the range shown in Fig. 12.15 (but do equally well in terms of seasonal growth and persistence), illustrates that functional growth analysis based on such features alone is of limited value (e.g. *Ranunculus glacialis* versus *Cerastium uniflorum*; Körner and Renhardt 1987; Körner 1991).

Most of these altitudinal trends could be simulated by growing plants at a low, rather than high

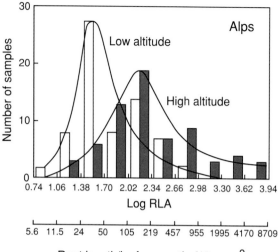

Fig. 12.18. The frequency distribution of the root length/leaf area ratio (RLA) in herbaceous perennial plant species collected at 600 m (five mature individuals of 22 species) and 3000 m altitude (5 mature individuals of 27 species) in the Tirolian Alps. (Körner and Renhardt 1987)

temperature (Table 12.5), although some inherent differences do exist as well. As can be seen from Table 12.5, low temperature influences a suite of alterations in carbon investments. A nice example of these interacting temperature influences was documented for *Trifolium repens* provenances (Boller 1980). At low temperature, alpine ecotypes invested more into stolons, which also accumulate reserve carbohydrates at the expense of leaves. This change of allocation was found to be responsible for reduced growth, despite increased leaf photosynthetic rates.

With current knowledge it is not possible to explain the significance of the greater mean mass fraction and length of fine roots in alpine compared with lowland forbs, and how these changes may influence the plant carbon balance. Future comparative research will have to combine data for specific root metabolism with actual root zone temperatures. We need quantitative data for both root turnover and carbon export by roots to either the rhizosphere or to mycorrhiza under field conditions.

Table 12.5. Summary of trends in mobile carbohydrates and structural investments observed in alpine and lowland plant species grown at 10 and 20 °C in daylight growth chambers (solar domes). For comparison, the natural high versus low altitude differences are shown. From references mentioned in the text, plus unpublished results for a set of alpine and lowland forbs. (M. Schober and Ch. Körner, unpubl.)

Parameter	Diff. between cold versus warm grown		Natural diff. high/low altit.
	alpine	lowland plants	
TNC	+	+	+
Sugars	++	++	++
Starch	–	–	–
Fructans		+	
Total lipids			+
Energy content			+
SLA	–	–	–
SRL	+	+	++
LMR	./–	./–	.
SMR	–	–	––
SOR			.
RMR	+	+	++
LAR	–	–	–
S/R	–	–	–
RLA	+	+	+++

TNC total non structural carbohydrates, *SLA* specific leaf area, *SRL* specific root length, *LMR* leaf mass ratio, *SMR* shoot mass ratio, *SOR* Storage organ mass ratio, RMR fine root mass ratio, *LAR* leaf area ratio, S/R above-/belowground ratio, *RLA* fine root length/leaf area ratio. All mass ratios except S/R are per total plant dry mass. Empty space = no data, "." = no difference.

The carbon investment data presented here suggest that carbon is not a particularly limiting resource for alpine plant growth. Such a conclusion seems to be supported by the observations by Diemer (1996) and Blumer and Diemer (1996) that alpine plants, including those from extreme high altitudes are able to "invest" a substantial fraction of their annual leaf crop in natural herbivores with no obvious drawbacks on performance.

13   Growth dynamics and phenology

The previous chapters have illustrated that rates of carbon assimilation by leaves are relatively high in alpine plants. Structural investments favor roots rather than stems, but the biomass fraction invested in leaves is the same as in comparable morphotypes at low altitude. However, interspecific variation in these traits is far greater than the mean elevational difference for communities. Non-structural carbon compounds (carbohydrates, lipids) in tissues are so abundant – and often not even recovered at senescence – that carbon supply seems unlikely to limit growth and the **size of alpine plants** (Table 13.1). Despite the existence of some alpine "giants" (Smith 1980, 1994), the overall elevational trend certainly is a reduction of plant and plant organ size (Fig. 13.1), which contrasts with the high metabolic capacity of tissues, especially leaves.

This chapter will consider three aspects of plant growth and vegetative development. It will start first by considering seasonal and diurnal growth dynamics, followed by a consideration of growth rates of alpine plants under controlled conditions, and will close with an overview of functional durations of alpine leaves and roots. Reproductive growth and development will be discussed in Chapter 16.

Seasonal growth

The growth process itself has received much less attention than, for instance, leaf photosynthesis. A number of simple but important **questions** have hardly been addressed, although they may be equally important for carbon gain as leaf gas exchange. Do alpine plants produce new structures (cells) during the whole season or only during part of it? When do alpine plants produce new structures – both day and night, or only during the day? Are there any thermal thresholds for growth? When growing at equal temperatures, do alpine plants grow more slowly than lowland plants? How long does it take leaves or roots to mature and senesce? Do leaf dynamics differ between seasonal and non-seasonal (tropical) alpine zones? These are some of the questions addressed below from the point of view of vegetative growth.

Season length and photoperiodism

The length of the period of vegetative plant activity, often termed the growth period or **growing season**, varies from 5–6 weeks in snowbed communities of the temperate and arctic-alpine zone to the full year at the equator. In the temperate zone, closed alpine vegetation commonly experiences a 10–12 week growing season, i.e. utilizes not more than one fourth of the year, which is related to the duration of the snow free period (see Chap. 5).

In the temperate and subarctic zone, measurable plant growth commonly starts, depending on species, within 10 days before or after snow melt, and reaches peak rates very fast. The weight of environmental factors which control growth dynamics differs between the beginning and the end of the season. Inherent, weather independent controls – largely **photoperiodism** – assure that growth does not commence during, and does not

Table 13.1. A survey of size parameters of alpine and lowland herbaceous plant species in the Central Alps. Means for n species (in brackets). (Körner et al. 1989a)

Size parameter	600 m	3000 m
Maximum plant height (cm)	29 (17)	4 (33)
Leaf area per individual (cm²)	143 (16)	16 (23)
Area of individual leaves (cm²)	10.3 (22)	0.9 (33)
Length of leaf petiole (cm)	7.5 (16)	1.1 (24)

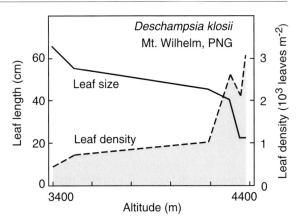

Fig. 13.1. Altitudinal variation of leaf length and leaf density per tussock ground area in tropical alpine *Deschampsia klosii* on Mt. Wilhelm, Papua New Guinea. Note the steady decrease in leaf size. Treeline is around 3850 m and small patches of forest are found up to 4100 m in this area. (Hnatiuk 1994)

continue into periods which, by "evolutionary experience", would be fatal for newly grown plant tissue. Photoperiodic controls of the beginning of growth vary with microhabitat, but generally become less stringent as late winter progresses and thermal conditions become more influential. For instance, Ram et al. (1988) found that growth initiation closely tracks spring temperature in 142 grassland species in the central Himalayas (3250–4200 m). Controls of the late season termination of growth are under much stronger photoperiodic, and less thermal control, and are much less variable, for obvious reasons.

Soil and canopy warming by solar radiation during snow-free periods in winter must not, under any conditions, be mistaken by plants as signs of the beginning of spring. Breaking **dormancy** and activation of meristems would sensitize plants to freezing temperatures and almost certainly lead to the loss of tissue during following cold periods. How stringent these controls of winter dormancy are depends on the predictability of snow cover. Plants inhabiting "safe sites" (e.g. snowbed plants, so called chionophytes) are more opportunistic (i.e. less endogenously controlled) than plants living in open habitats. Hence, there is a wide range, from almost no to very conservative control. Excellent examples are the two evergreen Ericaceae shrubs *Rhododendron ferrugineum* and *Loiseleuria procumbens*, inhabiting the extreme ends of winter snow cover gradients in the Alps. *Rhododendron*, commonly buried by deep snow, becomes active whenever snow disappears (or

is removed) and temperatures permit, whereas *Loiseleuria*, found on windswept ridges, often bare of any snow, remains fully dormant while experiencing winter sunshine and daytime canopy warming above 20 °C (see Fig. 4.6). Figure 13.2 illustrates the pronounced photoperiod control of development in alpine and lowland genotypes of *Taraxacum officinale*. In contrast, no such control exists in *Luzula alpino-pilosa*, a snowbed species, which synchronizes its development with lowland relatives when grown at a common (warm) growth temperature at low elevation (unpublished data).

Once a specific photoperiod threshold has been surpassed, the onset of spring growth depends on snow cover and weather and may vary from year to year by more than a month. Not so the late season termination of vegetative activity. Waiting for "cool weather" signals for the completion of the seasonal activity cycle would be dangerous and would often be too late for the recovery of mobile resources from deciduous tissue. Therefore, most alpine plants from seasonal climates initiate senescence and winter bud formation largely independently of weather conditions, but tightly coupled to the photoperiod. As a consequence, the **effective**

Fig. 13.2. Example of photoperiodic control of development in alpine (2500 m) and lowland (350–550 m) subspecies of *Taraxacum officinale* grown in a common garden at low altitude in Basel (April). Despite warm spring temperatures, the alpine sub-species (arrows) still "waits" for long days (in its natural habitat, the season starts at the end of June). The same is true for the arctic-alpine *Oxyria digyna*

length of the growing season at a given site is largely determined by the beginning of the season (snow melt). Prolonged periods of favorable weather late in the season have comparatively little influence on the effective growth period. It should also be noted that latitudinal differences in photoperiod reach a maximum by the end of June (end of December in the southern hemisphere), when most alpine species start to grow. The latitudinal differences pass through zero by the end of September (or end of March), at or soon after the end of the extratropical growing season.

Mooney and Billings (1961) grew latitudinal provenances of alpine *Oxyria digyna* in a common greenhouse and noted that the whole series of events from breaking of dormancy, through flowering and the formation of perennial buds was under strong origin-specific influence of photoperiod, subject to modifications by actual tem-

perature. Heide (1992) documented a very high resolution (temporal precision) of photoperiodic sensitivity in arctic-alpine grass populations. Transplantation of plants to either warmer or equally cold but photoperiodically different environments can demonstrate the strength of these controls. A reciprocal cross-continental **transplant experiment** with temperate- (47°N lat) and arctic-alpine (68°N lat) provenances of herbaceous species showed delayed senescence in plants transplanted from south to north and accelerated senescence in those transplanted from north to south, compared with local populations (sites in Tirol, Austria, and Abisko, north Sweden; Prock and Körner 1996). As part of the same experiment, temperate-alpine genotypes of *Ranunculus glacialis* were grown at subarctic lowland conditions, i.e. under the combined influence of prolonged photoperiod and higher temperature.

Under these conditions plants started to grow so early, and continued flowering so long, that up to three instead of the normal single flower cohort were produced in one season. Temperate-alpine populations grown at the arctic-alpine site commenced growth whenever snow disappeared, i.e. together with the local population, but continued to remain green for at least 2 more late season weeks after local populations. Until now this delayed senescence had no negative effect on fitness. In 1996, 9 years after transplantation, temperate-alpine provenances in fact did better than local populations, and produced a second generation of vital seedlings under these harsh subarctic fellfield conditions.

In none of the species tested did the beginning of vegetative growth reveal strong origin effects, but followed local spring warming. However, the initiation of flowering and senescence in particular, reflected origin, except for representatives of the Rosaceae family (species of *Geum* and *Potentilla*), which continued producing new leaves during the late season at all sites, with late season leaf cohorts overwintering green.

Aside from the timing of growth, photoperiod does also influence **dry matter allocation** and specific leaf area in active plants. A simulation of long days by photosynthetically ineffective red light (while providing a constant white light photoperiod) caused growth rates of grasses to increase, despite a reduction of photosynthetic capacity per unit leaf area, as was shown for *Poa alpina* (Fig. 13.3) and nordic provenances of *Poa pratensis* (Hay and Heide 1983). Using the same method with 25 species from 3000 m in the Alps, confirmed that long days alone can increase SLA (F. Keller and Ch. Körner, unpubl.).

Seasonal growth dynamics

Outside the tropics, the **seasonal course of plant growth** commonly commences by rapid expansion of preformed tissue with a gradual transition into a phase of new tissue production and maturation, followed by a phase of stable above-ground

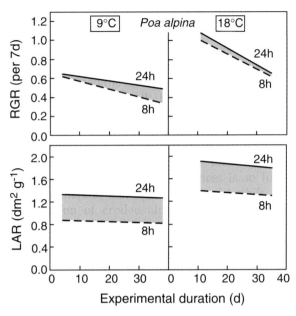

Fig. 13.3. Effects of photoperiod and temperature on relative growth rate (RGR) and leaf area ratio (LAR) in subarctic-alpine *Poa alpina*. Both long (24 h) and short day (8 h) treated plants received the same dose of photosynthetic active radiation (for 8 h); long days were simulated by a very low intensity signal from incandescent lamps. Long day increases LAR and RGR at both growth temperatures. The LAR effect is hardly duration-dependent. (Solhaug 1991)

biomass (formation of buds for the next season), and finally a phase of leaf dieback. In the temperate zone the period of measurable shoot enlargement is relatively short, commonly not exceeding 4–5 weeks irrespective of the duration of the snow-free period. Bliss (1956) compared growth dynamics of arctic and alpine dwarf shrubs, graminoids and forbs. In northern Alaska he found periods of shoot extension of 3–6 weeks, mostly 4 weeks (only one grass species continued to grow for 8 weeks). In the alpine zone of the Medicine Bow Mountains of Wyoming he observed 2 to 5 weeks, again 4 weeks were most common.

In the humid tropics, leaf and shoot production is continuous. Hnatiuk (1994) reports that he never observed any pronounced seasonal death of leaves in tussock grasses on Mt. Wilhelm, New Guinea, even though he noted some rhythms associated with the change from the wet to the less wet season in this ever humid tropical-alpine climate. Diemer (1998a, b) observed year round leaf production in 16 herbaceous species and five shrubs in the Ecuadorian páramos at 4000 to 4600 m altitude.

The mean **duration of leaf unfolding** (from first sign of a leaf to the end of lamina extension) in herbaceous alpine plant species in the Alps is 25 ± 3 days versus 27 ± 1 days in comparable lowland forbs (12 alpine and 16 lowland species; Prock 1994). Thus, the dynamics of leaf development are very similar despite the fact that final lamina length and leaf area differ by factors of 3 and nearly 10 respectively. Prock also found that the duration of leaf extension varies only little among different leaf cohorts or with total leaf life span (see later). According to Diemer et al. (1992) and Prock's work, **leaf dieback** commonly takes half as long as leaf expansion (mean 11–12 days), but may be accelerated or interrupted by bad late season weather. No altitudinal variation of the duration of leaf dieback was found in these studies. Figure 13.4 illustrates the seasonal course of shoot biomass in *Carex curvula*, which again reflects the relatively short period of aboveground tissue production in the early season.

For obvious reasons, much less is known for alpine **root dynamics** and the limited evidence available to date is largely of an indirect nature, because direct observation of living roots in situ is very difficult. It has long been known that any restriction of root growth has immediate consequences for shoot growth and that these interactions are not simply a consequence of reduced nutrient acquisition by roots. Plants may become dwarfed even when for instance, spatially restricted roots (in small pots) are flushed with nutrients and hormonal signals have been shown to be involved (e.g. Carmi and Heuer 1981). Clearly root system expansion is a function of soil temperature which influences both growth and developmental processes (e.g. Kaspar and Bland 1992). One of the major problems in the interpretation of root responses to their environment is the steep **thermal gradient** often found in soils, alpine ones

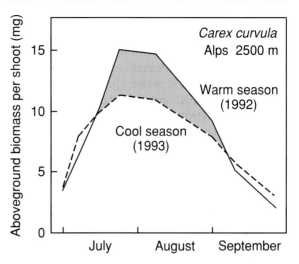

Fig. 13.4. The seasonal course of shoot biomass of *Carex curvula* at 2500 m in the Swiss Alps. Note that individual shoots of this sedge live several years and bear 3–4 leaves (see Figs. 8.4 and 11.13). Their belowground apex produces one to two new leaves every year. Leaves die back to the ground surface each fall but resume growth for two to three more seasons until their basal meristem dies. Hence, annual aboveground shoot biomass production is a combination of birth of new leaves and re-growth and dieback of older leaves. The diagram illustrates the in situ amplitude of growth between a rather warm (1992) and a cool season (1993). Fertilization (data not shown) caused total biomass per unit land area to double and delayed dieback. (Schäppi 1996)

in particular (see also Chaps. 4 to 6). The very same root with its active tips exploring different soil horizons may pass through a soil temperature profile from 25 °C at the top to a few degrees above zero in the deeper profile. Also, opposite gradients may occur with the topsoil chilled and warmth still stored at greater depth. In view of such gradients thermal acclimation and activity of roots are difficult to predict, and a better understanding of the functioning of alpine roots certainly awaits much more sophisticated experimental approaches.

Callaghan et al. (1991) reviewed the available evidence on cold climate root dynamics with an emphasis on tundra soils, but some of their conclusions are likely to be relevant for roots of alpine plants as well. While permafrost, water-logging and soil frost heaving are less abundant in alpine compared with arctic environments (see Chap. 6), short season, low early and late season soil temperatures, soil acidification and shallow profiles are common attributes. Because soil surfaces are mostly less wet and less densely covered by mosses, lichens or litter at alpine compared with low tundra habitats, soil heat flux is greater and soil warming may be substantial at mid season.

Root growth in arctic soils had been documented at temperatures approaching 0 °C. Chapin (1974) reported optima for root initiation of 5 °C in *Eriophorum* and optima for **root elongation** between 5 and 10 °C for several wet tundra graminoids. Under controlled conditions plants were observed to grow at −0.5 °C (Billings et al. 1977) and ranges of elongation of 0.2 to 2.5 mm d^{-1} have been observed below 5 °C (Bliss 1956). Root dynamics in the field appear to have been hardly studied. Mähr and Grabherr (1983) installed root windows in an alpine sedge community and noted a range of mean daily root extension rates between 1.0 and 3.7 mm, with a major peak at the beginning of the season (when soils are still cool), low rates at mid season (when soils reach peak temperatures) and a second smaller peak in the late season (Table 13.2). Hence, soil temperature per se appears to be an unreliable predictor of seasonal growth dynamics in alpine roots, and seasonally varying temperature responses of root growth are to be expected, as is known for conifers from cold climates.

Table 13.2. Dynamics of root length growth (mm d^{-1}) in an alpine sedge community at 2500 m altitude in the Alps (root window observations). (Mähr and Grabherr 1983)

	1977	1978[a]
Early season	3.7 ± 2.5	1.5 ± 0.6
Peak season	1.2 ± 0.5	1.0 ± 0.4
Late season	2.3 ± 1.1	1.6 ± 0.4

[a] The generally lower rates in 1978 possibly reflect a late disturbance effect due to the installation of the root observation box.

Diurnal leaf extension

This is a subject hardly ever touched, but of great importance for the understanding of alpine plant life. Do leaves of alpine plants grow (expand) at night? If not, they lose an important fraction of time for carbon investments compared with lowland plants. A number of investigations at low altitude revealed a strong influence of temperature on instantaneous leaf and shoot extension (for references see Körner and Woodward 1987), but to my knowledge, **auxanometers** have been used only twice to monitor growth dynamics in alpine altitudes with sufficient resolution. Senn (1922) used Pfeffer's classical mechanical auxanometer at 2450 m altitude in the Swiss Alps on *Hieracium* and *Arnica*, and he could not detect any night-time growth in leaves, while day-time growth dynamics yielded no consistent picture for the two dicots he studied, although expansion was clearly detectable with this simple device. In a second attempt, in situ growth dynamics were recorded with electronic displacement transducers in the Tirolian Alps (Woodward et al. 1986; Körner and Woodward 1987). The following results were obtained.

The rates of **leaf extension** growth in five species of *Poa* typical for the altitudes of four experimental sites between 600 and 3000 m altitude show a very pronounced dependency on meristem temperature (1–2 cm below ground surface). Lowland plants grew day and night, irrespective of temperature and night-time rates during warm summer weather were half the day-time rates (maximum temperature 26 °C, minimum 14 °C), and still reached one fifth of daytime rates during cool and rainy weather (maxima 14 °C, minima 11 °C). Periods of bright midsummer weather with relatively mild nights between 2600 and 3000 m altitude (maximum meristem temperature 24 °C, night-time minimum 6 °C) permitted night-time rates of growth of about one fifth of peak daytime rates. During bad weather conditions (temperatures <3 °C) almost no growth occurred.

In situ low **temperature thresholds** for leaf extension growth were between 5 and 7 °C at low altitude, and around 0 °C at highest alpine altitude. At low altitude, the critical temperature for zero extension growth never occurred in midsummer. However, the much lower threshold temperature of alpine plants was regularly reached in the field, and thus caused low temperature to block growth during adverse weather. At temperatures above 20 °C, rates of leaf extension in lowland species reach twice the rates of alpine species (Fig. 13.5, upper diagram). Clearly, alpine leaf growth is

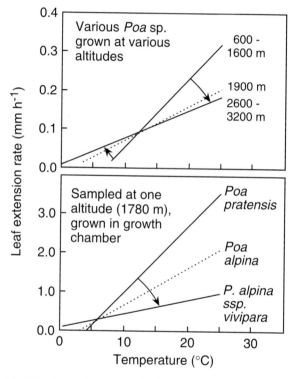

Fig. 13.5. *Upper diagram*: in situ leaf extension growth in *Poa* species native to different altitudes. Alpine species are able to grow at comparatively low temperatures, but growth rates are only half those of lowland species at high temperatures (Körner and Woodward 1987). *Lower diagram*: temperature responses of leaf extension rate of *Poa* species of different altitudinal preferences but collected at one site at medium altitude where these species coexist naturally. Plants were grown in a growth chamber at 15 °C and subjected to temperatures between 0 and 26 °C. Responses reflect the trend (*arrows*) seen in the *upper diagram* for field-grown plants. (Woodward and Friend 1988)

restricted much more by actual temperatures than is leaf growth in lowland plants.

A rather unique possibility for testing the **genetic nature** of such trends is on river banks at medium altitude where low and high altitude taxa coexist naturally. Under such conditions, species of high altitude origin exhibited low temperature thresholds of extension growth of between 2 and 3 °C compared with thresholds of between 6 and 7 °C for species commonly found at much lower altitudes (Woodward et al. 1986). This study included, besides *Poa* sp., species of the genera *Agrostis, Rumex, Polygonum, Achillea* and *Anthyllis*, hence illustrating that an altitude specific genetic differentiation does exist in both, grasses and dicotyledonous taxa. On these river banks, 300 m below the treeline, growth of alpine species was not significantly restricted by night-time temperatures, but that of lowland species was. In all cases leaf extension growth was tightly coupled to temperature, and thus closely followed the diurnal temperature course.

The above analyses of extension growth in the field revealed 20 °C leaf extension rates of lowland *Poa* of 0.25 mm h^{-1}, and 0.13 mm h^{-1} in *Poa* growing at highest altitudes (Körner and Woodward 1987). Alpine species always grew slower than lowland species unless temperatures dropped below 10 °C, where the difference was reversed (Fig. 13.5). Mean **daily sums of lamina length increments** in the main leaf cohort of herbaceous dicotyledonous species were found to be 1.2 mm d^{-1} at alpine sites (2600 m) compared with 2.8 mm d^{-1} at low altitude (across all weather conditions; manual measurements in the Tirolian Alps; Prock 1994).

These field data confirm results of earlier studies with alpine taxa obtained under growth chamber conditions (Woodward 1975, 1979a), and observations were also repeated in a subsequent growth chamber study, again with *Poa* (Fig. 13.5, lower diagram). Thus, alpine plants are able to maintain leaf extension growth at lower temperatures than lowland plants, but rates at high temperatures are much lower than in lowland plants. Night time or bad weather conditions at high alpine sites (unlike at lowland sites) severely constrain leaf growth, and thus the investment of carbon assimilates.

Rates of plant dry matter accumulation

Following from these phenological and biometric studies with high temporal resolution, growth analysis, using sequential harvest of alpine and lowland species grown under controlled conditions, revealed differences pointing in the same direction, but **alpine/lowland comparisons** are not straightforward. When a certain alpine species is compared with a certain lowland species, taxonomic, i.e. species specific effects, cannot be ruled out. Comparisons between altitudinal provenances within a species are also inadequate, because the comparison would always include individuals from marginal versus those from habitats central to the range of the species (see Chap. 1). Finally, such comparisons would only be valid if species or populations were investigated under their natural temperature regime and photoperiod. Ignorance of the latter could completely flaw results, as was shown by Heide and coworkers (see above). The only solution to this dilemma would again be broad screening and looking at frequency distributions of traits both under adequately controlled greenhouse or growth chamber conditions and in the field, but such data are not yet available. The following account will therefore rest on controlled environment studies.

Graminoids. Woodward (1979b) grew the lowland species, *Dactylis glomerata* and *Phleum bertolonii* and British upland provenances of *Phleum alpinum* and *Sesleria albicans* at 10 and 20 °C and optimal nutrient supply. Species specific responses of **relative growth rate** (RGR, the dry mass increment per day relative to the mass at the beginning of the day) of seedlings overshadowed any origin specific trends. While the two *Phleum* species and *Dactylis* did not significantly differ in RGR (a mean of 0.18 at 20 °C after 2 weeks), *Sesleria* grew only half as fast at any temperature (0.08

at 20 °C). After 2 weeks of growth, the Q_{10} of RGR was 2.0 and 2.2 for the lowland species and 1.7 and 1.8 for the upland species. Atkin et al. (1996a) compared RGR in hydroponically grown *Poa* species of contrasting altitudinal origin at 20 °C. While their three Dutch lowland species grew by a mean RGR of 0.21, the two Australian alpine species exhibited RGR of 0.12. It needs to be noted that the alpine species selected here form tussocks of festucoid leaves, morphologically different in every respect from those of the lowland species, which limits comparability. In both these studies, differences in RGR were explained by differences in specific leaf area (SLA) and not in photosynthesis (see Chap. 12). In a similar experiment, Atkin and Day (1990) compared RGR in a pair of Rocky Mountains *Luzula* species (alpine: *L. acutifolia* ssp.*nana*; lowland: *L. campestris*) and arrived at 0.067 for the alpine and 0.079 for the lowland species.

By comparing biomass accumulation in Swiss high altitude provenances of *Poa alpina* (2500 m) with lowland provenances of *Poa pratensis* (500 m) under both their respective native growth temperatures, and common temperatures in a low nutrient substrate over 2 months of peak season growth (A. Larigauderie and Körner, unpubl.) mean increments of 0.038 for *P. alpina* and 0.050 for *P. pratensis* were obtained. Remarkably, growth temperature had no effect on either species, i.e. acclimation was perfect (Q_{10} of mass increment = 1). However, this is not a valid expression of RGR because, by definition, RGR requires short observation intervals, which do not include ontogenetic effects, as the data here do. But taken together, Woodward's *Sesleria*, Atkin's Australian *Poa* and American *Luzula*, and the Swiss *Poa* data do suggest inherently slower growth rates at common 18–20 °C in alpine grasses, compared with lowland grasses, but the data base is clearly narrow and still inconsistent, and growth conditions were rather artificial.

Dicotyledonous species: Less is known about RGR for alpine versus lowland forbs. Relatively small differences were reported by Atkin and Day (1990) for pairs of alpine versus lowland *Ranunculus* (0.061 versus 0.084) and *Plantago* (0.081

versus 0.104), again grown at equally warm temperatures. RGR during the first month of life of seedlings of four alpine and five lowland forb species from the Tirolian Alps grown at close to natural growth conditions were 0.074 and 0.094 respectively (at ca. 18 °C), and were almost indifferent at 9 °C with RGRs of 0.051 and 0.055 (A. Schober and Körner, unpublished). *Arabis alpina* from 2600 m altitude and *Arabis hirsuta* from 600 m altitude, grown together from seed on a precisely controlled temperature gradient from 9 to 22 °C, both showed no germination or growth

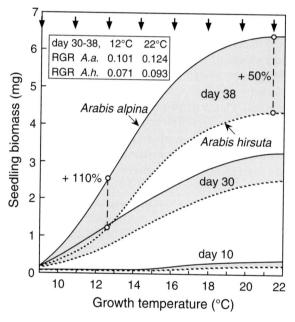

Fig. 13.6. The response of seedling growth of *Arabis alpina* (seed from 2600 m altitude in the Alps) and *Arabis hirsuta* (seed from 600 m) to a controlled temperature gradient. Plants were grown in fertile substrate in pots inserted into U-shaped aluminum bars whose ends were kept at 9 and 22 °C. The diagram shows data from three consecutive harvests for eight different temperatures (*arrows at the top*). Germination did not significantly differ between species. Embryo dry weight, specific leaf area, leaf mass ratio and % leaf N were 76 µg, 6.9 dm² g⁻¹, 50% and 3.7% in *A. alpina* and 66 µg, 6.4 dm² g⁻¹, 63% and 4.5% in *A. hirsuta*. The alpine species always grew faster, but the difference is greatest at low temperatures (+110 vs +50%; unpubl.)

below 9 °C, and *A. alpina* grew faster at any temperature <9 °C (RGR at 20–22 °C of 0.124 in A.a. compared with 0.093 in A.h.; Fig. 13.6). However, at the cold end of the gradient the difference in carbon gain between *A. alpina* and *A. hirsuta* was greatest. This example illustrates that the short-term growth potential of alpine plants is not always smaller than that of lowland species and that temperature optima for growth may not always differ greatly.

Data for longer term **accumulative growth** at two temperatures are shown in Table 13.3. Plants were grown from established seedlings in solar domes (91% of full solar radiation) during peak season. These data show that effects of plant nutrition on growth override influences of origin or temperature. At very low nutrient supply, biomass differences between alpine and lowland species were dramatic, but were rather small at high nutrient supply. At high nutrient supply, alpine plants growing at 9 °C require one third more time to reach the same leaf plastochron (number of visible leaves) than at ca. 18 °C. Lowland plants are slowed down by low temperature even more.

These data for forbs (except those for *Arabis* in the thermal gradient study mentioned above) seem to be in line with those presented for graminoids as long as only fertile and warm growth conditions are considered. In both cases, alpine species grow slightly slower than lowland species. Much larger origin related differences may be found under nutrient shortage, and differences can become reversed at low temperatures (i.e. alpine species growing faster than lowland species). Given the strong **interaction between nutrient supply, origin and temperature response** of growth, as illustrated by Table 13.3, it appears that laboratory data obtained under fertile conditions and only one (high) temperature can tell us little about the growth controls in alpine plants. A common denominator of the studies of leaf expansion and whole plant growth is that at high temperatures lowland plants grow faster than alpine plants and at very low temperatures, alpine plants grow faster than lowland plants. At some intermediate temperature, both groups may

Table 13.3. Mean biomass accumulation (mg of dry weight) in alpine and lowland herbaceous species (seeds from 2600 m and 600 m altitude in the Alps of Tirol) grown in solar domes at close to natural field temperatures (T) and at reciprocal temperatures. Numbers in brackets refer to the number of days elapsed since germination. (A. Schober and Ch. Körner, unpubl.)

	Low nutrients		High nutrients	
	Low T	High T	Low T	High T
Alpine species (n = 4)	17 (71)	22 (56)	643 (100)	569 (70)
Lowland species (n = 8)	83 (71)	147 (50)	784 (105)	849 (69)

Low T 8–11 °C; *High T* 17–20 °C (daily minimum and maximum soil temperatures; on sunny days maximum leaf temperatures may be 1–2 K warmer). *Low nutrients* 1 : 1 : 1 sand, peat and low N garden soil, 0.3 l per pot; *High nutrients* the same mix but with 100 g Osmocote complete slow release fertilizer pellets per m³ of soil. Mean mass of seeds was 0.45 mg in high and 0.94 mg in low altitude species. Plants from different temperatures within each nutrient treatment were harvested at the same developmental stage (equal plastochron).

be found growing at equal rates. Hence, results of the study of inherent growth characteristics of alpine and lowland plants depend on the common experimental temperatures chosen.

Functional duration of leaves and roots

As mentioned above, the life spans of leaves and roots co-determine the carbon and nutrient balance of plants. Functional **leaf life span** is defined here as the time elapsed between appearance ("birth") of a leaf and the end of its functional duration, i.e. the time when senescence has progressed so far that more than 5% of the leaf area becomes chlorotic. Total life span includes the continued duration of leaf senescence until 95% browning or abscission. Within the active part of a leaf's life there are two distinct phases: leaf expansion (until full lamina size is reached) and the mature life phase. Since the speed of senescence strongly reflects late season weather conditions, only the functional leaf duration is considered here (not meaning that the final part of leaf senescence has no function).

The mean functional leaf duration in herbaceous alpine plant species in the **temperate zone** (Alps) was found to vary, depending on year and habitat, between 68 ± 4 days (Diemer et al. 1992, 16 species) and 87 ± 7 days (Prock and Körner 1996, 14 species), statistically indifferent from the 71 ± 4

days (Diemer et al. 1992, 13 species) and 78 ± 4 days (Prock and Körner 1996, 16 species) found in comparable wild lowland species. The year to year variation of leaf life spans was within ±15%, and means of leaf duration range from 41 days in *Polygonum viviparum* to 93 days in *Ranunculus glacialis* (with a similar range in low altitude species). Plants grown in isolation in garden plots have longer leaf life spans, an effect enhanced when fertilizer is added (Prock 1994). Hence, if data from natural habitats are compared, mean life span of leaves, interspecific variability and the range of life spans of leaves were not different between these populations separated by 2000 m of altitude. A comparable group of arctic-alpine herbaceous species from 1040 m altitude in northern Sweden averaged at 67 ± 4 days leaf duration (seven species; Prock and Körner 1996). Thus, herbaceous alpine plants from these strongly seasonal mountains have a functional leaf duration of about 10 weeks.

Late **snowbeds** are a special case where plants of the same species are exposed to variable season length. According to Kudo's (1996a) analysis in Japan, leaf life span in species of *Peucedanum*, *Potentilla* and *Sieversia* (all herbaceous) decreases in proportion to the length of the snow-free period (leaf life spans of 40 to 84 days within total snow-free periods of between 67 and 117 days). With decreasing leaf duration, leaf N concentration and specific leaf area increase, similar to what is observed along latitudinal gradients of season length (Chap. 10). The opposite is seen in ever-green shrubs, which extend their leaf duration (as measured by number of seasons), the shorter the snow-free period (Kudo 1992).

A fractionation of alpine leaf duration into the phase of leaf expansion and the **mature phase** revealed that leaf duration is a function of the mature life span (Prock 1994). As mentioned above, the durations of leaf expansion (25–27 days) and leaf dieback (ca. 12 days) are rather invariable and only loosely correlated with total life span. The functional duration of temperate zone herbaceous alpine leaves correlates weakly with leaf nitrogen concentration. The longer leaves remain active, the lower is the mean concentration of nitrogen per unit of dry weight. However, since longer-lived leaves tend to be thicker, there is no correlation between leaf life span and leaf nitro-gen content per unit leaf area (Diemer et al. 1992). The small range of leaf life spans among temper-ate zone herbacous alpine plants is one of the explanations as to why correlations with other leaf traits (%N, SLA) are poor. However, when put into a global context (Reich 1993) leaves of alpine plants match the general relationship between leaf duration, leaf metabolic activity and leaf allome-try (Reich et al. 1997).

Longevity of **"evergreen" leaves** in arctic-alpine and temperate-alpine species varies between 1.4 and 3.8 seasons (16 species, Karlsson 1992) and does not reveal any altitudinal trend, but longevi-ties (when measured by number of seasons) tend to be longer at polar, compared with temperate lat-itudes. However, it is difficult to compare leaf longevities between climates of different season length, because it is unclear how the dormant season should be treated. Counting years of dura-tion is certainly misleading, but counting seasons is also problematic if season length differs. Func-tional duration might be better counted in weeks or months, but how does one account for the dif-ferent "qualities" of weeks or months? Early and late season periods, when herbaceous plants have not yet emerged or have senesced, have different weight compared with midseason periods. When Karlsson found no altitudinal difference in leaf

longevity in terms of numbers of seasons, this meant that the functional duration was in fact shorter (7–14 months) at high than at low altitude (11–30 months) in the ericaceous shrubs he studied. It is well known that conifers retain needles much longer at high altitudes. But a simple calculation illustrates that the functional duration does not necessarily differ. In the Alps, *Picea abies* near the treeline with a 5 month season, retains needles 10–12 years, i.e. ca. 55 months compared with 6–8 years at the 8 month season at low alti-tude, which corresponds to ca. 56 months.

An interesting "natural experiment" for testing the influence of season length on alpine leaf traits is the latitudinal comparison of alpine species growing at similar temperature conditions. Is alpine leaf longevity determined by (1) season length, (2) the need for a new leaf position in the canopy (because of intra- or interspecific leaf shading), (3) nutrient shortage and the need for re-allocating mobile leaf compounds, or (4) because of leaf damage (wearing, pathogens, epiphylls etc.)? Points (2) and (3) are linked.

The most detailed study of alpine leaf dynam-ics, which included the **equatorial tropics**, was that by Diemer (1998a, b). The mean leaf duration (in this case including the phase of leaf senescence) of 16 **herbaceous species** between 4000 and 4600 m altitude in Ecuador was 193 ± 19 days compared with 80–90 days for comparable forbs in the Alps. Hence, life spans are more than doubled in the tropics, despite the fact that the tropical-alpine temperatures and radiation are very similar to the growing season means of the temperate zone (Diemer 1996, see Chap. 4). Diemer also noted that the interspecific variabilty of leaf longevity increased substantially with decreasing latitude (Fig. 13.7). Hence, it appears that leaf duration becomes increasingly season dependent, the shorter the season (the above point 1). If there is no season length constraint, plant life strategy determines leaf longevity, which apparently permits a broad suite of options, reflected in the wide ranges of alpine leaf life spans seen in the alpine tropics. Apart from the endless season, greater leaf longevities in the tropics appear to be

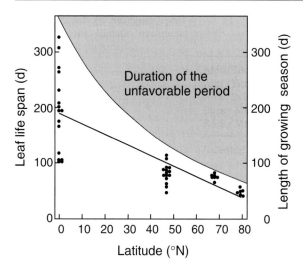

Fig. 13.7. The latitudinal variation of total leaf life span from leaf appearance to leaf death, i.e. including a mean duration of leaf senescence (which is known to last ca. 11–12 days in extratropical latitudes) in herbaceous alpine plants. Note the association with season length at high latitude and the greater range at low latitude. (Data from Diemer et al. 1992; Prock and Körner 1996; Diemer 1998a)

associated with the greater carbon investments per unit of leaf area (Körner 1989b). Also, the shorter daylength of only 12 h in the tropics compared with 15 h in the temperate zone alpine summer and 24 h experienced by arctic-alpine plants diminishes the photosynthetically effective latitudinal difference in mean leaf longevity (Diemer 1998a).

For **shrubs** at 4060 m altitude Diemer (1998b) reports leaf life spans of 4–22 months. He concluded that it is inappropriate to distinguish deciduous and evergreen habits in tropical alpine plants. Leaf longevity correlated with mass based nitrogen concentration but not with nitrogen content per unit leaf area, once more, because long lived leaves are thicker (have lower SLA) than short lived leaves. Among the sample of Ecuadorian shrubs studied by Diemer, leaf persistence was only very vaguely related to particular leaf traits such as nitrogen concentration or specific leaf area. Disregarding the dormant season, the 7–19 months of active leaf duration in evergreen ericaceous shrubs mentioned above for higher latitudes falls midway in the tropical range.

No immediate correlation was observed between leaf life span and the rate of **leaf initiation** in the tropics. On average, 1.6 herbaceous leaves were initiated per month in Ecuador (Diemer 1998a). In the temperate zone, leaf appearance is pulsed, so that the initiation rate per month would be 3–4 per month at the beginning of the season, and close to zero later on. Diemer calculated a mean for the whole temperate season in the Alps of 1.4 leaves per month, similar to the continuous rate he found in the tropics.

In **summary**, alpine leaf duration shows a wide spectrum in the tropics and a much narrower range in the temperate and arctic-alpine zone. Within the tropics, and within the high latitude regions for which altitudinal data exist, there appears to be nothing unusual about leaf duration in alpine plants. Neither herbaceous, nor woody alpine species appear to have leaf durations significantly different from those also found at lower altitudes of the respective regions (for lowland tropics see Reich 1993).

At temperate and higher latitudes the pattern seen for herbaceous plants and evergreen dwarf shrubs differs from the one seen in deciduous trees and shrubs, where species from genera such as *Betula, Sorbus, Larix, Vaccinium, Ribes, Lonicera* etc. have to complete their annual cycle with much shorter leaf duration at high rather than low altitudes.

Data from the temperate zone indicate that alpine species respond to season length by adjusting leaf duration. In contrast, the longer duration of growth in herbaceous lowland species of higher latitudes is associated with the production of multiple leaf cohorts. Late snowbed species at high latitudes are a very special case; they have to survive with extremely short leaf duration, similar to what may be seen in hot desert ephemerals. In herbaceous plants such short life spans have been shown to coincide with very low carbon investment per unit leaf area and high leaf N concentration (Körner 1989b; Kudo 1996a).

Root turnover depends on life form (shrub, forb, graminoid), species, and type of root. Massive tap roots may live as long as the plant lives, whereas fine roots may be replaced annually. Some cold climate species have been shown to replace two thirds of their fine roots every year (e.g. the arctic grass *Phippsia algida*; Bell and Bliss 1978), while in others most roots survive 5–6 years (Shaver and Billings 1975). Data for temperate alpine herbaceous species by Bell and Bliss suggest fine root survival for 3–4 years (e.g. *Ranunculus, Alopecurus*) and survival of 7 to 10 years was estimated for the adventituous roots of some *Carex* and *Luzula* species. Schäppi and Körner (1996) found the annual increment of fine root biomass in ingrowth cores in alpine heath dominated by *Carex curvula* to represent one sixth of the total root biomass in undisturbed soil. By comparing different methodological approaches, Mähr and Grabherr (1983) arrived at 20 years of root duration in this species, and they concluded that main roots are possibly as old as the rhizome section from which they emerge. In Fig. 12.6 it was shown that the main roots of *Carex curvula* retained an undiluted carbon isotope label 2 years after labeling. Active, 11-year-old roots were found by Jònsdòttir and Callaghan (1988) in *Carex bigelowii* and Callaghan et al. (1991) refer to a number of studies in which even longer-lived roots in shrubs and tap-rooted species were observed (see the reports of alpine **root morphology** by Daubenmire 1941 and Holch et al. 1941).

Although turnover of younger, more active side- or adventitous roots is much faster, one can safely conclude that overall life spans of roots in cold alpine soils exceed those in warmer regions. This greater root persistence is the most plausible explanation for the greater root masses found in arctic and alpine, compared with temperate lowland plants (see Chaps. 12 and 15). As was mentioned in Chapter 12, roots of such cold soils are generally thinner than in warm soils, although low temperature tends to enhance thickening within a given species or ecotype. Hence, it appears that alpine root dynamics contrast with alpine leaf dynamics, a difference obviously associated with the contrasting chances of survival during the cold season above and below the ground.

In **conclusion**, the data presented in this chapter (which concentrated on individual plants and their leaves and roots) do provide some, but not exhaustive explanations for small plant size and low community productivity in alpine altitudes. Even when compared under optimized growth conditions, alpine plants often grow slower than lowland plants. There is no obvious growth-related reason why tropical-alpine plants are not taller than comparable temperate-alpine and sub-arctic-alpine plants. Although leaf dynamics appear to be increasingly linked to the length of the growing season the shorter the season becomes, neither the rate of leaf initiation nor the rate of leaf expansion seem to be latitude (season length) specific.

Whether rates of leaf or root expansion and biomass accretion are slower in alpine compared with lowland plants depends on the temperature range considered. At common high temperatures, alpine plants mostly grow slower, but at a common low temperature they grow faster than lowland plants. It seems that growth controls other than those conventionally included in functional plant growth analysis must be considered to explain alpine dwarfism, the topic of the next chapter.

14   Cell division and tissue formation

Cellular aspects of growth have traditionally not been part of alpine plant ecology. It was perhaps taken as self-evident that once plants are in an active phase of life and sufficient carbon assimilates are present at meristems, plants would grow new cells. But this is not necessarily so. The formation of new cells involves a suite of synthetic steps, all potentially as sensitive, or even more so, to low temperature than the production of the raw materials for cell formation, photosynthates and amino acids in particular.

To grow new tissue requires:

1. Multiplication of genetic **information** and of embryonic cells,
2. Accretion of cell **mass**, in essence the formation and differentiation of a cell wall and functional cell organelles,
3. The **energy** to fuel these synthetic processes, provided by growth respiration.

At a certain low temperature, none of these processes will operate. The question is whether all three are similarly limited by low temperature, and whether these limits are higher or lower than those for photosynthesis. The process with the greatest temperature sensitivity will be the **bottleneck of growth** at low temperature. This brief account of ecological aspects of developmental biology of alpine plants will be based on very few facts, but should point at some of the least understood aspects of plant growth in cold climates.

Cell size and plant size

Why are alpine plants, their leaves in particular, so small? Do they consist of smaller cells or less cells? The answer is short and simple: the cells are not smaller than in lowland plants (Fig. 11.3), in fact, there is even a trend in some very high altitude species to have unusually big cells making up their tissues (Körner et al. 1989a). Consequently, alpine plants and their leaves consist of a smaller number of cells than plants from low altitude; they produce less cells. Even the most extreme forms of natural **alpine plant dwarfism** (Fig. 14.2), but also artificially dwarfed plants such as Bonsais do not exhibit reduced cell size (Körner et al. 1989c). Cell size is a most conservative characteristic and is maintained (within certain bounds) irrespective of whether plants are phenotypically or genotypically small, and irrespective of whether plant dwarfism is a response to drought, nutrient shortage, high light stress or low temperature (Körner and Pelaez Menendez-Riedl 1989). As an example, Fig. 14.2 shows quite "normal" cell size in what are possibly the smallest mature terrestrial higher plants on earth, collected in the high Andes. Stress-induced or genotypic variations in leaf size within a given higher plant species or across species are largely controlled by the number of cells produced, and not, or to lesser extent, by variations in the final size of cells.

In Chapter 13 it was reported that **leaf formation** in herbaceous species takes roughly 3–4 weeks from first appearance to mature size, both at high alpine and lowland elevations in the temperate and subarctic zone. Most of that time is required for step two of tissue growth (see above),

Fig. 14.1. The world of dwarfs. What causes small size and slow growth in plants at such high elevations? The photograph shows a valley at 4200 m altitude in the Cumbres Calchaquies Mountains of northwest Argentina where some of the world's tiniest angiosperms grow. The two miniature specimen of *Oxalis* and *Geranium* (and others in *Nototriche*, *Muehlembergia*, *Draba* and *Crassula* not shown), are fully mature and produce seeds (for cell size see Fig. 14.2). These plants were collected on the rocky outcrops in the center of the valley (a site discovered by S. Halloy)

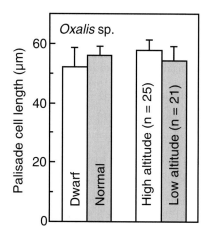

Fig. 14.2. The cell size in leaves of the dwarf *Oxalis* species from 4200 m elevation shown in Fig. 14.1 is not different from that in a large congeneric species collected at only 3000 m elevation in the same area, and neither are different from the mean for all alpine and lowland forbs examined in Fig. 11.3. Alpine dwarfism is not associated with small cell size

because cell division is completed when leaves have reached 30–40% of their final size, and most cells are produced during the earliest phase of leaf life (<10% of final size, Dale and Milthorpe 1983; Fig. 14.3). Since alpine plants produce similar (or smaller) numbers of leaves per flush, and since the mean area of a single leaf is about ten times smaller in alpine compared with lowland forbs (Körner et al. 1989a), the average alpine leaf contains up to ten times fewer cells (depending on leaf thickness), and also produces up to ten times fewer mature cells per unit time.

The size of leaves is determined at a very early stage of leaf growth (Fig. 14.3), and it has been known for many years that the termination of **epidermal** cell production, i.e. the formation of the **envelope** of the final organ, is a key factor in size control (Dale 1988). Alpine plants do not seem to be an exception. Epidermal cell division (except for stomata formation) has been found to cease long before mesophyll cell division in both alpine

Leaf development in *Geum reptans*, Alps 2600 m

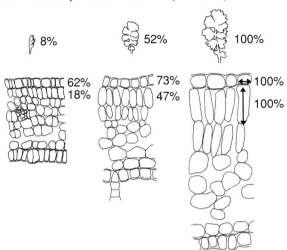

Fig. 14.3. Cross sections of leaves of *Geum reptans* at different developmental stages (samples taken from a site at 2600 m in the Alps). Note that most cells are present when the leaf has reached 8% of its final leaf area, so leaf enlargement to 100% size is largely a result of cell expansion and formation of intercellular spaces (Körner and Pelaez-Menendez Riedl 1989)

and lowland forbs (Körner and Pelaez-Menendez Riedl 1989). If cell numbers, and thus leaf size, are determined at such an early stage of leaf development, it cannot be the rate of resource supply (e.g. photo-assimilates) that controls the mature size of an individual leaf (resource demand is very small at the time when cell number is fixed). Rather, a program must be carried out that reflects the overall resource supply status of the whole plant within the limits of its genotypic size control. As mentioned earlier, cell size variation is relatively small, and cannot explain the mean tenfold size difference in alpine plants and their leaves compared with related low altitude plants.

An interesting aspect worth mentioning here is the reported correlation between **cell size and genome size**. Grime (1983) observed that plants whose leaves expand in early spring at low altitude (when temperatures are low) have larger cells and greater haploid genome size than plants active during warmer periods. According to Grime, the advantage of bigger cells is a reduced dependence on cell division during early leaf development. On the other hand, genome size has been found to correlate with the duration of the cell cycle (Francis and Barlow 1988; Dale 1992; Creber et al. 1993). Cells with bigger genomes appear to take longer to duplicate. Rated by cell size, alpine plants do not seem to have adopted this spring plant strategy observed by Grime, although a few species have been found with rather large cells (e.g. *Oxyria digyna*). As Bennett (1987) noted, genome size is much more variable between species than cell size. Hence control of cell size must be more subtle than a simple maintenance of certain DNA/cell volume ratios.

Mitosis and the cell cycle

I am not aware of any published work on the relative limitations of tissue growth under low temperatures by the three components mentioned in the introduction to this chapter. So what follows is an account derived from indirect evidence and is partly speculative. If resource supply for either ATP formation or synthesis of cell walls and cell organelles limits growth at low temperatures, it is most likely step 2 (see introduction to this chapter) which is critical, because this is where most of the resources and their processing are required. In terms of fuel and mass investment, cell division itself is the "cheapest" part of the process. As will be discussed later, there are also other reasons why the rate of production of embryonic cells is not likely to be the bottleneck of alpine plant growth (if there is one).

The separation of step 1 and step 2 processes is only meant to underline the fundamental difference between the genome replication (the formation of tiny "water bubbles" with packages of DNA) and subsequent mass investment (cell enlargement and differentiation). In reality, organ formation is a concerted action, with one step not possible without the other (Fig. 14.4). It is obvious that if there were no cell division, there would also

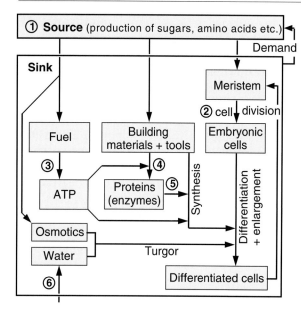

Fig. 14.4. A schematic presentation of the major processes involved in tissue growth. *Numbers* indicate where low temperature effects could play a role and thus limit the overall process, given that all partial processes are interlinked (see the text). For the sake of clearness, fueling and substrate provision to cell division itself (2) was omitted in this scheme, because the quantities required are very small compared with those for subsequent supplies for growth and differentiation of the emergent "embryonic" cell populations

be nothing to differentiate. But would there be any **cell division** without **cell differentiation**? Not for long, because the tissue would fall apart. Hence the rate of differentiation, the most costly part of plant tissue formation, will feed back on cell division in one way or the other.

What are the critical steps in Fig. 14.4 where low temperature could become effective and limit alpine plant growth? From the previous three chapters, supply limitations by primary products (1) seem most unlikely. Similarly, water supply in cold climates was shown to be not critical under most conditions (Chap. 9). Hence, **three groups of limitations** remain: the direct limitation of cell division (2), ATP provision to synthetic processes (3), and the synthetic processes themselves: (4) synthesis and turnover of enzymes and (5) for-

mation of cell walls and organelle differentiation. (3), (4) and (5) may follow similar kinetics and thus may be represented by one common metabolic function, as was suggsted by Criddle et al. (1997), but the mechanisms are not the same, and the distinction between "fuel-" and "process-" limitation seems important.

From what is known, it does not seem to be the amount of fuel present at the sites of synthesis which becomes limiting at low temperature, because sugars are very abundant (see Chap. 12), and their concentration has even been shown to increase in developing tissue at low temperatures (Francis and Barlow 1988). Yet the often observed change in leaf shape with growth temperature was suggested to be related to diffusional barriers for substrate at the inter-veinal scale, which may build up as temperatures decline (Fischer 1960). Whether the use of sugars for ATP production is limiting as Dahl (1986) had suggested, and as is assumed by Criddle et al. (1997), is hard to say, because of the potential co-limitation by other processes. It may be assumed that the reduction of mitochondrial activity at low temperatures is partly compensated by their increased number (Miroslavov and Kravkina 1991). As a consequence, rates of tissue respiration measured at low temperatures in cold habitats have often been found to be similar to rates measured at warmer temperatures in warm habitats, as a result of acclimation (see Chap. 11). With processes (4) and (5), we enter pure guessing, since no alpine plant seems to have been investigated, but it is clear that such thermal limitations of protein metabolism and synthesis of cellular structures must exist. In the following I will concentrate on cell division (2) for two reasons. First, because cell division can be documented in the field, and second, because one can safely assume that rates of cell division also reflect rates of cell differentiation since both processes are tightly coupled through feedbacks. Thus, cell division is a gateway to the dynamics of tissue formation.

Cell division includes phases of activity which are visible under the microscope (mitosis, the separation and sorting of chromosomes) and invisible phases during which the genome is copied (the synthesis or S phase) or "gaps" during which early researchers could not see anything spectacular happening (before and after the S phase, the G1 and G2 phases). Meanwhile it was discovered that quite significant things happen during G-phases. Together, all these steps are called the **cell cycle**. However, not all cells in a meristematic region are actively dividing ("cycling" through the above phases), but some "rest", and cells do not continue to cycle infinitely, but will leave the cell cycle lineage in an orderly manner so that the number of cycling cells does not increase exponentially, but achieves a sort of steady state, at least for certain developmental stages. Furthermore, not all cycling cells are cycling at the same speed. Hence, to be precise, one has to distinguish between actual measured rates of cycling in discrete cell cohorts and the statistical mean across all cells in a meristematic region, which is termed the **cell doubling time**. In the older literature this distinction was not always made, and the two terms have been used as synonyms. For practical purposes, the cell doubling time is a very useful quantity, but one has to be aware that (depending on the region of a meristem considered) it includes components other than just cell cycle dynamics.

The challenge is to obtain a picture of these dynamic processes through the collection of a series of static pictures. Whether the number of cells in a mitotic (i.e. visible) stage is low because of low numbers of cycling cells or because cells are cycling slowly is hard to know. However, tricks are available to find this out. A classic one is to block the separation of chromosomes of a mitotic cell (the beginning of the formation of two daughter nuclei). This can be done, for instance by adding colchicine, the substance which makes the geophyte *Colchicum autumnale* so poisonous. The result of this selective interference with the cell cycle is the gradual (over some hours, commonly linear) accumulation of all cycling cells in the metaphase, because they cannot get any further (the "metaphase arrest" method). By measuring the rate of appearance of metaphase cells in the treated tissue over the linear part of the response,

one can obtain an approximate rate of cell doubling or the cell doubling time.

Do cells divide more slowly at low alpine temperatures? Although published work (mostly with root tips of onion, bean or cereals in growth chambers) is largely for warm temperatures, in a few cases temperatures at or below 10 °C have been included. A literature review yielded the response shown in Fig. 14.5. These data suggest a dramatic slowing of the cell doubling time when temperatures drop below 20 °C. While cell doubling times of around 24 h are common at 20 °C, the regression suggests that this time is more than doubled at 10 °C, and approaches infinity near 0 °C. The shape of the curve resembles a mirrored temperature response of dark respiration (see Chap. 11). The diagram also illustrates that the **duration of mitosis** follows the response of the cell doubling time in a constant proportion of about one tenth (note the different scales in the diagram). Since prevailing night-time temperatures in the alpine

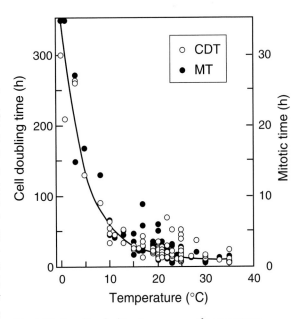

Fig. 14.5. Results of a literature survey of temperature responses of cell doubling time (*CDT*) and its mitotic phase (*MT*) in root tip meristems of herbaceous plants grown in controlled environments (ca. 50 publications by 1990; unpubl. review by S. Pelaez-Riedl)

zone are around 5 °C during the growing season, mean cell doubling times of several days might be expected from this response for non cold-adapted lowland plants.

The data collected in Fig 14.5, but also the vast agronomic literature on cold season growth of cold adapted crops indicates that commonly there is no measurable growth at or below 0 °C, which also holds for alpine plants (see Chap. 13). In most cases, growth ceases at several degrees above zero. According to the data in Fig. 14.5, cell doubling time becomes extremely long below 5 °C. In contrast, photosynthetic CO_2-fixation may reach one third of full capacity at 5 °C, and all alpine plants tested photosynthesized intensively at 0 °C. As was shown in Chapter 11, photosynthesis in active leaves stops only once intercellular ice forms, which happens at several degrees below zero (see Chap. 8). Hence, under all conditions, one can expect the cessation of meristematic acitivity to precede the cessation of carbon acquisition, as was concluded for overall plant growth (see Chap. 12).

What do we really know about cell division in alpine plants in their natural environment? Very little indeed. Creber et al. (1993) in their elegant study with latitudinal and altitudinal provenances of *Dactylis glomerata* investigated the possibility of ecotypic differentiation of cell cycle responses to temperature, but their plants were all from relatively low elevations. The two major results of their study may still indicate trends which might be observed in alpine plants as well: (1) cell division in root tips of their hydroponically grown seedlings was less sensitive to changes in temperature in provenances from cooler, compared with warmer climates; (2) when temperature caused the cell cycle to slow, the number of cycling cells increased in a compensatory manner.

To my knowledge, no data of this sort for alpine plants have been published. In 1988, Francis and Barlow concluded that nothing was known about the relationship between temperature and the cell cycle in shoot and leaf meristems in general, because all work was with root tips. In the following I will therefore present some original work which provides some initial ideas about what

might happen in developing leaves under alpine field conditions. Leaves (5–10% of final size) were selected, simply because root tips are difficult to identify in mixed plant communities and are not accessible in the field for regular observation without severe disturbance. Leaves also exhibit more dramatic size reductions with altitude than do roots.

The first and simplest (though inconclusive) approach is to check meristematic regions for mitotic cells, count these, and relate their number to the total number of cells in the observed meristematic region. This is called the **mitotic index (MI)**. Numbers for 24–48 h sampling periods collected at high elevations in the Alps, subarctic Sweden and the Ecuadorian Andes suggest that there is nothing special in alpine plants compared with their lowland relatives. MIs in embryonic alpine leaves are between 5 and 10%, mostly around 7%, and there are no obvious elevational differences or diurnal changes that could be related to the course of temperature (Fig. 14.6 and unpublished data for other species). There is perhaps a slight trend of somewhat higher MI in the greenhouse grown *Ranunculus* maintained at a constant 18 °C, but *Hypochoeris* in the Andes has an MI of 9% at temperatures of between 0 and 3 °C. We had thought that this diurnal constancy of overall MI masks significant shifts in the relative abundance of pro-, meta-, ana- and telophase, but this was hardly the case. Only in one out of several 24-h cycles in *Ranunculus glacialis* in the Alps did we find some indication of complimentarity between metaphase and ana- plus telophase fractions (Fig. 14.7), but overall, proportions did not change a lot during the day, irrespective of temperature, species and site.

With some care, we may conclude that the "whole wheel" of the cell cycle turns in a concerted way, so that no phase is limited by temperature more than another one. Otherwise we should see significant shifts in the abundance of mitotic phases and significant overall changes in MI throughout the day. As Francis and Barlow (1988) pointed out, plants are expected to lengthen the G1 phase relative to other stages of the cell cycle, in

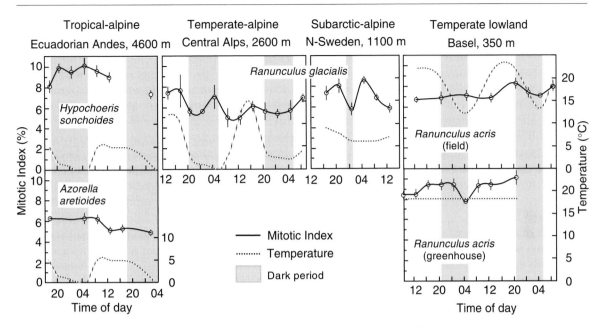

Fig. 14.6. The in situ diurnal course of mitotic index (percentage of cells in mitosis of all meristematic cells) in embryonic leaves of alpine plants from contrasting alpine climates and comparable data for a low altitude forb. *Dashed lines* indicate bud temperature. (S. Pelaez-Riedl and Ch. Körner, unpubl.)

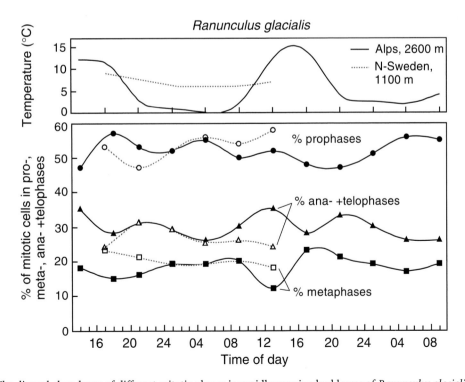

Fig. 14.7. The diurnal abundance of different mitotic phases in rapidly growing bud leaves of *Ranunculus glacialis* in the Austrian Alps (2600 m) and in the subarctic alpine zone of northern Sweden (1050 m). Means for 5–8 leaves at each sampling point. One hundred percent refers to the total number of mitotic cells found. (S. Pelaez-Riedl and Ch. Körner, unpubl.)

situations of stress, freezing temperatures included. This would delay the arrival of prophase cells in the "mitotic window", unless subsequent S and G2 phases caught up significantly. Pronounced **diurnal cycles** of mitotic index are absent despite considerable variation of temperatures. Cells are found at all mitotic stages at growth temperatures between 0 and 7 °C on subpolar, temperate and tropical mountains. Despite some insight into cell cycle dynamics that may be obtained by just looking at the abundance of various mitotic stages, such observations cannot tell us how fast the "wheel" is turning.

In situ determinations of cell doubling time are not an easy task, particularly not in alpine terrain. One needs a lot of similarly aged plants, six to ten specimens per sampling time, with hourly sampling over more than 10 hours. **Colchicine application** must follow a precise time schedule for each individual, and each plant can only be treated once. Every single embryonic leaf must be fixed, stained with the Feulgen reagent and analyzed under the microscope. So it is not surprising that this is not a very popular field of alpine ecology. Fortunately, soaking for 1 h of very young developing leaves (Fig. 14.8) with 0.5% colchicine solution (plus a wetting agent) does work. Also, cochicine impregnated filterpaper attached to such leaves helps. It was possible to halt the cell cycle in *Ranunculus glacialis,* and count the **metaphase accumulation** rate (Fig. 14.9). From this single experiment, it appears that leaf meristematic cells of *Ranunculus glacialis* have cell doubling times almost twice as long at 11–12 °C (a relatively warm day) as *R. acris* at low altitude at 22–23 °C. This would suggest a "Q_{10}" for the cell doubling of somewhere around 1.8 if data from the two different species are compared (not the way in which Q_{10} should be calculated, but still an interesting relationship).

With this first ever look into the heart of the growth process of an alpine plant species, generalizations cannot be made. Compared with the data in Fig. 14.5 it seems, however, that the cell doubling time in *Ranunculus glacialis* falls within the range of numbers reported in the literature for 11–12 °C in non-alpine plants. If this can be confirmed, it would suggest that thermal respon-

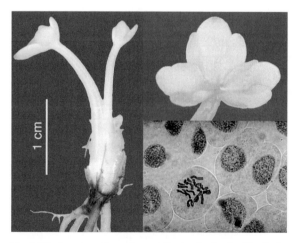

Fig. 14.8. A preparation of a rapidly developing bud leaf of *Ranunculus glacialis*, soon after snow melt at 2500 m elevation in the Swiss Alps. Leaves of this size (ca. 5% of final size) have been used in situ for the metaphase arrest method (see Fig. 14.9). *Scale bar* for the plantlet to the left only.

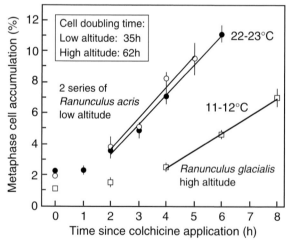

Fig. 14.9. *In situ* cell doubling time, calculated from the linear rate of metaphase cell accumulation after colchicine application to bud-leaves of *Ranunculus glacialis* at 2500 m in the Swiss Alps (July), compared with data for *Ranunculus acris* studied in the botanical garden and (under constant temperature) in the greenhouse in Basel (June). (S. Pelaez-Riedl and Ch. Körner, unpubl.)

ses in these very elementary processes of growth in alpine plants are not fundamentally different from those in plants in general. At first sight, it seems, this alpine plant species is not any "better" with respect to cell doubling time at its low growth temperatures than the crop plants for which data are shown in Fig. 14.5. Taken alone, these data might suggest *Ranunculus glacialis* produces less cells per unit of meristem area and time than plants from warmer low elevation sites (no acclimation). However, it remains uncertain as to whether the rate of cell production is indeed lower, because the **total number of cycling cells** per growing leaf is unknown.

Given the observation that the mitotic index was not very different between the low and high altitude *Ranunculus* examined, but the concurrent cell doubling time per unit of inspected meristem area was almost twice as long in the alpine species, theoretically, three answers are possible (with the first one being most plausible):

- Mitosis occupied a constant fraction of the cell cycle duration, hence was also taking twice as long at the alpine site. In this case, the chance of finding cells in a mitotic stage per unit of examined meristematic tissue would be equal at both sites, provided similar numbers of cells were cycling.
- The mitotic phase in the alpine species occupied a longer section of cell cycle duration, hence was relatively more restricted by lower temperature than S, G1 or G2. This would mean that less cells were actually cycling.
- Mitosis occupied a shorter fraction of the overall slower cell cycle, reducing the chances of being seen in a snapshot. This would require more cells to cycle to arrive at equal MI.

None of these possibilities can be rejected. But in a first approximation, a constant mitotic time fraction appears most strongly supported by the available literature (see Fig. 14.5), which would mean that the fraction of cycling cells was similar at both elevations. By simply adjusting the fraction of cycling cells (the so-called growth fraction), meristems could mitigate rate-limitations at the level of the single cell. This is what was actually found in the above mentioned study by Creber et al. (1993). But it needs to be re-emphasized that conclusions and interpretations of such data are full of pitfalls. A number of aspects cannot be accounted for in such a simple field trial, and even highly controlled laboratory experiments, such as the one by Creber et al., soon approach limits of interpretation.

For instance, assuming the cycling fraction of cells was indeed similar in the two *Ranunculi*, one could have expected *R. glacialis* with its smaller genome to have a shorter cell doubling time than *R. acris*, even at equal temperatures. The 1C-value for *R. glacialis* is 3.3 pg (1n = 8 chromosomes) and 5.3 pg for *R. acris* (1n = 7), both with ploidy levels of 2 (Bennett et al. 1982), with the latter confirmed for the populations studied here. With this in mind, the actual observed cell doubling time in the field of *R. glacialis* appears even slower.

The most critical unknown is the total size of the meristematic cell population per leaf, which depends in the duration of dividing cell lineages. Adjustments of activities among populations of meristematic cells can theoretically overrule large limitations experienced by individual dividing cells (see below).

From meristem activity to growth control

Given the limited data base, it does not make much sense to carry the above discussion any further. The purpose of showing and discussing these data was mainly to illuminate this largely neglected field of research. It needs to be repeated that such data do not permit a causal separation of limitations operating directly on the cell cycle, or indirectly, via feedbacks on cell division from subsequent cell growth. The relevant experimental literature (largely with agronomic plants) reflects a strong focus on cell cycle limitations themselves, which does not seem to be justified in the light of the far greater synthetic efforts and resource requirements during cell enlargement and differentiation. Given this uncertainty, it seems inappropriate to conclude (as for instance Francis and Barlow 1988 did) that "the rate of cell division determines the rates of organ and cell growth". With similar justification, one could conclude that the rate of tissue maturation (demand) determines the rate of cell division. In view of the various options a meristematic region has for producing more embryonic cells, if required (faster cycling, more cells cycling, cells cycling more often, all with exponential consequences), it seems that **feedback effects from tissue maturation** deserve greatest attention in order to find out how low temperatures limit plant growth. Hence the study of limitations at points 4 and 5 in Fig. 14.4 is needed most urgently. This is not a field were ecologists are normally experts.

In summary, if alpine plants (like any plant) were short of embryonic cells to be fed into the differentiation process, they have several response options:

• Have a given number of meristematic cells divide faster,
• Increase the initial number of dividing cells,
• Increase the fraction of cycling cells per unit of meristem tissue,
• Retain cells longer in a cell cycle lineage (increase the size of the meristematic zone).

Imagine the geometric increase in cell number by starting with 12 instead of 10 initials in an embryonic leaf, replicating consecutively over only three cell cycles (ca. 6 days at 10 °C) – 20 736 versus 10 000 new cells! The consequences for cell production of only the slightest re-arrangements in the pool of potentially dividing cells are so enormous that a plant should almost always be able to meet the demand in embryonic cells for tissue construction. The process of cell division is most unlikely to limit the growth and size of alpine plants. From the above it seems logical to conclude that it is the construction, the differentiation and maturation of cells that most likely represent the bottleneck of growth under thermal limitation, not the provision of raw materials, including the number of undifferentiated cells. Whether the total number of cycling cells is indeed larger at any given size of an alpine versus a lowland leaf remains to be explored. As illustrated by the above calculation, minute differences, possibly too small to be detected, would be sufficient to compensate for physiological limitations of cell duplication in a cold climate. That such compensatory processes do occur, was demonstrated by Creber et al. (1993).

This chapter closes the series of considerations related to a functional understanding of **alpine plant growth**. After considering physical constraints for survival, limitations by external resources such as water and mineral nutrients were explored. Via discussing potential constraints of carbon fixation and the carbon balance, constraints of carbon investments and growth dynamics, we arrived at the rather subtle question of meristematic processes in the cold. In part, this chapter ends like the previous ones, by finding little if any physiological evidence or reason which sufficiently explains alpine growth limitations. While photosynthesis and the carbon balance are most unlikely candidates for a physiological bottleneck of plant growth at high elevation, there are also no obvious indications that limitations of cell division are critical. This chapter ends by pointing to the potentially crucial role of synthesis of structures in the maturing cell, the cell wall in particular – an open field for biochemical ecology at high elevations. What if, in a couple of years, researchers find, perfect thermal adjustment in the processes involved, and once more no obvious physiological reason for a specific bottleneck?

Perhaps we would then take a fresh look at the data presented in the following chapter, which illustrate that alpine plant production is in fact not necessarily low, if expressed per unit of growing period duration rather than per year. However, we would still be left with the problem that tropical alpine plants don't grow any faster or larger despite being unlimited by time. Perhaps we would then realize that small size is also part of a plan, a plan for survival and fitness, rather than the result of immediate limitations by a hostile environment.

Alpine plants are **selected for small size** and can thereby cope with alpine life conditions (Körner and Larcher 1988). At all levels of their physiology alpine plants have been found to be perfectly adapted to grow where they do. It seems, that alpine plants grow in such a way that the resources available per unit land area and time are efficiently utilized, and that they avoid coupling to the unfavorable climate as much as possible by creating their own climate. Dwarfs on purpose!

In a more general sense, alpine plants are just another extreme example of how stature, size, allocation and growth are tuned so that a minimum of functional variability occurs at the cellular level. Alpine plants produce fewer and smaller organs, but these are well equipped. This means that chances are limited of obtaining an understanding of growth constraints and stature of alpine plants by examining primary metabolic processes (those associated with photosynthesis). Developmental controls appear to overrule many potential constraints by metabolism and resource use. Yet, by looking at the diverse appearances and functioning of alpine plants, it is fascinating to see how many different ways there are of being fit and successful in an environmentally rather demanding world.

Plants commonly don't grow alone. Thus, the growth of an individual leaf or an isolated plant – the topic of the previous three chapters – needs to be placed in the context of interactions, the growth of many other leaves and plants. A slowing down of one individual may be balanced by enhanced growth of another, hence only a consideration of all members of a plant community will reveal what may be the typical biomass production for a certain environment. Plants may respond in quite different ways if growing together, rather than alone.

The bulk of leaves and their integrated **activity per unit of land area** determine how much solar energy will be trapped by photosynthesis, how much water and mineral nutrients will be consumed, and how much biomass will be accreted and decomposed. In contrast to the previous three chapters, this chapter will consider whole plant communities, their structure, biomass production, and overall dry matter accretion, thus addressing land area based aspects.

The stucture of alpine plant canopies

Low stature in plant individuals, as is typical for most alpine plants, is not necessarily associated with less photosynthetic leaf area per unit of land area, a smaller **leaf area index** (LAI, $m^2 m^{-2}$). However, equal LAI at a low canopy height means greater **leaf area density** (LAD, $m^2 m^{-3}$). Closed alpine plant communities hold the world record for LAD by accumulating LAIs between ca. 2 and 5 in canopies of only 2 to 10 cm height (LAD 50–100 $m^2 m^{-3}$). For comparison, closed humid tropical forests averaging at maximum LAI of ca. 8 over a height of 40 m operate with LAD of ca. 0.2, similar to that found in temperate deciduous forests (LAI of ca. 5 over 25 m). Although leaves are not evenly distributed over the full canopy profile in such forests, the most active crown layers still measure 10 m or more in depth, driving layer specific LAD perhaps to 0.5. Lowland grassland may reach LAIs of 8 over a height of 0.5–1 m, yielding LADs between 4 and 8. Hence, dense clustering of leaves is an unbeaten feature of the alpine vegetation. Table 15.1 provides an overview of LAIs reported for closed alpine plant communities. LAI reaches high values up to the lower alpine zone, diminishes to less than half the value of low altitude meadows in typical alpine grassland (often around 2; Caldwell et al. 1974; Fig. 15.1), and below 1, once vegetation opens at higher elevations.

Canopy closure is an important (though not easily defined) criterion for applying the LAI concept with respect to light utilization. In fragmented vegetation with bare ground in between, LAI still provides a useful measure of land coverage by leaves, but functional links (e.g. plant efficiency of radiation use per unit land area, reflection of radiation, fluxes of water vapor etc.) are not straightforward. At the microscale "LAI" of isolated graminoid tussocks may still be similar to lowland meadows, but the most compact of all life forms, the prostrate cushion, is characterized by LAIs of less than 2 per unit of cushion surface (Körner and DeMoraes 1979), an obvious trade-off of condensing the active leaf layer to a few millimeters. Closed dwarf-shrub communities are not very different from graminoid communities, and

Table 15.1. Peak season canopy structure of alpine plant communities as compared with examples of low altitude vegetation (data for humid regions only). (Klug-Pümpel 1989; Cernusca and Seeber 1989; Larcher 1977)

Vegetation type	Plant canopy height (m)	Leaf height (m)	GAI $(m^2 m^{-2})$	PAI $(m^2 m^{-2})$
Alpine grassland				
Carex dominated (A1)	0.14	0.06	2.3–2.6	9
Carex dominated (CH)	0.12	0.06	1.6	–
Sesleria dominated (A1)	0.20	0.12	2.5	15
Luzula dominated (A1)	0.12	0.05	1.4	–
Alpine dwarf shrubs				
Loiseleuria dom. (A2)	0.05	0.02	–	3.3
Vaccinium dom. (A2)	0.25	0.25	–	3.4
Lower elevation vegetation				
Pasture near treeline (A3)	0.5	0.12	6–7	9–10
Montane hay field (A4)	0.9	0.35	8	9
Lowland hay fields	1.2	0.5	6–8	6–8
Mixed temperate forest	ca. 30	ca. 30	5–6	6–7
Humid tropical forest	ca. 40	ca. 40	6–8	7–9

A1 Hohe Tauern, 2300 m, *A2* Mt. Patscherkofel, 2000 m, *A3* and *A4* Hohe Tauern 1900 and 1600 m, all in Austria; *CH* Furka Pass, 2500 m, Swiss Alps; lowland references are means from various sources.

may reach LAIs of between 3 and 5 in the lower alpine belt.

A discussion of the consequences of canopy structure for radiation interception by plants needs to consider all absorbing structures, and hence needs to include non-leaf live plant material such as stems and flowers, the total **green area index** (GAI), as well as dead plant material in the canopy, which yields **plant area index** (PAI) i.e. the sum of the projected area of all structures per unit land area.

The inclusion of the area of dead structures causes alpine vegetation to come into line with lowland vegetation, because the amount of dead plant material increases with altitude almost as much as LAI decreases. As a consequence, photosynthetically useful quantum flux density (QFD) declines down the canopy profile of alpine vegetation more than it would if only LAI were considered (Fig. 15.1). Table 15.2 summarizes the results of an analysis of the interception of QFD in alpine plant canopies in the Alps, which illustrate that **extinction coefficients** of 0.45 are most common if based on PAI (the only meaningful basis for

describing radiation extinction). Self-shading of leaves – dead or alive – in closed alpine plant canopies is as significant as it is in lowland communities, which explains the strong dependency of photosynthesis on canopy structure (see Fig. 11.13).

Primary productivity of alpine vegetation

Three terms need to be distinguished:

- **Productivity**, which is the rate of new biomass formation (commonly per day),
- **Biomass production**, the amount of new biomass accumulating over a longer period of time (e.g. a growing season or a year), and
- The **pool of plant dry matter** per unit land area at any given time (biomass or phytomass, see below).

The first two will be considered in this section, although their conventional estimation by sequential harvesting depends on biomass estimates.

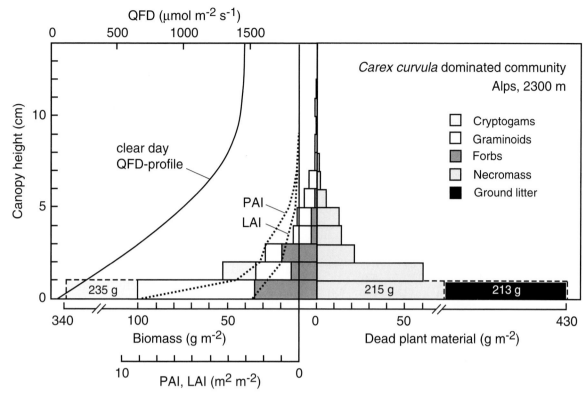

Fig. 15.1. Canopy structure and vertical distribution of quantum flux density (QFD) in a typical alpine *Carex* dominated grassland sward in the Alps. *Left* Biomass (i.e. live structures), *right* dead plant material. *PAI* Cumulative plant area index; LAI green leaf area index of higher plants. (Cernusca and Seeber 1989)

Since some newly produced biomass may have been lost between two harvest dates, actual production (and productivity calculated from this) is usually underestimated. In contrast, differences in aboveground biomass, sampled during the most active early season growth, can greatly overestimate true productivity, because apparent initial "growth" may largely result from allocation from belowground organs ("stored" growth). This hints at **fundamental problems of growth analysis** in perennial vegetation, which requires a few clarifications to start with.

Although widely reported, nobody has ever measured true net primary production (NPP) of a natural ecosystem, because its generally accepted definition as GPP-R = NPP makes it nearly impossible to determine to reasonable accuracy (GPP =

gross primary production, i.e. the result of cumulative net photosynthetic production by leaves; R, the sum of all respiratory losses of plants). None of the published numbers on **NPP** based on harvest data match this definition because the inter-harvest carbon exports of plants remain largely unknown (hence, I will use "NPP"). Above the ground, new litter may be accounted for, but fine root turnover, root exudation, supplies to mycorrhiza and above- and belowground macro- and microherbivory losses (which may be substantial – see the last section of this chapter), remain unaccounted for, although all were "produced". In shorter-lived plants, which start from seed every season (e.g. in agriculture), such errors may be smaller, but in perennial late successional vegetation such as most alpine types, annual

Table 15.2. Daily means of extinction coefficient k for quantum flux density during clear and overcast weather using a Lambert-Beer's law analogue: $QFD_i = QFD_o\ e^{-k\ LAI}$, where the indices i and o stand for below the canopy (ground level) and above the canopy, respectively. Examples are for grasslands from three elevations in the Alps, studied during clear and overcast weather in July and August. GAI, the green area index (representing >90% of LAI), PAI, index of total area of plant structures (including all standing dead material). Note, PAI-based k is relatively stable across altitudes, while GAI-based k increases with altitude due to the reduction of green structures compared with dead structures. Canopy reflection of QFD is ca. 5–12% of incoming QFD. (Cernusca and Seeber 1989)

Type of vegetation	GAI clear	PAI clear	PAI overcast
Montane hay field 1600 m	0.43	0.36	0.34
Pasture near treeline 1900 m	0.61	0.43	0.46
Alpine *Carex* grassland 2300 m	1.11	0.41	–
Alpine *Sesleria* grassland 2300 m	1.53	0.46	0.41

"NPP" in reality equals the sum of these losses, and it is only by defining certain sampling intervals and categories of plant matter that one can arrive at comparable "NPP" numbers.

Having said this, it is obvious that what follows reflects certain conventions, which assure some comparability among alpine sites of seasonal climates, although comparisons with production data from elsewhere are problematic. When there is no growing season and no pulsed growth, as in the tropics, conventional estimates of plant production do not work, but would require tedious monitoring of litter production. This is the reason why so little is known about biomass production of tropical alpine vegetation.

During the International Biological Program (IBP, late 1960s) a **terminology convention** was established which helps avoid confusion among various plant dry matter fractions:

- **Biomass** is defined as live plant mass (irrespective of specific tissue activity, thus including wood of live shoots).
- **Necromass** is dead plant material still attached to the plant ("standing" dead, but also belowground attached dead material).

- **Phytomass** is the sum of bio- and necromass.
- **Litter** (both above- and belowground) is a separate category, neither belonging to phytomass nor to soil organic matter, and includes all loose dead plant parts, big enough to be identified as structured plant material by eye.

Hence, biomass by definition, is live; phytomass, by definition, includes live and dead structures. Data not specified in this respect are of little comparative value and were disregarded here. A further problem with production estimates arises from asynchrony of plant phenology within a community. Some individuals or species may start to die back while others are still expanding their green mass. Harvests repeated at short intervals, and sorted by species and by live and dead matter are thus required to correctly integrate at least aboveground biomass production across a longer time interval such as a full season (instantaneous "book-keeping" of the various matter pools). Much more sophisticated analysis would be required to achieve similar resolution for root production (root windows, mini-rhizotrons, labeling techniques etc.). Problems of "stored growth" and overestimates of early growth have already been mentioned above. Given these numerous uncertainties, only a small fraction of the published literature on alpine biomass production is readily comparable and complete. I explain these difficulties also because some of these problems could be avoided in future data collection. The problem of the distinction of what is above- and belowground was discussed in chapter 12. For the sake of this re-assessment, I tried to be not too rigorous, but the reader should know, that the numbers presented for alpine "NPP" are likely to be underestimates, whereas those for aboveground productivity, when disregarding the late growing season, are overestimates.

Seasonal net primary production ("NPP")

Based on a number of studies, **whole season net biomass production above the ground** may be anywhere between 100 and 400 g m^{-2} a^{-1} for closed alpine vegetation in the temperate zone, with $200\,g\,m^{-2}\,a^{-1}$ representing a useful mean. Including **belowground** plant biomass production, $400\,g\,m^{-2}\,a^{-1}$ (±200) seems a realistic estimate. Since season length for sites from which such data are available varies between 2 and 4 months, alpine "NPP" converted to a full year would reach 800–$1600\,g\,m^{-2}\,a^{-1}$, with local values exceeding $2000\,g\,m^{-2}\,a^{-1}$, similar to productive low altitude vegetation (see Table 15.5). One of the reasons for such high productivities and the possible limitation of the validity of such extrapolations is the pulsed availability of nutrients during a short period of time, accumulated over the remaining 8–10 months of the year. Thus, much lower productivities are to be expected for non-seasonal tropical alpine vegetation. The only estimate I am aware of is $200 \pm 70\,g\,m^{-2}$ for above ground biomass in páramo grassland in Colombia, estimated by Hofstede et al. (1995a). This number is indeed low, given the year-round growing season. Since seasonal biomass production in the temperate and sub-arctic alpine zone (predominantly for herbaceous vegetation) is commonly treated as a synonym for peak season biomass, "production data" will not be presented here in detail and the reader is referred to the biomass section and the following discussion of productivities also derived from the biomass data.

Daily dry matter accumulation

The total dry matter difference between two successive harvests is now assumed to approximate a net production (but see above). Divided by the number of days elapsed, mean daily increments, i.e. **productivity** can be calculated. Considering aboveground rates of matter accumulation for closed vegetation only, data for North American alpine grasslands summarized by Bliss (1966) indicate rates largely between 1.6 and $2.8\,g\,d.m.\,m^{-2}\,d^{-1}$ (extremes 1.1 and 4.0; see Scott and Billings 1964; Table 15.3), for growing seasons of 60 to 70 days. Similar numbers are reported by Klug-Pümpel (1989) for graminoid communities in the Alps (1.3–$2.5\,g\,m^{-2}\,d^{-1}$ for season lengths between

77 and 142 days). For the lower alpine belt in the southern central Himalayas Rikhari et al. (1992) report harvest data from which 1.4–$3.1\,g\,m^{-2}\,d^{-1}$ can be estimated for ungrazed closed alpine vegetation (ca. 90–120 day season). Comparable numbers for tropical alpine vegetation do not seem to be available, for reasons mentioned above. If one divides Hofstede et al's (1995a) estimate of annual net biomass production of $200\,g\,m^{-2}$ by 365 days of tropical season, one would arrive at ca. $0.5\,g\,m^{-2}\,d^{-1}$, by all standards a low mean, which may also reflect unaccounted litter production.

Productivity calculated as observed difference in biomass divided by number of days, differs a lot when calculated for **different parts of the season** rather than for the whole growing season. When restricted to the initial flush after snow melt, pseudo-productivities (utilization of belowground reserves, see above) as high as twice the season means may be found, which was illustrated for instance by Klug-Pümpel (1989) and Brzoska (1973b). Efficiencies of energy conversion calculated for such periods are thus meaningless.

Conversion of photosynthetically active solar radiation into energy bound by plant dry matter has been described as "**efficiency of energy utilization** by plants". If only aboveground plant matter is considered, several authors arrived at "efficiencies" between 0.4 and 0.8% (according to Bliss 1966, 0.2–1.5%) for closed alpine vegetation (Table 15.3). Accounting for belowground biomass production total necromass production and dry matter losses not considered in Table 15.3, true "efficiencies" may range between 1 and 2%, again similar to numbers reported in the literature for lowland vegetation and crops. Bliss concluded that it may be generalized "that alpine plants (sensu vegetation) are as efficient in energy conversion and dry matter production as are many temperate region plants". If one accounts for the smaller LAI, the actual efficiency of energy capture per unit green matter is in fact larger than in lowland vegetation. Hence, there seems to be little in addition to the short duration of the growing season that limits alpine primary productivity, compared with other biomes. This conclusion is perfectly in line with those of Chapters 11–14.

Table 15.3. Estimates of aboveground primary productivity of alpine vegetation (live matter, i.e. biomass only). (From Bliss 1966; Larcher 1977; Galland 1986; Klug-Pümpel 1989; Rikhari et al. 1992; Brzoska 1973 a,b)

Type of vegetation		Length of season (days)	Aboveground productivity ($g\,m^{-2}\,d^{-1}$)	Energy fixation (%)
Alpine grassland with >70% cover				
Medicne Bow Mts.	wet *Carex-Deschampsia*	60	1.9	–
	snow bed, outer edge	55	2.5	–
Central Rocky Mts.	moist swales	65	2.1	–
	alpine turf	65	1.1	–
	moist *Carex-Deschampsia*	60	1.7	–
Mt. Washington	sedge-rush heath	70	1.6–2.9	0.5
	heath-rush meadow	70	1.8	–
	snow bank	70	2.7–2.8	0.5
	heath	70	4.0–4.3	0.8
Central Alps	*Carex curvula* heath	120	1.2–1.5	0.3–0.5
	Carex firma mats	75	1.1–2.0	–
	Deschampsia meadow	90	2.5	0.8
	Luzula snowbed	75	2.5	–
	Salix sedge snowbed	80	1.3	0.4
	Loiseleuria dwarf shrub heath	140	2.3	0.7
	Vaccinium dwarf shrub heath	130	3.8	0.9
	Open scree vegetation <10% cover	45	0.5–0.7	0.1–0.4
South Himalayan	*Danthonia* meadow	120	3.1	–
	Kobresia meadow	90	1.4	–
Approximate mean for all 18 sites with full cover		84	2.2 ± 0.9	

All numbers rounded up; length of growing season to nearest 5 or 10.

Surprisingly, the trend seen in closed alpine vegetation becomes enhanced at extremely high altitudes. Brzoska's (1973b) data for scree communities in the Alps (3150 m) with species like *Ranunculus glacialis*, *Androsace alpina* and *Primula glutinosa* yield extraordinary productivity if converted from 8–10% land cover to 100% land cover. For actual cover, Brzoska calculated 0.5–0.7 $g\,m^{-2}\,d^{-1}$ for a 40 to 50 day season, i.e. 5 to 7 $g\,m^{-2}\,d^{-1}$ if all ground were covered with this vegetation (and if the rate of growth were the same), not accounting for changes in belowground plant mass. The energy use efficiency of aboveground productivity alone of 0.09–0.38% for actual cover of 8–10% would scale up to 0.9–3.8% for full cover of 1 m^2 of land! Even allowing for large errors, there seems to be no doubt that alpine productivity and efficiency of solar energy utilization is not low on a daily basis, at least in the temperate zone mountains, for which such data are available.

The **range of aboveground productivities** shown in Table 15.3 appear to strongly reflect differences in soil fertility. Maximum productivity is always found in moist (not wet) places which permit season-long soil microbial activity (e.g. Scott and Billings 1964). As discussed in Chapters 9 and 10, soil moisture itself is unlikely to be a significant driver of alpine plant growth, but its indirect influence on nutrient availability in soils is of great importance.

Belowground productivity and seasonal belowground dry matter production of alpine vegetation are hardly known because the seasonal dynamics of root growth and root turnover (root litter production) are unknown. Annual regrowth of new roots gets drowned in a large matrix of old roots of mostly unknown longevity (see Chapt. 13). One reasonable approach to alpine fine root production might be to assume a constant leaf mass to fine root mass ratio and similar rates of above- and belowground turnover. Since both fine root to green leaf mass ratio and the stem to belowground storage organ ratio (including thick roots) are roughly 1:1 in herbaceous alpine plants (see Chap. 12), belowground production may be assumed to equal aboveground production, as a first approximation. In long-lived tussock grasses, cushion plants, dwarf shrubs and tap-rooted species, a leaf mass to fine root-mass ratio of 1:1 may still be valid, because the bulk of non-leaf or non-fine-root mass turns over very slowly. The only alternative approaches are very detailed root studies of the sort performed by Mähr and Grabherr (1983) with root windows, or by other tracing techniques.

Plant dry matter pools

The total amount of plant material per unit land area varies with season, with the aboveground live fraction in herbaceous vegetation close to zero at the beginning and end of the growing period in seasonal climates. Since details for belowground plant mass such as those in Fig. 15.2 are commonly not provided, the data compiled in Table 15.4 for peak season pools include a mix of various belowground biomass compartments. Where possible live and dead dry matter data are reported separately (see also the compilations by Franz 1979).

According to Table 15.4, **aboveground phytomass** of closed or almost closed alpine vegetation may be anywhere between 200 and 3500 g m^{-2}, a variation largely resulting from the wide range of necromass accumulation and the inclusion of woody plants. *Carex firma* mats in the Alps, *Chionochloa* tussocks in New Zealand, and the Colombian páramo grassland (Fig. 15.3) are examples of enormous necromass accumulation. Similarly, large necromass accumulation is reported from the high arctic (Henry et al. 1990). Apart from such extremes most herbaceous communities fall between 200 and 800 g m^{-2} aboveground phytomass with no latitudinal trend. Live aboveground plant mass ("biomass") of closed vegetation varies less, i.e. from 100 to 400 g m^{-2} (extremes of 600) with most values around 200.

Belowground phytomass data are much less abundant, and the range of 70 to 3600 g m^{-2} is similar to aboveground phytomass, but numbers above 500 appear to be more common. The variation seen in belowground phytomass cannot be explained because dead roots, dead rhizomes and belowground dead leaf sheets can accumulate to >90% of some of these total phytomass numbers.

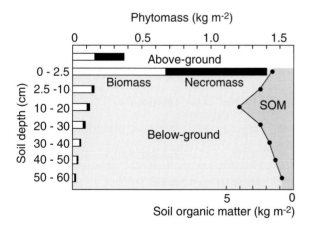

Fig. 15.2. Biomass, necromass, litter and raw humus: an example of carbon distribution in a late successional alpine grassland dominated by *Carex curvula*, 2500 m in the central Alps. (Unpubl.)

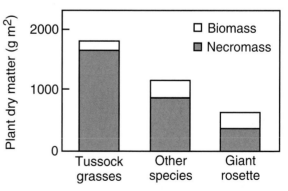

Fig. 15.3. Live and dead plant dry matter in a high Andean páramo vegetation, at ca. 4000 m elevation in Colombia (aboveground material only). Giant rosettes are *Espeletia* sp. Note the enormous stocks of dead material accumulating to 28 t ha^{-1}. (Hofstede et al. 1995a)

Table 15.4. Above (A) and below ground (B) phytomass (Ph), biomass (Bio) and ground litter. Because reported data are partly inconsistent with respect to the inclusion or exclusion of necromass, peak season biomass data have often been used as a measure of seasonal dry matter accumulation, hence some of the data listed here are redundant with productivity x duration from **Table** 15.3. For the shortcomings of such estimates of "primary production", see the text

Area/vegetation	Elev. (m)	Cover (%)	A-Ph	B-Ph	A-Bio	Litter	Ref.[a]
					(g d.m. m^{-2})		
Alps Scree veg.	3000	<15	26–75	–	–		1
Carex heath	2550	>90	750–800	–	300–470	55–165	2
Carex heath	2300	100	720–800	–	170	330–390	3
Sesleria meadow	2280	100	380	–	200	310	3
Deschampsia m.	2280	>80	420	–	210	240	3
Luzula meadow	2280	>70	240	–	190	240	3
Carex cushion	2160	100	1250	–	250	1900	4
Sesleria heath	2150	100	–	–	260	1100	4
Loiseleuria heath	2170	>90	1310	–	1070(w)	1100	5
Vaccinium heath	1980	100	1400	–	1020(w)	850	5
Central Caucasus							
Carex heath	2600	100	510	–	150	–	6
Alp. pasture	2000	100	500–750	2600	–	–	7
Alp. meadow	2000	100	660	240	400	–	7
Rhododendron	2200	100	1900	–	1820(w)	–	8
New Zealand Alps							
Tussock grassland	1980	>90	3540	1740	–	–	9
Mt. Washington							
Heath vegetation	1800	>70	–	540–3640	200–300	–	10
Medicine Bow Mts.							
–	3000	>70	–	–	110–350	–	10
Central Rocky Mts.							
Fellfield	3650	–	–	–	240	–	11
Dry meadow	3650	–	–	–	230	–	11
Moist meadow	3650	–	–	–	220	–	11
Wet meadow	3650	–	–	–	160	–	11
Snowbed	3650	–	–	–	100	–	11
South-central Himalayan							
Danthonia grassl.	3550	–	700	1500	390	250	12
Kobresia grassl.	3750	–	200	950	110	50	12
Rhododendron	3750	–	400	750	350(w)	70	12
Venezuelan Páramos							
Grassland	3550	>70	210–650	70–390	150–430	–	13
Colombian Páramos							
Tussock grassland							14
Papua New Guinea							
Tussock grassland	3600	>70	–	–	440–630	–	15

[a] 1, Brzoska (1969); 2, Grabherr et al. (1978); 3, Klug-Pümpel (1982); 4, Rehder (1976); 5, Schmidt (1977); 6, A. Cernusca (pers. comm.); 7, Nachutsrisvili (1975); 8, Tappeiner et al. (1989); 9, Meurk (1978); 10, Bliss (1966); 11, Walker et al. (1994); 12, Rikhari et al. (1992); 13, Smith and Klinger (1985); 14, Hofstede et al. 1995a; 15, Hnatiuk (1978); (*w*) means woody (dwarf shrub) vegetation.

Extremely high fractions of below ground phytomass of >95% of the total are reported for the alpine semi-desert of the Pamirs (Agakhanyantz and Lopatin 1978), but much of this may be non-functional remains.

The limited amount of information for low latitudes does not permit a conclusion with respect to the effect of seasonality per se. The data reported by Smith and Klinger (1985) and Hnatiuk (1978) are within the range found outside the tropics. Much of the variation within the **tropics** appears to be related to the presence or absence of tussock grasses, which tend to accumulate more biomass and, in particular, more necromass than other life forms. The missing separation of belowground organs by functional units (fine roots, thick roots, rhizomes, leaf bases) seriously limits the interpretation of these often large dry matter pools. Long life spans of thick roots and rhizomes contribute to the often high belowground plant mass (see Chap. 12). However, such structures may have a similar function to woody stems in shrubs, certainly not equivalent to the active fine root fraction, and of doubtful function with respect to storage. Retarded senescence and slow decay seem to be the most plausible explanation. Below-versus aboveground dry matter ratios were discussed in chapter 12.

I want to close this section with two aspects of biomass pools: the question of spatial and of year-to-year variation within a given region. A dense net of biomass study sites exists for the Alps. Figure 15.4 illustrates the **altitudinal variation of aboveground biomass** (live plant material only) in herbaceous or graminoid dominated vegetation of the central Alps. Peak season biomass between lowland and treeline elevations does not necessarily differ. Above the treeline, biomass per unit land area drops significantly, and only 300 m above the treeline, less than half of the typical lowland standing crop can be harvested at the time of maximum development of the vegetation. The further reduction of biomass with elevation reflects the reduction of land cover and not biomass per unit of ground area covered by plants. It needs to be noted that none of the grasslands

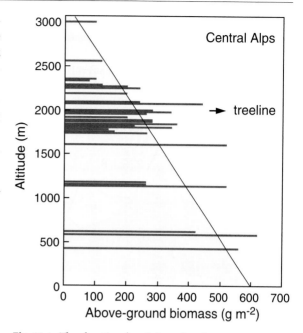

Fig. 15.4. The elevational variation of peak season aboveground biomass in herbaceous vegetation (largely grassland) in the central Alps. (compiled from various sources)

below the treeline are natural and those at lowest elevations are harvested three times a year for hay production.

Several authors have reported sequential harvests of alpine vegetation over more than one year (e.g. 3–5 years Bliss 1966; Klug-Pümpel 1989). The longest observation series is now available for Niwot Ridge in the Rocky Mountains' Front Range (Walker et al. 1994). Such studies help answer questions about how much the variation seen in Table 15.5 reflects **year-to-year variation** of growth conditions. For clearness, Fig. 15.5 depicts only three of the five types of vegetation studied, among which peak season biomass varied between 100 and 240 g in any year. The year-to-year variation was much smaller than the community differences, and only a small fraction (15–40%) could be explained by climate (a suite of parameters was tested). Hence, part of the variation seems to be related to phenological (reproductive?) rythms, winter conditions, storage cycles or random phenomena. Overall these data suggest that single year harvests of biomass can be considered as a reasonable site-typical estimate.

In **conclusion**, this brief assessment of alpine canopy structure, productivity and plant dry

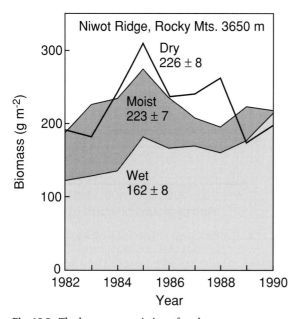

Fig. 15.5. The long-term variation of peak season aboveground biomass in alpine grassland on Niwot Ridge, Front Range (Rocky Mountains). The three meadow types are characterized as dry, moist and wet. Numbers are all-year means ± SE. Note the relatively small year-to-year variation and the slight trend to decreasing biomass from dry to wet. (Walker et al. 1994)

matter pools underlines that the length of the growing season and the percent of ground cover are the two major determinants of alpine plant biomass and its rate of production. Per unit covered ground area and per unit of time during the growing season, alpine plant communities produce as much or more as most communities at lower elevations, including forests (Table 15.5). These numbers are complementary to the more physiological data presented in earlier chapters. There is no reason to consider alpine vegetation as having low productivity, its productivity is not at all low. Its accumulative biomass production per year is low, but this is not because of a "stressful environment", "low temperatures", "adverse life conditions" etc., which were frequently supposed to limit productivity during the growing season, but simply because of short season and partly because of incomplete ground cover (rocks, washes, scree, patchy soil). Evidently the picture drawn in earlier assessments deserves a revision. Since the productivity of tropical, i.e. non-seasonal alpine vegetation is still largely unknown (but see Hofstede et al. 1995a,b) it remains to be seen whether such data from the tropics will deviate from the seasonal mountains on which the conclusions from this chapter are based. I would predict their productivity to be lower, because of continuous nutrient demand and the missing "pulse-effect" in seasonal climates as was discussed above.

Table 15.5. A comparison of net biomass production (above- plus belowground) in major types of global vegetation, calculated either per year or re-calculated per month of growing season (approximate ranges in brackets; see the caveats on "NPP" in the text). (Schulze 1982 and the alpine data presented here.) The mean for all six biomes is $210 \pm 30 \, g \, m^{-2} \, month^{-1}$

Biome	Annual net primary production ($kg \, m^{-2} \, a^{-1}$)	Length of growing season (month)	Monthly net primary production ($g \, m^{-2} \, month^{-1}$)
Humid tropical forest	2.5 (1.8–3.0)	12	210 (150–250)
Temperate deciduous forest	1.2 (1.0–1.5)	5	240 (110–300)
Boreal forest	1.1 (0.3–2.0)	5	210 (60–300)
Tropical grassland	2.5 (0.2–4.0)	10	250 (70–400)
Temperate grassland	1.0 (0.2–1.5)	6	170 (70–280)
Alpine vegetation of the northern temperate zone	0.4 (0.2–0.6)	2	200 (100–300)

Original calculations are for site specific season length, hence means and ranges per month do not exactly match those per year. The listed length of growing season is only a rough estimate for the major active part of the year.

Biomass losses through herbivores

Plant consumers are everywhere, even at the uppermost extremes of higher plant life (Halloy 1991; Swan 1992). Depending on the habitat and the type of vegetation, a substantial fraction of alpine plant biomass is harvested annually by natural herbivores, and they harvest selectively, hence affecting the species composition of plant communities. In many alpine regions of the world, large natural herbivores were replaced in historic times by domestic animals, sheep and goats in particular. These aspects of human land use will be treated in Chapter 17. Here I will briefly consider natural herbivores, often less conspicuous, but not necessarily less influential. For quantitative data on alpine animal communities see the reviews by Franz (1979, 1980). Since the evidence on natural alpine grazing appears to be rather thin, this account will rest on only a handful of case studies.

Starting at the lower end of alpine vegetation, Galen (1990) reported the massive influence of **elk** browsing on inflorescences of the **herb field** perennial *Polemonium viscosum* near the Rocky Mountains treeline in Colorado (3550 m). Due to obligatory flower bud preformation in the previous year, plants have no leeway for compensatory responses, and thus lose the entire seed crop of the current season. They also lose the potential crops of the following two years, because of reserve exhaustion – a rather severe impact on **reproduction**. Galen noted that herbivore impact decreases with elevation, and was small in the summit region (4050 m) . Probably elk prefer the lower elevations. This altitudinal pattern matches predictions by Grime (1979) that herbivory should be reduced in more stressful environments. As will be shown below, this generalisation does not seem to withstand testing.

In the heart of the **alpine grassland** belt, at ca. 2500 m elevation in the Swiss Alps, Blumer and Diemer (1996) documented substantial biomass harvesting by **grasshoppers** (*Melanopus frigidus* and *Aeropus sibirica*). With daily peak season biomass removal rates of about 0.4 g d.m. m^{-2}, these animals were responsible for 19% and 30% removal of the life standing crop in a *Carex*

curvula and a *Carex foetida* dominated community respectively (removal is significantly greater than consumption, because both species first cut leaves and only then nibble at their edges). This amount is at least three times the estimated annual consumption through midsummer cattle visits at this altitude (S. Schneiter and Ch. Körner, unpubl.). An interesting aspect of this study is the food preference of the two typical alpine grasshopper species. Both species were feeding on graminoids, which have relatively low leaf nitrogen concentrations (between 2 and 2.5%). However, the much more nutritious (2.6–4% N) forbs, were only consumed my *Melanopus*. But this species also avoided about half of the forb species present, most noteworthy the second most important species after *Carex curvula*, the rosette forb *Leontodon helveticus* (3.2% N), presumably because of latex content. The overall biomass losses due to grasshopper grazing would not be included in a conventional productivity assessment through biomass harvests, a substantial error given the amount of plant mass removed. The soils under these alpine turfs host a rich phytophagous larval fauna (e.g. Chironomidae) which could consume equally significant amounts of belowground biomass, but no data seem to exist.

Natural alpine herbivory is also very species specific in the high **Andean short grassland** of northwest Argentina (4250 m, Cumbres Calchaquies). A late season census in five forb species (from the genera *Hypochoeris*, *Werneria*, *Perezia*, *Geranium* and *Calycera*) revealed that *Hypochoeris* and *Calycera* had 26 and 29% of their leaves partly or fully eaten, *Perezia* 8% and the other two species were undamaged (unpublished data from March 1988). The dominant grazers in these high elevation planes are **Guanacos** (a small relative of the Lama) and rodents.

Biomass losses to herbivores in the open **fellfield** vegetation at higher elevations may be substantial as well, and increasing environmental stress does not seem to preclude plant life from herbivory pressure (Oksanen and Oksanen 1989; Oksanen and Ranta 1992; Virtanen et al. 1997). Data from the Alps and from northern Scandinavia suggest that **rodents** are major consumers, and once more, very selective ones. They clearly go for the high protein food, such as leaves of **Ranunculus glacialis** and *Oxyria digyna*, both species with 3–4% N in their leaves. Järvinen (1987) showed that flowering rhythms observed in *Ranunculus glacialis* in subarctic-alpine fellfields (Kilpisjärvi, 860 m) had to do with population dynamics (and feeding dynamics) of lemmings (*Lemmus lemmus*). An 8-year exclosure of low density lemming polulations from snowbed communities in the same area caused substantial alterations in community composition, particularly among mosses, which started to suppress various phanerogam species (Virtanen et al. 1997). Their observations are in line with earlier work in the arctic, which showed that moderate grazing maintains higher species diversity and is not detrimental to productivity (Henry and Svoboda 1994; see also Chap. 17). In a recent study on the impact of pikas (*Ochotona collaris*, small rodents) on an alpine *Kobresia* meadow in the Ruby Range (Canada, 61°N) annual foliage removal of 60% was well tolerated (McIntire and Hik 2002). Plants responded to persistent heavy grazing by producing shorter leaves and delayed leaf senescence. For further references on N-American alpine herbivory see Dearing (2001).

For highest elevations in the central Alps, Diemer (1996) showed that **snow vole** (*Microtus*) which are found up to 3900 m elevation (1700 m above treeline), are major consumers of *Ranunculus glacialis*. At three of four sites between 2600 and 3300 m elevation, 15 to 26% of all individuals showed herbivory damage at a single inspection early in the season (*Microtus* usually harvests whole inflorescences and whole leaves and moves this material to the nest). The one site with no damage was an isolated peak with permanent snowfields around. A detailed full season census on two sites (2600 m and 3150 m) revealed much

Fig. 15.6. *Ranunculus glacialis*, a preferred food for snow vole near the upper limits of higher plant life in the Alps. Voles prefer inflorescences and also harvest whole leaves, with no existential impact on these plants

greater losses: 60–65% of all individuals were affected, with most of them having lost their large inflorescences (Fig. 15.6). Of all these individuals, 36 to 58% also lost at least two of their ca. five to ten leaves. Surprisingly, this had no effect on survival of plants (all survived) and 36% of those damaged flowered quite normally the year after.

It seems that *Ranunculus glacialis*, growing at the upper limits of higher plant life in these mountains, copes well with its major herbivore, and both the carbon and nutrient balance of these plants seem to be all but tight, permitting substantial provisions to the rodent's life even above the snowline. This adds to the conclusions of Chapter 12, and re-emphasizes that great care is advised in projecting our human perception of what is stressful and limiting onto adapted organisms (see Chap. 1).

The alpine flora of a given mountain region commonly contains 200–300 different species of higher plants (Chap. 2). How do they manage to **maintain** their **presence** and expand their range into new open land? How do they ensure the maintenance of intra-population diversity required for sustained ground cover in a harsh and ever changing environment? There are three ways:

- Invest in seed production and seed dispersal – sexual reproduction,
- Invest in vegetative propagules – clonal reproduction,
- Stay where you are as long as you can – the space-holder strategy.

These three strategies may be combined in various ways, as will be discussed later. Plants may periodically switch between strategies 1 and 2 or use both simultaneously. Independently of the predominance of either of these strategies, plants may in addition (once successfully established) adopt strategy 3. Strategy 1 will always be involved, but often contributes little recruitment in late successional stages. The following sections will first consider sexual reproduction of alpine plants, namely flowering, pollination, seed development, germination and seedling establishment, which are then followed by an account of clonal propagation and alpine plant age.

Flowering and pollination

In the tropics, plants can theoretically flower year-round, and according to Hedberg (1957) and Coe (1967), there are no distinct flowering seasons in the afro-alpine life zone, with most species bearing flowers throughout a 10–12 month period. However, pronounced seasonal cycles of precipitation can narrow the flowering period for the majority of species to a few months in the tropics too, as is the case in parts of the Andean páramos; (e.g. Berry and Calvo 1994). Irrespective of precipitation, flowering seasons get shorter with increasing latitude, until there is very little temporal choice left. However, it is still possible to distinguish three types of flowering regimes in such short-seasoned climates:

- Early flowering – at or shortly after snow melt or soil thawing,
- Mid season flowering – at the peak of vegetative expansion,
- Late flowering – after the main part of the growing period is over (sometimes only prolonged flowering).

All early, and most midseason flowering arctic and alpine species perform **flower preformation**, which means inflorescences are initiated and differentiated to variable degree in the previous or even pre-previous season (to a variable degree; Resvoll 1917; Rübel 1925; Sörensen 1941; Billings and Mooney 1968; Larcher 1980; Dahl 1986). In flowers emerging first, a fully "pre-fabricated" organ is unfolded. In species flowering 1 or 2 weeks later, inflorescence expansion involves some current season differentiation growth, and species flowering at mid season often start from preformed floral primordia only, with substantial current season dry matter investment. In some of the mid and late season flowering herbaceous species, particularly those which produce multistemmed shoot systems with comparatively small flowers (many Caryophyllaceae), flowers are current season products. As always, there are exceptions to such generalizations. For instance, among species with extreme floral preformation, *Polygonum viviparum* flowers late, whereas *Ranunculus glacialis* flowers early.

These two species are classic examples of flower bud preformation. In *R. glacialis*, one to two future cohorts of flowers are initiated as plants enter the winter (Resvoll 1917 p. 89ff; Moser et al. 1977; Fig. 16.1). When transplanted to a warmer long-day climate, up to three generations of flowers may emerge within one season (Prock and Körner 1996). In *P. viviparum*, three seasons of preparation are required for a flowering event in the fourth season, as was documented with SEM micrographs by Diggle (1997) in specimens collected at Niwot Ridge (3750 m, Rocky Mts.). Indeed, these are long periods during which environmental conditions may influence the size and time of appearance of such inflorescences. However, the last, most critical steps, meiosis and gametogenesis are included in preformation. At least this is what Erschbamer et al. (1994) noted in alpine *Carex* in which ovules and pollen (and meiosis) were completed only immediately before anthesis.

Like vegetative development (Chap. 13), the **timing of seasonal flowering** is under tight **environmental control** with temperature and photoperiod exerting most important influences. From the few species carefully examined (mainly generalist grass species such as *Poa alpina* or *Phleum alpinum*) it appears that vernalization by low **winter temperature** is essential. The eventual appearance of flowers requires a secondary induction, either another cold winter or a certain photoperiod (50% of 25 species tested in the Alps require long days; Keller and Körner 2003). Very early flowering species seem to be opportunists and flower whenever snow releases them (e.g. *Soldanella alpina* and other Primulaceae; *Ranunculus nivalis*, *R. glacialis*, *R. adoneus*; *Saxifraga oppositifolia*; most alpine *Carex* and *Luzula* species). This explains why small (experimental or environmental) increases in temperature can induce

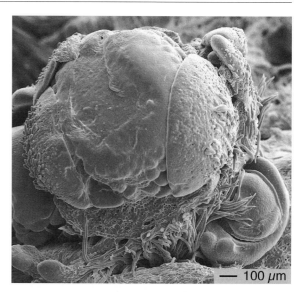

— 100 μm

Fig. 16.1. Flower bud preformation is employed by many alpine species in order to make maximum use of the short growing season. Here as an example, floral initials (terminal and lateral flower) for the next season (1999) in *Ranunculus glacialis* (SEM, 60x). Potential primordia for the year 2000 are hidden. The sample was taken from a flowering individual on July 10, 1998. Together with the actual flowers, three generations of reproductive structures may be present at one time

earlier flowering and accelerated overall development in such species (Alatalo and Totland 1997; Henry and Molau 1997; Stenström et al. 1997; Suzuki and Kudo 1997), in a similar way to the effect of shorter snow cover duration (Walker et al. 1995).

Species flowering only once the annual leaf crop has been produced (my impression is that this is the case in more than half of all alpine species in the Alps), may use **photoperiodic triggers** to initiate anthesis. At least this was shown for *Poa alpina* and *Phleum alpinum* (Heide 1990, 1994; Pahl and Darroch 1997 and references therein). In obligatory late flowering species, transplantation to warm low elevation growth conditions can illustrate photoperiodic inhibition of development until the "time is right" (see Fig. 13.2). In temperate and higher latitude alpine climates, only very few species commence flowering later than two months after snow melt. According to Resvoll

(1917), the extremes she found were 33 days in an alpine ecotype of *Taraxacum officinale* and 40 days in *Polygonum viviparum* (Dovrefjell in Norway at ca. 1500 m, and in north Sweden at ca. 1000 m). Commonly Primulaceae, Ranunculaceae and Cyperaceae species are among the ones **flowering earliest**, and species of Asteraceae and Campanulaceae most often belong to the group **flowering latest**. *Gentiana* and *Gentianella* species are often either among the first or very last ones to flower (Fig. 16.2).

Although flowering periods are highly species specific, and not at all synchronized within communities, despite the short season (not even within closely related species e.g. Pickering 1995), the overlap of flowering periods (Fig. 16.2) leads to pronounced seasonal peaks of **flower abundance** in all seasonal climates (Fig. 16.3 and 16.4). More than 70% of all species present can be found flowering during mid July in the central Himalayas, and by the end of July in the alpine regions of temperate zone mountains. Peaks seem to be broader in the mediterranean, summer-dry climate of Chile, compared with the monsoonal climate in the Himalayas, and slope exposure clearly becomes more important with elevation.

A schematic summary of trends across latitudes is attempted in Fig. 16.5, and illustrates the extreme narrowing of the flowering period at higher latitudes. This compression of the blooming period is the main reason why alpine vegetation at these higher latitudes often appears more attractive (colorful) to people than tropical alpine vegetation.

Why is there still so much variation in phenology, despite the extremely short season at higher latitudes? Molau (1993) and Kudo (1992, 1996b) have related this to differences in **life history** strategies. Molau proposed, based on a data set for 137 species from the subarctic-alpine life zone (68°N lat, 1000 m, in northern Sweden) that early flowering species show a low relative reproductive success (mean RRS of 0.3; RRS = fruits/flowers × seed/ovule), but those seeds that have been set have a very high chance of maturing (Fig. 16.6). In contrast, late flowering species have a high seed output per genet (mean RRS of 0.7), but compar-

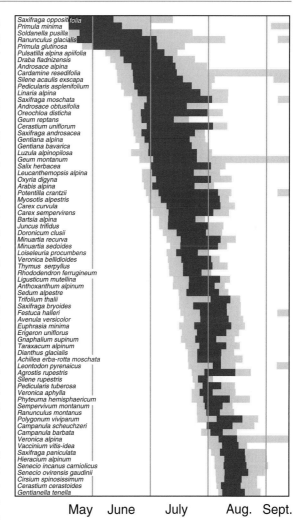

Fig. 16.2. Flowering spectra observed in 79 alpine plant species in the summit area of Mt. Glungezer (2600 m, 600 m above treeline, Tirol, Austria). The survey area covers 4000 m² with rock-, scree- and snowbed-vegetation and fragments of alpine grassland. The vegetation covered ground area was ca. 1100 m². 4 species of a total of 83 species did not flower, two of which are viviparous. Note the broad overlap of spectra by the end of July, which leads to the 78% flowering peak in Fig. 16.3. (Bahn and Körner 1987)

atively lower chance of reaching maturity. Complementary to this reproductive success is the **breeding system**: high outbreeding rates (<20% selfing) in early flowering species and a great fraction of selfing (>60%), apomixis and vivipary in

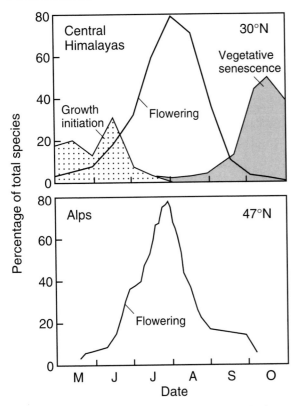

Fig. 16.3. Flowering phenology in alpine vegetation in the central Himalayas (ca. 4000 m, 30°N lat) and in the Alps (2600 m, 47°N lat). The diagrams show the relative fraction of species flowering as % of all species (142 species in the Himalayas and 79 in the Alps). (Bahn and Körner 1987; Ram et al. 1988)

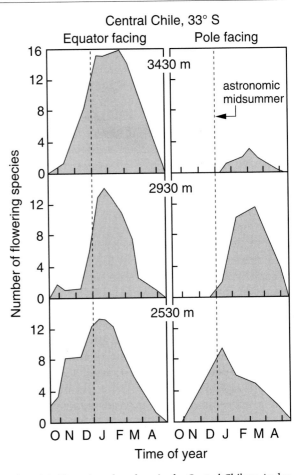

Fig. 16.4. Flowering phenology in the Central Chilean Andes (33°S lat) at various elevations, and pole versus equator facing slopes. The lowest altitude site is in the Andean dwarf shrub belt, the middle altitude site falls in cushion plant vegetation, and the uppermost altitude site is in open fellfields. (Arroyo et al. 1981)

late flowering species. Finally, there is a characteristic difference in ploidy: of the one third of all early flowering species which are not diploid, the majority are tetra or hexaploid. In contrast, 45% of all the late flowering species are more than diploid, and among these, species with 8 to 12 sets of chromosomes represent one third (such high ploidy levels are completely missing in the early flowering group). Molau concludes that early and late flowering is associated with two different risks:

- A pollen-loss risk (too cold for successful pollination, but sufficient time for seed maturation) and

- A seed-loss risk (safe pollination, but not enough time to mature seeds).

The truly functional determinant of seed/ovule ratios is the breeding system (greater fertilization failure in heterozygotes versus more homozygotes), whereas the link to flowering phenology is of secondary or indirect nature, with the two risks as possible resultants. It is not clear why this seasonal separation of breeding systems occurs, and it needs to be added that the high frequency of

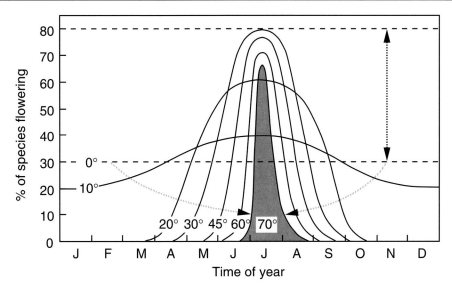

Fig. 16.5. A schematic presentation of flowering phenology in the alpine zone at different latitudes. Peaks in temperate and subarctic vegetation do not reach up to the subtropical peak, because of reduced flowering duration per species and the termination of flowering in many very early flowering species by the time the bulk of species reaches the peak. The two *dashed lines* for the tropics do not account for single florets which might be found at any time of the year in any species, which would bring these lines up to 100%. Note the most dramatic change in phenology in the tropical-subtropical transition. For simplicity, a Northern Hemisphere seasonal calendar was employed

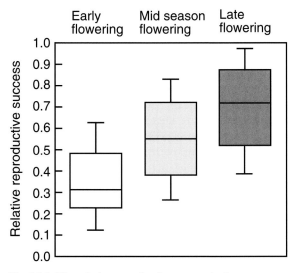

Fig. 16.6. The relative reproductive success in three groups of subarctic-alpine plant species differing in flowering phenology (median, and the ranges from 25–75 and 10–90 percentiles). The data set includes 21, 51 and 32 species for the early-, mid season- and late-flowering group. (Molau 1993)

higher ploidy levels in late species is contributed by species from very late thawing snowbed habitats, in which selfing appears essential. Nevertheless, either of the two breeding systems may be more successful, depending on season and microhabitat, hence their common representation in a given flora. It is important that these two "strategies" are seen as end points of a continuum of possibilities, although there is some indication (e.g. with autogamy) that frequency distributions may be bimodal – either/or – with intermediates less common.

Kudo (1992, 1993, 1996b), who in essence arrived at similar conclusions to Molau for **snow duration gradients** in Japan, emphasized pollinator availability as a generally important factor. According to his observations, early flowering individuals/species suffered from a lack of pollinator activity, whereas later flowering individuals/species had a similar problem but for other reasons: they were in competition for pollinators with many other species, produced a lot of flowers, but often did not produce mature seed, either because of this pollination problem or because of the short remaining season. Despite the suggested pollinator shortage at the beginning of the season, all these studies (including the ones in the extremely early flowering *Saxifraga oppositifolia* by Stenström and Molau 1992 and Gugerli 1998), demonstrated that early flowering species were predominantly outbreeding and hence pollinator-dependent (see also Zoller et al. 2002). Perhaps, the pollinator presence and their diurnal migration is underestimated or the flower/pollinators number is such that early pollinators are short in flowers rather than flowers being short in pollinators?

Information on breeding systems in **tropical alpine** plants are scarce and most literature relates to giant rosettes, which according to the work by Sobrevila (1989) in *Espeletia schultzii*, are largely self-incompatible (empty achenes in selfed flowers), but remarkably, the chances for seed maturation declined with elevation even in this weakly seasonal climate. Berry and Calvo (1994) in their review of the *Espeletia* breeding system list self-compatibility indices between 0.0 and 0.15, with most values close to zero. These authors refer to several other studies which suggest that, counter to common prediction, autogamy is not or only very slightly enhanced at high tropical elevations. *Argyroxiphium* in Hawaii is also reported to be self-incompatible (Carr et al. 1986).

The fact that autogamy and apomixis were found to correlate with higher ploidy levels, which in turn are more frequently found in late flowering species (see above) may have contributed to the view that alpine species in general have higher **ploidy levels** than lowland species (reviews by Bliss 1971 and Packer 1974). However, if any, these trends are small, and may rather reflect the different representations of breeding systems at different elevations. The suggestion that polyploids are more successful under stressful conditions may in fact relate to the perpetuance of successful genets through inbreeding and apomixis in such climates, rather than to ploidy as such. According to Favarger (1954, 1961) the abundance of **polyploids** in the uppermost part of the alpine zone in the

Alps is not significantly different from surrounding lowlands (about half of all species). A recent isoenzyme and DNA comparison in populations of *Anthoxanthum odoratum* from different elevations in the Alps revealed no reduction in genetic variability, which might have been expected if inbreeding played a greater role at high altitude (Felber et al. 1996).

Although none of the hermaphroditic species tested by Molau (1993; except for apomicts) are strictly self-incompatible, outbreeding appears to be the dominant breeding system in alpine climates (see also the critical review by Packer 1974), thus **pollination** is critical. Rated by biomass or abundance of species (Poaceae, Cyperaceae) wind pollination is by far the most important. Rated by species number, insect pollination is much more important. Rated by number of species which produce seedlings, insect pollination becomes even more significant, because the wind pollinated alpine graminoids predominantly propagate clonally. Which insects pollinate alpine flowers and how much do alpine plants invest in their flowers?

While leaves of alpine plants measure only 1/10 the size of lowland plants (Table 13.1; Körner et al. 1989a) the mean size of flowers does not differ across a 2-km elevation distance in the Alps (mean diameter and mass of flowers is $2.2/1.8\,cm^2$ and 14.7/11.5 mg for low/high altitude herbaceous species (n = 15/22), excluding pseudanthia of Asteraceae, the size and mass of which also does not differ significantly. Th. Fabbro and Ch. Körner, unpubl.). In terms of mass, the most expensive investment for flowering in lowland forbs often is the flowering stalk, an organ which is drastically reduced at high elevations (Körner and Renhardt 1987; Bauert 1993). Transplant experiments by Clausen et al. (1948) and Neuffer and Bartelheim (1989) have demonstrated that this shortening of inflorescences includes a strong genotypic component, which is closely related to elevation of seed origin. Disproportionately big flowers in large numbers often completely shade the leaf canopy for 2–3 weeks (see color Plate 1 at the end of the book), so that the actual **costs of flowering** substantially exceeds their own structural and metabolic costs. Rated by flower duration, as much as 10% of the potential seasonal photosynthetic carbon gain may be sacrificed through self shading by plants during flower display (e.g. *Silene acaulis*, *Androsace alpina*, *Phlox* sp.). These diminished gains may balance the savings in stalks compared with lowland species. Such examples illustrate the high priority given to reproduction versus carbon gain in many of these alpine species.

A much debated topic is **flower coloration** at high elevations. Flowers are often more intense in color, but not always, as exemplified by the rich alpine flora of New Zealand, which in essence is a "white" flora (with few exceptions; Mark and Adams 1979). Also, the equatorial alpine florae I have seen generally were not very colorful, with exceptions such as the outstanding "paint brushes" of *Castillea* and lupines in the Andes and the many tiny, blue or pink Gentianaceae, Geraniaceae or Malvaceae found in almost any tropical alpine region. It seems that coloration reaches a peak at the high latitude edge of the subtropics and in the temperate zone and declines somewhat at higher latitudes (although this has not been quantified to my knowledge). So elevation by itself cannot be the sole cause, but may enhance the phenomenon, perhaps through effects of high solar radiation (no evidence that UV is involved as is often claimed, see Chap. 8) on pigmentation. It seems intuitively plausible that the high fragmentation of alpine habitats, combined with a short flowering season and often windy climate should select for showy flowers, provided the selective force, the right pollinators, are available (see below).

There are also characteristic **seasonal changes in the dominant coloration** of flowers ("aspect"). For instance, in the central Himalayas, the season starts with blue (*Gentiana*, *Primula*), changes into yellow (*Ranunculus*, *Taraxacum*) and then white at early peak season (*Anemone*) merging into a mix of yellow and red when the monsoonal season starts in July (*Polygonum*, *Potentilla*, *Geum*). In August, red and blue take over (*Polygonum*, *Cyanthus*) followed by a white September (*Selinum*, *Anaphalis*), and the season closes in blue again with *Gentiana* and *Cyanthus* (Ram et al. 1988). It

is likely that such periodicity is also associated with pollinator preferences, but it may simply reflect the dominance of certain taxa at certain times without any additional function. Coloration phasing is not that pronounced in the alpine zone of the Alps, although it is well known from montane hayfields, which often go through sharp transitions from yellow to white and then to mixtures including pink and blue at the flowering peak. In alpine grasslands in the Alps, white is common early in the season, whereas the second half of the summer is dominated by yellow (*Leontodon*, *Hieracium*). At the same time, a suite of different colors may be found in open scree vegetation.

Given the difficulties of successful pollination at low temperatures (see below), the widespread phenomenon of **solar tracking** by alpine flowers ("heliotropism") has been discussed as an adaptive means. Totland (1996) examined this in alpine *Ranunculus acris* by not allowing flowers to track and by comparing reproductive success with controls, but he could not see any effect in this species, although flower centers were slightly warmer in those allowed to track the sun (see also Kevan 1975; Luzar and Gottsberger 2001). It is noteworthy that solar tracking in alpine flowers was found to be spectra sensitive (Stanton and Galen 1993) with blue light being most effective in yellow flowers.

If one browses the more general alpine literature, one gets the impression that abundance and activity of **pollinators** decreases with elevation, so that alpine plants have a problem. Neither the above trends in breeding systems, nor the more detailed work by specialists on alpine pollination (starting with Müller's 1881 documentation) supports this view (e.g. Kalin-Arroyo et al. 1982; Philipp et al. 1990; Totland 2001).

For instance, Erhardt (1993) noted that the edelweiss (*Leontopodium alpinum*) receives flower visits from 29 different insect families, with flies being most important. The pollinator spectrum definitely does changes with elevation. Butterflies and beetles become much less important, whereas bumblebees (*Bombus* spp.) and flies become more important (in the Alps, flies of the genus

Rhynchotrichops are very important pollinators according to Franz 1979 p. 253). Working with *Campanula rotundifolia*, Bingham and Orthner (1998) demonstrated that actual pollination does not differ across a wide range of elevations, although visiting rates to flowers decreased. They resolved this discrepancy by showing that:

- the period of stigma receptivity was significantly longer, and
- the deposition of pollen per visit was significantly larger at alpine altitudes.

Their low elevation populations of *Campanula* were receptive for 1.46 days, the high altitude populations for 2.36 days. Prolonged **flower duration** seems to be common at high elevations. Arroyo et al. (1981) determined means across many species of 4.1 days for 2310 m versus 9.0 days at 3550 m elevation in the Chilean Andes. In addition, these authors found the mean total flowering period per species to increase from 3.2 weeks to 10.8 weeks along this Andean transect. Hence, the overall duration of receptive flowers increases with altitude. In line with the observations by Bingham and Orthner, Gugerli (1998) also found no support for the pollinator limitation hypothesis in high versus lower altitude populations of the early flowering *Saxifraga oppositifolia* in the Alps.

At low elevation, less efficient solitary bees accounted for most flower visits in Bingham and Orthner's study, whereas **bumblebees** were the dominant pollinators at high elevation, and these carry a lot more pollen. These authors re-emphasized the great importance of bumblebees for alpine plants. Bergmann et al. (1996) provided some climatological and behavioral explanations for this. According to these authors, bumblebees were observed flying at temperatures as low as 4 °C (6 °C for butterflies) and at wind speeds of up to 8 m s^{-1} (6 m s^{-1} for butterflies). Butterflies were only on flowers in 0.9% of the observations, compared with 69% in bumblebees. Bumblebees were also much less fancy about specific flowers: of the 18 observed plant species, they visited 17, while butterflies visited 4, with two thirds of all their visits to a single species, *Silene acaulis*. A census of

flower visitations in alpine grassland (site 1 in Fig. 4.3) during a student field course in early July (flowering peak) and during bright weather, did not suggest a similar significance of bumblebees as discribed for the Scandes. Flies dominated (as reported by Totlanol 1993 for alpine S Norway), and there were lots of different butterfly species, but hardly any bees or bumblebees.

Elevational trends of pollinator abundances also occur in the subtropics and **tropics**: less pollination by birds and butterflies, increasing importance of flies, bees and bumblebees (e.g. Berry and Calvo 1994; Loope and Medeiros 1994). According to Berry and Calvo, *Bombus* species are the most common pollinators of *Espeletias*, but two species of humming birds also visited *Espeletia* between 3900 and 4300 m elevation, and one of these was active even during light rain and snow flurries, when no insect activity was recorded. These authors also report experiments with mesh bags which suggest that some high altitude *Espeletia* species may, at least partly, be wind pollinated.

Seed development and seed size

Once pollination is successful, seed development is the next critical step. Like the ecological importance of developmental processes, which in general are poorly researched, alpine **embryology** is a widely neglected field. Given the long acknowledged difficulties many alpine species have with seed ripening this is surprising indeed.

For a fertilized ovule to become a viable seed it requires three, partly overlapping **developmental steps**:

- Cell division (histogenesis of the testa and the future storage tissue, mostly endosperm),
- Differentiation of embryo tissue and testa, and filling of the endosperm with reserves, and
- Dehydration of the seed to ca. 15% moisture content and hardening of the testa.

Parallel to seed development, the ovary turns into a **fruit**, which also requires investments because of growth in size, formation of special surface structures for dispersal, development of an attractive pericarp for dispersers etc. Energetic and carbon costs for the formation of the seeds' envelope can substantially exceed those for seed formation. Hence there must be enough time, enough thermal energy and enough resources for both the fruit and its seeds to mature. Somehow, the plant must control its reproductive output with respect to resource availability, which means adjusting the number of fertilized ovules per fruit (in multi-seeded fruits) which will enter the seed development lineage, and thereby also determine final fruit size (first detailed discussion by Söyrinki 1938). A significant fraction of the overall **control of reproductive investment** has already occurred, when the size of the inflorescence and the number of flowers were determined. At this second step, the control runs via **abortion** of surplus ovules, which must happen rapidly after fertilization, before significant investments in seed development have occurred. The causes for and the control of these adjustments are extensively discussed in the lit-erature. For alpine plants I refer to the in-depth consideration by Stöcklin and Favre (1994). The net result of these processes is the total plant **seed to ovule ratio**, which is rarely 1.0, and more commonly somewhere between 0.3 and 0.7. Akhalkatsi and Wagner (1996) and Wagner and Reichegger (1997) for instance, report values of 0.5 to 0.9 for *Gentianella caucasea*, and 0.5 to 0.6 for two alpine *Carex* species (see below).

Although the greatest investments in seeds in terms of assimilates occur during step 2, available evidence suggests that it is step 1 which is most sensitive to temperature and thus critically determines the success of seed formation. According to the detailed studies in **Carex curvula** and **Carex firma** in the Alps by Wagner and Reichegger (1997), **seed development** starts with an 8–16 day lag with hardly any measurable change in size or increase in cell number. During this phase, either no or only the very first cell divisions occur within the embryo sac as they were documented by Erschbamer et al. (1994). Such lags include impor-

tant developmental processes, given the fact that alpine plants have no time to waste. The major, almost linear process of **histogenesis**, i.e. the creation of the future seed's cell number and tissue volume, takes between 30 and 40 days in these sedges, depending on the local thermal regime. The sensitivity of this process to temperature was demonstrated by comparing durations of histogenesis on slopes of contrasting exposures, which caused the means to differ by 8 to 10 days for north versus south or west slopes. As can be seen from Fig. 16.7, it is largely the **endosperm** tissue which requires this long formation period and which accounts for >90% of the final seed volume. Most of the size increment of the **embryo** occurs during the last two weeks of histogenesis. Once the tissue volume is completed, endosperm filling with reserves and seed

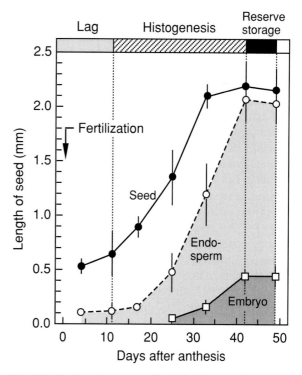

Fig. 16.7. In situ time course of seed development in *Carex curvula* expressed as length increment of total seed and its endosperm and embryo fractions (central Alps, summit of Mt. Patscherkofel, 2240 m). (Wagner and Reichegger 1997)

maturation are a matter of less than 10 days. Altogether, these two early flowering species take between 49 and 56 (in one special case up to 69) days for developing their seeds.

Much more rapid seed development was measured by Akhalkatsi and Wagner (1996) and Wagner and Mitterhofer (1998) in two of the latest flowering alpine species known, the closely related annuals *Gentianella caucasea* (central Caucasus, 2200–2700 m) and *Gentianella germanica* (central Alps, 2000 m; Fig. 16.8). The study in the Caucasus revealed no developmental lag. The endosperm nucleus started to divide immediately after fertilization and only once a mass of 128 free nuclei were formed cell wall formation began and the endosperm became cellular. The first division of the zygote occurred when endosperm nuclei entered their forth division cycle. Sixteen to 20 days after anthesis, mature seeds were released (about half the size of seeds in the two *Carex* species). The *Gentianella* populations studied in the Alps were separated into late and less late-flowering genotypes. The less late ones took 22 days to produce seeds, the very late ones took 33 days, but these days included cold October weather. Cummulated **thermal time** (degree days) were very similar for both genotypes (ca. 250 degree days). While endosperm filling and maturation occupied a small fraction of seed development in *Carex*, it represented about half of the development time in *Gentianella*, but again, turned out to be more robust against temperature differences than histogenesis. Thus the acceleration of seed development compared with *Carex* was due to the absence of a lag phase, faster histogenesis and smaller seed size. With 0.95, the seed/ovule ratio reached an extremely high value in both genotypes studied in the Alps, indicating very high reproductive success, despite the extraordinary late date of flowering (some plants started to flower in September) and possibly reflects a high degree of selfing.

Comparing phenophases and seed development in *Festuca rubra* near the treeline and at low elevation, Larcher (1996) concluded that the delay seen at high altitude was largely the result of prolonged differentiation of flowers and seed- plus

embryo-tissue, and was due to a lesser extent to a slowing of expansion growth and tissue maturation (seed filling). Rate differences were largely explained by different heat sums. A 9-year record of germinability (i.e. quality) of seeds collected in the subarctic alpine belt by Laine et al. (1995) illustrated substantial year to year variation (including zero values in several species), which could also be explained by heat sums during respective growing periods. Given that the deviation of the mean June to August air temperature in this region varied by ±1.5 K during these 9 years, this is perhaps the strongest evidence for the sensitivity of seed development to temperature in such cold regions.

The two case studies for early (*Carex*) and late flowering alpine species (*Gentianella*) illustrate the likely amplitude of development duration of seeds in temperate zone mountains (extremes 16 and 69 days). A most important message from these investigations is that it is tissue (cell) formation rather than tissue filling with reserves which appears to constrain the overall rate of develop-

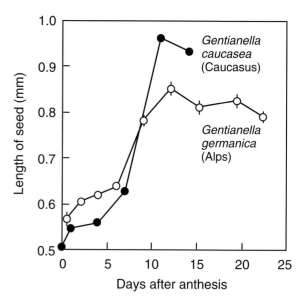

Fig. 16.8. In situ time course of seed development in two extremely late flowering *Gentianella* species, which started to flower in mid August (*G. caucasea*, central Caucasus, 2200–2700 m; *G. germanica*, central Alps 2000 m). (Akhalkatsi and Wagner 1996; Wagner and Mitterhofer 1998)

ment in seeds in the cold. In line with the conclusions on vegetative growth (Chaps. 12 and 13), formative processes once more appear to limit growth in cold climates to a greater degree than assimilate provision. The caveat to this is that the overall investments in sexual reproduction by alpine plants do reflect their vegetative vigor and size, which are co-controlled by resource availability, as they are in plants in general (Stöcklin and Favre 1994).

The product of these processes are **mature seeds**. Given that alpine plants are so much smaller than lowland plants, one might have expected that they also produce much smaller diaspores. However, this does not seem to be the case (Table 16.1; see also Thompson and Rabinovitz 1989).

Although **diaspore weight** may be anywhere between 2 and 2000 µg, mean weights of random samples of comparable lowland and alpine forbs from various latitudes are surprisingly similar and average around 470 µg (excluding the microscopic seeds of orchids and diaspores with massive pericarp involvement like wings or parachutes). The survey presented in Table 16.1 is by no means exhaustive, and a more careful examination of a specific flora may reveal some interesting trends, but these data at least suggest that it is unlikely that massive elevational difference occur in diaspore weight, as they do exist in leaf and plant size.

Because the list in Table 16.1 includes many genera in which diaspores resemble compound units of seed and fruit (achenes, nuts with various

Table 16.1. Diaspore weights (µg) of alpine forbs from various latitudes and from temperate lowland grassland (original data except for[a,b,c])

Origin	Number of species	Smallest seed	Heaviest seed	Mean ± s.e.	
Lowland meadow, 550 m Tirol, Austria, 47°N	17	50	1450	525	114
Lowland meadow, 370 m Abisko, Sweden 68°N	9	2	980	415	110
Montane meadow, 1700 m Tirol, Austria, 47°N	19	3	2000	635	130
Alpine grassland, 2600 m Tirol, Austria, 47°N	18	4	960	442	80
Alpine glacier forefield, 1960 m Swiss Alps, 47°N	9[a]	40	1220	369	123
Alpine vegetation, (2300–3000 m) Swiss Alps 47°N	12[b]	40	1380	598	95
Subarctic-alpine fellfield, 1150 m Abisko, Sweden, 68°N	7	4	840	339	110
Sub-tropical alpine, 4250 m Cumbres Calchaquies, Argentina, 26°S	20	8	1540	411	92
Total means across all sites (n = 8)	–	2	2000	467	38
Reference: British grassland	53[c]	20	2550	818	143

[a] From Stöcklin and Bäumler (1996).

[b] From Urbanska and Schütz (1986), only Caryophyllaceae and Asteraceae, a single number for *Cirsium spinosissimum* of 2340 µg was disregarded.

[c] From Grime et al. (1988). Only those dicotyledonous, herbaceous species were included in this comparison which grow in sunny places, on well drained ground and in relatively undisturbed and unfertilized habitats. *Anthriscus silvester* (5180 µg), *Centaurea scabiosa* (7460 µg) and *Heracleum spondylum* (5520 µg) where omitted from the sample, because of their extraordinary diaspore sizes. For a discussion of this mean, see the text.

structures for dispersal attached), means would come out slightly lower for seed-only diaspores. An examination of the raw data for Table 16.1 indicates that means are strongly affected by the inclusion of species with very light diaspores (mostly seeds) as they occur among alpine Caryophyllaceae, Saxifragaceae, and some Asteraceae (e.g. *Gnaphalium* and *Antennaria*), and species with very heavy diaspores, common in some Asteraceae, Ranunculaceae, Rosaceae and particularly in Apiaceae, Fabaceae and Polygonaceae (e.g. *Rumex*). Heaviest true seeds are commonly those of legumes (not included in Table 16.1), irrespective of altitude or latitude, a family getting increasingly rare at higher elevations.

The mean for comparable species (and life conditions) extracted from Grime et al.'s (1988) compilation of British grassland species is higher than in the various samples considered in Table 16.1, even when some extreme cases are omitted. The mean seed weight of the six grassland legumes in the British sample of 1380 ± 330 (s.e.), adds to that difference, but cannot fully explain it. Urbanska and Schütz (1986) report an even higher mean for eight alpine legumes of $2180 \pm 480 \mu g$. The mean for six comparable British *Carex* species (not included in the above mean) is $1088 \pm 203 \mu g$ (s.e.) with a range from 370–$1880 \mu g$, which overlaps with the $420 \mu g$ and $1620 \mu g$ reported for *Carex firma* and *Carex curvula* by Wagner and Reichegger (1997). No data seem to be available for typical alpine grasses, but the mean for caryopses of the 24 Poaceae listed by Grime et al., which fulfill the criteria used in Table 16.1, is similar to that in the British forbs, namely $853 \pm 170 \mu g$ (s.e.; range 20–$2400 \mu g$) if *Bromus erectus* with $4230 \mu g$ is disregarded. This would be the mean against which caryopses of alpine grasses could be compared with, although variable contributions of awns could substantially bias such a comparison.

A functional **interpretation of the diaspore sizes** observed in the alpine flora seems difficult at this stage. The classical concept that pioneer vegetation is small seeded whereas late successional vegetation is heavy seeded (Kerner 1871; Salisbury 1974) can only be tested within taxonomically related groups. Within alpine Asteraceae and Rosaceae, such a trend seems possible (very small seeds in some snowbed species), but overall the sample size available is too small and family is too high a taxon for testing this. There also seems to be a trend in late flowering alpine species of producing smaller seeds, but many of these are also late successional, and some very early flowering species also produce very small seed (e.g. in Saxifragaceae) contrasting with the commonly heavy diaspores of early *Ranunculus*. Obviously, this is a field that requires more systematic research. Overall, diaspore weight is a genus and species characteristic (Urbanska and Schütz 1986; Thompson and Rabinovitz 1989), and only at this level might one see ecologically significant elevational trends. However, according to these authors, the chances are small because seed weight is a rather conservative plant trait. Landolt's (1967) assessment of taxonomically related lowland and upland species revealed no significant elevational difference, but a trend of increasing rather than decreasing diaspore weight in some of the groups. For Asteraceae Schütz and Stöcklin (2001) report 1.6 mg for alpine ($n = 88$) and 1.2 mg for lowland species ($n = 92$). Baker (1972) in his large survey of Californian species, including alpines, concluded an overall decrease in herbaceous seed size with elevation and he demonstrated this at genus level, within *Penstemon* and *Trifolium*. However, the *Trifolium* data shown in Fig. 16.9 illustrate why this trend occurred. It is only due to the occurrence of a few species with very heavy seeds from low elevation, whereas the minimum seed size observed actually increases with elevation. If these heavy seeded species were from dry lowland habitats (which is likely in California), then the comparison is invalid, because it includes a gradient which is drought, not elevation specific (see Chap. 3). A study of *Plantago asiatica* indicated a trend from larger numbers of small seeds at lower elevations to smaller numbers of larger seeds at higher elevations (Sawada et al. 1994).

Taking all this information together, mean diaspore weights for non-woody alpine plant species

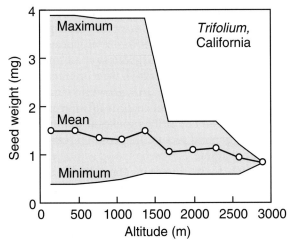

Fig. 16.9. Elevational trends of seed size in the genus *Trifolium* in California (see the discussion in the text). (Baker 1972)

are either similar or, for some groups of species, moderately smaller, for others even slightly bigger than in comparable groups of lowland species, but the bias of **taxonomic relatedness** makes the comparison very difficult. It can safely be concluded that extremely heavy diaspores (>4 mg) as found in some lowland forbs, do not occur at high elevation. The comparison of diaspore weights in subtropical-, temperate- and subarctic-alpine plants suggests that season length has little influence on weight spectra and means. From this it may be concluded that alpine seed or diaspore size are not closely linked to climate or resource availability, but more arctic and tropical alpine data would be needed to verify this statement.

Germination

Whenever tested, seeds of the majority of alpine plant species were found to germinate readily if they had gone through adequate **dormant periods** and/or cold pre-treatment. The first broad screening tests of alpine seed germinability were those by Braun (1913), Lüdi (1933) and Söyrinki (1938).

At the time Braun examined almost the complete above snowline flora of the Alps, it was not known that enforced quiescence and/or low temperature pretreatment can accelerate germination, so he often arrived at low **germination rates,** but he noted the much greater success when seeds were collected in spring instead of fall. Overall, these authors noted the most essential points for seasonal alpine climates:

- No (or little) current season (pre-winter) germination,
- High germinability (>50%) in the majority of the species after experiencing winter,
- Very rapid onset of germination following snow melt,
- Remarkable temporal spreading of succeeding (later) germination events.

These observations refer to germination under controlled, mostly warm conditions, which will be considered first, and which reflect the potential for recruitment. Actual germination and seedling survival in the field will be discussed later. From the most extensive of these germination experiments, the ones by Söyrinki with species from the subarctic-alpine zone (he experimented with 91 species of the 197 he observed in the field), the following results can be summarized: storage at winter temperatures was not essential, but greatly increased germinability in the majority of species. Later experience (see below) suggests that in most (not all) cases it is the several months **quiescence** as such, and not necessarily low temperature which is responsible for the much greater germination success in spring compared with fall. Quiescence can be extended by dry storage. Stored at 4 °C, alpine seeds were found to remain viable over several years (e.g. Weilenmann 1981), in some cases (*Hutchinsia, Arabis, Dryas*) germinability even increased with time. At freezing temperatures, seed longevity seems almost unlimited (Billings and Mooney 1968).

After experiencing winter, only 12 out of the 91 species studied by Söyrinki did not germinate, but this either had to do with unripe seed or other peculiarities, so that one can safely conclude that more than 90% of the species germinated. In half of these, more than 60% of all seeds germinated (mostly >80%). Lowest germination rates (1–20%) were found in *Viola*, *Pinguicula* and in three of the ten *Carex*, and three of the seven Ericaceae species tested. Typical snowbed species ranked in the top group. Similar patterns were observed by Braun (1913) in the Alps who concluded that environmental conditions high above the snowline do not preclude seed maturation and that these outposts of higher plant life do not depend on seed import from lower elevations.

Extremely fast germination (within 1 week after moistening) occurred in Söyrinki's tests in *Oxytropis campestris*, *Oxyria digyna*, *Cerastium alpinum*, *Cardamine resedifolia*, *Silene acaulis*, *Taraxacum officinale*, *Dryas octopetala*, *Gnaphalium norwegicum* and a few others. Among the fastest grasses were *Poa alpina* and *Festuca ovina*. After a first rapid wave of germination, almost a third of the species exhibited additional germination events delayed over several weeks. An interesting observation was that a number of species (e.g. six of the ten *Carex*) germinated in two waves separated by a whole year, obviously reducing the risk of complete failure of recruitment in a bad season. Söyrinki noted that a few species which germinated very badly under his warm test conditions, did so rapidly in the field, notably *Luzula* and *Pedicularis*. Like Söyrinki, Lüdi (1933) was also impressed by the very high percentages of germinating seeds. Of his 86 species from the alpine grassland belt only two (*Agrostis capillaris* and *Calamagrostis varia*), did not germinate on wet blotting paper in a warm room. All three of these early investigators noted the high **species specificity** of germination behavior. These few notes should suffice to illustrate the type of responses also likely to be seen in other seasonal alpine vegetation.

All more recent studies from the temperate and sub-arctic zone added to the enormously varied picture of germination characteristics in alpine plants (e.g. reviews by Amen 1966; Billings and Mooney 1968; Bliss 1971; Fossati 1980; Urbanska and Schütz 1986; Chambers et al. 1987, Wildner-

Eccher 1988; newer work for instance by Laine et al. 1995; Stöcklin and Bäumler 1996; Schütz 2002). It almost seems like this is the aspect of alpine plant life which bears the greatest of all functional variability. The complicated web of interactions between genotypic features, growth conditions of seeds, imposed quiescence, time of snow melt and actual seedbed conditions makes it impossible to draw any general picture that could be considered "typically alpine".

In fact, there does not seem to be anything special compared with low elevation vegetation at these higher latitudes – perhaps except for a more pronounced quiescence requirement (winter) in the majority of alpine taxa before they germinate, but even this was questioned as a result of some studies in North America (Billings and Mooney 1968). Winter quiescence was suggested by these authors to be largely imposed by the environment, rather than being seed controlled or required. This conclusion largely rests on germination success under warm test conditions, which may overrule dormancy, compared with seedbed conditions in the field (Schütz 2002). Early flowering and early seeding species in particular, would run into serious problems if they germinated in warm late summer weather (see the discussion by Urbanska and Schütz 1986). Söyrinki's data discussed above, clearly demonstrated the comparatively poor (in many species zero) autumn germinability. A few species seem to need the **experience of low temperatures** before they germinate, and thus cannot be fooled by dry storage alone – at least their germination is greatly enhanced after freezing instead of warm-dry storage (e.g. *Luzula spicata*, *Viola biflora*). Among other possibilities, seed coat scarification during freeze-thaw cycles in nature has been discussed as a likely mechanism that can break dormancy in such species (Bliss 1971 and references therein). Seeds of tropical *Espeletia* would not germinate unless cold pretreated (Pannier 1969, see below). It was also observed that it makes a difference to germinability whether the cold storage is dry or wet, with wet stored seeds germinating much faster (Table 16.2).

Table 16.2. Days required for 50% of final germination to be attained in alpine seeds stored for 90 days at either cold and dry or cold and wet conditions. Seed beds of alpine soil maintained at 18/4 °C day/night temperature. Species ranked by wet storage success (samples from Beartooth Plateau, 3050 m, Montana). From Chambers et al. (1987; their data pooled across successional stage of species and light regime, which on average had comparatively small or irregular effects)

Species	Wet cold storage	Dry cold storage
Festuca idahoensis	1	10
Artemisia scopulorum	4	10
Deschampsia caespitosa	5	9
Potentilla diversifolia	5	10
Geum rosii	5	16
Calamagrostis purpurascence	6	15
Sibbaldia procumbens	7	14
Polemonium viscosum	10	18
Mean ± SD	5.4 ± 2.6	12.8 ± 3.4

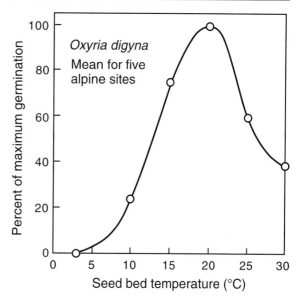

Fig. 16.10. Percentage of germinated seeds of *Oxyria digyna* at day 14 after moist exposure to different temperatures. The maximum number of germlings was observed at 20 °C which was used as the 100% reference. Means of log transformed percentages for seed families from five alpine sites in the Rocky Mts. and the Sierra Nevada of California calculated from data in Mooney and Billings (1961)

Above, the potential rate of germination under standardized, warm conditions was discussed. Which component of the natural environment can limit the actual germination process once dormancy is broken? There is consensus in the literature that the influence of **temperature** exceeds any other potential influence, and moist alpine seeds commonly do not care for a special light regime, although there is occasional evidence that in some species light/dark treatments have a modulating effect of unpredictable direction (e.g. Chambers et al. 1987). Also, the influence of various natural substrates have been tested, but there were no effects on germination (Fossati 1980; Weilenmann 1981).

The first exact experimental approach to the question of thermal influences on the progress of germination was the provenance screening by Mooney and Billings (1961) in *Oxyria digyna*. This typical cold-adapted species native to arctic and alpine environments was unable to germinate at 3 °C within 2 weeks, and reached a maximum at 20 °C (Fig. 16.10). Remarkably, the rate of germination was less at 10 °C than at 30 °C. When the 3 °C pretreated seeds were moved to 20 °C, their rate of germination quickly exceeded the rate of those which were at 20 °C from the beginning. Thus, surprisingly high temperatures are required for optimal germination, and deep cycle variation of temperature during germination accelerated the process, rather than delayed it.

Subsequent experiments by various authors and with various arctic and alpine species confirmed the relatively high temperature requirements for germination in cold climate plants (cf. Billings and Mooney 1968, Wildner-Eccher 1988; Schütz 2002). A constant temperature of 5 °C, Amen (1966) even claimed 10 °C, seems to be a low threshold, but oscillations between 2–4 °C and 9–13 °C, as they occur in nature, led to positive results. Seeds of **tropical-alpine** species seem to have similarly high thermal requirements for optimal germination. Pannier (1969) tested temperature responses of *Espeletia schultzii* achenes. *Espeletia* did not germinate unless treated for 30 days with 2 °C. Except for 17 °C, constant seed bed temperatures of 5, 12, 22, 24, 27 and 30 °C without cold pre-treatment led to zero germination and at 17 °C the success was 25% of viable seeds. However, cold pretreated achenes germinated rapidly with a peak at 17 °C when kept in light (Fig. 16.11). In complete darkness, germination was reduced (zero germination below 17 °C) and the peak occurred at 24 °C. Repeated surface drying and rewetting enhanced germination significantly in comparison to constant moisture.

Including further published data, Bliss (1971) concluded that **alternating day-night tempera-tures** between 10 and 20 °C lead to highest germi-nation rates. From the observations by Pannier (above), one may extend this to say that envi-ronmental variation (in a physiological range) in general is stimulative. Billings and Mooney (1968) refer to agronomic literature for grass and winter cereal germination for which similar or even lower threshold temperatures for germination were reported. Thus germination, one of the most ele-mentary developmental processes in the life cycle of alpine plants, does not seem to differ in its thermal requirements from what is known from cool adapted low elevation plants from temperate latitudes. Soil warming above ambient tempera-tures by solar radiation is therefore an important pre-requisite of alpine plant recruitment. At the same time, this is one of the greatest dangers for seedlings, because following topsoil desiccation may rapidly become fatal. Also, soil heating by itself may be deadly for unprotected seedlings on bare ground (see Chaps. 4 and 8). Rapid germi-nation in spring allows plants to utilize the gap between cold limitation and these alternative dangers of a progressing season. Billings and Mooney (1968) had suggested that the high thermal threshold for germination increases the probability that the initial life phase of a seedling falls in a most favorable part of the year.

Alpine seed banks and natural recruitment

In situ observations of seed germination are less abundant than laboratory tests of germinability as was discussed above. However, throughout the literature (starting with Söyrinki's 1938 survey) authors noted ample seedling presence in most alpine plant communities. Due to high mortality, only a very small fraction of seedlings survived. In order to illuminate this critical part of the life cycle under natural conditions, I will present some data on seed production and dispersal, and then comment on alpine seed banks and recruitment. In the following, "seed" is used as a synonym for all sorts of diaspores.

Kerner's (1871) rather modern account of **seed dispersal strategies** in alpine plants contains the most essential points. He distinguished between seed attributes for short and long distance trans-port, with the former strategy more often found in late successional and the latter more often in early successional communities. He predicted higher seed rain densities from pioneer communities

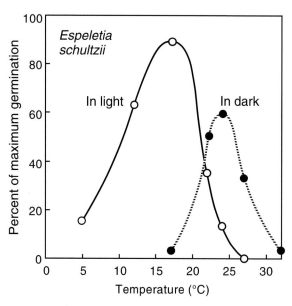

Fig. 16.11. The temperature response of germination in the light in 2 °C-stratified achenes of *Espeletia schultzii* of the Venezuelan páramos. One hundred percent corresponds to total number of viable seeds as determined by the tetrazolium method. (Pannier 1969)

than from old turf vegetation, but was surprised that the bulk of the seed rain he collected from old snow closely reflected community structure on adjacent young moraines. Hence, long distance transport was not as "long" as might be expected from the frequency of parachutes and wings etc. attached to seeds in those pioneer communities. Aliens were missing in his collections. Kerner's aeronautic experiments with seeds of 1 mg weight suggested that the mean distances winged seeds reach at equal **wind** velocity are only two to three times those reached by wingless, when both are launched from the same platform. Kerner concluded that most seed dispersal in alpine vegetation is of relatively short distance, in line with modern dispersal and seed bank analysis (but see Molau and Larsson 2000 for rare long-distance dispersal events).

Species spectra in **seed rain studies** with water filled seed traps distributed over ca 1 km^2 of foreland of the Swiss Morteratsch glacier by Stöcklin and Bäumler (1996) confirmed the very close association with nearby vegetation (see Welling and Laine 2000, and Erschbamer et al. 2001 for more recent examples). During 95 days of the fruiting season, these authors trapped between 125 and 2000 seeds m^{-2} depending on the successional age of the nearby community, which ranged from ca. 5 to more than 45 years since release from ice (34 species trapped in total). The highest seed number was trapped near 45-year-old vegetation (still an early successional assemblage of species).

The species abundance in the seed rain (long distance dispersal) reflected the effectiveness of propagation devices and was highly skewed. Of all trapped seeds, 65% were parachutists (*Epilobium*, *Hieracium*), 30% had wings (*Oxyria*, *Rumex*) and only ca. 5% had no such devices, but were extremely small seeded (*Saxifraga*). Hence, seed rain, while closely resembling the species spectrum of the region, was far from reflecting the actual species abundance. These authors also directly tested **short distance dispersal** of seeds in six pioneer species differing in "aeronautic fitness" (the spectrum ranging from species with very small seeds and a long plume to species with heavy round seeds; Table 16.3). Except for *Epilobium fleischeri* with extremely light and plumed seeds, very few seeds were trapped at more than 40 cm from the seed source, most seeds were trapped within 14 cm.

A rather detailed analysis of alpine seed dispersal was conducted by Scherff et al. (1994) for the Rocky Mountains snowbed species *Ranunculus adoneus*. Applying the adhesive **seed trap** method, they documented that virtually all achenes of this species remained within 16 cm of the maternal parent – similar to what Stöcklin and Bäumler observed in their five non-parachutist species. Secondary dispersal through snow gliding, melt water or rain etc. added 10 cm (downhill) to the primary dispersal distance. Scherff et al. explain this rather conservative dispersal strategy by the specific spatial limitations of the

Table 16.3. Short distance dispersal of seeds in glacier forefield species (% of trapped seeds). Seeds were trapped by adhesive on Petri dishes arranged in four circles with increasing radius. Percentages per circle were calculated from the total number of seeds trapped in each circle, corrected for increasing perimeter (Swiss Alps, 2000 m). From Stöcklin and Bäumler (1996)

Species	Distance from source (cm)				Seed weight (µg)
	14	39	114	214	
Trifolium pallescence	100	–	–	–	688
Oxyria digyna	94	6	–	–	907
Linaria alpina	92	5	3	–	253
Achillea moschata	83	17	–	–	232
Saxifraga aizoides	68	22	10	–	40
Epilobium fleischeri	32	47	5	16	132[a]

Fringe/wing/plume size increase from top to bottom.

[a] Including the long plume.

snowbed habitat and the pronounced mycorrhizal dependence of this species. Infections outside the specific "life niche" seem unlikely. The text by these authors is a highly recommended review of dispersal strategies and niche theory, relevant for high mountain situations. Taken together, these two studies suggest that short distance dispersal is not only a snowbed peculiarity, but appears to be a more common phenomenon in alpine vegetation.

The above examples were for young communities. Spence (1990) compared seed rain in four types of late successional alpine communities in the Craigieburn Range of New Zealand and obtained mean annual inputs of between 340 seeds m^{-2} (fellfield) and 5000 (snowbed), with tussock grassland and herb fields taking middle positions (1050 to 1630 seeds m^{-2}). Out of 10 to 32 species occurring in these communities, 5 to 13 showed up in seed rain (mean 40–50%, with dispersal distances largely <50 cm). Chambers (1993) conducted seed rain studies in both late successional alpine turf (90% cover, 54 species) and an adjacent 35 year old borrow area (25% cover, 34 species, mostly species which also occur in the late successional plots) in the Beartooth Mts., Montana. Her data for 3 years average at 5200 seeds $m^{-2} a^{-1}$ in the turf and $10\,700 \, m^{-2} a^{-1}$ in the young vegetation on disturbed flats, which is in stark contrast to cover. The species spectrum found in traps was similar. The overall difference in seed number

resulted largely from the 4100 instead of 340 seeds from *Deschampsia* and 1400 versus 200 seeds in *Cerastium* (plus some further differences in pioneer graminoids). In the turf, seed abundance in seed rain did not reflect actual species abundance, because the dominant long-lived clonal species were highly underrepresented, whereas in the disturbed area, the shorter lived dominants were also the major seed source. Hence, **species life history** was the major determinant of seed rain composition. This is also reflected in the seed bank data presented below, in particular those by Semenova and Onipchenko (1994), who studied seed banks which turned out to be dominated by seed of non-dominant species.

The net result of seed rain, seed predation or other losses, and germination is the dormant **seed bank**, from which a community derives its recruitment or is driven along a successional lineage (Table 16.4). Alpine seed banks strongly reflect seed rain composition, confirm predominant short distance dispersal and rarely contain alien seed. To balance the annual input of seed, a similar number of seeds m^{-2} must disappear from the seed bank every year. For three alpine annuals, Reynolds (1984a) found that the major loss of individuals occurred between seed dispersal and germination (the latter included).

Disregarding extremes of only a few hundred seeds m^{-2} in a lichen heath, and >10 000 m^{-2} for an alpine bog reported by Semenova and Onipchenko

Table 16.4. Examples for emergent seed bank sizes in alpine vegetation (rounded numbers from germination trials in warm greenhouses). Species numbers include all, even very rare species. Aliens never exceeded 3 species

Type of vegetation, site	Number of all species (seed bank only)	Seed bank (n m^{-2})	Ref.[a]
Dolomite grassland, Alps (2400 m)	35 (11)	1400	1
Acidic silicate grassland, Alps (2450 m)	44 (15)	1900	1
Open grassland/fellfield, Alps (2600 m)	57 (33)	1350	2
Carex firma community, Alps (2300 m)	56 (22)	1500	2
Festuca grassland, Caucasus (2750 m)	55 (24)	1190	3
Tall Herbfield, Caucasus (2700 m)	38 (16)	3850	3
Snowbed, Caucasus (2700 m)	22 (19)	2800	3
Geum turf, Beartooth Mts. (3200 m)	54 (44)	3800	4

[a] 1, Hatt (1991); 2, Diemer and Prock (1993); 3, Onipchenko and Blinnikov (1994); 4, Chambers (1993).

(1994, not shown in Table 16.4), numbers for emergent seed banks of alpine vegetation with high ground cover are surprisingly similar across a wide spectrum of habitats, and their magnitude matches known numbers for lowland grassland (Thompson 1978). Semenova and Onipchenko (1994) compared emergent seed banks obtained from greenhouse and field trials and also investigated the vertical distribution of the seed bank in soil profiles (Table 16.5). The field trials revealed only 9–22% of the seed numbers obtained in the greenhouse, except for the very poor lichen-heath seed bank, where field germination reached 78% of the greenhouse result. Of all seeds, 70 to 95% came from the top 2 cm of the soil profile, the remainder from 2–10 cm depth, with hardly any from below 6 cm (10% from deeper than 6 cm in Chamber's 1993 study). In the field, 68–96% of all seeds germinated in the first year, the majority in spring. However, the longevity of buried seeds increases with elevation (Cavieres and Kalyn-Arroyo 2001).

In summary, these observations suggest that at most, only about half of the species composing the actual vegetation are represented in emergent seed banks. The other half either have deeply dormant seeds, very little seed production or very high seed mortality. Seeds of the species emerging through warm greenhouse treatment would germinate only in small percentages under field conditions. In all seed banks tested, total seed numbers strongly reflected the presence of very few species,

in late successional vegetation mostly subdominant species with very small seeds. Thus, data on species presence and species specific seed abundance need to be reported.

Seedling emergence and establishment are the next crucial steps. This is the life phase during which most genets are lost at all elevations. Unless seedlings manage to rapidly anchor deep into the substrate, they will die. As mentioned above, topsoil desiccation, heat on bare soils and night-time needle ice heaving are detrimental, and cryogenic processes in the soil during winter in seasonal climates permit only few seedlings to survive. Along a transect from lower montane (2300 m), across the treeline (3500 m) to lower alpine grassland (3700 m) in the Rocky Mts., Jolls and Bock (1983) found the number of seedlings of *Sedum lanceolatum* per m^2 to decline from 25 to 2 to almost zero. However, high numbers of adults were present at all elevations, suggesting that seedling abundance and survival are due to different factors than those which control adult distribution. The loss of seedlings was not constant during the season, but was highest at the beginning and, most surprisingly, the relative survival of individuals did not differ among elevations. Once more, this example warns against the use of absolute numbers of propagules for explaining the fate of populations. Full probability matrices and flowcharts such as the ones developed by Rusterholz et al. (1993, Fig. 16.12) and Chambers (1995a) are needed to model population dynamics.

Belowground investments are the key to survival. In *Oxyria* for instance, hardly any shoot growth occurs until the second year and the first growing season is devoted primarily to **root system establishment** (Billings and Mooney 1968). There are several reports in the literature that **seedling mortality** due to physical stress is reduced in the shelter of other plants, and cushion plants were even considered as a "nurse plants" ("facilitation"; Bliss 1971; Urbanska and Schütz 1986; Scherff et al. 1994). The often observed clumped distribution of seedlings may also reflect shelter effects of micro-relief or neighbor plants (Järvinen 1984; Diemer 1992), although neighbors

Table 16.5. A comparison of greenhouse and field emergence of soil seed banks and vertical seed distribution in four alpine plant communities of the northwest Caucasus. Greenhouse data as in **Table 16.4**; species numbers in brackets. (Semenova and Onipchenko 1994)

Site	% In top 2 cm	Green-house	Field
Lichen heath	88	350 (21)	272 (20)
Festuca grassland	93	1190 (24)	108 (12)
Herb field	70	3850 (16)	846 (15)
Snowbed	95	2810 (19)	256 (12)

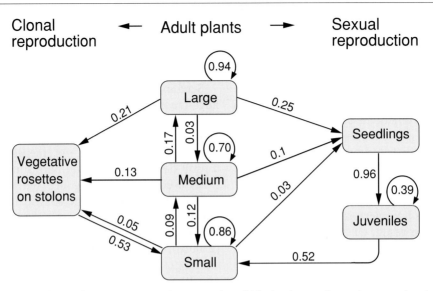

Fig. 16.12. Life-cycle scheme of *Geum reptans*, a pioneer species, which simultaneously employs sexual and clonal propagation on glacier forefields in the Alps. Numbers are stage-specific transition probabilities or seedling and vegetative rosette production on a 1-year basis (T Weppler and J Stöcklin, pers. commun.; see also an earlier attempt by Rusterholz et al. 1993)

could have negative effects as well, for instance by screening solar radiation and reducing seedbed warming (Moen 1993). According to Diemer's (1992) 6-year permanent quadrate observations in *Ranunculus glacialis*, fertilizer addition does not contribute to seedling survival, but particle size at the soil surface is critical (Chambers 1995b).

There is no rule as to whether the first growing season or the first winter are responsible for greatest seedling mortality. Of the six habitats with complete cover compared by Urbanska and Schütz (1986), four had the greatest overall seedling losses in **summer**, one in **winter** and one had none. In *Biscutella levigata*, seedling mortality was more or less evenly spread over the first growing season, the first winter and the second growing season, with no clear habitat effect. Stöcklin and Bäumler's (1996) studies in glacier forefield communities of varying age revealed small summer and high winter mortality in moist sites near the retreating glacier and similar mortalities in summer and over winter at

older, better drained sites. Seedling mortalities reported in the literature cover the full range from none to all, but total 12 month losses commonly exceed 50% in large seeded and 99% in small seeded species (see Fig. 16.12).

Given that in a single season one average *Ranunculus glacialis* individual produces 500 large achenes, and one *Epilobium fleischeri* individual 60 000 tiny plumed seeds (Järvinen 1994; Stöcklin and Bäumler 1996), there seems to be sufficient leeway for such losses. Several authors have reported considerable year to year variation of seedling abundance, not paralleled by significant variations in juvenile or adult populations (e.g. Diemer 1992), reflecting substantial buffering capacity of unfavorable seasons at the whole population level. Most of the authors mentioned above expressed their surprise about the effectiveness of the reproductive system in alpine plants and emphasized the (often underestimated) significance of sexual reproduction even in very harsh alpine habitats.

Clonal propagation

Outstanding flowering with rich colors (in most regions) makes alpine vegetation worldwide so attractive. However, periodically, flowering and seed production can be a rather risky mode of propagation at high altitudes, as was discussed above, and a suite of alternatives have been adopted by alpine plants in order to bypass the dangers of recruitment failure. These alternatives – various ways of vegetative multiplication – generally tend to increase in importance with increasing elevation (Bliss 1971; Billings 1974).

However, as Söyrinki (1938) in his classical study of reproductive ecology in the Scandinavian alpine zone has pointed out, the visual impression may be misleading. Often, it is the dominant alpine species in terms of biomass which primarily use **vegetative propagation**, but not the majority in terms of species numbers. Of the total high altitude flora of northeast Lapland investigated by Söyrinki (197 species of higher plants), only 52% of the species had some mode of clonal growth, in 44% this was clearly expressed, but the majority of these species represent boreal or montane floristic elements. Among exclusive alpine species, particularly those inhabiting very hostile places such as snowbeds (with obligatory seeders such as in the genera *Sibbaldia*, *Veronica* and *Gnaphalium*), clonal propagation is not particularly pronounced in this nordic-alpine region. Of the 58 rock and scree species listed by Söyrinki, only 20 exhibited no obvious seedling re-cruitment, among these six *Carex* species and four Ericaceae dwarf shrubs, all obligatory clonal plants. These observations are in line with findings from the arctic (Grulke and Bliss 1985), and Billings (1974) has suggested that clonal propagation is even less abundant in alpine than arctic environments. In contrast, some alpine grasslands and scree vegetation in the Alps were characterized as 80 to 90% clonal (Hartmann 1957; Stöcklin 1992). It appears there is an abundance peak of clonal growth in the lower alpine belt, and a reduction in frequency at higher elevations (perhaps related to more fragmented habitats and reduced com-petition). However, such statistics also strongly depend on vegetation type and on **what is** considered **clonal propagation** and what not, which deserves a brief discussion.

Plants, unlike most animals, are modular organisms, in which each module (or segment) which bears at least one bud can in principle grow into a new plant. In some cases not even a bud is required, and new apical meristems may crystallize from homogenous tissue regions, including roots. Hence, in this widest sense, all plants would be potential clones, but not all species materialize these options. A more practical use of the term clone refers to situations where modules or groups of modules gain a certain degree of independence from a "mother-ramet", and thereby become centers of further ramet production. Through this proliferation, a single genet may cover large areas of ground.

Daughter-**ramets** may remain physically connected to the mother ramet or become completely separated. Also groups of ramets may form highly integrated plant units, IPUs (see the review for arctic-alpine clonal plants by Jónsdóttir et al. 1996). The degree of independence and separation of ramets or integrated ramet groups is commonly used as a criterion for distinguishing clonal versus non-clonal plants. Thereby independence is not necessarily the disconnection of all physical ties, but rather a **functional independence**, which means ramets become self-supported by roots and are fully photo-autotrophic. If connections are maintained, these may become functional during disturbance (e.g. herbivory) or patchy resource supply and connected ramets may "help each other out" – one of the reasons why clonal plants are so robust in the face of perturbations, and have stabilizing effects on ecosystems. However, without disturbance, priority seems to be given to fast ramet independence (a test of 19 clonal plants from the central Alps by Tschurr 1992). Given this rough definition, clonal forms of propagation in the alpine zone range from slowly migrating tussock-fronts to vegetative bulbils, which are dispersed like seed.

A typology of clonal propagation in the alpine life zone

There have been several attempts to categorize clonal growth in alpine plants. Hartmann (1957; Fig. 16.13), for instance, distinguished 20 types of clonal growth on the basis of morphology. These types can be lumped into two major groups (distinct by the absence or presence of a persistent main root of the mother ramet) and contrasted with non-clonal species (Table 16.6). *Cyperaceae*, *Poaceae* and *Asteraceae*, the three most dominant alpine plant families, almost exclusively belong to the "obligatory clonal" group. In contrast, *Fabaceae* are largely non-clonal and tap-rooted. Stöcklin's (1992) categorization combines mor-

Table 16.6. Vegetative propagation of alpine plants. A classification of 228 species of the Swiss central Alps into three major categories. (Hartmann 1957)

Plant family (n species)	A	B	C
Cyperaceae	9	0	0
Juncaceae	6	0	0
Poaceae	20	2	0
Asteraceae	29	5	1
Scrophulariaceae	7	1	2
Ranunculaceae	4	2	0
Primulaceae	6	6	0
Saxifragaceae	4	9	0
Gentianaceae	2	5	2
Rosaceae	1	11	0
Campanulaceae	0	6	0
Caryophyllaceae	0	14	6
Brassicaceae	0	9	6
Fabaceae	0	1	8
Additional minor families pooled	6	33	5
Total for major families (184 spp.)	88	71	25
Total for all species (228 spp.)	94	104	30
Major families (%)	47.8	38.6	13.6
All species (%)	41.2	45.6	13.2

Major families include five or more species, minor families one to four species. Families are ranked by the relative fraction of species falling in A, and then B. A obligatory clonal species, primary root system not retained; B partially clonal species, primary root system retained, but adventitious roots and satellite ramets are formed; C not clonal or only accidentally (e.g. if buried), mostly tap rooted species.

phological traits with aspects of plant life strategy, and Halloy (1990) included architectural (size) aspects, but his system covers all alpin plant types, hence is more general. In simple terms, one may distinguish the eight main forms of clonal propagation listed below, with all sorts of intermediate forms and special types existing as well (the letters given below refer to Hartmann's typology given in Figure 16.13 where appropriate, but neglect the root properties).

1. Tussock Graminoids (A, B, C). Grasses, sedges and rushes which form dense clusters of ramets (shoots) with new ramets produced at one end, while old ones senesce at the other (phalanx-type of clonal growth; Figs. 16.14 and 16.15, color Plate 1b at the end of the book). Depending on species, this may – after an initial phase of establishment – lead to rings, fragmenting clusters or garlands, very slowly moving through the landscape (often by less than 1 mm per year). This is possibly the most important strategy in late successional alpine vegetation around the world. Examples are found in alpine species of *Carex*, *Kobresia*, *Juncus*, *Danthonia*, *Deschampsia*, *Chionochloa*, *Festuca*, *Nardus*, *Poa*, *Stipa*.

2. Stoloniferous Graminoids (D, E, L, N, W). Graminoids with belowground stolons from which widely spaced ramets emerge. These may either explore new open terrain (alpine ruderals) or invade existing assemblages of plants (guerrilla type; Fig. 16.15). Some species perform both phalanx- and guerrilla-type clonal growth (they may switch between the two). Alpine examples are found again in *Carex*, but also in *Luzula*, *Agrostis*, *Festuca* and others.

3. Mat-Forming Forbs (F, G, H, I, M, O, R, S). Clusters of herbaceous rosettes (Fig. 16.16) which, by slow centrifugal spread, may disintegrate into separate fragments of the genet. Examples exist both in ruderal and late successional habitats in *Antennaria*, *Celmisia*, *Helichrysum*, *Werneria*, *Hypochoeris*, *Gentiana*, *Plantago*, *Veronica*.

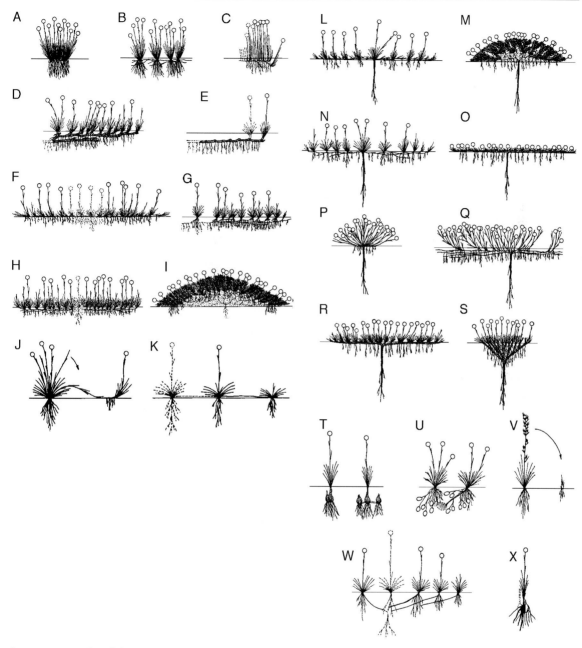

Fig. 16.13. Examples of clonal propagation in alpine plants after Hartmann (1957). The first group of examples (**A–K**) is for species which either never had (graminoids), or lost the main root of the mother ramet. The second group (**L–X**) is for species which Hartmann considers partially clonal, because they largely remain centered around a main (tap) root. This distinction, though logical, is difficult to ascertain in the field, hence was not employed in the simpler categorization provided in the text (types 5, 6 and 8 are not represented among these examples). The leaf display is schematic and does not necessarily distinguish graminoids and dicots. References to these examples by letters in the text ignore the root characteristics

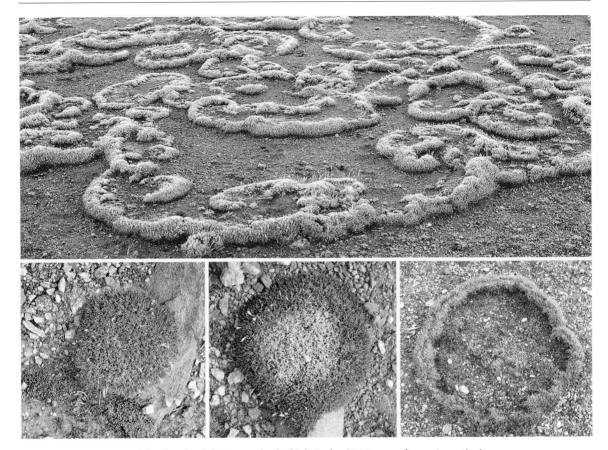

Fig. 16.14. The genesis of clonal garlands in *Festuca* in the high Andes (4250 m, northwest Argentina)

4. *Stoloniferous or Rhizomatous Forbs* (*J, K, L, N, W*). Isolated ramets or groups of ramets, at least initially connected by below- or aboveground stolons/runners can form widely spread nets of a single genet (Fig. 16.17). Examples are *Gentiana, Primula, Soldanella* in late successional communities and both these genera plus *Geum, Doronicum, Saussurea, Senecio, Rumex*, and a large group of other genera, often found on less compact, more inorganic substrates. A very special group of alpine species in this group are scree slope plants with long and elastic belowground shoots and very flexible internode length, well suited to cope with an ever mobile substrate. Ramet separation in this case is mostly induced by sheer forces in the ground. Examples are found in *Campanula, Linaria, Thlaspi*.

5. *Creeping Dwarf Shrubs.* Woody-stemmed species creeping within the topsoil/raw humus layer with buds at or just above the soil surface (Fig. 16.18). Note, not all species of this type perform clonal propagation. Many still depend on a primary stem and root stock. Clonal fragments become self-supporting by adventitious roots emerging from older stems. But this may strongly depend on the moisture of the substrate (the moister, the greater the likelihood of adventitious root development). Some flat cushion plants, normally not a clonal growth form (commonly

Fig. 16.16. Mat-forming forbs: *Celmisia* sp. (Rock and Pillar Range, 1150 m, South Island, New Zealand)

Fig. 16.15. The guerilla (*top*) and phalanx type (*bottom*) of clonal growth in alpine graminoids: a stoloniferous clonal fragment of *Carex bigelowii* (N-Sweden, 1050 m) and a decapitated tussock of *Nardus stricta* (Alps, 2000 m; *arrows* indicate the active fronts)

tap-rooted), may also disintegrate and build self-supporting sub-units, the sizes of which thus are not correlated with genet age. Examples in *Coprosma*, *Drapetes*, *Dryas*, *Loiseleuria*, *Pernettya*, *Salix*.

6. Prostrate Dwarf Shrubs. Alpine dwarf shrubs with above-ground branched stems and buds between 5 and 50 cm above the soil surface, which tend to layer and get buried in raw humus and litter (Fig. 16.19). New shoots emerge from buried buds. Adventitious roots develop very late (again depending on moisture) and genet disintegration may take a very long time if it does occur at all. Examples in genera of *Empetrum*, *Gaultheria*,

Rhododendron, *Vaccinium*, *Hebe*, *Hypericum*, *Salix*, *Styphelia*.

7. "Viviparous" Plants (V). A number of alpine plants produce vegetative propagules on inflorescences (Fig. 16.20). The term viviparous is in common use, but is misleading because propagules are not the result of sexual reproduction and "birth", but develop from mother tissue instead of zygotes (Sarapultsev 2001). In some cases little plantlets emerge, as in the often studied *Poa alpina* ssp. *vivipara*. In other cases, as for instance in *Polygonum viviparum*, bulbils, i.e. compressed, seed-like shoots emerge instead of flowers. "Vivipary" causes rapid and, and in the case of seed-like bulbils, very widespread distribution of a genet.

8. "Accidental" clonal plants. A number of alpine plants can propagate clonally under certain circumstances, but normally do not do so. One example is *Ranunculus glacialis*, the basal segment of flowering stalks of which tend to layer, and if they get buried by loose substrate may produce adventitious roots and become autonomous. In

Fig. 16.17. Stoloniferous forbs: *Petasites paradoxus* (Lechtal Alps, Austria, 2100 m)

most regions *Oxyria digyna* also belongs to this group.

This list is not exhaustive and relates partly to the discussion of plant growth forms in Chapter 2. In the following, I will give some examples of environmental triggers, internal controls and long-term consequences of some of these modes of vegetative spreading. It is important to remember that clonal growth is not a substitute for sexual reproduction, but an addition. Temporal gaps between years with successful sexual recruitment may be zero or millennia. Clonal propagation commonly runs parallel to flowering and fruiting and secures the persistent presence of a species in a given community, irrespective of reproductive success. Energetically there is a trade-off between the two modes of propagation. The following examples were also selected to illustrate these two

aspects, the benefits for persistence and the conflict between sexual and vegetative investments.

Case studies of clonal propagation in alpine plants

The first example is for a phalanx-type tussock graminoid, as these occur in almost all alpine floras. With some species specific modification, such tussocks have to solve similar "geometric" problems, the understanding of which requires some insight into morphology and module turnover. In *Carex curvula*, a dominant sedge in the Alps (Fig. 8.4 and 11.13), single shoots of a clonal system take approximately 9 years to mature and flower (and then die, or die without having flowered). Three and six years after its birth, a shoot produces a daughter shoot (the

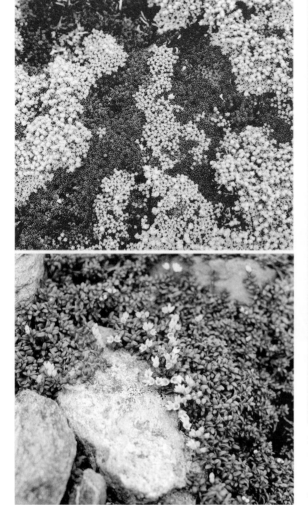

Fig. 16.19. Dwarf shrubs: *Rhododendron ferrugineum* (Austrian Alps 2000 m)

Fig. 16.18. Creeping dwarf shrubs: *Raoulia hectori, Drapetes lyallii* (*top*, Old Man Range, New Zealand, 1600 m) and *Loiseleuria procumbens* (*bottom*, Swiss central Alps, 2600 m)

horizontal angle between the two is 120°) each of which again goes through this shoot life cycle (determined from annual leaf scar production; C. Heid and Ch. Körner, unpubl.). Shoots grow less than 0.5 mm in length per year (in other places 0.9 mm were found, Grabherr et al. 1978), so before they die, they may have expanded the clone by 4 mm in their growth direction. Since the terminal third of the shoot with the inflorescence is lost, the spatial net gain will in fact be less. Also, the 120°

branching and non-centrifugal spread reduces the radial expansion rate. The above direct measurements of age match with the 11% annual shoot mortality reported by Erschbamer et al. (1994) for this species.

Hence, on average, each vital shoot gives birth to two new shoots and, in the ideal case, to one inflorescence. Evidently, more shoots than only those dying through flowering must be eliminated from the lineage in order to maintain a stable ramet population once space is occupied (in this case ca. 85 ramets per 100 cm^2 including all interclonal gaps; only 3% of all shoots at a time flower). Such a clonal system has only two options, namely to expand into empty space (if available) or suffer from massive **intra-clone competition**. In old com-

Fig. 16.20. Viviparous clonal plants: *Polygonum viviparum* and *Poa alpina ssp. vivipara* (Alps, 2500 m)

munities of this type the pressure for ramet space is so high that the topsoil develops into an almost impenetrable and rapidly dehydrating felt of roots where *Carex* seedlings appear to be unable to establish (see below), although viability of seeds on wet filter paper is very high. Despite the 120° branching angle (which would cause the clone to disperse in all directions), compact phalanx fronts consisting of 20 to 50 shoots can be seen, hence abortion of superfluous shoots is not random. During an unknown period of site colonization (perhaps 10 000 years ago, when glaciers released the terrain) an unknown number of genets per unit of land area may have established, expanded, and then started to interact with each other. The current alpine turf reflects the result of this millennial battle for space, but we cannot see the boundaries of clones, whether they intermingled or remained separated, and how many clones survived this interaction. Perhaps only one most successful genet, a single "mega-clone"?

An attempt to **map genets** of *Carex curvula* in the field using DNA analysis revealed most interesting results (Steinger et al. 1996). Even on a relatively small scale of a few meters, several genetically different clones could be identified. Hence, their is substantial genetic diversity, though possibly not as much as was recently reported for more easily fragmenting clones of alpine *Hylocomium* moss in northern Sweden by Cronberg et al. (1997). According to the DNA data, the area of a single *Carex* clone (see Fig. 11.13) can exceed 1 m² and intermingling among clones is moderate. Over at least 90 cm (several thousands of ramets) only one clone seems to be present. Taking the maximum spatial extension of the same genet (at least 1.6 m) and realistic annual spreading rates (see above), **clonal ages** of several thousands of years appear plausible. This is an example where clonal propagation leads to persistent space occupancy and where successful sexual reproduction appears to have been restricted to the colonizing phase. Over thousands of years flowers and seeds were produced of which no genetic precipitate can be found in interclonal gaps. Yet, some forbs do manage to regularly recruit cohorts of seedlings in this tussock matrix, most successfully *Leontodon helveticus* the second most important species of this community.

A very different strategy of clonal growth is adopted by **Carex bigelowii**, a widely spread arctic-alpine species (Fig. 16.15). Many years of research by Callaghan and coworkers (above 1000 m elevation in northern Sweden) showed that this species has a much more opportunistic strategy, and switches between phases of slow or rapid expansion, purely vegetative or reproductive growth (e.g. Callaghan 1976; Carlsson and Callaghan 1990; Jónsdóttir et al. 1996; Jónsdóttir and Callaghan 1990).

Carex bigelowii, produces guerrilla and phalanx tillers together, but the relative proportion depends on habitat and resource availability. Guerrilla tillers have higher vegetative reproduction, whereas phalanx tillers are more likely to produce flowers. The periodic stoloniferous "outbreaks" are under the combined control of reserve accumulation and flowering. It was observed that both shoot mortality and floral initiation are highly correlated with shoot size, and flowering events were found to depend on the climate of two successive growing seasons. Due to resource exhaustion, ramet population size actually decreases after a year with massive flowering and seed production, hence undergoes cyclic changes. It was suggested that flowering directly controls growth by restricting meristems for photosynthetic tissue. By carbon isotope labeling it was shown that functional clonal connections remained intact in this sedge over at least 9 to 11 years and the minimum size of a successful, physiologically functional unit was around five interconnected tiller generations. There was a clear hierarchy among tillers, hence apical dominance effects also had control over the rhizome system. It is surprising that also in this much more openly growing sedge, recruitment from seedlings was not observed in the alpine zone.

In contrast to the two previous examples of late successional clonal species, which continued sexual reproduction at a high cost, while actual propagation is to 100% clonal, **Geum reptans**, represents a pioneer strategy where both modes of propagation are employed simultaneously and successfully. Rusterholz et al. (1993) investigated the breeding system of this species in a glacier forefield in the Alps (2450 m). Flowering efforts and investments in seeds are very high in this species (30 000 heavy seeds per 100 m²), but only 5% of all new plants established in a population came from seeds, because 86% of all seeds failed to germinate, and from those which did, merely 1% survived the first growing season (Fig. 16.12). Hence, in this species the population was again predominantly increasing due to clonal growth, although only 20% of all stoloniferous propagules did successfully establish (root). However, given the longevity of genets (which may be at least several decades), a 5% fraction of reproduction through seed guarantees high genetic diversity and rapid colonization of new ground (*G. reptans* is an obligatory outbreeder). Since stolon production, on average, "consumes" twice as much of annual aboveground biomass production as flowering, a failure of 80% of these new ramets indicates a high cost of clonal propagation. This may explain why the investment policy of *Geum reptans* prioritizes belowground structures, followed by flowering and seeding, and only in a third phase are significant resources allocated to stolons. Most stolons come from old and well established plants. A similar strategy was found in alpine *Epilobium* (Stöcklin and Bäumler 1996).

The last two examples are for **viviparous plants** (Fig. 16.20). Their viviparous nature increases with elevation or latitude, hence becomes more pronounced the harsher the environment gets (one exception is mentioned below).

Polygonum viviparum produces both flowers (terminal) and, lower down on the same inflorescence, bulbils, which are dispersed in a similar way to normal seeds. Bauert (1993) demonstrated a negative correlation between the relative abundance of flowers and bulbils on the same stalk, as if there were a certain fraction of the length of a flowering stem available for positioning either of the two organs at a set spacing regime (internode length). In an extreme (rare) case this leads to inflorescences with only flowers or only bulbils, but commonly both types of organs are present at all elevations (Fig. 16.20), with the bulbil fraction increasing with elevation.

Once more, predominant clonal propagation at high alpine elevations does not preclude substantial genetic variability among populations, as Bauert showed, using isoenzyme assays. Minor sexual reproduction is maintained, although this may depend on particularly favorable growing seasons. Diggle et al. (1994) found that it takes one or several seasons before the preformation of an inflorescence is completed in this species (see the first Section). They also found substantial genotype variation in alpine populations, but phenotypic variation within genotypes was significant as well. The above mentioned trade-off between vegetative and sexual organs within the same inflorescence finds a parallel in the viviparous *Saxifraga cernua*, in which Wehrmeister and Bonde (1977) showed that alpine ecotypes have larger bulbils than arctic ecotypes, whereas pollen is viable in arctic, but not in the alpine populations they had studied in Colorado.

Poa alpina ssp. vivipara was given the rank of a subspecies before it became known that vivipary is a phenotypic feature in this species. Schwarzenbach (1956), Bachmann (1980) and Heide (1994) demonstrated that vivipary is influenced by temperature and day length (for further references see Hermesh and Acharya 1987; Pahl and Darroch 1997). A high proportion of normal flowering individuals can be obtained in habitually viviparous species by optimal primary and secondary floral induction (warm winter, and – depending on origin – variably long days). Also, high soil moisture favors bulbil versus seed production. Thus sexuality is by no means entirely suppressed in this viviparous species, but is under environmental control.

Bulbils, in this case in fact little green plantlets, develop from both somatic and generative tissue of the inflorescence (first described in detail by Resvoll 1917; see review by Bachmann 1980) and are propagated mainly by layering of the mature "inflorescence". Long flowering stalks seem to have a selective advantage, since they allow wider spacing between the mother and daughter ramets. After layering, the stalk becomes a functional stolon and supports bulbils until adventitious rooting is successful (only on wet ground). One can find whole offspring circles around such viviparous *Poa alpina*. The "viviparous" clonal potential of some *Poa alpina* provenances is so high that commercial mass production of bulbils for revegetation of eroded alpine terrain has been suggest-ed (Grabherr 1995). Vivipary occurs in several other *Poa* species from cold climates and also in species of *Deschampsia* and *Festuca* (Bachmann 1980). Another, rather curious type of "vivipary" is the production of plantlets on leaf margins. This is employed by an evergreen Liliaceae in Japan (*Heloniopsis orientalis*) which preferentially grows at lower altitudes, but can also be found in alpine meadows and snowbeds, where this mode of reproduction is, however, substantially reduced (Kawano and Masuda 1980).

A very special form of asexual reproduction is **apomixis**, where seeds are formed asexually, and thus represent copies of the maternal genome. Apomixis plays a greater role in alpine compared with low altitudes and seems to be more frequent in polyploid genotypes, but the actual processes involved are complicated and may vary a lot among species (Richards 1997). It has often been believed that apomixis is a "dead end" of evolution, but this does not seem to be the case. Although full apomixis is a cloning process, genetically different apomictic "clones" have been identified among populations, and apomictic species are very successful, perhaps through the rapid dispersal of most successful genets. Asexually produced seeds preserve successful genotypes. Apomictic species are not only known to be particularly successful seeders, they also may occasionally produce seeds sexually, hence the observed genetic diversity. Apomixis is found mainly within Asteraceae and Poaceae, the two most important plant families in alpine vegetation worldwide. Plant species like *Poa alpina* (25 publications alone between 1930 and

1974, according to Bachmann 1980) and species of *Hieracium*, *Taraxacum* and *Antennaria* are apomicts. Recently, apomixis was shown in *Nardus stricta* and many more such discoveries are to be expected (cf. Molau 1993; Pahl and Darroch 1997; Richards 1997).

Finally, it should be added that clonal propagation is part of normal life in all **bryophytes** and **lichens**, groups of organisms which are able to inhabit the most hostile alpine habitats. Plant propagation at its upper elevational limits is only clonal.

Alpine plant age

The above section on clonal growth makes it clear that age becomes a problematic term once vegetative propagation of genets occurs. Some of these **clones** may be of very old age. In *Carex curvula*, DNA analysis of clonal expansion suggests genet ages of several thousands of years, older than any of the most often quoted patriarch trees. The same possibly applies to Ericaceae dwarf shrubs such as *Empetrum nigrum* or *Loiseleuria procumbens*, or clonal systems of creeping *Salix*. These plants may be functionally immortal.

On the other hand, a few alpine **annuals** are found with life cycles of less than 3 months (e.g. Bliss 1971; Jackson and Bliss 1982; Reynolds 1984b). In temperate and subarctic mountains these rarely exceed 1 or 2% of the local alpine flora (Brassicaceae, Gentianaceae and Poaceae being most successful). Annuals may reach higher abundance in the mediterranean alpine zone, for instance in the Sierra Nevada of California (Bliss 1971) or in subtropical mountains with longer growing seasons (Ram et al. 1988, report a fraction of 12% for the central Himalayas). Often such numbers strongly depend on whether disturbed areas in the treeline ecotone and in treeless upper montane altitudes are included or not.

In the remaining non-clonal and non-annual plants, age ranges from 2 to several hundreds of years, but there is a clear **tendency for higher plant age** at higher elevations. A good example are the many bi-annual species found at low elevation

which form a rosette and a tap root in the first year and reproduce the second year. At alpine elevations it may take 5 to 10 years for such hapaxantic (i.e. monocarpic) herbaceous species to complete their life cycle. Only rarely is the life cycle completed in less than 3 years (*Arabis alpina* may be an example). For the majority of the taller, tap-rooted or clonal alpine forbs, ages between 30 and 50 years seem a reasonable estimate (Schweingruber and Dietz 2001).

Agakhanyantz and Lopatin (1978) review literature for alpine plants in the Pamirs, with some examples of very old plant ages, as for instance 400 years in *Acantholimon diapensoides*, and 100–300 years for several other **tap rooted, slow growing species**. They also concluded that plant age increases with elevation in the Pamirs. As with leaf life span (Chap. 13), the altitudinal reduction of the length of the growing season in seasonal climates introduces some bias in such calculations when years are counted. A more appropriate measure would be the number of active months, disregarding periods of complete quiescence. Yet, even when accounting for active periods only, the resultant ages are impressive.

Cushion plants can reach very old ages. They are the prototype species of long-term space occupancy, for reasons largely related to plant nutrition and soil stability (see Chaps. 4, 6 and 9). With expansion rates of only a few millimeters per year, some of the famous giant cushions (e.g. *Raoulia* sp.) of New Zealand may be many hundreds of years old. This also applies to the "hard as wood" cushions found in the southern Andes (*Azorella* sp.) whose extensions over several meters of ground area (up to 1 m high) suggest ages of several thousand years (Plate 2f). Because of their age and steady increase in size, cushion plant species such as *Silene acaulis* (Plate 1d) have been used to estimate the minimum age of land released from glaciers or rivers (Benedict 1989).

A group of alpine plants in which individuals can reach very old ages are the tropical **giant rosettes** which obviously take a long time to become "giant" (Fig. 7.6). According to Rundel and Witter (1994), individuals of the Hawaiian alpine giant rosette *Argyroxiphium sandwicense* live up to 90 years, commonly not starting to flower before they are 20 years old and most flowering individuals have an age of about 30–50 years. *Lobelia telekii* on Mt. Kenya flowers and then dies at 40–70 years of age. In *L. keniensis*, first flowering occurs after 30 to 60 years of vegetative growth. New ramets emerge from the stem base which continue the flowering cycle well over 100 years, but possibly not more than 300–400 (Young 1994). Also *Senecio kenyodendron* is reported to reach several hundred years of age when plants become 4–5 m tall (Smith and Young 1994).

Long life cycles are a typical feature of alpine plant life and represent the most important reason why alpine vegetation is more **vulnerable to any sort of disturbance** than vegetation at low elevation. Clonal growth and space occupancy for long periods are two ways of compensation. However, long life is only possible in the alpine environment through great resistance to the mechanical forces of snow and erosion and climatic extremes. To withstand these impacts requires costly structures (hence the often slow seasonal growth) and compact growth forms. One should remember that these growth forms were not selected to resist mountain boots or ski-edges, nor to withstand excessive grazing and trampling by domestic animals. It is also good to remember that a single step on a cushion or dwarf shrub mat could break connections to the ground which were built up over a century, with no chance of repair – aspects which will be discussed in Chap. 17.

Community processes

Numerous plant–plant interactions determine the ultimate outcome of alpine plant reproduction. For a target individual such interactions can be positive (**facilitation**) or negative (**competition**).

Given that small plant stature and dense ground cover facilitate substantial decoupling from the often harsh above-canopy climate (Chap. 4; Fig. 9.17), the aggregation of any above-ground biomass diminishes climatic impact. Not surpris-

ingly, isolating individuals by removing neighbours has been found disadvantageous for some alpine species (Callaway et al. 2002). Although confounded by site differences in canopy size and moisture, this study shows the expected loss of such positive neighbour effects once **shelter** provides no climatic advantage at lower elevations. A more differentiated picture emerges, with both, positive (dominant species) and negative effects of removals (subordinate species), if one accounts for the nature of target species (Aksenova et al. 1998; Pavlov et al. 1998; Choler et al. 2001). Unvegetated ground impairs seedling establishment by overheating in the sun (Fig. 4.8) and by ground freezing (Fig. 6.4) during clear nights. Dominant, commonly clonal alpine species, produce a dense ground cover for their own and their smaller neighbours sake. Similarly, nutrient-poor, loose substrate can be stabilized and enriched with humus by specialist species (clonal plants, cushion plants), which then can "nurse" communities of other species (Fig. 4.13; Plate 2b; e.g. Cavieres et al. 1998; Kikvidze and Nakhutsrishvili 1998; Nunez et al. 1999; see paragraph on seedling emergence). Facilitative **nutrient effects** have also been identified for plants with high quality litter (Steltzer and Bowman 1998), legumes (Theodose et al. 1996; Jacot et al. 2000a,b) and, surprisingly, also for an arctic-alpine hemiparasite (Press 1998).

Competitive effects can be associated with soil nutrients (e.g. Jonasson 1992; Havström et al. 1993; Onipchenko and Blinnikov 1994; Thomas and Bowman 1998), acidic raw humus production (well known for Ericaceae and Cyperaceae) impairing seedling establishment in most species (Welling and Laine 2002), or with spatial dominance of a dense root felt (many tussocks) or clonal growth in general. In early successional communities, competition tends to become less frequent and facilitative effects become more important with elevation (Moen 1993). However, in late successional communities, **clonal dominance** enhances competition and often becomes exclusive for other species at high altitude.

Conditions for plants have always changed and will always change, everywhere on the globe, and high elevation vegetation is no exception (Barry 1990). However, these natural changes are commonly rather slow and of a largely physical nature. The current human-induced changes are rather rapid, include chemical influences (CO_2, soluble N, acid rain) as well as land surface management, causing unprecedented impacts (Messerli and Ives 1984, 1997; Körner 1992, 1994, 2000, Körner and Spehn 2002; Price 1995a,b). Intensification or the rapid abandonment of alpine land use are the most immediate dangers, followed by potential consequences of altered atmospheric composition. Climatic changes exceeding those seen in the recent past (warming, more intense rain events, reduced snow cover) may also become critical in places.

The **centerpiece** of any consideration of the impact of global changes on alpine ecosystems is the stability of **alpine soils**. On slopes, soils persist only as long as vegetation persists. A persistent ground cover with intact root systems is thus the criterion by which the risk of all anthropogenic influences on high elevation ecosystems are to be rated. It should be recalled that 40% of the world's population benefit from mountains, and 10% are directly dependent (Messerli and Ives 1984). This chapter is a brief account of important global change risks for the alpine life zone.

Alpine land use

Alpine land use is possibly as old as human presence in mountain foothills and forelands (Fig. 17.1). **Hunting and pasturing** have influenced alpine vegetation in temperate zone mountains for at least 7000 years (Patzelt 1996), and possibly much longer in warmer regions. "The man in the ice", a Bronze Age hunter or shepherd found with largely intact mountaineer outfit, released from retreating ice above 3000 m in the central Alps of Tirol is a most obvious proof of man's active presence in alpine environments, long before heli-skiing (Eijgenraam and Anderson 1991; Spindler et al. 1995; Bortenschlager and Oeggl 1998). Remarkably, this ancient Tirolian was found in an area where contemporary farmers still herd their flocks from the south across the glacier to summer pastures on the north of the main divide.

Except for extreme elevations or inaccessible rock terraces, it is a safe assumption that all alpine vegetation has undergone some influence of anthropogenic land use. Ungulate wild herbivores have been diminished by hunting (in the case of Ibex in the Alps, in fact eliminated during the 19th century, and only recently re-introduced) and were gradually replaced by seasonal pasturing with **domestic animals** (Fig. 17.2). Traditional, man-made alpine pasture land near the treeline (see color Plate 4 at the end of the book) is common in all mountainous regions with permanent settlements (all over Eurasia, parts of South America) but is missing in areas with a historically predominant nomadic life style (North America, Australia). It is largely for these pasture lands that the Alps, the Carpathians, the Caucasus, the Hindukush and Himalayas have been praised by travellers for their colorful alpine mats. These pastures, with their wooden fences and stone walls, dwellings and shrines, drainage and irrigation systems, specific soil dynamics and very special flora represent a unique cultural heritage that is about to be lost (Werner 1981). Because of

Fig. 17.1. Land use in the Alps. Highly diverse grasslands near the treeline and above have been used for millennia for hay production and grazing, and represent an ecologically stable natural as well as cultural heritage (1680 m, Lechtal Alps, Austria; see also color Plate 4 at the end of the book)

sustainable management, these forms of land use only exceptionally lead to erosion. Across the Alps, high elevation pasturing supplied approximately a million people with food until very recently. Measures to secure **sustainable agriculture** in traditional alpine pasture land near the treeline are urgently needed in many parts of the world for three reasons:

- To maintain a healthy, unpolluted food source for future generations,
- To conserve biologically highly diverse, stable and attractive plant communities,
- To retain a millennia old cultural heritage.

In pastures near the treeline, things can go wrong in three ways: (1) uncontrolled, non-traditional (i.e. patchy) grazing causing spot-impacts under otherwise low stocking rates, (2) stocking beyond the carrying capacity or introduction of too heavy animals (Fig. 17.2) or (3) sudden abandonment of pastures. All three may affect soils and induce erosion (Körner 2000).

In the case of **abandonment**, this has to do with sudden occlusion of drainage systems (over-saturation of soils), and with turf-erosion caused by creeping late winter snow, frozen to over-long grass. The risky transition period, back to self-sustainable ground cover, may take at least half a century (Cernusca 1978), but sensitivity varies a lot with slope and vegetation type (Gigon 1984). A future reversal of such post-abandonment shrub and tree invasions (which eliminate the product of manual work from many generations), will most likely be unaffordable, hence the loss will be finite.

Among direct impacts of pasturing on alpine vegetation, trampling effects are much more severe than grazing effects as such. Late successional alpine turf with a dense root felt is rather robust, whereas dwarf shrub communities appear extremely sensitive (Körner 1980). Adequate grazing regimes in alpine grassland can substantially improve hydroelectric **catchment value**. According to measurements and calculations by Körner et al. (1989c) the added hydroelectric value

Fig. 17.2. Traditional grazing by domestic animals has shaped the alpine vegetation over millennia in many parts of the world (*above*, Hohe Tauern National Park, 2300 m, Austria). Destructive alpine land consumption by overgrazing (in this case by tourist horses) is followed by soil erosion (*below*, near Tafí del Valle, 2500 m, northwest Argentina)

may be in the order of € or $150 per ha per year at current exchange rates (1 l gasoline in Europe costs ca. 1 €). In contrast, inappropriate grazing in previously ungrazed, grazing sensitive alpine vegetation negatively affects catchment value (Costin 1958). This insight paved the way to the foundation of Kosciusko National Park in the Snowy Mountains of southeast Australia.

A particular field of conflict are alpine tussock grasslands found in many parts of the world. The grazing value of the commonly rigid leaves of the dominant species is poor, hence regular **burning** is practiced in order to stimulate palatable regrowth and promote non-tussock grasses (Mark and Holdsworth 1979; Mark et al. 1980; Hofstede et al. 1995). As a consequence, catchment value declines,

nutrients are leached, and the vegetation cover becomes reduced in the long-term.

As mentioned above, "adapted" traditional **grazing** may have no negative effects on alpine vegetation. A study in the Swiss Alps (Körner 2000) revealed unexpected effects of six seasons of cattle exclosure from "natural" alpine *Carex curvula* dominated grassland at 2500 m elevation (200–300 m above the treeline): total (live) above-ground phanerogam biomass at peak growing season (just before the first cattle visits), was reduced in the fenced area by 15%, and there were clear indications of reduced abundance of rare forbs. Hence, what is described in text books as one of the most typical natural alpine grasslands, seems to benefit from sporadic cattle presence, both in terms of productivity and biodiversity. Dung deposition was found to create patch dynamics in this system with a statistical ca. 50 year rotation time. A positive biomass response to traditional alpine grazing was also observed in the Garhwal Himalaya (Sundriyal 1992).

Other, more localized, but rather severe forms of alpine land use are the construction of **ski runs** and **transport routes, summer tourism** (Fig. 17.3), **hydroelectric installations** and **mining.** Given their extent of alpine land consumption, ski runs and their infrastructure are regionally causing substantial perturbations. Service roads may sometimes create even more damage than the smoothed ski run itself (Cernusca 1977). Unfortunately, the insight that sustainable re-vegetation of machine graded terrain above the climatic treeline is almost impossible, is rather recent. Initially promising "green" often disappears in following seasons, because of mal-adaptation. Rather sophisticated (and expensive) **re-vegetation** procedures may help in places (e.g. Schiechtl 1988; Urbanska 1988; Grabherr 1995), but will not re-establish the stability of naturally evolved, deeply rooted soil. Measures for snow allocation (e.g. snow fences) create new microhabitats, with new snow cover dependent vegetation. Sudden removal of such structures or breakdown (e.g. after bankruptcy of a lift company) exposes sensitive vegetation, normally found only in protected areas, to the harsh winter climate of ridges, leading

Fig. 17.3. Alpine tourism: heavy local impact in often sensitive areas, but also contributing to local villagers' income and the wider public's awareness of the need of conservation. *Top*, tourists on the fragile cinder slopes of Mt. Fuji, Japan; *center*, midsummer at Grimsel Pass, Switzerland; *bottom*, wood transport along tourist routes, Langtang, Nepal

to death with subsequent soil erosion risk. **Artificial snow** has also been described to have adverse effects (Cernusca et al. 1990), but often also protects the turf from damage under otherwise insufficient snow cover. There is a lot of long term responsibility involved in the management and alteration of alpine terrain and its microclimate, given the rather low self-repair capacity and overall slow responses of alpine vegetation.

With a wider perspective, the ecological judgement of such touristic intrusions into the alpine landscape is not straightforward. If there were no income from tourism, many of the afore mentioned, biologically precious, man-made parts of the alpine landscape near the treeline were already gone. The appreciation of the beauty and recreational value of the alpine ecosystem by lowland visitors contributed significantly to the success of conservation initiatives. One has also to bear in mind that the most intense types of land use by tourism are commonly small in area, compared with agricultural land use, but on the other hand often affect rather sensitive terrain, which would be better left untouched (e.g. dwarf shrub vegetation in the lower, and fragmented vegetation in higher alpine zone, on steep slopes in particular). Figure 17.4 illustrates the extent of historical deforestation for pasture land in the Alps.

The impact of altered atmospheric chemistry

Although commonly far away from urban agglomerations and industrial emission, atmospheric pollution does reach alpine ecosystems to a variable degree, and atmospheric CO_2 enrichment is a global phenomenon. No doubt, many alpine regions in the Northern Hemisphere are currently receiving a multiple of the pre-industrial rate of soluble **nitrogen deposition** (Fig. 17.5; Table 17.1) and alpine plants have been shown to be very responsive to N addition (Chap. 10, Fig. 17.6). More vigorously growing plants tend to be more receptive to atmospheric fertilizer. Commonly, such fast growing species are not very robust when facing physical stress. Their increasing abundance under

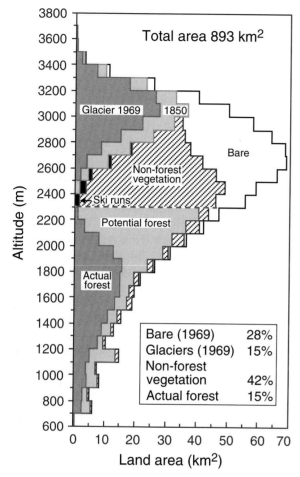

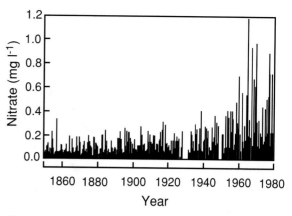

Fig. 17.5. Measured concentrations of nitrate in an ice core from Monte Rosa (Switzerland 4400 m) as a function of ^{210}Pb dated age. (Döscher et al. 1995)

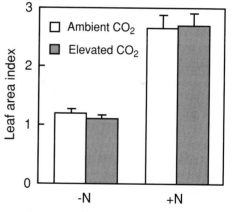

Fig. 17.4. The contribution of certain land surface types to total land area in the Ötztal catchment (893 km²; Tirol, Austria). The altitudinal range of this catchment is comprised between 700 and 3700 m, with 50% of the area above 2500 m. The climatic treeline is at ca. 2200 m. Note the land area released from glaciers between 1850 and 1969 (80 km² or 9% of catchment area) and the difference between the potential and actual forest area (160 km² or 18%), largely resulting from land claims during millenia of pasturing activities. Settlements are restricted to 5% of the catchment area because of natural risks. Machined ski runs cover ca. 1% of the area, but the area actually influenced through construction activities and skiing is substantially larger, especially if work roads and associated rock or scree slides are considered (Patzelt 1996)

Fig. 17.6. The influence of soluble N deposition on alpine grassland. The addition of 40 kg N ha⁻¹ a⁻¹ (the current lowland deposition in central Europe), causes leaf area index (LAI) of an alpine turf at 2500 m elevation to increase 2.5 times by the forth season of treatment. Biomass doubled. For current deposition see Table 17.1. For comparison, 100–300 kg N ha⁻¹ a⁻¹ are common applications in agriculture. LAI: orginal data; biomass: Schäppi and Körner (1997)

enhanced N input could weaken the overall robustness of ecosystems (Welker et al. 2001).

According to Bowman (1992) and Williams et al. (1995) the largest fraction of annual wet deposition of N is stored in the seasonal snowpack and maximum runoff concentrations occur during the first part of snow melt. The comparatively low rates of N deposition in the Rocky Mts. and the Sierra Nevada of California correspond to approximately one third of annual N mineralisation in soils, but the current rates reported for some parts of the Alps exceed natural N mineralisation.

Table 17.1. Soluble nitrogen deposition above the climatic treeline of Europe and North America

Area	N deposition ($kg\,N\,ha^{-1}\,a^{-1}$)	Reference
Alps (Tirol)[a]	7	Psenner and Nickus (1986)
Alps (Tirol)	8–13	Smidt and Mutsch (1993)
Alps (Swiss)	14–21	Graber et al. (1996)
Rocky Mts. (Co.)	6	Bowman (1992)
Rocky Mts. (Co.)	4	Baron and Campbell (1997)

[a] Central part, all other data from front range catchments.

Anthropogenic **acid deposition** is significant at high elevation, in some cases exceeding rates found at lower altitudes (Psenner and Nickus 1986; Lovett and Kinsman 1990; Döscher et al. 1995). Rusek (1993) reports for the Tatra Mts. that acid deposition since 1977 became particularly effective in alpine grassland, where a drop by 1.5 pH units was observed in places with snow accumulation. Spruce forests started to die back from higher elevations. Among all compounds deposited in alpine ecosystems, soluble nitrogen deserves greatest attention, because of the key role of nitrogen for plant metabolism and its immediate influence on plant growth and biodiversity (see Chap. 10). Acid rain may impose particularly severe changes in aquatic systems (Psenner and Nickus 1986; Psenner and Schmidt 1992).

Elevated CO_2 has been expected to become most effective in plants at high elevations, because of the already reduced availability due to lower partial pressure (see Chap. 11). Results of gas exchange studies have indeed supported the idea of instantaneous (Billings et al. 1961; Mooney et al. 1966; Körner and Diemer 1987; Ward and Strain 1997) and prolonged (Diemer 1994; Körner and Diemer 1994) strongly positive effects of elevated CO_2 on alpine photosynthesis, in line with long-term adjustments to life at high elevation as reflected in carbon isotope discrimination (Körner et al. 1991). However, results of four seasons of in situ simulation of a double-CO_2 atmosphere in alpine grassland in Switzerland did not support this view (Schäppi and Körner 1996; Körner et al. 1997; see color Plate 4h at the end of the book). Aboveground plant biomass remained completely unaffected and belowground effects were small (+12%, P = 0.09). Not even legumes and their symbiotic N_2 fixation were stimulated. These observations are in line with those by Tissue and Oechel (1987) for graminoid arctic tundra.

CO_2-enrichment did, however, reveal some species specific responses (stimulation of faster growing, currently very rare species), affected tissue quality of plants (less protein more carbohydrates; Schäppi and Körner 1997), and did influence herbivore behavior (significantly increases consumption by grasshoppers; see Körner et al. 1997). These more subtle effects may translate into significant ecosystem effects in the very long term. However, the net effect of CO_2 enrichment on the ecosystem C balance approached zero after four seasons (see the synthesis in Körner et al. 1997 and Körner and Hättenschwiler 1998). Overall the influences of doubling elevated CO_2 are much smaller than those imposed by the application of amounts of soluble nitrogen, which are currently contained in the annual rainfall of many European lowland regions.

Climatic change and alpine ecosystems

A most important feature of alpine plant life is "change" – rapid and small scale change of climatic and soil conditions. Climatic global change scenarios appear almost negligible compared with those natural changes and contrasts alpine plants master today, as illustrated in Chapter 4. However, an overall **warming** and associated changes in **precipitation** patterns and **snow cover** will influence alpine vegetation (Guisan et al. 1995; Theurillat and Guisan 2001; Welker et al. 2001). Current predictions suggest that warming-only effects will be minimal in the tropics and maximal at high latitudes. Twentieth century trends in the Alps match global warming trends (Fig. 17.7). Changes appear to be most pronounced in Western Europe and parts of Asia (Diaz and Bradley 1997). In the Alps, minimum temperatures increased much more than means (by 2 K; Beniston et al. 1997). As was discussed in Chapter 8, winter minima per se will have little effect on alpine plants, but if they affect snow cover they can become effective. However, there is evidence for the Swiss Alps and the Rocky Mountains of Colorado, that effects on snow cover generally diminish as elevation increases (Beniston 1997; Inouye et al. 2000). Most sensitive areas are below the treeline, and are not considered here. At the treeline itself we see significant growth stimulation in the Alps over the past 120 years, with most dramatic changes in the recent 40 years (Paulsen et al. 2000). An advance of the treeline is, however, not evident so far, but increased young tree establishment in forest gaps near the treeline can be seen in many places unless a dense ground cover of dwarf shrubs or alder krummholz inhibits this process. Evidence is accumulating that as heavy rainfall events (associated with warming, but not necessarily extreme events)

become more frequent, erosion is likely to be enhanced (e.g. Rebetez et al. 1997).

Predictions based on models which assume current **plant-climate coupling** to persist, and plants to migrate with isotherms (e.g. Ozenda and Borel 1990; Guisan et al. 1998) may overestimate change at the level of vegetation belts (which are partly linked to soils and relief, and depend on slow clonal spreading, hence are conservative), but may underestimate effects on single species, the level at which migration appears to happen, according to fossil records (e.g. Ammann 1995; see also the discussion by Theurillat 1995). Holten (1993) and Saetersdal and Birks (1997) suggested that species of narrow thermal ranges (often rare species) should be affected first.

The evidence that alpine plants are likely to respond to climate warming comes from two sides: **simulation** experiments with small open top greenhouses such as the ones used in the International Tundra Experiment (ITEX; e.g. Henry and Molau 1997; Totland 1997) and alpine site **re-visitation** (Hofer 1992; Gottfried et al. 1994; Grabherr and Pauli 1994). The last authors calculated mean 10-year up-slope moving rates of plant species of 1 to 4 m in elevation for various mountain tops in the Alps. They thus confirmed trends already suggested by Braun-Blanquet (1956), who noted increased plant species presence above 3000 m elevation in 1947–1955, compared with 1812–1835.

Temperature increases in the range of interest here (1–3 K) do not necessarily enhance **plant metabolism** because cold climate plants commonly adjust their respiration to prevailing temperatures (see Chap. 11 and Criddle et al. 1994; Larigauderie and Körner 1995), though perhaps not completely in all cases. The most common observation during in situ warming experiments in temperate and subarctic latitudes (e.g. ITEX) was little vegetative effect (growth, biomass production), but substantial **phenological acceleration** both in arctic and alpine areas (e.g. earlier flowering; Wookey et al. 1994; Havström et al. 1995; Alatalo and Totland 1997; Stenström et al. 1997; Suzuki and Kudo 1997; Molau 1997; Stenström and Jonsdottir 1997; tests for *Carex, Cassiope, Saxifraga, Silene*). This is in line with conclusions by

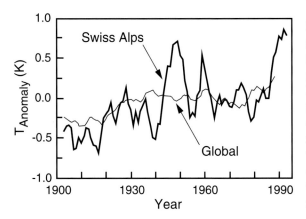

Fig. 17.7. Climate warming: a comparison of global trends in mean surface temperature anomalies with those averaged for eight high elevation sites in the Alps (smoothed with a 5-year filter). (Beniston et al. 1997)

Prock and Körner (1996) that early season development of cold climate plants is often opportunistic (according to unpublished data by F. Keller and Ch. Körner, the onset of flowering is photoperiod sensitive in only about half of the uppermost alpine flora in the Alps), while late season phenology is deterministic (photoperiod controlled; see Chap. 13). Hence, climatic warming will largely affect early season development, the key to which is the possibility of earlier **snow melt** (Guisan et al. 1995, 1998; Guisan 1996; Theurillant and Guisan 2001), not what had actually happened during the recent warm years in the Alps and the Rocky Mountains, because of enhanced snow pack at high elevation (Beniston 1997; Inouye et al. 2000). In regions with potential topsoil desiccation and thus drought induced nutrient shortage in midsummer, accelerated snow melt may, however, have negative effects (Walker et al. 1995). As a plausible alternative to bulk up-slope migration, Grabherr et al. (1995) have suggested local niche filling and re-arrangements in vegetation mosaics, driven by patterns of snow distribution and microclimate. In regions where the moisture regime changes, quite unexpected effects might be seen. For instance, ongoing long-term reductions of precipitation on Kilimanjaro had not only affected the famous glacier but enhanced fire frequency in the zone above the montane rainforest and led to a lowering of the actual treeline (A. Hemp, pers. commun.).

Enhanced UV-B is unlikely to exert major effects on alpine plants since they are well adjusted to cope with a varying degree and high intensity of UV-B compared with doses considered critical at low elevation (see Chap. 8). Experimental evidence from high latitudes also suggests little if any effect (Sonesson et al. 1996; Van de Staaij et al. 1997; Weih et al. 1998; tests for genera *Betula, Hylocomium, Silene*), although ericaceous dwarf shrubs in open tundra woodland have been shown to be sensitive to experimental UV-B enhancement (Johanson et al. 1995; *Vaccinium* sp.).

In **summary**, rising growing season temperatures, longer season length and increased nitrogen supply alone or in combination will reduce some of those constraints dominating alpine plant life as was discussed in this book. Changes in cloudiness, snow pack and precipitation may enhance or diminish such trends. In interaction with elevated CO_2, plant diversity and abundance of certain species will change. "Lessening" environmental "limitations" (see Chap. 1) will open alpine terrain for **invaders** from lower elevations and create pressure for upward migration of alpine species. Actual **migration** will always strongly depend on migration corridors and whether mountains provide high enough escapes. Whether rates of migrations will track current rapid changes seems doubtful. Most alpine species perform some sort of clonal, rather slow mode of propagation and thus retain space occupancy irrespective of such climatic variation. For instance, it is thought that clones of *Carex curvula* have persisted several thousand years on the very same spot at 2500 m in the Alps (Steinger et al. 1996) while the climate has undergone substantial variation. Obligatory seeders and alpine "**ruderals**", restricted to open high elevation terrain, are thus the most likely components of the alpine flora to exhibit fast responses. However, even these may operate at remarkably conservative population dynamics despite climatic warming, as was demonstrated in a 12-year permanent plot study with *Ranunculus glacialis* in the Austrian Alps (Diemer 2002). Late successional, closed vegetation will change very slowly if at all (with N deposition bearing the greatest influence).

Compared with these atmospheric changes, **direct human influences** are much more severe and immediate. The current rapid and worldwide deterioration of the lower alpine vegetation belts and traditional pasture land near the treeline calls for rapid intervention (Körner and Spehn 2002). On the other hand, wild plants inhabiting high alpine terrain bear a great potential for **bio-monitoring** atmospheric influences on a global scale, because such cold climate wilderness habitats are unique in occurring at all latitudes (Fig. 2.1). Hence, well documented historic research sites in alpine areas represent invaluable reference points for global change research.

References

(The numbers in the brackets denote the chapters in which the reference appears.)

Abbott RJ, Chapman HM, Crawford RMM, Forbes DG (1995) Molecular diversity and deviations of populations of *Silene acaulis* and *Saxifraga oppositifolia* from the high Arctic and southern latitudes. Mol Ecol 4:199–207 [2]

Abdaladze OG (1987) CO_2-gas exchange in plants of subalpine belt of the central Caucasus. Bot Zhurn 72:1042–1049 [11]

Agakhanyantz OE, Breckle SW (1995) Origin and evolution of the mountain flora in middle Asia and neighbouring mountain regions. In: Chapin FS III, Körner Ch (eds) Arctic and alpine biodiversity: Patterns, causes and ecosystem consequences. Ecological studies, vol 113. Springer, Berlin Heidelberg New York, pp 63–80 [1, 2]

Agakhanyantz OE, Lopatin IK (1978) Main characteristics of the ecosystems of the Pamirs, USSR. Arct Alp Res 10:397–407 [1, 9, 15, 16]

Akhalkatsi M, Wagner J (1996) Reproductive phenology and seed development of *Gentianella caucasea* in different habitats in the central Caucasus. Flora 191:161–168 [16]

Aksenova AA, Onipchenko VG, Blinnikov MS (1998) Plant interactions in alpine tundra: 13 years of experimental removal of dominant species. Ecoscience 5:258–270 [16]

Alatalo JM, Totland O (1997) Response to simulated climatic change in an alpine and subarctic pollen-risk strategist, *Silene acaulis*. Glob Change Biol 3:74–79 [16, 17]

Amen RD (1966) The extent and role of seed dormancy in alpine plants. Quart Rev Biol 41:271–281 [16]

Ammann B (1995) Paleorecords of plant biodiversity in the Alps. In: Chapin FS III, Körner Ch (eds) Arctic and alpine biodiversity: Patterns, causes and ecosystem consequences. Ecological studies, vol 113. Springer, Berlin Heidelberg New York, pp 137–149 [2, 17]

Anchisi E (1985) Quatrieme contribution à l'étude de la flore valaisanne. Bull Murithienne 102:115–126 [2]

Andrews CJ (1996) How do plants survive ice? Ann Bot 78:529–536 [5, 8]

Archibold OW (1995) Ecology of world vegetation. Chapman and Hall, London [1]

Armand AD (1992) Sharp and gradual mountain timberlines as a result of species interaction. In: Hansen AJ, di Castri F (eds) Landscape boundaries. Consequences for biotic diversity and ecological flows. Ecological studies, vol 92.360–378, Springer, Berlin Heidelberg New York, pp 360–378 [7]

Arno SF (1984) Timberline. The Mountaineers, Seattle [7]

Arnone J, Hirschel G (1997) Does fertilizer application alter the effects of elevated CO_2 on *Carex* leaf litter quality and in situ decomposition in alpine grassland? Acta Oecol 18:201–206 [6]

Arnone JA III (1997) Indices of plant N availability in an alpine grassland under elevated atmospheric CO_2. Plant Soil 190:61–66 [10]

Arnone JA, Körner C (1997) Temperature adaption and acclimation potential of leaf dark respiration in two species of *Ranunculus* from warm and cold habitats. Arct Alp Res 29:122–125 [11]

Arroyo MK, Medina E, Ziegler H (1990) Distribution and $\partial^{13}C$ values of Portulaceae species of the high Andes in northern Chile. Bot Acta 103:291–295 [1, 9]

Arroyo MTK, Armesto JJ, Villagran C (1981) Plant phenoloical patterns in the high Andean cordillera in central Chile. J Ecol 69:205–223 [16]

Atkin OK (1996) Reassessing the nitrogen relations of arctic plants: a mini review. Plant Cell Environ 19:695–704 [10]

Atkin OK, Botman B, Lambers H (1996a) The causes of inherently slow groth in alpine plants: an analysis based on the underlying carbon economies of alpine and lowland *Poa* species. Funct Ecol 10:698–707 [11, 12, 13]

Atkin OK, Botman B, Lambers H (1996b) The relationship between the relative growth rate and nitrogen economy of alpine and lowland *Poa* species. Plant Cell Environ 19:1324–1330 [11]

Atkin OK, Collier DE (1992) Relationship between soil nitrogen and floristic variation in late snow areas of the Kosciusko alpine region. Aust J Bot 40:139–149 [10]

Atkin OK, Cummins WR (1994) The effect of nitrogen source on growth, nitrogen economy and respiration of two high arctic plant species differing in relative growth rate. Funct Ecol 8:389–399 [5, 10]

Atkin OK, Day DA (1990) A comparison of the respiratory processes and growth rates of selected Australian alpine and related lowland plant species. Aust J. Plant Physiol 17:517–526 [11, 13]

Atkin OK, Westbeek MHM, Cambridge ML, Lambers H, Pons TL (1997) Leaf respiration in light and darkness. A comparison of slow- and fast-growing Poa species. Plant Physiol 113:961–965 [11]

Aulitzky H (1961) Lufttemperatur und Luftfeuchtigkeit. Mitt Forstl Bundes Versuchsanst Mariabrunn (Wien) 59:105–125 [7]

Aulitzky H (1963) Grundlagen und Anwendung des vorläufigen Wind-Schnee-Ökogrammes. Mitt Forstl Bundes Versuchsanst Mariabrunn (Wien) 60:763–834 [4]

Aulitzky H (1984) Die Windverhältnisse an einer zentralalpinen Hangstation und ihre ökologische Bedeutung. Centralbl Forstwes 101:193–232 [5]

Azocar A, Rada F, Goldstein G (1988) Freezing tolerance in Draba chionophila, a "miniature" caulescent rosette species. Oecologia 75:156–160 [8, 12]

Bachmann MA (1980) Oekologie und Breeding System bei Poa alpina L. (PhD Thesis, University of Zürich). Mitt Bot Mus Univ Zürich 318 [16]

Bahn M, Körner Ch (1987) Vegetation und Phänologie der hochalpinen Gipfelflur des Glungezer in Tirol. Ber Natwiss Med Ver Innsbr 74:61–80 [16]

Baig MN, Tranquillini W (1980) The effects of wind and temperature on cuticular transpiration of Picea abies and Pinus cembra and their significance in desiccation damage at the alpine treeline. Oecologia 47:252–256 [4, 7]

Baker HG (1972) Seed weight in relation to environmental conditions in California. Ecology 53:997–1010 [16]

Ballard TM (1972) Subalpine soil temperature regimes in southwestern British Columbia. Arct Alp Res 4:139–146 [7]

Barbour MG, Berg NH, Kittel TGF, Kunz ME (1991) Snowpack and the distribution of a major vegetation ecotone in the Sierra Nevada of California. J Biogeogr 19:141–149 [4]

Barclay AM, Crawford RMM (1982) Winter desiccation stress and resting bud viability in relation to high altitude survival. Flora 172:21–34 [7]

Barnola LG, Montilla MG (1997) Vertical distribution of mycorrhizal colonization, root hairs and below-ground biomass in three contrasting sites from the tropical high mountains, Merida, Venezuela. Arct Alp Res 29:206–212 [10]

Baron JS, Campbell DH (1997) Nitrogen fluxes in a high elevation Colorado Rocky Mountain basin. Hydrol Process 11:783–799 [17]

Barry RG (1978) H-B de Saussure: the first mountain meteorologist. Bull Am Meteorol Soc 59:702–705 [3]

Barry RG (1981) Mountain weather and climate. Methuen, London [3, 4, 7]

Barry RG (1990) Changes in mountain climate and glaciohydrological responses. Mount Res Dev 10:161–170 [17]

Barthlott W, Lauer W, Placke A (1996) Global distribution of species diversity in vascular plants: towards a world map of phytodiversity. Erdkunde 50:317–327 [1]

Baruch Z (1979) Elevational differentiation in Espeletia schultzii (Compositae), a giant rosette plant of the Venezuelan Páramos. Ecology 60:85–98 [1, 9, 11, 12]

Baruch Z (1982) Patterns of energy content in plants from the Venezuelan Páramos. Oecologia 55:47–52 [12]

Bauert MR (1993) Vivipary in Polygonum viviparum: an adaptation to cold climate? Nord J Bot 13:473–480 [2, 12, 16]

Bauert MR, Kälin M, Baltisberger M, Edwards PS (1998) No genetic variation detected within isolated relict populations of Saxifraga cernua in the Alps using RAPD markers. Mol Ecol 7:1519–1527 [2]

Baumgartner A (1980) Mountain climates from a perspective of forest growth. In: Benecke U, Davis MR (eds) Mountain environments and subalpine tree growth. New Zealand Forest Service, Wellington, pp 27–40 [7]

Baumgartner A, Reichel E, Weber G (1983) Der Wasserhaushalt der Alpen. Oldenbourg, München [9]

Baxter R, Bell Sa, Sparks TH, Ashenden TW, Farrar JF (1995) Effects of elevated CO2 concentrations on three montane grass species. III. Source leaf metabolism and whole-plant carbon partitioning. J Exp Bot 46:917–929 [12]

Beaman JH (1962) The timberline of Iztaccihuatl and Popocatepetl, Mexico. Ecology 43:377–385 [7]

Beck E (1988) Plant life on top of Mt. Kilimanjaro (Tanzania). Flora 181:379–381 [2]

Beck E (1994) Cold tolerance in tropical alpine plants. In: Rundel PW, Smith AP, Meinzer FC (eds) Tropical alpine environments. Cambridge University Press, Cambridge, pp 77–110 [1, 4, 8, 12]

Beck E (1994) Turnover and conservation of nutrients in the pachycaul Senecio keniodendron. In: Rundel PW, Smith AP, Meinzer FC (eds) Tropical alpine environments. Cambridge University Press, Cambridge, pp 215–221 [10]

Beck E, Nauke P, Fetene M (2002) Hotspots der Biodiversitätsentwicklung: Tropische Hochgebirge. Biologie in unserer Zeit 32:82–888 [1]

Beck E, Rehder H, Pongratz P, Scheibe R, Senser M (1981) Ecological analysis of the boundary between the afro-alpine vegetation types "Dendrosenecio woodlands" and "Senecio brassica- Lobelia keniensis" community on Mt. Kenya. J East African Nat Hist Soc Natl Mus 172:1–11 [9]

Beck E, Scheibe R, Schulze ED (1986) Recovery from fire: observations in the alpine vegetation of western Mt. Kilimanjaro (Tanzania). Phytocoenologia 14:55–77 [7]

Beck E, Schlutter I, Scheibe R, Schulze ED (1984) Growth rates and population rejuvenation of East African giant groundsels (Dendrosenecio keniodendron). Flora 175:243–248 [16]

Beck E, Senser M, Scheibe R, Steiger HM, Pongratz P (1982) Frost avoidance and freezing tolerance in afroalpine giant rosette plants. Plant Cell Environ 5:215–222 [4, 9]

Becwar MR, Rajashekar C, Hansen-Bristow KJ, Burke MJ (1981) Deep undercooling of tissue water and winter hardiness limitations in timberline flora. Plant Physiol 68:111–114 [8]

Bell KL, Bliss LC (1978) Root growth in a polar semidesert environment. Can J Bot 56:2470–2490 [13]

Bell KL, Bliss LC (1979) Autecology of *Kobresia bellardii*: why winter snow accumulation limits local distribution. Ecol Monogr 49:377–402 [4, 5, 8, 12]

Benecke U (1972) Wachstum, CO$_2$-Gaswechsel und Pigmentgehalt einiger Baumarten nach Ausbringung in verschiedene Höhenlagen. Angew Bot 46:117–135 [7]

Benecke U, Havranek WM (1980) Gas-exchange of trees at altitudes up to timberline, Craigieburn Range, New Zealand. N Z Forest Serv Tech Pap 70:195–212 [7]

Benecke U, Schulze ED, Matyssek R, Havranek WM (1981) Environmental control of CO$_2$-assimilation and leaf conductance in Larix decidua Mill. I. A comparison of contrasting natural environments. Oecologia 50:54–61 [7]

Benedict JB (1984) Rates of tree-island migration, Colorado Rocky Mountains. Ecology 65:820–823 [6]

Benedict JB (1989) Use of *Silene acaulis* for dating: the relationship of cushion diameter to age. Arct Alp Res 21:91–96 [16]

Benedict JB (1990) Lichen mortality due to late-lying snow: results of a transplant study. Arct Alp Res 22:81–89 [5]

Beniston M (1997) Variation of snow depth and duration in the Swiss Alps over the last 50 years: links to changes in large-scale climatic forcings. Clim Change 36:281–300 [17]

Beniston M, Diaz HF, Bradley RS (1997) Climatic change at high elevation sites: An overview. Clim Change 36:233–251 [17]

Bennett MD (1987) Variation in genomic form in plants and its ecological implications. New Phytol 106:177–200 [14]

Bennett MD, Smith JB, Heslop-Harrison JS (1982) Nuclear DNA amounts in angiosperms. Proc R Soc Lond Biol 216:179–199 [14]

Berger-Landefeldt U (1936) Der Wasserhaushalt der Alpenpflanzen. Bibl Bot 115:38–40 [9]

Bergmann P, Molau U, Holmgren B (1996) Micrometeorological impacts on insect activity and plant reproductive success in an alpine environment, Swedish Lapland. Arct Alp Res 28:196–202 [16]

Bergweiler P (1987) Charakterisierung von Bau und Funktion der Photosynthesemembranen ausgewählter Pflanzen unter den Extrembedingungen des Hochgebirges. PhD Thesis, University of Köln [5, 8, 11]

Bernoulli M, Körner Ch (1999) Dry matter allocation in treeline trees. Phyton 39:7–12 [7]

Berry PE, Calvo R (1994) An overview of the reproductive biology of *Espeletia* (Asteraceae) in the Venezuelan Andes. In: Rundel PW, Smith AP, Meinzer FC (eds) Tropical alpine environments. Cambridge University Press, Cambridge, pp 229–249 [16]

Biddington NL (1985) A review of mechanically induced stress in plants. Sci Hortic 36:12–20 [4]

Bilan MV (1967) Effect of low temperature on root elongation in loblolly pine seedlings. Proc 14 IUFRO-Congress, München. Intern Union Forest Res Org 4/23:74–82 [7]

Billings WD (1957) Physiological ecology. Annu Rev Plant Physiol 8:375–391 [1]

Billings WD (1973) Arctic and alpine vegetations: similarities, differences, and susceptibility to disturbance. Bioscience 23:697–704 [2, 5]

Billings WD (1974) Arctic and alpine vegetation: plant adaptations to cold summer climates. In: Ives JD, Barry RG (eds) Arctic and alpine environments. Methuen, London, pp 403–443 [16]

Billings WD (1974) Adaptations and origins of alpine plants. Arct Alp Res 6:129–142 [1, 10, 12]

Billings WD (1978) Alpine phytogeography across the great basin. Great Basin Nat 2:105–117 [2]

Billings WD (1979) High mountain ecosystems. Evolution, structure, operation and maintenance. In: Webber PJ (ed) High altitude geoecology. AAAS Select Symp 12:97–125 [1, 2]

Billings WD (1987) Constraints to plant growth, reproduction, and establishment in arctic environments. Arct Alp Res 19:357–365 [1]

Billings WD (1988) Alpine vegetation. In: Barbour MG, Billings WD (eds) North American terrestrial vegetation. Cambridge University Press, Cambridge, pp 392–420 [1, 2]

Billings WD, Bliss LC (1959) An alpine snowbank environment and its effects on vegetation, plant development, and productivity. Ecology 40:388–397 [4, 5]

Billings WD, Clebsch EEC, Mooney HA (1961) Effect of low concentrations of carbon dioxide on photosynthesis rates of two races of *Oxyria*. Science 133:1834 [1, 11, 17]

Billings WD, Clebsch EEC, Mooney HA (1966) Photosynthesis and respiration rates of Rocky Mountain alpine plants under field conditions. Am Midl Nat 75:34–44 [11]

Billings WD, Mooney HA (1968) The ecology of arctic and alpine plants. Biol Rev 43:481–529 [1, 11, 16]

Billings WD, Peterson KM, Shaver GR, Trent AW (1977) Root growth, respiration, and carbon dioxide evolution in an arctic tundra soil. Arct Alp Res 9:129–137 [13]

Billings WD, Peterson KM, Shaver GR (1978) Growth, turnover, and respiration rates of roots and tillers in tundra graminoids. In: Tieszen LL (ed) Vegetation and production ecology of an Alaskan arctic tundra. Ecological studies, vol 29. Springer, Berlin Heidelberg New York, pp 415–434 [12]

Billings WD, Shaver GR, Trent AW (1976) Measurement of root growth in simulated and natural temperature gradients over permafrost. Arct Alp Res 8:247–250 [4]

Bingham RA, Orthner AR (1998) Efficient pollination of alpine plants. Nature 391:238–239 [16]

Bird MI, Haberle SG, Chivas AR (1994) Effect of altitude on the carbon-isotope composition of forest and grassland soils from Papua New Guinea. Global Biogeochem Cycles 8:13–22 [11]

Björkman O, Holmgren P (1961) Studies of climatic ecotypes in higher plants. Leaf respiration in different populations of *Solidago virgaurea*. Ann R Agric Coll 27:297–304 [11]

Blagowestschenski WA (1935) Über den Verlauf der Photosynthese im Hochgebirge des Pamirs. Planta 24:276–287 [1, 11]

Bliss LC (1956) A comparison of plant development in microenvironments of arctic and alpine tundras. Ecol Monogr 26:303–337 [4, 12, 13]

Bliss LC (1964) Leaf water content in two alpine plants on Mt. Washington, New Hampshire. Ecology 45:163–166 [9]

Bliss LC (1966) Plant productivity in alpine microenvironments on Mt. Washington, New Hampshire. Ecol Monogr 36:125–155 [12, 15]

Bliss LC (1971) Arctic and alpine plant life cycles. Annu Rev Ecol Syst 2:405–438 [1, 10, 16]

Bliss LC (1985) Alpine. In: Chabot BF, Mooney HA (eds) Physiological ecology of North American plant communities. Chapman and Hall, London, pp 41–65 [1]

Blöschl G, Kirnbauer R, Gutknecht D (1991) Distributed snowmelt simulations in an alpine catchment. I. Model evaluation on the basis of snow cover patterns. Water Resour Res 27:3171–3179 [9]

Blum G (1926) Untersuchungen über die Saugkraft einiger Alpenpflanzen. Beihefte Bot Centalbl 43 Abt I:1–100 [8]

Blumer P, Diemer M (1996) The occurrence and consequences of grasshopper herbivory in an alpine grassland, Swiss Central Alps. Arct Alp Res 28:435–440 [10, 12, 15]

Blumthaler M, Ambach W, Huber M (1993) Altitude effect of solar UV radiation dependent on albedo, turbidity, and solar elevation. Meteorol Z N F 2:116–120 [3, 8]

Boller BC (1980) Bestandesphotosynthese und Assimilatverteilung bei Ökotypen von Weißklee (*Trifolium repens*L.) unter verschiedenen Temperaturen und Tageslängen. PhD Thesis ETH Zürich [12]

Bonnier G (1890a) Cultures expérimentales dans les hautes altitudes. C R Acad Sci Paris 110:363–365 [1]

Bonnier G (1890b) Influence des hautes altitudes sur les fonctions des vegetaux. C R Acad Sci (Paris) 111:377–380 [11]

Bonnier G (1895) Recherches expérimentales. L'adaptation des plantes au climat alpin. Ann Sci Nat 7th Ser Bot 19:219–360 [1, 11]

Bortenschlager S (1977) Ursachen und Ausmaß postglazialer Waldgrenzschwankungen in den Ostalpen. In: Frenzel B (ed) Dendrochronologie und postglaziale Klimaschwankungen in Europa. Erdwiss Forsch 13:260–266 [7]

Bortenschlager S (1993) Das höchst gelegene Moor der Ostalpen, "Moor am Rofenberg", 2760 m. Festschrift Zoller. Diss Bot 196:329–334 [7]

Bortenschlager S, Oeggl K (1998) The iceman and its natural environment. The man in the ice 4. Springer, Vienna [17]

Bowman WD (1992) Inputs and storage of nitrogen in winter snowpack in an alpine ecosystem. Arct Alp Res 24:211–215 [10, 5, 17]

Bowman WD (1994) Accumulation and use of nitrogen and phosphorus following fertilization in two alpine tundra communities. Oikos 70:261–270 [10]

Bowman WD, Conant RT (1994) Shoot growth dynamics and photosynthetic response to increased nitrogen availability in the alpine willow *Salix glauca*. Oecologia 97:93–99 [10, 11]

Bowman WD, Theodose TA, Fisk MC (1995) Physiological and production responses of plant growth forms to increases in limiting resources in alpine tundra: implications for differential community response to environmental change. Oecologia 101:217–227 [9, 10, 11]

Bowman WD, Theodose TA, Schardt JC, Conant RT (1993) Constraints of nutrient availability on primary production in two alpine tundra communities. Ecology 74:2085–2097 [1, 10]

Bowman WD, Turner L (1993) Photosynthetic sensitivity to temperature in populations of two C4 *Bouteloua* (Poaceae) species native to different altitudes. Am J Bot 80:369–374 [11]

Bowman WD, Schardt JC, Schmidt SK (1996) Symbiotic N_2-fixation in alpine tundra: ecosytem input and variation in fixation rates among communities. Oecologia 108:345–350 [10]

Bowman WD, Seastedt TR (eds) (2001) Structure and function of an alpine ecosystem – Niwot Ridge, Colorado. Oxford University Press, Oxford [1, 10]

Boysen Jensen P (1949) Causal plant-geography. Biol Medd (Kopenhagen) 21:1–19 [7]

Braun J (1913) Die Vegetationsverhältnisse der Schneestufe in den Rätisch-Lepontischen Alpen. Ein Bild des Pflanzenlebens an seinen äussersten Grenzen. Neue Denkschr Schweiz Naturforsch Ges 48:1–347 [16]

Braun-Blanquet J (1923) Über die Genesis der Alpenflora. Verh Naturforsch Ges Basel 35:243–261 [2]

Braun-Blanquet J (1956) Ein Jahrhundert Florenwandel am Piz Linard (3414 m). Bull Jard Bot Brux 26:221–232 [17]

Braun-Blanquet J, Jenny H (1926) Vegetations-Entwicklung und Bodenbildung in der alpinen Stufe der Zentralalpen. Denkschr Schweiz Naturforsch Ges 63 [6]

Breckle SW (1973) Mikroklimatische Messungen und ökologische Beobachtungen in der alpinen Stufe des afghanischen Hindukusch. Bot Jahrb Syst 93:25–55 [8, 9]

Brewer CA, Smith WK (1993) The natural occurrence of leaf surface wetness: structural and functional implications for native plants in the central Rocky Mountains. Bull Ecol Soc Am Suppl 74:173 [11]

Brewer CA, Smith WK (1997) Patterns of leaf surface wetness for montane and subalpine plants. Plant Cell Environ 20:1–11 [11]

Brockmann-Jerosch H (1919) Baumgrenze und Klimacharakter. Pflanzengeographische Kommission der Schweiz Naturforschenden Gesellschaft, Beiträge zur geobotanischen Landesaufnahme, vol 6. Rascher, Zürich [7]

Broll G (1998) Diversity of soil organisms in alpine and arctic soils in Europe. Review and research needs. Pirineos (Jaca) 151–152:43–72 [6]

Brooks PD, Schmidt SK, Willims MW (1997) Winter production of CO$_2$ and N$_2$O from alpine tundra: environmental controls and relationships to inter-system C and N fluxes. Oecologia 110:403–413 [11]

Brooks PD, Williams MW, Schmidt SK (1996) Microbial activity under alpine snowpacks, Niwot Ridge, Colorado. Biogeochemistry 32:93–113 [10]

Brown WV (1977) The Kranz syndrome and its subtypes in grass systematics. Mem Torrey Bot Club 23:1–97 [11]

Brzoska W (1969) Stoffproduktion und Energiehaushalt der Vegetation auf hochalpinem Standort unter besonderer Berücksichtigung von Ranunculus glacialis L. PhD Thesis, University of Innsbruck [15]

Brzoska W (1973) Stoffproduktion und Energiehaushalt von Nivalpflanzen. In: Ellenberg H (ed) Ökosystemforschung. Springer, Berlin Heidelberg New York, pp 225–233 [12]

Brzoska W (1973a) Stoffproduktion und Energiehaushalt von Nivalpflanzen. In: Ellenberg H (ed) Ökosystemforschung. Springer, Berlin Heidelberg New York, pp 225–233 [15]

Brzoska W (1973b) Dry matter production and energy utilization of high mountain plants in the Austrian Alps. Oecol Plant 8:63–70 [15]

Burga CA (1988) Swiss vegetation history during the last 18 000 years. New Phytol 110:581–602

Burger R, Franz H (1969) Die Bodenbildung in der Pasterzenlandschaft. Wissenschaftl Alpenverein Hefte 21:253–265 [5]

Cabido M, Ateca N, Astegiano ME, Anton AM (1997) Distribution of C3 and C4 grasses along an altitudinal gradient in central Argentina. J Biogeogr 24:197–204 [11]

Cabrera HM, Rada F, Cavieres L (1998) Effects of temperature on photosynthesis of two morphologically contrasting plant species along an altitudinal gradient in the tropical high Andes. Oecologia 114:145–152 [11]

Caine N (1974) The geomorphic processes of alpine environments. In: Ives JD, Barry RG (eds) Arctic and alpine environments. Methuen, London, pp 721–748 [6]

Caldwell MM (1968) Solar ultraviolet radiation as an ecological factor for alpine plants. Ecol Monogr 38:243–268 [1, 3, 5, 8]

Caldwell MM (1970) The wind regime at the surface of the vegetation layer above timberline in the Central Alps. Centralbl Gesamte Forstwes 87:65–74 [4]

Caldwell MM (1971) Solar UV irradiation and the growth and development of higher plants. Photophysiology 6:131–177 [8]

Caldwell MM, Robberecht R, Billings WD (1980) A steep latitudinal gradient of solar ultraviolet-B radiation in the arctic-alpine life zone. Ecology 61:600–611 [8]

Caldwell MM, Robberecht R, Nowak RS (1982) Differential photosynthetic inhibition by ultraviolet radiation in species from the arctic-alpine life zone. Arct Alp Res 14:195–202 [8]

Caldwell MM, Teramura AH, Tevini M (1989) The changing solar ultraviolet climate and the ecological consequences for higher plants. Tree 4:363–367 [3, 5, 8]

Caldwell MM, Teramura AH, Tevini M, Bornmann JF, Björn LO, Kulandaivelu G (1995) Effects of increased solar ultraviolet radiation on terrestrial plants. Ambio 24:166–173 [8]

Caldwell MM, Tieszen LL, Fareed M (1974) The canopy structure of Tundra plant communities at Barrow, Alaska, and Niwot Ridge, Colorado. Arctic Alp Res 6:151–159 [15]

Callaghan TV (1976) Growth and population dynamics of Carex bigelowii in an alpine environment. Oikos 27:402–413 [1, 16]

Callaghan TV, Headley AD, Lee JA (1991) Root function related to the morphology, life history and ecology of tundra plants. In: Atkinson D (ed) Plant root growth, an ecological perspective. Special publication of the British Ecological Society, no 10. Blackwell, Oxford, pp 311–340 [6, 13]

Callaway RM, Brooker RW, Choler P, Zaal K, Lortie CJ, Michelet R, Paolini L, Pugnaire FI, Newingham B, Aschehoug ET, Armas C, Kikodze D, Cook BJ (2002) Positive interactions among alpine plants increase with stress. Nature 417:844–848 [16]

Campbell JS (1997) North American alpine ecosystems. In: Wielgolaski FE (ed) Polar and alpine tundra. Ecosystems of the world. Elsevier, Amsterdam, pp 211–261 [1]

Carlsson BA, Callaghan TV (1990) Effects of flowering on the shoot dynamics of Carex bigelowii along an altitudinal gradient in Swedish Lapland. J Ecol 78:152–165 [16]

Carmi A, Heuer B (1981) The role of roots in control of bean shoot growth. Ann Bot 48:519–527 [13]

Carr GD, Powell EA, Kyhos DW (1986) Self incompatibility in the Hawaiian Madiinae (Compositae): an exception to Bakers' rule. Evolution 40:430–434 [16]

Cartellieri E (1940) Über Transpiration und Kohlensäureassimilation an einem hochalpinen Standort. Sitzungsber Akad Wiss (Wien) Math Naturwiss Kl Abt I 149(3–6):95–143 [11]

Castrillo M (1995) Rubilose-1,5-bis-phosphate carboxylase activity in altitudinal populations of Espeletia schultzii Wedd. Oecologia 101:193–196 [11]

Cavieres LA, Kalyn-Arroyo MT (2001) Persistent soil seed bank in Phacelia secunda J.F. Gmel. (Hydrophyllaceae): experimental detection of variation along an altitudinal gradient in the Andes of central Chile. J Ecol 88:31–39 [16]

Cavieres LA, Penaloza A, Papic C, Tambutti M (1998) Efecto nodriza del cojin Laretia acaulis (Umbelliferae) en la zona alto-andina de Chile central. Rev Chil Hist Nat 71:337–347 [16]

Cernusca A (1976) Bestandesstruktur, Bioklima und Energiehaushalt von alpinen Zwergstrauchbeständen. Oecol Plant 11:71–102 [1, 4, 8, 13]

Cernusca A (1977) Bestandesstruktur, Mikroklima, Bestandesklima und Energiehaushalt von Pflanzenbeständen des alpinen Grasheidegürtels in den Hohen Tauern. In: Cernusca A (ed) Alpine Grasheide Hohe Tauern. Veröffentlichungen des österr MaB-Hochgebirgsprogramms Hohe Tauern, vol 1. Wagner, Innbruck, pp 25–45 [4, 9]

Cernusca A (1977) Schipisten gefährden die Gebirgslandschaft. Umschau 77:109–112 [17]

Cernusca A (1978) Ökologische Analysen von Almflächen im Gasteiner Tal. Veröff Österr MaB-Hochgebirgsprogramms Hohe Tauern, vol 2. Wagner, Innsbruck [17]

Cernusca A (ed) (1989) Struktur und Funktion von Graslandökosystemen im Nationalpark Hohe Tauern. Veröffentlichungen des Österreichischen MaB-Programmes vol 13. Österr Akad Wiss Wien and Wagner, Innsbruck, pp 33–47 [1]

Cernusca A, Angerer H, Newesely C, Tappeiner U (1990) Ökologische Auswirkungen von Kunstschnee – Eine Kausalanalyse der Belastungsfaktoren. Verh Ges Ökol 19:746–757 [17]

Cernusca A, Decker P (1989a) Faktorenabhängigkeit der respiratorischen Kohlenstoffverluste einer alpinen Grasheide (Caricetum curvulae) in 2300 m MH in den Hohen Trauern. In: Cernusca A (ed) Struktur und Funktion von Graslandökosystemen im Nationalpark Hohe Tauern. Veröffentlichungen Österr MaB-Hochgebirgsprogramms Hohe Tauern, vol 13. Österr Akad Wiss (Wien) and Wagner, Innsbruck, pp 371–396 [11]

Cernusca A, Decker P (1989b) Vergleichende Atmungsmessungen an Graslandökosystemen entlang einer Höhenstufenabfolge zwischen 1612 und 2528 m NN in den Alpen. In: Cernusca A (ed) Struktur und Funktion von Graslandökosystemen im Nationalpark Hohe Tauern. Veröffentlichungen Österr MaB-Hochgebirgsprogramms Hohe Tauern, vol 13. Österr Akad Wiss (Wien) and Wagner, Innsbruck, pp 405–418 [11]

Cernusca A, Seeber MC (1981) Canopy structure, microclimate and the energy budget in different alpine plant communities. In: Grace J, Ford ED, Jarvis PG (eds) Plants and their atmospheric environment. Symp Brit Ecol Soc. Blackwell, Oxford, pp 75–81 [1, 9]

Cernusca A, Seeber MC (1989) Mesoklimatische Hinweise und Beschreibung vonWitterungsablauf und Phänologie während der Oekosystemstudie "Höhentransekt" in den Hohen Trauern. In: Cernusca A (ed) Struktur und Funktion von Graslandökosystemen im Nationalpark Hohe Tauern. Veröffentlichungen Österr MaB-Hochgebirgsprogramms, vol 13. Österr Akad Wiss (Wien) and Wagner, Innsbruck, pp 311–330 [15]

Chabot BF, Chabot JF, Billings WD (1972) Ribulose-1,5-diphosphate carboxylase activity in arctic and alpine populations of Oxyria digyna. Photosynthetica 6:364–369 [11]

Chambers JC (1993) Seed and vegetation dynamics in an alpine herb field: effects of disturbance type. Can J Bot 71:471–485 [16]

Chambers JC (1995a) Disturbance, life history strategies, and seed fates in alpine herbfield communities. Am J Bot 82:421–433 [16]

Chambers JC (1995b) Relationships between seed fates and seedling establishment in an alpine ecosystem. Ecology 76:2124–2133 [16]

Chambers JC, MacMahon JA, Brown RW (1990) Alpine seedling establishment: the influence of disturbance type. Ecology 71:1323–1341 [5]

Chambers JC, McMahon JA, Brown RW (1987) Response of an early seral dominant alpine grass and a late seral dominant alpine forb to N and P availability. Reclam Reveg Res 6:219–234 [10, 12, 16]

Chapin DM, Bliss LC (1988) Soil-plant water relations of two subalpine herbs from Mount St. Helens. Can J Bot 66:809–818 [9]

Chapin DM, Bliss LC, Bledsoe LJ (1991) Environmental regulation of nitrogen fixation in a high arctic lowland ecosystem. Can J Bot 69:2744–2755 [10]

Chapin FS III (1987) Environmental controls over growth of tundra plants. Ecol Bull 38:69–76 [10]

Chapin FS III (1974) Phosphate absorption capacity and acclimation potential in plants along a latitudinal gradient. Science 183:521–522 [13]

Chapin FS III (1989) The cost of tundral plant structures: evaluation of concepts and currencies. Am Nat 133:1–19 [11]

Chapin FS III, Chapin MC (1981) Ecotypic differentiation of growth processes in Carex aquatilis along latitudinal and local gradients. Ecology 62:1000–1009 [12]

Chapin FS III, Körner Ch (1995) Patterns, changes, and consequences of biodiversity in arctic and alpine ecosystems. In: Chapin FS III, Körner Ch (eds) Arctic and alpine biodiversity: patterns, causes and ecosystem consequences. Ecological studies vol 113. Springer, Berlin Heidelberg New York, pp 313–320 [8]

Chapin FS III, Körner Ch (eds) (1995) Arctic and alpine biodiversity: Patterns, causes and ecosystem consequences. Ecological studies, vol 113. Springer, Berlin Heidelberg New York [1]

Chapin FS III, Moilanen L, Kielland K (1993) Preferential use of organic nitrogen for growth by a non-mycorrizal arctic sedge. Nature 361:150–153 [10]

Chapin FS III, Oechel WC (1983) Photosynthesis, respiration, and phosphate absorption by Carex aquatilis ecotypes along latitudinal and local environmental gradients. Ecology 64:743–751 [10, 11]

Chapin FS III, Shaver GR (1989) Lack of latitudinal variations in graminoid storage reserves. Ecology 70:269–272 [8, 12]

Chapin FS III, Vitousek PM, Van Cleve K (1986) The nature of nutrient limitation in plant communities. Am Naturalist 127:48–58 [10]

Chapin FS, Schulze E-D, Mooney HA (1990) The ecology and economics of storage in plants. Annu Rev Ecol Syst 21:423–447 [12]

Chapin SF III (1978) Phosphate uptake and nutrient utilization by Barrow tundra vegetation. In: Tieszen LL (ed) Vegetation and production ecology of an Alaskan arctic tundra. Ecological studies, vol 29. Springer, Berlin Heidelberg New York, pp 483–507 [10]

Chatterton NJ, Harrison PA, Bennett JH, Asay KH (1989) Carbohydrate partitioning in 185 accessions of Gramineae grown under warm and cool temperatures. J Plant Physiol 134:169–179 [12]

Choler P, Michalet R, Callaway RM (2001) Facilitation and competition on gradients in alpine plant communities. Ecology 82:3295–3308 [16]

Clausen J, Keck DD, Hiesey WM (1948) Experimental studies on the nature of species. III. Environmental responses of climatic races of *Achillea*. Carnegie Inst Wash Publ 581:1–125 [1, 11, 16]

Clements FE, Martin EV, Long FL (1950) Adaptation and origin in the plant world. The role of environment in evolution. Waltham, MA [1]

Coe J (1967) The ecology of the alpine zone of Mount Kenya. Junk, The Hague [16]

Collier DE (1996) No difference in leaf respiration rates among temperate, subarctic, and arctic species grown under controlled conditions. Can J Bot 74:317–320 [11]

Combes MR (1910) L'eclairement optimum pour le developpement des vegetaux. C R Acad Sci (Paris) 150: 1701–1702 [11]

Comes HP, Kadereit JW (1998) The effect of Quarternary climatic changes on plant distribution and evolution: a molecular perspective. Trends Plant Sci 3:432–438 [2]

Costin AB (1958) The grazing factor and the maintenance of catchment values in the Australian Alps. CSIRO Div Plant Ind Techn Pap 10:3–13 [17]

Costin AB (1966) Management opportunities in Australian high mountain catchments. Proceedings of the International Symposium on Forest Hydrology, Pennsylvania 1965. Pergamon Press, Oxford, pp 565–577 [1, 9]

Costin AB, Gray M, Totterdell CJ, Wimbush DJ (1979) Kosciusko alpine flora. CSIRO and Collins, Melbourne [2]

Courtin GM, Mayo JM (1975) Arctic and alpine plant water relations. In: Vernberg FJ (ed) Physiological adaptation to the environment. Intext Educational, New York, pp 201–224 [9]

Cowan IR, Farquhar GD (1977) Stomatal function in relation to leaf metabolism and environment. In: (ed) Integration of activity in the higher plant. Cambridge University Press, Cambridge. pp 471–505 [9]

Crawford RMM (1992) Oxygen availability as an ecological limit to plant distribution. Adv Ecol Res 23:93–185 [5]

Crawford RMM, Chapman HM, Hodge H (1994) Anoxia tolerance in high arctic vegetation. Arct Alp Res 26:308–312 [5]

Creber HMC, Davies MS, Francis D (1993) Effects of temperature on cell division in root meristems of natural populations of *Dactylis glomerata* of contrasting latitudinal origins. Environ Exp Bot 33:433–442 [14]

Criddle RS, Hopkin MS, McArthur ED, Hansen LD (1994) Plant distribution and the temperature coefficient of metabolism. Plant Cell Environ 17:233–243 [17]

Criddle RS, Smith BN, Hansen LD (1997) A respiration based description of plant growth rate responses to temperature. Planta 201:441–445 [14]

Cronberg N, Molau U, Sonesson M (1997) Genetic variation in the clonal bryophyte *Hylocomium splendens* at hierarchical geographical scales in Scandinavia. Heredity 78:293–301 [16]

Cuatrecasas J (1986) Speciation and radiation of the Espeletiinae in the Andes. In: Vuilleumier F, Monasterio M (eds) High altitude tropical biogeography. Oxford University Press, New York, pp 267–303 [2]

Cuevas JG (2000) Tree recruitment at the *Nothofagus pumilio* alpine timberline in Tierra del Fuego, Chile. J Ecol 88:840–855 [7]

Curl H, Hardy JT, Ellermeier R (1972) Spectral absorption of solar radiation in alpine snowfields. Ecology 53:1189–1194 [5]

Dahl E (1951) On the relation between summer temperature and the distribution of alpine vascular plants in the lowlands of Fennoscandia. Oikos 3:22–52 [4]

Dahl E (1951) On the relation between summer temperature and the distribution of alpine vascular plants in the lowlands of Fennoscandia. Oikos 3:22–52 [11]

Dahl E (1986) Zonation in arctic and alpine tundra and fellfield ecobiomes. In: Polunin N (ed) Ecosystem theory application. Wiley, London, pp 35–62 [1, 7, 14, 16]

Dahl E (1987) The nunatak theory reconsidered. Ecol Bull 38:77–94 [2]

Dahl E (1990) History of the Scandinavian alpine flora. In: Gjaerevoll O (ed) Alpine plants. The Royal Norwegian Society of Science and Tapir Publishers, Trondheim, pp 16–21 [2]

Dale JE (1988) The control of leaf expansion. Ann Rev Plant Physiol Plant Mol Biol 39:267–295 [14]

Dale JE (1992) How do leaves grow? Advances in cell and molecular biology are unraveling some of the mysteries of leaf development. Bioscience 42:423–432 [14]

Dale JE, Milthorpe FL (1983) The growth and functioning of leaves. Cambridge Univ Press, Cambridge [14]

Däniker A (1923) Biologische Studien über Baum- und Waldgrenze, insbesondere über die klimatischen Ursachen und deren Zusammenhänge. Vierteljahresschr Naturforsch Ges Zürich 68:1–102 [7]

Danneberg OH, Jenisch HS, Richter E (1980) Der Humushaushalt eines alpinen Pseudogleys unter Curvuletum. In: Franz H (ed) Untersuchungen an alpinen Böden in den Hohen Tauern 1974–1978, Stoffdynamik und Wasserhaushalt. Veröffentlichungen Österr MaB-Hochgebirgsprogramms Hohe Tauern, vol 3. Wagner, Innsbruck, pp 109–129 [6, 10]

Daubenmire RF (1941) Some ecologic features of the subterranean organs of alpine plants. Ecology 22:370–378 [13]

Daubenmire RF (1954) Alpine timberlines in the Americas and their interpretation. Butler Univ Bot Stud 2:119–136 [7]

Davis J, Schober A, Bahn M, Sveinbjoernsson B (1991) Soil carbon and nitrogen turnover at and below the elevational treeline in northern Fennoscandia. Arct Alp Res 23:279–286 [6]

Day TA (1993) Relating UV-B radiation screening effectiveness of foliage to absorbing-compound concentration and anatomical characteristics in a diverse group of plants. Oecologia 95:542–550 [8]

Day TA, DeLucia EH, Smith WK (1989) Influence of cold soil and snow cover on photosynthesis and leaf conductance in two Rocky Mountain conifers. Oecologia 80:546–552 [7]

Day TA, Vogelmann TC, DeLucia EH (1992) Are some plant life forms more effective than others in screening out ultraviolet-B radiation? Oecologia 92:513–519 [8]

Dearing D (2001) Plant–herbivore interactions. In: Bowman WD, Seastedt WD (eds) (2001) Structure and function of an alpine ecosystem – Niwot Ridge, Colorado. Oxford University Press, Oxford pp 266–281 [10, 15]

Decker JP (1959) Some effects of temperature and carbon dioxide concentration on photosynthesis of mimules. Plant Physiol 34:103–106 [1, 11]

DeLucia EH, Day TA, Vogelman TC (1992) Ultraviolet-B and visible light penetration into needles of two species of subalpine conifers during foliar development. Plant Cell Environ 15:921–929 [8]

Deshmukh I (1986) Ecology and tropical biology. Blackwell, Oxford [7]

Diaz HF, Bradley R (1997) Temperature variations during the last century at high elevation sites. Clim Change 36:253–279 [17]

Diemer M (2002) Population stasis in a high-elevation herbaceous plant under moderate climate warming. Basic Appl Ecol 3:77–83 [17]

Diemer M (1996) The incidence of herbivory in high-elevation populations of Ranunculus glacialis: a re-evaluation of stress-tolerance in alpine environments. Oikos 75:486–492 [11, 12, 15]

Diemer M (1998a) Life span and dynamics of leaves of herbaceous perennials in high-elevation environments – "news from the elephant's leg". Funct Ecol 12:413–425 [13]

Diemer M (1998b) Leaf lifespans of high-elevation aseasonal Andean shrub species in relation to leaf traits and leaf habitat. Global Ecol Biogeogr 7:457–465 [13]

Diemer M, Körner Ch (1996) Lifetime leaf carbon balances of herbaceous perennial plants from low and high altitudes in the central Alps. Funct Ecol 10:33–43 [11]

Diemer M, Körner Ch, Prock S (1992) Leaf life spans in wild perennial herbaceous plants: a survey and attempts at a functional interpretation. Oecologia 89:10–16 [12, 13]

Diemer M, Prock S (1993) Estimates of alpine seed bank size in two central European and one Scandinavian subarctic plant communities. Arct Alp Res 25:194–200 [16]

Diemer M, Körner Ch (1998) Transient enhancement of carbon uptake in an alpine grassland ecosystem under elevated CO2. Arct Alp Res 30:381–387 [11]

Diemer MW (1992) Population dynamics and spatial arrangement of Ranunculus glacialis L., an alpine perennial herb, in permanent plots. Vegetatio 103:159–166 [10, 16]

Diemer MW (1994) Mid-season gas exchange of an alpine grassland under elevated CO2. Oecologia 98:429–435 [11, 17]

Diemer MW (1996) Evidence of microclimatic convergence of high elevation aseasonal tropical páramo and seasonal temperate zone alpine environments. J Veg Sci 7:821–830 [3, 4, 8, 13]

Diggle PK (1997) Extreme preformation in alpine Polygonum viviparum: an architectural and developmental analysis. Am J Bot 84:154–169 [16]

Diggle PK, Lower SS, Ranker TA (1994) Clonal diversity and phenotypic plasticity in three alpine populations of Polygonum viviparum (Polygonaceae). Am J Bot Suppl 81:22 [16]

Dirmhirn I (1964) Das Strahlungsfeld im Lebensraum. Akademische Verlagsgesellschaft, Frankfurt [3]

Döscher A, Gäggeler HW, Schotterer U, Schwikowski M (1995) A 130-year deposition record of sulfate, nitrate and chlorine from a high-alpine glacier. Water Air Soil Pollut 85:603–609 [17]

Douglas GW, Bliss LC (1977) Alpine and high subalpine plant communities of the north Cascades Range, Washington and British Columbia. Ecol Monogr 47:113–150 [4]

Duguay CR (1994) Remote sensing of the radiation balance during the growing season at the Niwot Ridge long-term ecological research site, Front Range, Colorado, USA. Arct Alp Res 26:393–402 [4]

Earnshaw MJ, Carver KA, Gunn TC, Kerenga K, Harvey V, Griffiths H, Broadmeadow MSJ (1990) Photosynthetic pathway, chilling tolerance and cell sap osmotic potential values of grasses along an altitudinal gradient in Papua New Guinea. Oecologia 84:280–288 [8]

Earnshaw MJ, Carver KA, Gunn TC, Kerenga K, Harvey V, Griffiths H, Broadmeadow MSJ (1990) Photosynthetic pathway, chilling tolerance and cell sap osmotic potential values of grasses along an altitudinal gradient in Papua New Guinea. Oecologia 84:280–288 [11]

Eckel O, Thams C (1939) Untersuchungen über Dichte, Temperatur und Strahlungsverhältnisse der Schneedecke von Davos. Geol Schweiz Hydrol 3:275–340 [5]

Eddelman LE, Ward RT (1984) Phytoedaphic relationships in alpine tundra of north-central Colorado, USA. Arct Alp Res 16:343–359 [4]

Ehleringer JR, Miller PC (1975) Water relations of selected plant species in the alpine tundra, Colorado. Ecology 56:370–380 [1, 9]

Eijgenraam F, Anderson A (1991) A window on life in the Bronze Age. Science 254:187–188 [7, 17]

Elias SA (2001) Paleoecology and late Quarternary environments of the Colorado Rockies. In: Bowman WD, Seastedt TR (eds) (2001) Structure and function of an alpine ecosystem – Niwot Ridge, Colorado. Oxford University Press, Oxford, pp 285–303 [7]

Ellenberg H (1958) Wald oder Steppe? Die natürliche Pflanzendecke der Anden Perus, I u. II. Umschau 22:645–648 [7]

Ellenberg H (1963) Vegetation Mitteleuropas mit den Alpen in kausaler, dynamischer und historischer Sicht. Ulmer, Stuttgart [7]

Ellenberg H (1996) Páramos und Punas der Hochanden Südamerikas, heute grossenteils als potentielle Wälder anerkannt. Verh Ges Ökol 25:17–23 [7]

Engler A (1913) Einfluß der Provenienz des Samens auf die Eigenschaften der forstlichen Holzgewächse. Mitt Schweiz Centralanst Forstl Versuchswesen 10:190–386 [1]

Enquist BJ, Ebersole JJ (1994) Effects of added water on photosynthesis of *Bistorta vivipara*: the importance of water relations and leaf nitrogen in two alpine communities, Pikes Peak, Colorado, USA. Arct Alp Res 26:29–34 [9, 11]

Erhardt A (1993) Pollination of the edelweiss, *Leontopodium alpinum*. Bot J Linn Soc 111:229–240 [16]

Erschbamer B (1990) Substratabhängigkeit alpiner Rasengesellschaften. Flora 184:389–403 [10]

Erschbamer B (1996) Wachstumsdynamik und Nährstoffgehalt der alpinen Segge, *Carex curvula* subsp. *rosae*, auf unterschiedlichen Substraten. Flora 191:121–129 [10]

Erschbamer B, Kneringer E, Niederfriniger-Schlag R (2001) Seed rain, soil seed bank, seedling recruitment, and survival of seedlings on a glacier foreland in the central Alps. Flora 196:304–312 [16]

Erschbamer B, Winkler J, Wagner J (1994) Vegetative und generative Entwicklung von drei *Carex curvula*-Sippen in den Zentralalpen. Flora 189:277–286 [16]

Espinosa R (1933) Oekologische Studien über Kordillierenpflanzen. Englers Botanische Jahrb 65:121–211 [9]

Evans GR (1980) Phytomass, litter, and nutrients in montane and alpine grasslands, Craigieburn Range, New Zealand. In: Benecke U, Davis MR (eds) Mountain environment and subalpine tree growth. New Zealand Forest Service, Wellington, pp 95–110 [10]

Eviner VT, Chapin FS (1997) Nitrogen cycle – plant-microbial interactions. Nature 385:26–27 [10]

Fahey BD (1974) Seasonal frost heave and frost penetration measurements in the Indian Peaks region of the Colorado Front Range. Arct Alp Res 6:63–70 [6]

Farquhar GD, Ehleringer JR, Hubick KT (1989) Carbon isotope discrimination and photosynthesis. Ann Rev Plant Physiol Plant Mol Biol 40:503–537 [9]

Farquhar GD, Richards RA (1984) Isotopic composition of plant carbon correlates with water-use efficiency of wheat genotypes. Aust J Plant Physiol 11:539–552 [11]

Farrar JF (1988) Temperature and the partitioning and translocation of carbon. Symp Soc Exp Biol 42:203–235 [12]

Favarger C (1954) Sur le pourcentage des polyploides dans la flore de l'étage nival des Alpes Suisses. 8th Int Bot Congr (Paris) Sect 9–10:51–56 [16]

Favarger C (1961) Sur l'emploi des nombres des chromosomes en géographie botanique historique. Ber Geobot Inst ETH (Stiftung Rübel, Zürich) 32:119–146 [2, 16]

Felber F, Zhao G.-F, Küpfer P (1996) Étude de la variabilité génétique de la flouve alpine (*Anthoxanthum odoratum* A. & D. Löve) et du mélèze (*Larix decidua* Miller) dans le'écocline subalpin-alpin. Bull Murithienne 114:179–185 [16]

Fetene M, Gashaw M, Nauke P, Beck E (1998) Microclimate and ecophysiological significance of the tree-like life-form of *Lobelia rhynchopetalum* in a tropical alpine environment. Oecologia 113:332–340 [4]

Field C, Chiariello N, Williams WE (1982) Determinants of leaf temperature in California *Mimulus* species at different altitudes. Oecologia 55:414–420 [9]

Fischer FJF (1960) A discussion of leaf morphogenesis in *Ranunculus hirtus*. N Z J Sci 3:685–693 [14]

Fisk MC, Schmidt SK (1995) Nitrogen mineralization and microbial biomass nitrogen dynamics in three alpine tundra communities. Soil Sci Soc Am J 59:1036–1043 [10]

Flenley JR (1979) The equatorial rain forest: a geological history. Butterworth, London [7]

Flenley JR (1992) Ultraviolet-B insolation and the altidudinal forest limit. In: Furley PA, Proctor J, Ratter JA (eds) Nature and dynamics of forest savanna boundaries. Chapman and Hall, London, pp 273–282 [2]

Flenley JR (1995) Cloud forest, the Massenerhebung effect, and ultraviolet insolation. In: Hamilton LS, Juvik JO, Scatena FN (eds) Tropical montane cloud forests. Ecol Stud, Springer, New York, pp 150–155 [8]

Flint SD, Caldwell MM (1983) Influence of floral optical properties on the ultraviolet radiation environment of pollen. Am J Bot 70:1416–1419 [8]

Fliri F (1975) Das Klima der Alpen im Raume von Tirol. Wagner, Innsbruck [3]

Flohn H (1974) Contribution to a comparative meteorology of mountain areas. In: Ives JD, Barry RG (eds) Arctic and alpine environments. Methuen, London, pp 55–71 [3]

Fonda RW, Bliss LC (1966) Annual carbohydrate cycles of alpine plants on Mt. Washington, New Hampshire. Bull Torrey Bot Club 93:268–277 [12]

Fossati A (1980) Keimverhalten und frühe Entwicklungsphasen einiger Alpenpflanzen. Veröff Geobot Inst ETH (Stiftung Rübel, Zürich) 73:1–193 [16]

Fox JF (1981) Intermediate levels of soil disturbance maximize alpine plant diversity. Nature 293:564–565 [4, 5]

Francis D, Barlow PW (1988) Temperature and the cell cycle. Symp Soc Exp Biol 42:181–201 [14]

Franz H (1979) Ökologie der Hochgebirge. Ulmer, Stuttgart [1, 3, 4, 5, 6, 7, 9, 15, 16]

Franz H (1980) Untersuchungen an alpinen Böden in den Hohen Tauern 1974–1978, Stoffdynamik und Wasserhaushalt. Veröffentlichungen Österr MaB-

Hochgebirgsprogramms Hohe Tauern, vol 3. Wagner, Innsbruck [6, 15]

French HM (1996) The periglacial environment. Longman, Harlow, 2nd ed [6]

Frey W (1977) Wechselseitige Beziehungen zwischen Schnee und Pflanze – eine Zusammenstellung anhand von Literatur. Mitt Eidgenöss Inst Schnee Lawinenforsch 34 [5, 7]

Friedel H (1961) Schneedeckendauer und Vegetationsverteilungen im Gelände. Mitt Forstl Bundes Versuchsanst Mariabrunn (Wien) 59:317–369 [1, 4, 5]

Friend AD, Woodward FI (1990) Evolutionary and ecophysiological responses of mountain plants to the growing season environment. Adv Ecol Res 20:59–124 [1, 11]

Friend AD, Woodward FI, Switsur VR (1989) Field measurements of photosynthesis, stomatal conductance, leaf nitrogen and $\partial^{13}C$ along altitudinal gradients in Scotland. Funct Ecol 3:117–122 [11]

Furrer G, Graf K (1978) Die subnivale Höhenstufe am Kilimandjaro und in den Anden Boliviens und Ecuadors. In: Troll C, Lauer W (eds) Geoecological relations between the southern temperate zone and the tropical mountains. Steiner, Wiesbaden, pp 441–457 [6]

Gale J (1972) Availability of carbon dioxide for photosynthesis at high altitudes: theoretical considerations. Ecology 53:494–497 [3, 9, 11]

Galen C (1990) Limits to the distributions of alpine tundra plants: herbivores and the alpine skypilot, *Polemonium viscosum*. Oikos 59:355–358 [15]

Galen C, Stanton ML (1995) Responses of snowbed plant species to changes in growing-season length. Ecology 76:1546–1557 [4, 5]

Galen C, Stanton ML (1999) Seedling establishment in alpine buttercups under experimental manipulations of growing-season length. Ecology 80:2033–2044 [5]

Galland P (1982) Recherches sur les sols des pelouses alpines au parc national suisse. Bull Bodenkundl Ges Schweiz 6:137–144 [10]

Galland P (1986) Croissance et strategie de reproduction de deux especes alpines: *Carex firma* L. et *Dryas octopetala* L. Bull Soc Neuchat Sci Nat 109:101–112 [15]

Gamalei YV, van Bel AJE, Pakhomova MV, Sjutkina AV (1994) Effects of temperature on the conformation of the endoplasmic reticulum and on starch accumulation in leaves with the symplastic minor-vein configuration. Planta 194:443–453 [12]

Gamper M (1981) Heutige Solifluktionsbeträge von Erdströmen und klimamorphologische Interpretation fossiler Böden. Ergeb Wiss Untersuchungen Schweiz Nationalpark 15:355–443 [6]

Gams H (1933) Der tertiäre Grundstock der Alpenflora. Der Begriff des alpigenen Florenelementes. Jahrb Verein Schutz Alpenpflanz Tiere 5:7–37 [2]

Gams H (1963) Die Herkunft der hochalpinen Moose und Flechten. Jahrb Verein Schutz Alpenpflanz Tiere 28 [2]

Gardes M, Dahlberg A (1996) Mycorrhizal diversity in arctic and alpine tundra: an open question. New Phytol 133:147–157 [10]

Garnier BJ, Ohmura A (1968) A method of calculating the direct short wave radiation income of slopes. J Appl Meteorol 7:796–800 [4]

Gauslaa Y (1984) Heat resistance and energy budget in different Scandinavian plants. Holarct Ecol 7:1–78 [1, 4, 8]

Geiger R (1965) The climate near the ground. Harvard University Press, Cambridge, MA [4, 5]

Geissler P, Velluti C (1996) L'écocline subalpin-alpin: approche par les bryophythes. Bull Murithienne 114:171–177 [2]

Germino MJ, Smith WK (1999) Sky exposure, crown architecture, and low-temperature photoinhibition in conifer seedlings at the alpine treeline. Plant Cell Environ 22:407–415 [7, 11]

Germino MJ, Smith WK (2000) Differences in microsite, plant form, and low-temperature photoinhibition in alpine plants. Arct Antarct Alp Res 32:388–396 [7, 11]

Geyger E (1985) Untersuchungen zum Wasserhaushalt der Vegetation im nordwestargentinischen Andenhochland. Diss Bot 88 [1, 9, 11]

Gigon A (1984) Typology and principles of ecological stability and instability. Mountain Res Dev 3:95–102 [17]

Gjaerevoll O (1956) The plant communities of the Scandinavian alpine snow-beds I. Bruns, Trondheim [4]

Gjaerevoll O (ed) (1990) Alpine plants. The Royal Norwegian Society of Sciences and Tapir Publishers, Trondheim [1, 2]

Gjaerevoll O, Ryvarden L (1977) Botanical investigations on J.A.D. Jensens Nunatakker in Greenland. K Nor Vidensk Selsk Skr 4:1–40 [2]

Gold WG, Bliss LC (1995) The nature of water limitations for plants in a high arctic polar desert. Ecosystems research report, vol 10. European Commission, Brussels, pp 149–155 [9]

Goldstein G, Drake DR, Melcher P, Giambelluca TW, Heraux J (1996) Photosynthetic gas exchange and temperature-induced damage in seedlings of the tropical alpine species *Argyroxiphium sandwicense*. Oecologia 106:298–307 [11]

Goldstein G, Meinzer F, Monasterio M (1985) Physiological and mechanical factors in relation to size-dependent mortality in an Andean giant rosette species. Oecol Plant 6:263–275 [1, 9]

Goldstein G, Meinzer FC, Rada F (1994) Environmental biology of a tropical treeline species, *Polylepis sericea*. In: Rundel PW, Smith AP, Meinzer FC (eds) Tropical alpine environments. Cambridge University Press, Cambridge, pp 129–149 [7]

Goldstein G, Rada F, Azocar A (1985) Cold hardiness and supercooling along an altitudinal gradient in Andean giant rosette species. Oecologia 68:147–152 [8]

Goldstein G, Rada F, Canales MO, Zabala O (1989) Leaf gas exchange of two giant caulescent rosette species. Acta Oecol Oec Plant 10:359–370 [9, 11]

Gonzalez JA (1985) El potencial agua en algunas plantas de altura y el problema del stress hidrico en alta montana. Lilloa 36:167–172 [9]

Gonzalez JA (1991) The annual TNC (total non-structural carbohydrates) cycles in two high mountain species: *Woodsia montevidensis* and *Calandrinia acaulis*. Botanica Acta Ci Venez 42:276–280 [12]

Gonzalez JA, de Riera MQ, de Israilev LA (1993) Chlorophyll concentration and flavonoids in the fern *Woodsia montevidensis* in different light regimes at two altitudes in northwestern Argentina. Acta Oecol 14:839–846 [1, 8]

Gottfried M, Pauli H, Grabherr G (1994) Die Alpen im "Treibhaus": Nachweis für das erwärmungsbedingte Höhersteigen der alpinen und nivalen Vegetation. Jahrb Ver Schutz Bergwelt 59:13–27 [17]

Gottfried M, Pauli H, Grabherr G (1998) Prediction of vegetation patterns at the limits of plant life: a new view of the alpine-nival ecotone. Arctic Alp Res 30:207–221 [4]

Graber WK, Siegwolf RTW, Nater W, Leonardi S (1996) Mapping the impact of anthropogenic depositions on high elevated alpine forests. EnvironSoftware 11:29–64 [10, 17]

Grabherr G (1976) Der CO_2-Gaswechsel des immergrünen Zwergstrauches *Loiseleuria procumbens* (L.) Desv. in Abhängigkeit von Strahlung, Temperatur, Wasserstress und phänologischem Zustand. Photosynthetica 11:302–310 [4, 5]

Grabherr G (1995) Renaturierung von natürlichen und künstlichen Erosionsflächen in den Hochalpen. Ber Reinh Tüxen Ges 7:37–46 [16, 17]

Grabherr G (1997) The high mountain ecosystems of the alps. In: Wielgolaski FE (ed) Ecosytems of the world, vol 3: Polar and alpine tundra. Elsevier, Amsterdam, pp 97–121 [1]

Grabherr G, Brzoska W, Hofer H, Reisigl H (1980) Energiebindung und Wirkungsgrad der Nettoprimarproduktivitat in einem Krummseggenrasen (Caricetum curvulae) der Otztaler Alpen, Tirol. Oecol Plant 1:307–316 [12]

Grabherr G, Cernusca A (1977) Influence of radiation, wind, and temperature on the CO_2 gas exchange of the Alpine dwarf shrub community Loiseleurietum cetrariosum. Photosynthetica 11:22–28 [11]

Grabherr G, Gottfried M, Gruber A, Pauli H (1995) Patterns and current changes in alpine plant diversity. In: Chapin FS III, Körner Ch (eds) Arctic and alpine biodiversity: patterns, causes and ecosystem consequences. Ecological studies, vol 113. Springer, Berlin Heidelberg New York, pp 167–181 [1, 2, 17]

Grabherr G, Mähr E, Reisigl H (1978) Nettoprimarproduktion und Reproduktion in einem Krummseggenrasen (Caricetum curvulae) der Ötztaler Alpen, Tirol. Oecol Plant 13:227–251 [15, 16]

Grabherr G, Pauli MGH (1994) Climate effects on mountain plants. Nature 369:448 [17]

Gracanin Z (1972) Die Böden der Alpen. In: R Granssen (ed) Bodengeographie mit besonderer Berücksichtigung der Böden Mitteleuropas. Köhler, Stuttgart, pp 172–195 [6]

Grace J (1977) Plant response to wind. Academic Press, London [3, 4, 7]

Grace J (1987) Climatic tolerance and the distribution of plants. New Phytol 106:113–130 [1]

Grace J (1988) The functional significance of short stature in montane vegetation. In: Werger MJA, Van der Aart PJM, During HJ, Verhoeven JTA (eds) Plant form and vegetation structure. SPB Academic Publishers, The Hague, pp 201–209 [7, 8]

Grace J (1989) Tree lines. Philos Trans R Soc London (Biol) 324:233–245 [1, 7]

Grace J (1990) Cuticular water loss unlikely to explain treeline in Scotland. Oecologia 84:64–68 [7]

Grace J, Norton DA (1990) Climate and growth of *Pinus sylvestris* at its upper altitudinal limit in Scotland: evidence from tree growth-rings. J Ecol 78:601–610 [7]

Graumlich LJ, Brubaker LB (1995) Long-term records of growth and distribution of conifers: integration of paleoecology. In: Smith WK, Hinckley TM (eds) Ecophysiology of coniferous forests. Academic Press, San Diego, pp 37–62 [7]

Gray DM, Male DH (eds) (1981) Handbook of snow. Principles, processes, management and use. Pergamon Press, Oxford [5]

Greer DH (1978) Comparative ecophysiology of some snow tussock (*Chionochloa* spp) populations in Otago. PhD Thesis, Dunedin, New Zealand [7]

Grime JP (1979) Plant strategies and vegetation processes. Wiley, Chichester [15]

Grime JP (1983) Prediction of weed and crop response to climate based upon measurements of nuclear DNA content. Aspec Appl Biol 4:87–98 [14]

Grime JP, Hodgson JG, Hunt R (1988) Comparative plant ecology. Unwin Hyman, London [16]

Gross M (1989) Untersuchungen an Fichten der alpinen Waldgrenze. Diss Bot 139 [7]

Gruber F (1980) Die Verstaubung der Hochgebirgsböden im Glocknergebiet. In: Franz H (ed) Untersuchungen an alpinen Böden in den Hohen Tauern 1974–1978, Stoffdynamik und Wasserhaushalt. Veröffentlichungen des Österr MaB-Hochgebirgsprogramms Hohe Tauern, vol 3. Wagner, Innsbruck, pp 69–90 [6]

Grulke NE, Bliss LC (1985) Growth forms, carbon allocation, and reproductive patterns of high arctic saxifrages. Arct Alp Res 17:241–250 [16]

Gugerli F (1998) Effect of elevation on sexual reproduction in alpine populations of *Saxifraga oppositifolia* (Saxifragaceae). Oecologia 114:60–66 [16]

Gugerli F, Eichenberger K, Schneller JJ (1999) Promiscuity in populations of the cushion plant *Saxifraga oppositifolia* in the swiss Alps as inferred from random amplified polymorphic DNA (APD). Mol Ecol 8:453–461 [2]

Guisan A (1996) Alplandi: évaluer la réponse des plantes alpines aux changements climatiques á travers la modélisation des distributions actuelles et future de leur habitat potentiel. Bull Murithienne 114:187–196 [17]

Guisan A, Holten JI, Spichiger R, Tessier L (1995) Potential ecological impacts of climate change in the Alps and Fennoscandian mountains (20 contributions). Conservatoir Jardin Botaniques (Genève), Publ hors-série 8, pp 1–184 [17]

Guisan A, Theurillat JP, Kienast F (1998) Predicting the potential distribution of plant species in an alpine environment. J Veg Sci 9:65–74 [17]

Gupta RK (1972) Boreal and arcto-alpine elements in the flora of western Himalaya. Vegetatio 24:159–175 [2]

Gurevitch J (1992) Differences in photosynthetic rate in populations of *Achillea lanulosa* from two altitudes. Funct Ecol 6:568–574 [11]

Hadley EB, Rosen RB (1974) Carbohydrate and lipid contents of *Celmisia* plants in alpine snowbank and herbfield communities on Rock and Pillar Range, New Zealand. Am Midl Naturalist 91:371–382 [12]

Hadley JL, Smith WK (1983) Influence of wind exposure on needle desiccation and mortality for timberline conifers in Wyoming, USA. Arct Alp Res 15:127–135 [7]

Hadley JL, Smith WK (1990) Influence of leaf surface wax and leaf area to water content ratio on cuticular transpiration in western conifers, USA. Can J For Res 20:1306–1311 [7]

Hadley KS (1987) Vascular alpine plant distributions within the central and southern Rocky Mountains, USA. Arct Alp Res 19:242–251 [2]

Hager J, Breckle SW (1985) Mikroklima der subalpinen Dornpolsterstufe Kretas. Verh Ges Ökol 13:671–676 [9]

Halloy SRP (1981) La presion de anhidrido carbonico como limitante altitudinal de las plantas. Lilloa 35:159–167 [11]

Halloy SRP (1982) Climatologia y edafologia de alta montana en relacion con la composicion y adaptacion de las comunidades bioticas (con especial referencia a las Cumbres Calchaquies, Tucuman, Argentina). PhD Thesis, San Miguel Tucuman. University Microfilm International, Ann Arbor, no 8502967 [1, 2, 4]

Halloy SRP (1989) Altitudinal limits of life in subtropical mountains: what do we know? Pacific Sci 43:170–184 [2]

Halloy SRP (1990) A morphological classification of plants, with special reference to the New Zealand alpine flora. J Veg Sci 1:291–304 [2, 10, 15, 16]

Halloy SRP (1991) Islands of life at 6000 m altitude: the environment of the highest autotrophic communities on earth (Socompa Volcano, Andes). Arct Alp Res 23:247–262 [1, 2, 15]

Halloy SRP, Gonzalez JA (1993) An inverse relation between frost survival and atmospheric pressure. Arct Alp Res 25:117–123 [8]

Halloy SRP, Mark AF (1996) Comparative leaf morphology spectra of plant communities in New Zealand, the Andes and the European Alps. J R Soc N Z 26:41–78 [4, 7, 9]

Hamerlynck EP, Smith WK (1994) Subnivean and emergent microclimate, photosynthesis, and growth in *Erythronium grandiflorum* Pursh, a snowbank geophyte. Arct Alp Res 26:21–28 [1, 5]

Harris C (1981) Periglacial mass-wasting. A review of research. Geogr Abstr (Norwich), Res Monogr Series 4 [6]

Hartman EL, Rottman ML (1987) Alpine vascular flora of the Ruby Range, West Elk Mountains, Colorado. Great Basin Nat 47:152–160 [2]

Hartmann H (1957) Studien über die vegetative Fortpflanzung in den Hochalpen. Jahresber Naturf Ges Graubündens (Switzerland) 86:3–168 [16]

Haselwandter K (1979) Mycorrhizal status of ericaceous plants in alpine and subalpine areas. New Phytol 83:427–431 [10]

Haselwandter K (1987) Mycorrhizal infection and its possible ecological significance in climatically and nutritionally stressed alpine plant communities. Angew Bot 61:107–114 [6, 10]

Haselwandter K, Hofmann A, Holzmann HP, Read DJ (1983) Availability of nitrogen and phosphorus in the nival zone of the alps. Oecologia 57:266–269 [5, 10]

Haselwandter K, Read DJ (1982) The significance of a root-fungus association in two *Carex* species of high-alpine plant communities. Oecologia 53:352–354 [10]

Hässler R (1982) Net photosynthesis and transpiration of *Pinus montana* on east and north facing slopes at alpine timberline. Oecologia 54:14–22 [7]

Hässler R, Streule A, Turner H (1999) Shoot and root growth of young *Larix decidua* in contrasting microenvironments near the alpine timberline. Phyton 39:47–52 [7]

Hatt M (1991) Samenvorrat von zwei alpinen Böden. Ber Geobot Inst ETH (Stiftung Rübel, Zürich) 57:41–71 [16]

Hättenschwiler S, Handa T, Egli L, Asshoff R, Ammann W, Körner C (2002) Atmospheric CO2 enrichment of alpine treeline conifers. New Phytol 156:363–375 [7]

Hättenschwiler S, Smith WK (1999) Seedling occurrence in alpine treeline conifers: a case study from the central Rocky Mountains, USA. Acta Oecol 20:219–224 [7]

Haunold E, Gludovatz A, Richter E (1980) Stickstoffdynamik in einem alpinen Pseudogley unter Curvuletum. In: Franz H (ed) Untersuchungen an alpinen Böden in den Hohen Tauern 1974–1978, Stoffdynamik und Wasserhaushalt. Veröffentlichungen Österr MaB-Hochgebirgsprogramms Hohe Tauern, vol 3. Wagner, Innsbruck, pp 131–153 [10]

Havranek W (1972) Über die Bedeutung der Bodentemperatur für die Photosynthese und Transpiration junger Forstpflanzen und für die Stoffproduktion an der Waldgrenze. Angew Bot 46:101–116 [7]

Havstroem M, Callaghan TV, Jonasson S (1993) Differential growth responses of *Cassiope tetragona*, an arctic dwafr-shrub, to environmental perturbations among three contrasting high- and subarctic sites. Oikos 66:389–402 [16]

Havström M, Callaghan TV, Jonasson S (1995) Effects of simulated climate change on the sexual reproductive effort of *Cassiope tetragona*. Ecosyst Res Rep (European Commission, Brussels) 10:109–114 [17]

Hay RKM, Heide OM (1983) Specific photoperiodic stimulation of dry matter production in a high-latitude cultivar of *Poa pratensis*. Physiol Plant 57:135–142 [13]

Headley EB, Bliss LC (1964) Energy relationships of alpine plants on Mt. Washington, New Hampshire. Ecol Monogr 34:331–357 [12]

Heber U, Bligny R, Streb P, Douce R (1996) Photorespiration is essential for the protection of the photosynthetic apparatus of C3 plants against photoinactivation under sunlight. Bot Acta 109:307–315 [11]

Hedberg I, Hedberg O (1979) Tropical-alpine life-forms of vascular plants. Oikos 33:297–307 [1, 2]

Hedberg O (1957) Afro-alpine vascular plants. Symb Bot Ups 15:1–411 [2, 16]

Hedberg O (1964) Features of afroalpine plant ecology. Acta Phytogeogr Suec 139 [1, 2, 7]

Hedberg O (1970) Evolution of the afro-alpine flora. Biotropica 2:16–23 [2]

Heer C, Körner Ch (2002) High elevation pioneer plants are sensitive to mineral nutrient addition. Basic Appl Ecol 3:39–47 [10]

Hegg O, Feller U, Dahler W, Scherrer C (1992) Long term influence of fertilization in a Nardetum. Vegetatio 103:151–158 [10]

Heide OM (1985) Physiological aspects of climatic adaptation in plants with special reference to high-latitude environments. In: Kaurin A, Junttila O, Nilsen J (eds) Plant production in the north. Norwegian University Press, Tromso, pp 1–22 [8]

Heide OM (1990) Dual floral induction requirements in *Phleum alpinum*. Ann Bot 66:687–694 [16]

Heide OM (1992) Flowering strategies of the high-arctic and high-alpine snow bed grass species *Phippsia algida*. Physiol Plant 85:606–610 [13]

Heide OM (1994) Control of flowering and reproduction in temperate grassland. New Phytol 128:347–362 [16]

Hellmers H, Genthe MK, Ronco F (1970) Temperature affects on growth and development of Engelmann spruce. For Sci 16:447–452 [7]

Helm D (1982) Multivariate analysis of alpine snow-patch vegetation cover near Milner Pass, Rocky Mountain National Park, Colorado, USA Arct Alp Res 14:87–95 [4]

Hemp A (2002) Ecology of the pteridophytes on the southern slopes of Mt. Kilimanjaro, I. Altitudinal distribution. Plant Ecol 159:211–239 [1]

Hemp A, Beck E (2001) *Erica excelsa* as a fire-tolerating component of Mt. Kilimanjaro's forests. Phytocoenologia 31:449–475 [3]

Henrici M (1918) Chlorophyllgehalt und Kohlensäure-Assimilation bei Alpen- und Ebenen-Pflanzen. Verh Naturforsch Ges Basel 30:43–136 [1, 5, 11]

Henrici M (1921) Zweigipflige Assimilationskurven. Mit spezieller Berucksichtigung der Photosynthese von alpinen phanerogamen Schattenpflanzen und Flechten. Verh Naturforsch Ges Basel 32:107–172 [11]

Henry GHR, Molau U (1997) Tundra plants and climate change: the International Tundra Experiment (ITEX). Glob Change Biol 3:1–9 [16, 17]

Henry GHR, Svoboda J (1994) Comparisions of grazed and non-grazed high-arctic sedge meadows. In: Svoboda J, Freedman B (eds) Ecology of a polar oasis. Captus Press Inc, New York, pp 193–194 [15]

Henry GHR, Svoboda J, Freedman B (1990) Standing crop and net production of sedge meadows of an ungrazed polar desert oasis. Can J Bot 68:2660–2667 [15]

Hermes K (1955) Die Lage der oberen Waldgrenze in den Gebirgen der Erde und ihr Abstand zur Schneegrenze. Kölner geographische Arbeiten, Heft 5. Geographisches Institut, University of Köln [2, 7]

Hermesh R, Acharya SN (1987) Reproductive response to three temperature regimes of four *Poa alpina* populations from the Rocky Mountains of Alberta, Canada. Arct Alp Res 19:321–326 [16]

Hiesey WM, Milner HW (1965) Physiology of ecological races and species. Annu Rev Plant Physiol 16:203–216 [1, 11]

Hikosaka K, Nagamatsu D, Ishii HS, Hirose T (2002) Photosynthesis–nitrogen relationships in species at different altitudes on Mount Kinabalu, Malaysia. Ecol Res 17:305–313 [10, 11]

Hnatiuk RJ (1978) The growth of tussock grasses on an equatorial high mountain and on two sub-antarctic islands. In: Troll C, Lauer W (eds) Geoecological relations between the southern temperate zone and the tropical mountains. Erdwiss Forschung 11, Steiner, Wiesbaden, pp 159–190 [1, 4, 12, 15]

Hnatiuk RJ (1994) Plant form and function in alpine New Guinea. In: Rundel PW, Smith AP, Meinzer FC (eds) Tropical alpine environments. Cambridge University Press, Cambridge, pp 307–318 [13]

Hnatiuk RJ, Smith JMB, McVean DN (1976) Mt Wilhelm Studies 2. The climate of Mt Wilhelm. Aust Nat Univ Press, Canberra [7]

Hoch G, Körner Ch (2003) The carbon charging of pines at the climatic treeline: a global comparison. Oecologia 135:10–21 [7]

Hoch G, Popp M, Körner Ch (2002) Altitudinal increase of mobile carbon pools in *Pinus cembra* suggests sink limitation of growth at the Swiss treeline. Oikos 98:361–374 [7]

Hofer HR (1992) Veränderungen in der Vegetation von 14 Gipfeln des Berninagebietes zwischen 1905 und 1985. Ber Geobot Inst ETH, Stiftung Rübel (Zürich) 58:39–54 [17]

Hofstede RGM, Chilito EJ, Sandovals EM (1995a) Vegetative structure, microclimate, and leaf growth of al páramo tussock grass species, in undisturbed, burned and grazed conditions. Vegetatio 119:53–65 [15]

Hofstede RGM, Mondragon Castillo MX, Rocha Osorio CM (1995) Biomass of grazed, burned, and undisturbed Páramo grasslands, Colombia. I. Aboveground vegetation. Arct Alp Res 27:1–12 [15, 17]

Holch AE, Hertel EW, Oakes WO, Whitwell HH (1941) Root habits of certain plants of the foothill and alpine belts of Rocky Mountain National Park. Ecol Monogr 11:327–345 [13]

Holten JI (1993) Potential effects of climatic change on distribution of plant species, with emphasis on Norway. In: Holten JI, Paulsen G, Oechel WC (eds) Impacts of climatic change on natural ecosystems, with emphasis on boreal and arctic/alpine areas. NINA, Trondheim, pp 84–104 [17]

Holtmeier FK (1974) Geoökologische Beobachtungen und Studien an der subarktischen und alpinen Waldgrenze in vergleichender Sicht. Steiner, Wiesbaden [7]

Holtmeier FK (1994) Ecological aspects of climatically-caused timberline fluctuations. Review and outlook. In: Beniston M (ed) Mountain environments in changing climates. Routledge, London, pp 220–233 [7]

Holtmeier FK, Broll G (1992) The influence of tree islands and microtopography on pedoecological conditions in the forest-alpine tundra ecotone on Niwot Ridge, Colorado Front Range, U.S.A. Arct Alp Res 24:216–228 [4, 6, 7]

Holzmann HP, Haselwandter K (1988) Contribution of nitrogen fixation to nitrogen nutrition in an alpine sedge community (Caricetum curvulae). Oecologia 76:298–302 [10]

Höner D, Breckle SW (1986) Untersuchungen zum Speicherstoff-Gehalt der Dornpolster in den Hochgebirgslagen Kretas (Griechenland). Flora 178:297–305 [12]

Hope GS (1976) Vegetation. In: Hope GS, Peterson JA, Radok U, Allison I (eds) The equatorial glaciers of New Guinea. Results of the 1971–73 Australian University expeditions to Irian Jaya: survey, glaciology, meteorology, biology and palaeoenvironments. Balkema, Rotterdam, pp 113–172 [6]

Hope GS, Peterson JA, Radok U, Allison I (1976) The equatorial glaciers of New Guinea. Results of the 1971–73 Australian University expeditions to Irian Jaya: survey, glaciology, meteorology, biology and palaeoenvironments. Balkema, Rotterdam, pp 113–172 [10]

Huber B (1956) Die Temperatur pflanzlicher Oberflächen. In: Ruhland W (ed) Handbuch der Pflanzenphysiologie, vol 3. Springer, Berlin Göttingen Heidelberg, pp 285–292 [4]

Huber F (1976) Respiratorischer Kohlenstoffverbrauch alpiner Zwergstrauchbestande. In: Müller P (ed) Verhandlungen der Gesellschaft für ökologie, Wien. NV Publ, The Hague, pp 31–35 [11]

Hübl E (1985) Zu den nordischen Beziehungen der Vegetation der Alpen. Flora 176:309–323 [2]

Hungerer KB, Kadereit JW (1998) The phylogeny and biogeography of Gentiana L. sect. Ciminalis (Adans.) Dumort.: a historical interpetation of distribution ranges in the European high mountains. Perspect Plant Ecol Evol Syst 1:121–135 [2]

Ingestad T (1981) Nutrition and growth of birch and grey alder seedlings in low conductivity solutions and at varied relative rates of nutrient addition. Physiol Plant 52:454–466 [10]

Inouye DW, Barr B, Armitage KB, Inouye BD (2000) Climate change is affecting altitudinal migrants and hibernating species. Proc Natl Acad Sci (US) 97:1630–1633 [17]

Inouye DW, McGuire AD (1991) Effects of snowpack on timing and abundance of flowering in Delphinium nelsonii (Ranunculaceae): implications for climate change. Am J Bot 78:997–1001 [5]

Isard SA (1983) Estimating potential direct insolation to alpine terrain. Arct Alp Res 15:77–89 [4]

Isard SA (1986) Factors influencing soil moisture and plant community distribution on Niwot Ridge, Front Range, Colorado, USA. Arct Alp Res 18:83–96 [4, 9]

Isard SA, Belding MJ (1989) Evapotranspiration from the alpine tundra of Colorado, U.S.A. Arct Alp Res 21:71–82 [9]

Ives JD (1978) Remarks on the stability of timberline. In: Troll C, Lauer W (eds) Geoecological relations between the southern temperate zone and the tropical mountains. Steiner, Wiesbaden 313–317 [7]

Ives JD, Barry RG (eds) (1974) Arctic and alpine environments. Methuen, London [1, 3]

Ives JD, Hansen-Bristow KJ (1983) Stability and instability of natural and modified upper timberline landscapes in the Colorado Rocky Mts, USA. Mt Res & Dev 3:149–155 [7]

Izmailova NN (1977) Wasserhaushalt kryophiler Polsterpflanzen im östlichen Pamir. Ekol Akad Nauk SSSR 2:17–22 (in Russian) [1, 9]

Jackson LE, Bliss LC (1982) Distribution of ephemeral herbaceous plants near treeline in the Sierra Nevada, California, USA. Arct Alp Res 14:33–42 [2, 16]

Jacot KA, Lüscher A, Nösberger J, Hartwig UA (2000a) Symbiotic N_2 fixation of various legume species along an altitudinal gradient in the Swiss Alps. Soil Biol Biochem 32:1043–1052 [10]

Jacot KA, Lüscher A, Nösberger J, Hartwig UA (2000b) The relative contribution of symbiotic N_2 fixation and other nitrogen sources to grassland ecosystems along an altitudinal gradient in the Alps. Plant Soil 225:201–211 [10, 16]

Jaeger CH, Monson RK (1992) Adaptive significance of nitrogen storage in Bistorta bistortoides, an alpine herb. Oecologia 92:578–585 [10]

Jaeger III CH, Monson RK, Fisk MC, Schmidt SK (1999) Seasonal partitioning of nitrogen by plants and soil microorganisms in an alpine ecosystem. Ecology 80:1883–1891 [10]

James JC, Grace J, Hoad SP (1994) Growth and photosynthesis of Pinus sylvestris at its altitudinal limit in Scotland. J Ecol 82:297–306 [7]

Järvinen A (1984) Patterns and performance in a Ranunculus glacialis population in a mountain area in Finnish Lapland. Ann Bot Fenn 21:179–187 [16]

Järvinen A (1987) Microtine cycles and plant production: what is cause and effect? Oikos 49:352–357 [15]

Jenny H (1926) Die alpinen Böden. Denkschr Schweiz Naturforsch Ges 63:295–344 [6]

Jerosch MC (1903) Geschichte und Herkunft der schweizerischen Alpenflora. Eine Übersicht über den gegenwärtigen Stand der Frage. Engelmann, Leipzig [2]

Jobbagy EG, Jackson RB (2000) Global controls of forest line elevation in the Northern and Southern Hemispheres. Global Ecol Biogeogr 9:253-268 [7]

Johanson U, Gehrke C, Björn LO, Callaghan TV (1995) The effects of enhanced UV-B radiation on the growth of dwarf shrubs in a subarctic heathland. Funct Ecol 9:713-719 [17]

Johnson DA, Caldwell MM (1975) Gas exchange of four arctic and alpine tundra plant species in relation to atmospheric and soil moisture stress. Oecologia 21:93-108 [9, 11]

Johnson DA, Caldwell MM (1976) Water potential components, stomatal function, and liquid phase water transport resistances of four arctic and alpine species in relation to moisture stress. Physiol Plant 36:271-278 [9]

Johnson DA, Rumbaugh MD (1986) Field nodulation and acetylene reduction activity of high-altitude legumes in the western United States. Arctic Alp Res 18:171-179 [10]

Johnson PL, Billings WD (1962) The alpine vegetation of the Beartooth Plateau in relation to cryopedogenic processes and patterns. Ecol Monogr 32:105-135 [2, 5, 6]

Jolls CL, Bock AH (1983) Seedling density and mortality patterns among elevations in Sedum lanceolatum. Arct Alp Res 15:119-126 [2, 16]

Jonasson S (1986) Influence of frost heaving on soil chemistry and on the distribution of plant growth forms. Geogr Ann 68 A:185-195 [6]

Jonasson S (1989) Implications of leaf longevity, leaf nutrient re-absorption and translocation for the resource economy of five evergreen plant species. Oikos 56:121-131 [10]

Jonasson S (1992) Plant responses to fertilization and species removal in tundra related to community structure and clonality. Oikos 63:420-429 [16]

Jonasson S, Callaghan TV (1992) Root mechanical properties related to disturbed and stressed habitats in the Arctic. New Phytol 122:179-186 [6]

Jonasson S, Havstrom M, Jensen M, Callaghan TV (1993) In situ mineralisation of nitrogen and phosphorus of arctic soils after perturbations simulating climate change. Oecologia 95:179-186 [10]

Jonasson S, Michelsen A, Schmidt IK, Nielsen EV, Callaghan TV (1995) Microbial biomass C, N and P in two arctic soils and responses to addition of NPK fertilizer and sugar: implications for plant nutrient uptake. Oecologia 106:507-515 [10]

Jonasson S, Sköld SE (1983) Influences of frost-heaving on vegetation and nutrient regime of polygon-patterned ground. Vegetatio 53:97-112 [6]

Jones HG (1992) Plants and microclimate. Cambridge University Press, Cambridge [11]

Jones HG, Flowers TJ, Jones MB (1989) Plants under stress. Cambridge University Press, Cambridge [8]

Jónsdóttir IS, Callaghan TV (1988) Interrelationships between different generations of interconnected tillers of Carex bigelowii. Oikos 52:120-128 [13]

Jónsdóttir IS, Callaghan TV (1990) Intraclonal translocation of ammonium and nitrate nitrogen in Carex bigelowii Torr. ex Schwein. using 15N and nitrate reductase assays. New Phytol 114:419-428 [16]

Jónsdóttir IS, Callaghan TV, Headley AD (1996) Resource dynamics within arctic clonal plants. Ecol Bull 45:53-64 [16]

Junttila O (1986) Effects of temperature on shoot growth in northern provenances of Pinus sylvestris L. Tree Physiol 1:185-192 [7]

Junttila O, Robberecht R (1993) The influence of season and phenology on freezing tolerance in Silene acaulis L., a subarctic and arctic cushion plant of circumpolar distribution. Ann Bot 71:423-426 [8]

Kainmüller Ch (1975) Temperaturresistenz von Hochgebirgspflanzen. Anz Math Naturwiss Kl Österr Akad Wiss (Wien) 7:67-75 [8]

Kalin-Arroyo MT, Primack R, Armesto J (1982) Community studies in pollination ecology in the high temperate Andes of central Chile. I. pollination mechanisms and altitudinal variation. Amer J Bot 69:82-97 [16]

Kappen L, Sommerkorn M, Schroeter B (1995) Carbon acquisition and water relations of lichens in polar regions – potentials and limitations. Lichenologist 27:531-345 [5, 11]

Karlsson PS (1985) Photosynthetic characteristics and leaf carbon economy of a deciduous and an evergreen dwarf shrub: Vaccinium uliginosum L. and V. vitis-idaea L. Holarct Ecol 8:9-17 [5]

Karlsson PS (1992) Leaf longevity in evergreen shrubs: variation within and among European species. Oecologia 91:346-349 [13]

Karlsson PS (1994) Photosynthetic capacity and photosynthetic nutrient-use efficiency of Rhododendron lapponicum leaves as related to leaf nutrient status, leaf age and branch reproductive status. Funct Ecol 8:694-700 [10]

Karlsson PS, Nordell KO (1996) Effects of soil temperature on the nitrogen economy and growth of mountain birch seedlings near its presumed low temperature distribution limit. Ecoscience 3:183-189 [7, 10]

Kaspar TC, Bland WL (1992) Soil temperature and root growth. Soil Sci 154:290-298 [13]

Kaurin A (1985) Effects of light quality on frost hardening in Poa alpina. In: Kaurin A, Junttila O, Nilsen J (eds) Plant production in the north. Norwegian University Press Tromsö, pp 116-126 [12]

Kawano S, Masuda J (1980) The productive and reproductive biology of flowering plants. VII. Resource allocation and reproductive capacity in wild populations of Heloniopsis orientalis (Thunb.) C. Tanaka (Liliaceae). Oecologia 45:307-317 [16]

Keller (1962) Gewässerhaushalt des Festlandes. Eine Einführung in die Hydrogeographie. Teubner, Leipzig [9]

Keller F, Körner Ch (2003) The role of photoperiodism in alpine plant development. Arct Antarct Alp Res, in press [16]

Kelley JJ Jr, Weaver DF, Smith BP (1968) The variation of carbon dioxide under the snow in the arctic. Ecology 49:358–361 [5]

Kern H (1975) Mittlere jährliche Verdunstungshöhen 1931–1960. Schriftenr Bayerisches Landesamt Wasserwirtschaft (München) 2 [9]

Kerner A (1869) Die Abhängigkeit der Pflanzengestalt von Klima und Boden. Festschrift der 43. Jahresversammlung Deutscher Naturforscher und Ärzte, Wagner, Innsbruck, pp 29–45 [1, 7, 8]

Kerner A (1871) Der Einfluss der Winde auf die Verbreitung der Samen im Hochgebirge. Z Dtsch Alpenverein, issue of 1871:144–172 [16]

Kerner A (1898) Pflanzenleben. Bibliographisches Institut Leipzig [12]

Kessler M, Hohnwald S (1998) Bodentemperaturen innerhalb und ausserhalb bewaldeter und unbewaldeter Blockhalden in den bolivianischen Hochanden. Erdkunde 52:54–62 [7]

Kevan PG (1975) Sun-tracking solar furnaces in high arctic flowers – Significance for pollination and insects. Science 189:723–726 [16]

Kikuzawa K, Kudo G (1995) Effects of the length of the snow-free period on leaf longevity in alpine shrubs: a cost-benefit model. Oikos 73:214–220 [1]

Kikvidze Z, Nakhutsrishvili G (1998) Facilitation in subnival vegetation patches. J Veg Sci 9:261–264 [16]

Killham K (1994) Soil ecology. Cambridge University Press, Cambridge [10]

Klein RM (1978) Plants and near-ultraviolet radiation. Bot Rev 44:1–127 [8]

Klikoff LG (1968) Temperature dependence of mitochondrial oxidative rates of several plant species of the Sierra Nevada. Bot Gaz 129:227–230 [11]

Klimes L, Klimesova J, Osbornova J (1993) Regeneration capacity and carbohydrate reserves in a clonal plant *Rumex alpinus*: effect of burial. Vegetatio 109:153–160 [12]

Klotz G (1990) Hochgebirge der Erde und ihre Pflanzen- und Tierwelt. Urania, Leipzig [1]

Klötzli FA (1991a) Dornpolster und Kissenpolster – zwei divergierende Adaptationen. Festschrift Zoller, Diss Bot 196:155–162 [2]

Klötzli FA (1991b) Niches of longevity and stress. In: Esser G, Overdieck D (eds) Modern ecology: Basic and applied aspects. Elsevier, Amsterdam, pp 97–110 [7]

Klug-Pümpel B (1982) Effects of microrelief on species distribution and phytomass variations in a Caricetum curvulae stand. Vegetatio 48:249–254 [5, 15]

Klug-Pümpel B (1989) Phytomasse und Nettoproduktion naturnaher und anthropogen beeinflusster alpiner Pflanzengesellschaften in den Hohen Trauern. In:

Cernusca A (ed) Struktur und Funktion von Graslandökosystemen im Nationalpark Hohe Tauern. Veröffentlichungen Österr MaB-Hochgebirgsprogrammes, vol 13. Wagner, Österr Akad Wiss (Wien) and Wagner, Innsbruck, pp 331–356 [15]

Kogami H, Hanba YT, Kibe T, Terashima I, Masuzawa T (2001) CO_2 transfer conductance, leaf structure and carbon isotope composition of *Polygonum cuspidatum* leaves from low and high altitudes. Plant Cell Environ 24:529–538 [11]

Komarkova V (1993) Vegetation type hierarchies and landform disturbance in arctic Alaska and alpine Colorado with emphasis on snowpatches. Vegetatio 106:155–181 [4]

Körner Ch (1976) Wasserhaushalt und Spaltenverhalten alpiner Zwergsträucher. Verh Ges Ökol 5:23–30 [4, 5, 9, 11]

Körner Ch (1977) Blattdiffusionswiderstände verschiedener Pflanzen im alpinen Grasheidegürtel der Hohen Tauern. In: Cernusca A (ed) Alpine Grasheide Hohe Tauern. Ergebnisse der Ökosystemstudie 1976. Veröffentlichungen Österr MaB-Hochgebirgsprogramms Hohe Tauern, vol 1. Wagner, Innsbruck, pp 69–81 [11]

Körner Ch (1977) Evapotranspiration und Transpiration verschiedener Pflanzenbestände im alpinen Grasheidegürtel der Hohen Tauern. In: Cernusca A (ed) Alpine Grasheide Hohe Tauern. Ergebnisse der Ökosystemstudie 1976. Veröffentlichungen Österr MaB-Hochgebirgsprogramms Hohe Tauern, vol 1. Wagner, Innsbruck, pp 47–68 [9]

Körner Ch (1980) Zur anthropogenen Belastbarkeit der alpinen Vegetation. Verh Ges Oekol 8:451–461 [8, 9, 11, 17]

Körner Ch (1982) CO_2 exchange in the alpine sedge *Carex curvula* as influenced by canopy structure, light and temperature. Oecologia 53:98–104 [7, 11]

Körner Ch (1984) Auswirkungen von Mineraldünger auf alpine Zwergsträucher. Verh Ges Oekol 12:123–136 [5, 10]

Körner Ch (1989a) Der Flächenanteil unterschiedlicher Vegetationseinheiten in den Hohen Tauern: eine quantitative Analyse großmaßstäblicher Vegetationskartierungen in den Ostalpen. In: Cernusca A (ed) Struktur und Funktion von Graslandökosystemen im Nationalpark Hohe Tauern. Veröffent Österr MaB-Programms, vol 13. Österr Akad Wiss and Wagner, Innsbruck, pp 33–47 [2]

Körner Ch (1989b) The nutritional status of plants from high altitudes. A worldwide comparison. Oecologia 81:379–391 [5, 7, 10, 13]

Körner Ch (1989c) Der CO_2-Gaswechsel verschiedener Pflanzen im alpinen Grasheidegürtel. II. Photosynthetische Kohlenstoffbindung des Bestandes. In: Cernusca A (ed) Struktur und Funktion von Graslandökosystemen im Nationalpark Hohe Tauern. Veröffentlichungen Österr MaB-Hochgebirgsprogramms, vol 13. Österr Acad Wiss (Wien) and Wagner, Innsbruck, pp 357–369 [11]

Körner Ch (1991) Some often overlooked plant characteristics as determinants of plant growth: a reconsideration. Funct Ecol 5:162–173 [1, 7, 12]

Körner Ch (1992) Response of alpine vegetation to global climate change. CATENA Suppl 22:85–96 [17]

Körner Ch (1993) Das "Ökosystem Polsterpflanze": Recycling und Aircondition. Biol Unserer Zeit 23:353–355 [4, 6, 9]

Körner Ch (1994) Biomass fractionation in plants: a reconsideration of definitions based on plant functions. In: Roy J, Garnier E (eds) A whole plant perspective on carbon-nitrogen interactions. SPB Academic Publishers, The Hague, pp 173–185 [2, 7, 9, 12]

Körner Ch (1994) Impact of atmospheric changes on high mountain vegetation. In: Beniston M (ed) Mountain environments in changing climates. Routledge, London, pp 155–166 [17]

Körner Ch (1995) Alpine plant diversity: a global survey and functional interpretations. In: Chapin FS III, Körner Ch (eds) Arctic and alpine biodiversity: patterns, causes and ecosystem consequences. Ecological studies, vol 113. Springer, Berlin Heidelberg New York, pp 45–62 [1, 2, 4, 8]

Körner Ch (1998a) A re-assessment of high elevation treeline positions and their explanation. Oecologia 115:445–459 [7]

Körner Ch (1998b) Alpine plants: stressed or adapted? In: Press MC, Scholes JD, Barker MG (eds) Physiological plant ecology. Blackwell Science, Oxford pp 297–311 [1]

Körner Ch (2000) Why are there global gradients in species richness? Mountains might hold the answer. Trends Ecol Evol 15:513–514 [2]

Körner Ch (2002) Mountain biodiversity, its causes and function: an overview. In: Körner Ch, Spehn EM (eds) Mountain biodiversity. A global assessment. Parthenon, New York, pp 3–20 [2]

Körner Ch, Allison A, Hilscher H (1983) Altitudinal variation in leaf diffusive conductance and leaf anatomy in heliophytes of montane New Guinea and their interrelation with microclimate. Flora 174:91–135 [1, 3, 4, 7, 8, 9, 11, 12]

Körner Ch, Bannister P, Mark AF (1986) Altitudinal variation in stomatal conductance, nitrogen content and leaf anatomy in different plant life forms in New Zealand. Oecologia 69:577–588 [3, 4, 7, 9, 12]

Körner Ch, Cochrane P (1983) Influence of plant physiognomy on leaf temperature on clear midsummer days in the Snowy Mountains, south-eastern Australia. Acta Oecol Oec Plant 4:117–124 [4, 7]

Körner Ch, Cochrane PM (1985) Stomatal responses and water relations of Eucalyptus pauciflora in summer along an elevational gradient. Oecologia 66:443–455 [7, 9, 10]

Körner Ch, De Moraes JAPV (1979) Water potential and diffusion resistance in alpine cushion plants on clear summer days. Oecol Plant 14:109–120 [4, 9, 15]

Körner Ch, Diemer M (1994) Evidence that plants from high altitudes retain their greater photosynthetic efficiency under elevated CO_2. Funct Ecol 8:58–68 [17]

Körner Ch, Diemer M, Schäppi B, Niklaus P, Arnone J (1997) The responses of alpine grassland to four seasons of CO_2 enrichment: a synthesis. Acta Oecol 18:165–175 [10, 12, 17]

Körner Ch, Diemer M, Schäppi B, Zimmermann L (1996) Responses of alpine vegetation to elevated CO_2. In: Koch GW, Mooney HA (eds) Carbon dioxide and terrestrial ecosystems. Academic Press, San Diego, pp 177–196 [6]

Körner Ch, Diemer M (1987) In situ photosynthetic responses to light, temperature and carbon dioxide in herbaceous plants from low and high altitude. Funct Ecol 1:179–194 [3, 11, 12, 17]

Körner Ch, Farquhar GD, Roksandic Z (1988) A global survey of carbon isotope discrimination in plants from high altitude. Oecologia 74:623–632 [11]

Körner Ch, Larcher W (1988) Plant life in cold climates. Symp Soc Exp Biol 42:25–57 [1, 3, 5, 7, 8, 9, 11, 12, 14]

Körner Ch, Farquhar GD, Wong SC (1991) Carbon isotope discrimination by plants follows latitudinal and altitudinal trends. Oecologia 88:30–40 [1, 2, 3, 9, 11, 17]

Körner Ch, Hättenschwiler S (1998) Die Alpen und das CO_2 Problem. Biologische Perspektiven. vdf, Zürich [17]

Körner Ch, Mayr R (1981) Stomatal behaviour in alpine plant communities between 600 and 2600 metres above sea level. In: Grace J, Ford ED, Jarvis PG (eds) Plants and their atmospheric environment. Blackwell, Oxford, pp 205–218 [7, 9]

Körner Ch, Neumayer M, Pelaez Menendez-Riedl S, Smeets-Scheel A (1989a) Functional morphology of mountain plants. Flora 182:353–383 [8, 9, 10, 11, 12, 13, 14]

Körner Ch, Meusel H (1986) Zur ökophysiologischen und ökogeographischen Differenzierung nah verwandter Carlina-Arten. Flora 178:209–232 [2]

Körner Ch, Paulsen J, Pelaez-Riedl S (2003) A bioclimatic characterisation of Europe's alpine areas. In: Nagy L, Grabherr G, Körner Ch, Thompson DBA (eds) Alpine biodiversity in Europe. Ecological studies, vol 167, Springer, Berlin Heidelberg New York, pp 13–30 [4]

Körner Ch, Pelaez Menendez-Riedl S (1989) The significance of developmental aspects in plant growth analysis. In: Lambers H, Cambridge ML, Konings H, Pons TL (eds) Causes and consequences of variation in growth rate and productivity of higher plants. SPB Academic Publishers, The Hague, pp 141–157 [1, 10, 11, 14]

Körner Ch, Pelaez Menendez-Riedl S, John PCL (1989b) Why are Bonsai plants small? A consideration of cell size. Aust J Plant Physiol 16:443–448 [8, 14]

Körner Ch, Pelaez-Riedl S, van Bel AJE (1995) CO_2 responsiveness of plants: a possible link to phloem loading. Plant Cell Environ 18:595–600 [7, 12]

Körner Ch, Renhardt U (1987) Dry matter partitioning and root length /leaf area ratios in herbaceous perennial plants with diverse altitudinal distribution. Oecologia 74:411–418 [8, 9, 10, 12, 16]

Körner Ch, Scheel JA, Bauer H (1979) Maximum leaf diffusive conductance in vascular plants. Photosynthetica 13:45–82 [11]

Körner Ch, Spehn EM (eds) (2002) Mountain biodiversity. A global assessment. Parthenon, New York

Körner Ch, Wieser G, Cernusca A (1989c) Der Wasserhaushalt waldfreier Gebiete in den österreichischen Alpen zwischen 600 und 2600 m Höhe. In: Cernusca A (ed) Struktur und Funktion von Graslandökosystemen im Nationalpark Hohe Tauern. Veröffentlichungen Österr MaB-Hochgebirgsprogramms Hohe Tauern, vol 13. Österr Akad Wiss (Wien) and Wagner, Innbruck, pp 13:119–153 [9, 17]

Körner Ch, Wieser G, Guggenberger H (1980) Der Wasserhaushalt eines alpinen Rasens in den Zentralalpen. In: Franz H (ed) Untersuchungen an alpinen Böden in den Hohen Tauern 1974–1978, Stoffdynamik und Wasserhaushalt. Veröffentlichungen Österr MaB-Hochgebirgsprogramms Hohe Tauern, vol 3. Wagner, Innsbruck, pp 243–264 [9]

Körner Ch, Woodward FI (1987) The dynamics of leaf extension in plants with diverse altitudinal ranges. II. Field studies in *Poa* species between 600 and 3200 m altitude. Oecologia 72:279–283 [13]

Krašan F (1882) Über den combinierten Einfluß der Wärme und des Lichtes auf die Dauer der jährlichen Periode der Pflanzen. Ein Beitrag zur Nachweisung der ursprünglichen Heimatzone der Arten. Bot Jahrbuch 3:74–128 [12]

Kraus G (1869) Einige Beobachtungen über den Einfluß des Lichts und der Wärme auf die Stärkeerzeugung im Chlorophyll. Jahrb Wiss Bot 7:511–531 [12]

Kudo G (1992) Effect of snow-free duration on leaf life-span of four alpine plant species. Can J Bot 70:1684–1688 [5, 13]

Kudo G (1992) Performance and phenology of alpine herbs along a snow-melting gradient. Ecol Res 7:297–304 [16]

Kudo G (1993) Relationship between flowering time and fruit set of the entomophilous alpine shrub, *Rhododendron aureum* (Ericaceae), inhabiting snow patches. Am J Bot 80:1300–1304 [16]

Kudo G (1996a) Intraspecific variation of leaf traits in several deciduous species in relation to length of growing season. Ecoscience 3(4):483–489 [12, 13]

Kudo G (1996b) Effects of snowmelt timing on reproductive phenology and pollination process of alpine plants. Mem Natl Inst Polar Res (Tokyo), Spec Issue 51:71–82 [16]

Kudo G, Ito K (1992) Plant distribution in relation to the length of the growing season in a snow-bed in the Taisetsu Mountains, northern Japan. Vegetatio 98:165–174 [4]

Kullman L (1989) Geological aspects of episodic permafrost expansion in North Sweden. Geogr Ann 71A:255–262 [6]

Kume A, Ino Y (1993) Comparison of ecophysiological response to heavy snow in two varieties of *Aucuba japonica* with different areas of distribution. Ecol Res 8:111–121 [5]

Laine K, Malila E, Siuruainen M (1995) How is annual climaic variation reflected in the production of germinable seeds of arctic and alpine plants in the northern Scandes? Ecosystems research report (European Commission, Brussels) 10:89–95 [16]

LaMarche VC, Mooney HA (1972) Recent climatic change and development of bristlecone pine (*P. longaeva* Bailey) krummholz zone, Mt. Washington, Nevada. Arct Alp Res 4:61–72 [7]

Lambers H, Van der Werf A (1989) Variation in the rate of root respiration of two *Carex* species: a comparison of four related methods to determine the energy requirements for growth, maintenance and ion uptake. In: Loughman BC et al (eds) Structural and functional aspects of transport in roots. Kluwer, Dordrecht, pp 131–135 [11]

Landolt E (1967) Gebirgs- und Tieflandsippen von Blütenpflanzen im Bereich der Schweizer Alpen. Bot Jb 86:463–480 [16]

Landolt E (1983) Probleme der Höhenstufen in den Alpen. Botanica Helv 93:255–268 [2]

Lange OL (1959) Untersuchungen über Wärmehaushalt und Hitzeresistenz mauretanischer Wüsten- und Savannenpflanzen. Flora 147:595–651 [8]

Lange OL (1961) Die Hitzeresistenz einheimischer immerund wintergrüner Pflanzen im Jahreslauf. Flora 56:666–683 [8]

Lange OL, Lange R (1963) Untersuchungen über Blattemperaturen, Transpiration und Hitzeresistenz an Pflanzen mediterraner Standorte (Costa Brava, Spanien). Flora 153:387–425 [8]

Lange OL, Metzner H (1965) Lichtabhängiger Kohlenstoff-Einbau in Flechten bei tiefen Temperaturen. Naturwissenschaften 52:191 [11]

Langlet O (1971) Two hundred years genecology. Taxon 20:653–722 [1]

Larcher W (1957) Frosttrocknis an der Waldgrenze und in der alpinen Zwergstrauchheide auf dem Patscherkofel bei Innsbruck. Veröff. Mus Ferdinandeum (Innsbruck) 37:49–81 [5, 9]

Larcher W (1963a) Zur spätwinterlichen Erschwerung der Wasserbilanz von Holzpflanzen an der Waldgrenze. Ber Naturwiss-Med Verein Innsbruck 53:125–137 [7]

Larcher W (1963b) Zur Frage des Zusammenhanges zwischen Austrocknungsresistenz und Frostharte bei Immergrünen. Protoplasma 57:569–587 [8]

Larcher W (1967) Die Berge – einzigartiges Versuchsfeld der Natur. Jahrb Vereins Schutze Alpenpflanzen Tiere 32: 1–7 [1]

Larcher W (1969) The effect of environmental and physiological variables on the carbon dioxide gas exchange of trees. Photosynthetica 3:167–198 [11]

Larcher W (1970) Aufgaben und Möglichkeiten ökophysiologischer Forschung im Gebirge. Mitt Ostalp Dinarische Ges Vegetationskd 11:95–100 [1]

Larcher W (1972) Der Wasserhaushalt immergrüner Pflanzen im Winter. Ber Dtsch Bot Ges 85:315–327 [9]

Larcher W (1975) Pflanzenökologische Beobachtungen in der Paramostufe der venezolanischen Anden. Anz Math Naturwiss Kl Österr Akad Wiss (Wien)11:194–213 [1, 7, 9]

Larcher W (1977) Ergebnisse des IBP-Projekts "Zwergstrauch-heide Patscherkofel". Sitzungsber Österr Akad Wiss (Wien) Math Naturwiss Kl Abt I 186:301–371 [1, 4, 6, 8, 9, 10, 11, 15]

Larcher W (1980) Klimastress im Gebirge – Adaptationstraining und Selektionsfilter für Pflanzen. Rheinisch Westfäl Akad Wiss (Düsseldorf) Naturwiss Vortr 291:49–88 [1, 4, 5, 8, 16]

Larcher W (1981) Resistenzphysiologische Grundlagen der evolutiven Kälteakklimatisation von Sprosspflanzen. Plant Syst Evol 137:145–180 [1]

Larcher W (1983) Ökophysiologische Konstitutionseigen-schaften von Gebirgspflanzen. Ber Dtsch Bot Ges 96:73–85 [1]

Larcher W (1985a) Winter stress in high mountains. Ber Eid-genöss Anst Forstl Versuchswes 270:11–20 [5, 7]

Larcher W (1985b) Six chapters on freezing resistance. In: Sorauer P (founder) Handbuch der Pflanzenkrankheiten, vol 1(5)7. Parey, Berlin, pp 107–287 [8]

Larcher W (1987) Stress bei Pflanzen. Naturwiss 74:158–167 [8]

Larcher W (1994) Hochgebirge: An den Grenzen des Wachstums. In: Morawetz W (ed) Ökologische Grundwerte in Österreich. Österr Akad Wiss Wien, pp 304–343 [1]

Larcher W (2003) Physiological plant ecology: ecophysiology and stress physiology of functional groups, 4rd edn. Springer, Berlin Heidelberg New York [8, 9]

Larcher W (1996) Das Verpflanzungsexperiment als Forsch-ungsansatz für phänologische Analysen: Reproduktive Entwicklung von Rotschwingelgras in 600 m und 1920 m Meereshöhe. Wetter Leben 48:125–140 [16]

Larcher W, Bauer H (1981) Ecological significance of resis-tance to low temperature. Encyclopedia of plant phy-siology, vol 12A(1). Springer, Berlin Heidelberg New York, pp 403–437 [8]

Larcher W, Siegwolf R (1985) Development of acute frost drought in Rhododendron ferrugineum at the alpine tim-berline. Oecologia 67:298–300 [4, 5, 8, 9]

Larcher W, Vareschi V (1988) Variation in morphology and functional traits of Dictyonema glabratum from contrast-ing habitats in the Venezuelan Andes. Lichenologist 20:269–277 [11]

Larcher W, Wagner J (1976) Temperaturgrenzen der CO_2-Aufnahme und Temperaturresistenz der Blätter von Gebirgspflanzen im vegetationsaktiven Zustand. Oecol Plant 11:361–374 [4, 8, 11]

Larcher W, Wagner J (1983) Ökologischer Zeigerwert und physiologische Konstitution von Sempervivum montanum. Verh Ges Ökol 11:253–264 [2, 11]

Larigauderie A, Körner Ch (1995) Acclimation of leaf dark res-piration to temperature in alpine and lowland plant species. Ann Bot 76:245–252 [11, 17]

Larson RA, Garrison WJ, Carlson RW (1990) Differential responses of alpine and non-alpine Aquilegia species to increased ultraviolet-B radiation. Plant Cell Environ 13:983–987 [8]

Lauer W (1988) Zum Wandel der Vegetationszonierung in den Lateinamerikanischen Tropen seit dem Höhe-punkt der letzten Eiszeit. In: Buchholz HJ, Gerold G (eds) Jahrbuch der Geographischen Gesellschaft zu Hannover, Lateinamerikaforschung, Hannover. Selbstver-lag der Geographischen Gesellschaft, Hannover, pp 1–45 [2, 7]

Lauscher F (1966) Die Tagesschwankung der Lufttemperatur auf Höhenstationen in allen Erdteilen. In: Steinhauser F (ed) 60.-62. Jahresbericht des Sonnblick-Vereins für die Jahre 1962–1964. Kommissionsverlag Springer, Wien, pp 3–17 [3]

Lauscher F (1976) Weltweite Typen der Höhenabhängigkeit des Niederschlags. Wetter Leben 28:80–90 [3]

Lauscher F (1977) Ergebnisse der Beobachtungen an den nordchilenischen Hochgebirgsstationen Collahuasi und Chuquicamata. Jahresbericht des Sonnblick-Vereines für die Jahre 1976–1977:43–67 [3, 7]

Lautenschlager-Fleury D (1955) Über die Ultraviolettdurch-lässigkeit von Blattepidermen. Ber Schweiz Bot Ges 65:343–386 [8]

Lavin M (1983) Floristics of the upper Walker River, California and Nevada. Great Basin Nat 43:93–130 [2]

Lawler DM (1988) Environmental limits of needle ice: a global survey. Arct Alp Res 20:137–159 [6]

Ledgard NJ, Baker NJ (1988) Mountain forestry. 30 years' research in the Craigieburn Range. Forest Research Insti-tute Christchurch, New Zealand, Bulletin no. 145 [7]

Lee DW, Lowry JB (1980) Solar ultraviolet on tropical moun-tains: Can it affect plant speciation? Am Nat 115:880–883 [8]

Lescia P, Antibus RK (1986) Mycorrhizae of alpine fell-field communities on soils derived from crystalline and calcareous parent materials. Can J Bot 64:1691–1697 [10]

Leuschner Ch (1996) Timberline and alpine vegetation on the tropical and warm-temperate oceanic islands of the world: elevation, structure and floristics. Vegetatio 123:193–206 [7]

Leuschner C, Schulte M (1991) Microclimatological investiga-tions in the tropical alpine shrubs of Maui, Hawaii: evi-dence for a drought-induced alpine timberline. Pac Sci 45:152–168 [9]

Li MH, Hoch G, Körner C (2002) Source/sink removal affects moblie carbohydrates in Pinus cembra at the Swiss treeline. Trees 16:331–337 [7]

Lipp CC, Goldstein G, Meinzer FC, Niemczura W (1994) Freezing tolerance and avoidance in high elevation Hawaiian plants. Plant Cell Environ 17:1035–1044 [1, 8]

Lipson DA, Monson RK (1998) Plant-microbe competition for soil amino acids in the alpine tundra: effects of freeze–thaw and dry-rewet events. Oecologia 113:406–414 [10]

Lipson DA, Schmidt SK, Monson RK (1999) Links between microbial population dynamics and nitrogen availability in an alpine ecosystem. Ecology 80:1623–1631 [10]

Litaor MI (1988) Soil solution chemistry in an alpine watershed front range, Colorado, USA. Arct Alp Res 20:485–491 [6]

Löffler E (1975) Beobachtungen zur periglazialen Höhenstufe in den Hochgebirgen von Papua New Guinea. Erdkunde Arch Wiss Geogr 29:285–292 [6]

Lohr PL (1919) Untersuchungen über die Blattanatomie von Alpen- und Ebenenpflanzen. PhD Thesis, University of Basel [9]

Loope LL, Medeiros AC (1994) Biotic interactions in Hawaiian high elevation ecosystems. In: Rundel PW, Smith AP, Meinzer FC (eds) Tropical alpine environments. Cambridge University Press, Cambridge, pp 337–354 [16]

Loris K (1981) Dickenwachstum von Zirbe, Fichte und Lärche an der alpinen Waldgrenze/Patscherkofel. Ergebnisse der Dendrometermessungen 1976–79. Mitt Forstl Bundes Versuchsanst Wien 142:416–441 [7]

Lösch R (1994) Photosynthetic characteristics of high-mountain endemic species from mid-Atlantic islands as functional adaptations to their peculiar habitat. Rev Valdotainne Hist Nat Suppl 48:405–409 [11]

Lösch R, Kappen L, Wolf A (1983) Productivity and temperature biology of two snowbed bryophytes. Polar Biol 1:243–248 [5]

Löve D (1970) Subarctic and subalpine: where and what? Arct Alp Res 2:63–73 [2, 7]

Lovett GM, Kinsman JD (1990) Atmospheric pollutant deposition to high-elevation ecosystems. Atm Environ 24A(11):2767–2786 [17]

Lüdi W (1933) Keimungsversuche mit Samen von Alpenpflanzen. Mitt Naturforsch Ges Bern (for 1932), pp 46–50 [16]

Lüdi W (1936) Experimentelle Untersuchungen an alpiner Vegetation. Ber Schweiz Bot Ges 46:632–681 [10]

Lüdi W (1938) Mikroklimatische Untersuchungen an einem Vegatationsprofil in den Alpen von Davos III. Ber Geobot Inst ETH, Stift Rübel, issue 1938:29–49

Lütz C (1987) Cytology of high alpine plants. II. Microbody activity in leaves of Ranunculus glacialis L. Cytologia 52:679–686 [8, 12]

Luzar N, Gottsberger G (2001) Flower heliotropism and floral heating of five alpine plant species and the effect on flower visiting in Ranunculus montanus in the Austrian Alps. Arct Antarct Alp Res 33:93–99 [16]

Mabberley DJ (1986) Adaptive syndromes of the afroalpine species of Dendrosenecio. In: Vuilleumier F, Monasterio M (eds) High altitude tropical biogeography. Oxford University Press, New York, pp 81–102 [2]

Mächler F, Nösberger J (1977) Effect of light intensity and temperature on apparent photosynthesis of altitudinal ecotypes of Trifolium repens L. Oecologia 31:73–78 [11]

Mähr E, Grabherr G (1983) Wurzelwachstum und -produktion in einem Krummseggenrasen (Caricetum curvulae) der Hochalpen. In: Böhm W, Kutschera L, Lichtenegger E (eds) Wurzelkökologie und ihre Nutzanwendung. Bundesanstalt Alpenländische Landwirtschaft, Gumpenstein-Irdning (Austria), pp 405–416 [13, 15]

Mahringer W (1964) Untersuchungen von Boden- und Felstemperaturen auf dem Hohen Sonnblick (3100 m). 60.-62. Jahresbericht des Sonnblick-Vereines, 1962–1964. Kommissionsverlag Springer, Wien, pp 17–31 [4]

Malyshev L (1993) Levels of the upper forest boundary in nothern Asia. Vegetatio 109:175–186 [7]

Mancinelli RL (1984) Population dynamics of alpine tundra soil bacteria, Niwot Ridge, Colorado Front Range, U.S.A. Arct Alp Res 16:185–192 [6]

Marchand PJ (1991) Life in the cold. 2nd edn. University Press of New England, Hannover [3, 4, 5, 7]

Marchand PJ, Chabot BF (1978) Winter water relations of tree-line plant species on Mt. Washington, New Hampshire. Arct Alp Res 10:105–116 [7]

Mariko S, Bekku Y, Koizumi H (1994) Efflux of carbon dioxide from snow-covered forest floors. Ecol Res 9:343–350 [5]

Mark AF (1970) Floral initiation and development in New Zealand alpine plants. NZ J Bot 8:67–75 [16]

Mark AF (1975) Photosynthesis and dark respiration in three alpine snow tussocks (Chionochloa spp.) under controlled environments. N Z J Bot 13:93–122 [1, 4, 11]

Mark AF (1994) Patterned ground activity in a southern New Zealand high-alpine cushionfield. Arct Alp Res 26:270–280 [6]

Mark AF, Adams NM (1979) New Zealand alpine plants (2nd edn). Reed, Wellington [2, 16]

Mark AF, Dickinson KJM, Hofstede RGM (2000) Alpine vegetation, plant distribution, life forms, and environments in a perhumid New Zealand region: oceanic and tropical high mountain affinities. Arct Antarct Alp Res 32:240–254 [2]

Mark AF, Holdsworth DK (1979) Yield and macronutrient content of water in relation to plant cover from the snow tussock grassland zone of eastern and central Otago, New Zealand. Progr Water Technol 11:449–462 [9]

Mark AF, Holdsworth DK (1979) Yield and macronutrient content of water in relation to plant cover from the snow tussock grassland zone of Eastern and Central Otago, New Zealand. Prog Water Technol 11:449–462 [17]

Mark AF, Rowley J, Holdsworth DK (1980) Water yield from high-altitude snow tussock grassland in Central Otago. Tussock Grassl Mt Lands Inst Rev (N Z) 38:21–33 [9, 17]

Markgraf V (1969) Moorkundliche und vegetationsgeschichtliche Untersuchungen an einem Moorsee an der Waldgrenze im Wallis. Bot Jahrb 89:1–63 [7]

Masuzawa T (1987) A comparison of the photosynthetic activity of Polygonum weyrichii var. alpinum under field conditions at the timberline of Mt. Fuji and in the laboratory. Bot Mag Tokyo 100:103–108 [1]

Matson PA, Harriss RC (1988) Prospects for aircraft-based gas exchange measurements in ecosystem studies. Ecology 69:1318–1325 [11]

Matsuoka N (1994) Continuous recording of frost heave and creep on a Japanese alpine slope. Arct Alp Res 26:245–254 [6]

Maupetit F, Wagenbach D, Weddeling P, Delmas RJ (1995) Seasonal fluxes of major ions to a high altitude cold alpine glacier. Atm Environ 29:1–9 [5]

Mawson BT, Cummins WR (1989) Thermal acclimation of photosynthetic electron transport activity by thylakoids of *Saxifraga cernua*. Plant Physiol 89:325–332 [11]

May DS, Villarreal HM (1974) Altitudinal differentiation of the Hill Reaction in populations of *Taraxacum officinale* in Colorado. Photosynthetica 8:73–77 [11]

McCracken IJ, Wardle P, Benecke U, Buxton RP (1985) Winter water relations of tree foliage at timberline in New Zealand and Switzerland. Ber Eidgenöss Anst Forstl Versuchswes 270:85–93 [7, 8]

McCutchan C, Monson RK (2001) Night-time respiration rate and leaf carbohydrate concentrations are not coupled in two alpine perennial species. New Phytol 149:419–430 [11]

McGraw JB (1985a) Experimental ecology of *Dryas octopetala* ecotypes: relative response to competitors. New Phytol 100:233–241 [5]

McGraw JB (1985b): Experimental ecology of *Dryas octopetala* ecotypes. III Environmental factors and plant growth. Arct Alp Res 17:229–239 [5]

McGraw JB (1995) Patterns and causes of genetic diversity in arctic plants. In: Chapin FS III, Körner Ch (eds) Arctic and alpine biodiversity: Patterns, causes and ecosystem consequences. Ecological studies, vol 113. Springer, Berlin Heidelberg New York, pp 33–43 [5]

McIntire EJB, Hik DS (2002) Grazing history versus current grazing: leaf demography and compensatory growth of three alpine plants in response to a native herbivore (*Ochotona collaris*). J Ecol 90:348–359 [15]

McNulty AK, Cummins WR, Pellizzar A (1988) A field survey of respiration rates in leaves of arctic plants. Arctic 41:1–5 [11]

Medina E, Delgado M (1976) Photosynthesis and night CO_2 fixation in *Echeveria columbiana* v. Poellnitz. Photosynthetica 10:155–163 [2, 9, 11]

Meinzer FC, Goldstein GH, Rundel PW (1985) Morphological changes along an altitude gradient and their consequences for an Andean giant rosette plant. Oecologia 65:278–283 [1, 9]

Meinzer FC, Goldstein G, Rada F (1994) Páramo microclimate and leaf thermal balance of Andean giant rosette plants. In: Rundel PW, Smith AP, Meinzer FC (eds) Tropical alpine environments. Cambridge University Press, Cambridge, pp 45–59 [4]

Merxmueller H (1952–1954) Untersuchungen zur Sippengliederung und Arealbildung in den Alpen. Jahrb

Verein Schutz Alpenpflanz Tiere 17:88–105, 18:135–158, 19:97–139 [2]

Merxmueller H, Poelt J (1954) Beiträge zur Florengeschichte der Alpen. Ber Bayer Bot Ges 30:91–101 [2]

Messerli B (1983) Stability and instability of mountain ecosystems: introduction to a workshop sponsored by the United Nations University. Mountain Res Dev 3:81–94 [2]

Messerli B, Ives JD (1984) Mountain ecosystems: stability and instability. Mountain Res Development 3 (2, Spec Issue) [17]

Messerli B, Ives JD (1997) Mountains of the world – a global priority. Parthenon, Carnforth [17]

Meteotest (1995) Meteonorm. Verzeichnis der Globalstrahlung und der Temperatur für die Gemeinden der Schweiz. Bundesamt für Energiewirtschaft, Bern

Meurk CD (1978) Alpine phytomass and primary productivity in Central Otago, New Zealand. N Z J Ecol 1:27–50 [15]

Michaelis P (1934) Ökologische Studien an der alpinen Baumgrenze. IV. Zur Kenntnis des winterlichen Wasserhaushaltes. Jahrb Wiss Bot 80:169 [7]

Michaelis P (1934) Ökologische Studien an der alpinen Baumgrenze. V. Osmotischer Wert und Wassergehalt während des Winters in den verschiedenen Höhenlagen. Jahrb Wiss Bot 80:336 [8]

Michelsen A, Jonasson S, Sleep D, Havström M, Callaghan TV (1996b) Shoot biomass, $\partial^{13}C$, nitrogen and chlorophyll responses of two arctic dwarf shrubs to in situ shading, nutrient application and warming simulating climatic change. Oecologia 105:1–12 [11]

Michelsen A, Schmidt IK, Jonasson S, Quarmby C, Sleep D (1996) Leaf ^{15}N abundance of subarctic plants provides field evidence that ericoid, ectomycorrhizal and non- and arbuscular mycorrhizal species access different sources of soil nitrogen. Oecologia 105:53–63 [10]

Miehe G (1989) Vegetation patterns on Mount Everest as influenced by monsoon and fohn. Vegetatio 79:21–32 [2, 7]

Miehe G (1991) Der Himalaya, eine multizonale Gebirgsregion. In: Walter H, Breckle SW (eds) Ökologie der Erde, vol 4: Spezielle Ökologie der gemäßigten und arktischen Zonen außerhalb Euro-Nordasiens. Fischer, Stuttgart, pp 181–230 [1, 2]

Miehe G, Miehe S (1994) Zur oberen Waldgrenze in tropischen Gebirgen. Phytocoenologia 24:53–110 [7]

Mikola P (1962) Temperature and tree growth near the northern timber line. In: Kozlowski TT (ed) Tree growth. Ronald, New York, pp 265–274 [7]

Miller AE, Bowman WD (2002) Variation in nitrogen$_{15}$ natural abundance and nitrogen uptake traits among co-occurring alpine species: do species partition by nitrogen form? Oecologia 130:609–616 [10]

Miller PC, Stoner WA, Ehleringer JR (1978) Some aspects of water relations of arctic and alpine regions. In: Tieszen LL (ed) Vegetation and production ecology of an Alaskan arctic tundra. Ecological studies, vol 29. Springer, Berlin Heidelberg New York, 343–357 [9]

Miroslavov EA, Kravkina IM (1991) Comparative analysis of chloroplasts and mitochondria in leaf chlorenchyma from mountain plants grown at different altitudes. Ann Bot 68:195–200 [11, 14]

Miroslavov EA, Kravkina IM, Bubolo LS (1991) Structural aspects of plant adaptation to high altitudes and Far North. Ecologia 4:35–42 [11]

Moen J (1993) Positive versus negative interactions in a high alpine block field: germination of Oxyria digyna seeds in a Ranunculus nivalis community. Arctic Alp Res 25:201–206 [16]

Molau U (1993) Relationships between flowering phenology and life history strategies in tundra plants. Arct Alp Res 25:391–402 [5, 6]

Molau U (1997) Responses to natural climatic variation and experimental warming in two tundra plant species with contrasting life forms: Cassiope tetragona and Ranunculus nivalis. Glob Change Biol 3:97–107 [17]

Molau U, Larsson EL (2000) Seed rain and seed bank along an alpine altitudinal gradient in Swedish Lapland. Can J Bot 78:728–747 [16]

Monasterio M (1986) Adaptive strategies of Espeletia in the Andean desert páramo. In: Vuilleumier F, Monasterio M (eds) High altitude tropical biogeography. Oxford University Press, New York, pp 49–80 [2, 7, 8, 9]

Monasterio M, Sarmiento L (1991) Adaptive radiation of Espeletia in the cold Andean tropics. TREE 6:387–392 [1]

Mooney HA, Billings WD (1960) The annual carbohydrate cycle of alpine plants as related to growth. Am J Bot 47:594–598 [12]

Mooney HA, Billings WD (1961) Comparative physiological ecology of arctic and alpine populations of Oxyria digyna. Ecol Monographs 31:1–29 [8, 10, 11, 13, 16]

Mooney HA, Billings WD (1965) Effects of altitude on carbohydrate content of mountain plants. Ecology 46:750–751 [8, 12]

Mooney HA, Billings WD, Hillier RD (1965) Transpiration rates of alpine plants in the Sierra Nevada of California. Am Midl Nat 74:374–386 [9]

Mooney HA, Strain BR, West M (1966) Photosynthetic efficiency at reduced carbon dioxide tensions. Ecology 47:490–491 [11, 17]

Mooney HA, Wright RD, Strain BR (1964) The gas exchange capacity of plants in relation to vegetation zonation in the White Mountains of California. Am Midl Nat 72:281–297 [7, 11]

Morales D, Jimenez MS, Iriarte J, Gil F (1982) Altitudinal effects on chlorophyll and carotinoid concentrations in gymnosperms leaves. Photosynthetica 16:362–372 [8]

Morecroft MD, Marrs RH, Woodward FI (1992b) Altitudinal and seasonal trends in soil nitrogen mineralization rate in the Scottish Highlands. J Ecol 80:49–56 [10]

Morecroft MD, Woodward FI (1990) Experimental investigations on the environmental determination of $\partial^{13}C$ at different altitudes. J Exp Bot 41:1303–1308 [11]

Morecroft MD, Woodward FI (1996) Experiments on the causes of altitudinal differences in the leaf nutrient contents size and $\partial^{13}C$ of Alchemilla alpina. New Phytol 134:471–479 [10, 11, 12]

Morecroft MD, Woodward FI, Marrs RH (1992a) Altitudinal trends in leaf nutrient contents, leaf size and $\partial^{13}C$ of Alchemilla alpina. Funct Ecol 6:730–740 [10]

Mörikhofer W (1932) Das Hochgebirgsklima. In: Loewy A (ed) Physiologie des Höhenklimas. Springer, Berlin Heidelberg New york, pp 12–65 [3]

Moser M (1966) Die ektotrophe Ernährungsweise an der Waldgrenze. Allg Forstz 77:120–127 [7]

Moser W, Brzoska W, Zachhuber K, Larcher W (1977) Ergebnisse des IBP-Projekts "Hoher Nebelkogel 3184 m". Sitzungsber Oesterr Akad Wiss (Wien) Math Naturwiss Kl Abt I 186:387–419 [1, 4, 5, 8, 11, 16]

Mullen RB, Schmidt SK (1993) Mycorrhizal infection, phosphorus uptake, and phenology in Ranunculus adoneus: implications for the functioning of mycorrhizae in alpine systems. Oecologia 94:229–234 [10]

Müller H (1881) Alpenblumen, ihre Befruchtung durch Insekten und ihre Anpassung an dieselben. Wilhelm Engelmann, Leipzig [16]

Müller W (1965) Zur Schätzung der Gebietsverdunstung im Gebirge. Arch Meteorologie, Geophys Bioklimatol Ser B 13:193–205 [9]

Munn LC, Buchnan BA, Nielsen GA (1978) Soil temperatures in adjacent high elevation forests and meadows of Montana. Soil Sci Am J 42:982–983 [7]

Nadelhoffer K, Shaver G, Fry B, Giblin A, Johnson L, McKane R (1996) ^{15}N natural abundances and N use by tundra plants. Oecologia 107:386–394 [10]

Nägeli W (1971) Der Wind als Standortsfaktor bei Aufforstungen in der subalpinen Stufe (Stillbergalp im Dischmatal, Kanton Graubünden). Mitt Schweiz Anstalt Forstl Versuchswes 47:35–147 [4]

Nakano T, Ishida A (1994) Diurnal variations of photosynthetic rates and xylem pressure potentials in four dwarf shrubs (Ericaceae) in an alpine zone. Proc NIPR Symp Polar Biol 7:243–255 [9, 11]

Nakhutshrisvili GS, Gamtsemlize SG (1984) Plant life under extreme high mountain conditions (exemplified by the central Causasus). Isdatjelstwo Nauka, Leningrad (in Russian) [1, 12]

Nakhutsrishvili G (1975) Ökologische Untersuchungen auf der Hochgebirgsstation Kasbegi. Publ ZK, KP Grus (USSR), Tbilissi [15]

Nakhutsrishvili G (1976) Plant life in extreme high mountain conditions of the Caucasus. Is Akad Nauk Gruz SSR Ser Biol, 2:132–140 (in Russian) [1]

Nakhutsrishvili G (1999) The vegetation of Georgia (Caucasus). Braun-Blanquetia 15 [1]

Nakhuzrishvili GS, Körner Ch (1982) Water relations of subalpine plants in the central Caucasus. Dokl Akad Nauk SSSR 267:243–245 (in Russian) [9]

Neales TF, Incoll LD (1968) The control of leaf photosynthesis rate by the level of assimilate concentration in the leaf: a review of the hypothesis. Bot Rev 34:117–121 [12]

Neuffer B, Bartelheim S (1989) Gen-ecology of Capsella bursa-pastoris from an altitudinal transsect in the Alps. Oecologia 81:521–527 [16]

Neuner G, Braun V, Buchner O, Taschler D (1999) Leaf rosette closure in the alpine rock species Saxifraga paniculata Mill.: significance for survival of drought and heat under high irradiation. Plant Cell Environ 22:1539–1548 [9]

Neuwinger I (1970) Böden der subalpinen und alpinen Stufe in den Tiroler Alpen. Mitt Ostalpin-Dinarische Ges Vegetationskd (Innsbruck) 11:135–150 [6, 7]

Neuwinger I (1972) Standortsuntersuchungen am Sonnberg im Sellrainer Obertal, Tirol. Mitt Forstl Bundes Versuchsanst Wien 96:177–207 [10]

Neuwinger I (1980) Erwärmung, Wasserrückhalt und Erosionsbereitschaft subalpiner Böden. Mitt Forstl Bundes Versuchsanst (Wien) 129:113–144 [5, 9]

Neuwinger I (1980) Oekologische Kennzeichnung von Bodenreliefserien längs eines Höhengradienten im Gebiet des Tauernbaches am Grossglockner (Hohe Tauern, Oesterreichische Alpen). In: Cernusca A (ed) Veröffentlichungen Österr MAB Hochgebirgsprogramms Hohe Tauern, vol 13. Österr Akad Wiss (Wien) and Wagner, Innsbruck, pp 50–93 [6]

Newesely C, Cernusca A, Bodner M (1994) Entstehung und Auswirkung von Sauerstoffmangel im Bereich unterschiedlich präparierten Schipisten. Verhandl Ges Oekol 23:277–282 [5]

Nicolussi K, Bortenschlager S, Körner Ch (1995) Increase in tree-ring width in subalpine Pinus cembra from the central Alps that may be CO_2-related. Trees 9:181–189 [7]

Niklaus PA, Körner C (1996) Responses of soil microbiota of late successional alpine grassland to long term CO_2 enrichment. Plant Soil 184:219–229 [10]

Nilsson O (1986) Nordisk fjällflora. Bonniers, Göteborg [2]

Noble IR (1993) A model of the response of ecotones to climate change. Ecol Appl 3:396–403 [7]

Nosko P, Bliss LC, Cook FD (1994) The association of free-living nitrogen-fixing bacteria with the roots of high artic graminoids. Arct Alp Res 26:180–186 [10]

Nunez CI, Aizen MA, Ezurra C (1999) Species associations and nurse plant effects in patches of high-Andean vegetation. J Veg Sci 10:357–364 [16]

O'Lear HA, Seastedt TR (1994) Landscape patterns of litter decomposition in alpine tundra. Oecologia 99:95–101 [5]

Oberbauer SF, Billings WD (1981) Drought tolerance and water use by plants along an alpine topographic gradient. Oecologia 50:325–331 [9]

Oberhuber W, Thomaser G, Mayr S, Bauer H (1999) Radial growth of Norway spruce infected by Chrysomyxa rhododendri. Phyton 39:147–154 [7]

Ohsawa M (1990) An interpretation of latitudinal patterns of forest limits in South- and East-Asian mountains. J Ecol 78:326–339 [7]

Oksanen L and Oksanen T (1989) Natural grazing as a factor shaping out barren landscapes. J Arid Environ 17::219–233 [15]

Oksanen L, Ranta E (1992) Plant strategies along mountain vegetation gradients: a test of two theories. J Veg Sci 3:175–186 [15]

Onipchenko VG, Blinnikov MS (eds) (1994) Experimental investigation of alpine plant communities in the northwestern Caucasus. Veröff Geobot Inst ETH (Stiftung Rübel, Zürich) 115:3–118 [1, 16]

Öquist G, Martin B (1986) Cold climates. In: Baker NR, Long SP (eds) Photosynthesis in contrasting environments. Elsevier, Amsterdam, pp 237–293 [11]

Osmond CB, Austin MP, Berry JA, Billings WD, Boyer JS, Dacey JWH, Nobel PS, Smith SD, Winner WE (1987) Stress physiology and the distribution of plants. The survival of plants in any ecosystem depends on their physiological reactions to various stresses of the environment. Bioscience 73:38–48 [8]

Osmond CB, Ziegler H, Stichler W, Trimborn P (1975) Carbon isotope disrimination in alpine succulent plants supposed to be capable of crassulacean acid metabolism (CAM). Oecologia 18:209–217 [11]

Ostler WK, Harper KT, McKnight KB, Anderson DC (1982) The effects of increasing snowpack on a subalpine meadow in the Uinta Mountains, Utah, USA. Arct Alp Res 14:203–214 [5]

Oswald H (1963) Verteilung und Zuwachs der Zirbe (Pinus cembra L.) der subalpinen Stufe an einem zentralalpinen Standort. Mitt Forstl Versuchswes Österr 60:437–499 [7]

Ott E (1978) Über die Abhängigkeit des Radialzuwachses und der Oberhöhen bei Fichte und Lärche von der Meereshöhe und Exposition im Lötschental. Schweiz Zeitschrift Forstwesen 129:169–193 [7]

Oulton K, Williams GJ III, May DS (1979) Ribulose-1,5-bisphosphate carboxylase from altitudinal populations of Taraxacum officinale. Photosynthetica 13:15–20 [11]

Ozenda P (1988) Die Vegetation der Alpen im europäischen Gebirgsraum. Fischer, Stuttgart [1, 2, 4]

Ozenda P (1993) Etage alpin et toundra de montagne: parente ou convergence? Fragm Florist Geobot Suppl (Krakow) 2:457–471 [2]

Ozenda P, Borel JL (1990) The possible responses of vegetation to a global climatic change. Scenarios for Western Europe, with special reference to the Alps. In: Boer MM, De Groot RS (eds) Landscape-ecological impact of climatic change. IOS Press, Amsterdam, pp 221–249 [17]

Packer JG (1974) Differentiation and dispersal in alpine floras. Arct Alp Res 6:117–128 [2]

Pahl M, Darroch B (1997) The effect of temperature and photoperiod on primary floral induction in three lines of alpine bluegrass. Can J Plant Sci 77:615–622 [16]

Pandey OP, Bhadula SK, Purohit AN (1984) Changes in the activity of some photosynthetic and photorespiratory enzymes in *Selinum vaginatum* Clarke grown at two altitudes. Photosynthetica 18:153–155 [11]

Pangtey YPS, Rawal RS, Bankoti NS, Samant SS (1990) Phenology of high altitude plants of Kumaun in Central Himalaya, India. Int J Biometeorol 34:122–127 [1, 5, 8]

Pannier F (1969) Untersuchungen zur Keimung und Kultur von *Espeletia*, eines endemischen Megaphyten der alpinen Zone ("Páramos") der venezolanischen-kolumbianischen Anden. Ber Dtsch Bot Ges 82:359–371 [16]

Pantis JD, Diamantoglou S, Margaris NS (1987) Altitudinal variation in total lipid and soluble sugar content in herbaceous plants on Mount Olympus (Greece). Vegetatio 72: 21–25 [8, 12]

Passama L, Ghorbal MH, Hamze M, Salsac L, Wacquant JP (1975) Sur quelques facteurs ecophysiologiques de differentiation entre calcicoles et calcifuges en milieu calcaire. Rev Ecol Biol Sol 12:309–327 [10]

Patzelt G (1996) Modellstudie Ötztal – Landschaftsgeschichte im Hochgebirgsraum. Mitt Österr Geogr Ges 138:53–70 [17]

Paulsen J, Körner Ch (2001) GIS analysis of tree-line elevation in the Swiss Alps suggests no exposure effect. J Veg Sci 12:817–824 [7]

Paulsen J, Weber UM, Körner Ch (2000) Tree growth near treeline: abrupt or gradual reduction with altitude? Arctic Antarct Alp Res 32:14–20 [7]

Pavlov VN, Onipchenko VG, Aksenova AA, Volkova EV, Zueva OI, Makarov MI (1998) The role of competition in Alpine plant communities (the Northwestern Causacus): an experimental approach. Zh Obshch Biol 59:453–476 [16]

Perez FL (1987) Soil moisture and the upper altitudinal limit of giant páramo rosettes. J Biogeogr 14:173–186 [9]

Perfect E, Miller RD, Burton B (1988) Frost upheaval of overwintering plants: field study of the displacement process. Arct Alp Res 20:70–75 [6]

Perkins TD, Adams GT (1995) Rapid freezing induces winter injury symptomatology in red spruce foliage. Tree Physiol 15:259–266 [7]

Péwé TL (1983) Alpine permafrost in the contiguous United States: a review. Arct Alp Res 15:145–156 [6]

Pfitsch WA (1988) Microenvironment and the distribution of two species of *Draba* (Brassicaceae) in a Venezuelan páramo. Arct Alp Res 20:333–431 [6]

Philipp M, Böcher J, Mattson O, Woodell SRJ (1990) A quantitative approach to the sexual reproductive biology and population structure in some arctic flowering plants: *Dryas integrifolia*, *Silene acaulis* and *Ranunculus nivalis*. Meddelelser Gronland, Biosci (Copenhagen) 34, 60 pp [16]

Pickering CM (1995) Variation in flowering parameters within and among five species of Australian alpine *Ranunculus*. Aust J Bot 43:103–112 [16]

Pignatti E, Pignatti S (1983) La vegetazione delle Vette di Feltre al di sopra del limite degli alberi. Stud Geobot 3:7–57 [2]

Pipp E, Larcher W (1987) Energiegehalte pflanzlicher Substanz: II. Ergebnisse der Datenverarbeitung. Sitzungsber Österr Akad Wiss (Wien) Math Naturwiss Kl Abt I 196: 249–310 [12]

Pisek A (1956) Der Wasserhaushalt der Meso- und Hygrophyten. In: Stocker O (ed) Handbuch der Pflanzenphysiologie, vol 3. Springer, Berlin Göttingen Heidelberg, pp 825–853 [6]

Pisek A (1960) Pflanzen der Arktis und des Hochgebirges. In: Ruhland W (ed) Handbuch der Pflanzenphysiologie, vol 5. Springer, Berlin Göttingen Heidelberg, pp 377–413 [1, 11]

Pisek A, Cartellieri E (1934) Zur Kenntnis des Wasserhaushaltes der Pflanzen III. Alpine Zwergsträucher. Jahr Wiss Bot 79:131–190 [9]

Pisek A, Cartellieri E (1941) Der Wasserverbrauch einiger Pflanzenvereine. Jahrb Wiss Bot 90:282–291 [9]

Pisek A, Larcher W (1954) Zusammenhang zwischen Austrocknungsresistenz und Frosthärte bei Immergrünen. Protoplasma 44:30–45 [1, 8]

Pisek A, Larcher W, Unterholzner R (1967) Kardinale Temperaturbereiche der Photosynthese und Grenztemperaturen des Lebens der Blätter verschiedener Spermatophyten. I. Temperaturminimum der Nettoassimilation, Gefrier- und Frostschadensbereiche der Blätter. Flora, Abt B 157:239–264 [5]

Pisek A, Larcher W, Unterholzner R (1967) Kardinale Temperaturbereiche der Photosynthese und Grenztemperaturen des Lebens der Blätter verschiedener Spermatophyten. I. Temperaturminimum der Nettoassimilation, Gefrier- und Frostschadensbereiche der Blätter. Flora Abt B 157:239–264 [8]

Pisek A, Larcher W, Vegis A, Napp-Zinn K (1973) The normal temperature range. In: Precht H, Christophersen J, Hensel H, Larcher W (eds) Temperature and life. Springer, Berlin Heidelberg New York, pp 102–194 [11]

Pisek A, Sohm H, Cartellieri E (1935) Untersuchungen über osmotischen Wert und Wassergehalt von Pflanzen und Pflanzengesellschaften der alpinen Stufe. Beihefte Bot Centralbl (Praha) 52:634–675 [9]

Pisek A, Winkler E (1958) Assimilationsvermogen und Respiration der Fichte (*Picea excelsa* Link) in verschiedener Hohenlage und der Zirbe (*Pinus cembra* L.) an der alpinen Waldgrenze. Planta 51:518–543 [7, 11]

Pisek A, Winkler E (1959) Licht- und Temperaturabhangigkeit der CO_2-Assimilation von Fichte (*Picea excelsa* Link), Zirbe (*Pinus cembra* L.) und Sonnenblume (*Helianthus annuus* L.). Planta 53:532–550 [7]

Polle A, Mossnang M, Schonborn A, Sladkovic R, Rennenberg H (1992) Field studies on Norway spruce trees at high altitudes. New Phytol 121:89–99 [8]

Pollock CJ, Lloyd EJ (1987) The effect of low temperature upon starch, sucrose and fructan synthesis in leaves. Ann Bot 60:231–235 [12]

Pollock KM (1979) Aspects of the water relations of some alpine species of *Chionochloa*. PhD Thesis, University of Dunedin, NZ [9]

Polunin O, Stainton A (1988) Flowers of the Himalaya. Oxford University Press, Oxford [2]

Posch A (1980) Bodenkundliche Untersuchungen im Bereich der Glocknerstrasse in den hohen Tauern. In: Franz H (ed) Untersuchungen an alpinen Böden in den Hohen Tauern 1974–1978, Stoffdynamik und Wasserhaushalt. Veröffentlichungen des Österr MaB-Hochgebirgsprogramms Hohe Tauern, vol 3. Wagner, Innsbruck, pp 90–107 [6]

Press MC (1998) Dracula or Robin Hood? A functional role for root hemiparasites in nutrient-poor ecosystems. Oikos 82:609–611 [16]

Price MF (1995a) Mountain research in Europe: an overview of MAB research from the Pyrenees to Siberia. Parthenon, Carnforth [17]

Price MF (1995b) Climate change in mountain regions: a marginal issue? Environmentalist 15:272–280 [17]

Price M, Heywood I (eds) (1994) Mountain environments and geographic information systems. Taylor and Francis, Basingstoke, UK [4]

Prock S (1994) Vergleichende ökologische Untersuchungen zur Phänologie, Blattlebensdauer und Biomasseallokation von krautigen Tal- und Gebirgspflanzen aus der Alpenregion, der Subarktis und der Arktis. PhD Thesis, University of Innsbruck [13]

Prock S, Körner C (1996) A cross-continental comparison of phenology, leaf dynamics and dry matter allocation in arctic and temperate zone herbaceous plants from contrasting altitudes. Ecol Bull 45:93–103 [1, 3, 5, 8, 12, 13, 16, 17]

Psenner R, Nickus U (1986) Snow chemistry of a glacier in the Central Eastern Alps (Hintereisferner, Tyrol, Austria). Z Gletscherk Glazialgeol 22:1–18 [10, 17]

Psenner R, Schmidt R (1992) Climate-driven pH control of remote alpine lakes and effects of acid deposition. Nature 356:781–783 [17]

Purohit AN, Nautiyal AR, Thapliyal P (1988) Leaf optical properties of an alpine perennial herb *Selinum vaginatum* Clarke grown at two altitudes. Biol Plant (Praha) 30:373–378 [1]

Pyankov VI, Kondratchuk AV, Shipley B (1999) Leaf structure and specific leaf mass: the alpine desert plants of the Eastern Pamirs, Tadjikistan. New Phytol 143:131–142 [11,12]

Pyankov VI, Mokronosov AT (1993) General trends in changes of the earth's vegetation related to global warming. Russian J Plant Physiol 40:515–531 [9]

Pyankov VI, Voznesenskaya EV, Kuzmin AN, Demidov ED, Vasilev AA, Dzyubenko OA (1992) C4 photosynthesis in alpine species of the Pamirs, Soviet. Plant Physiol 39:421–430 [1, 11]

Rabotnov TA (1987) The biocoenoses of alpine tundra (for example, the northwestern Caucasus) (russ). Istadelstwo Nauka, Moscow (in Russian) [1, 10]

Rada F, Azocar A, Briceno B, Gonzalez J, Garcia-Nunez C (1996) Carbon and water balance in *Polylepis sericea*, a tropical treeline species. Trees 10:218–222 [7]

Rada F, Azocar A, Gonzalez J, Briceno B (1998) Leaf gas exchange in *Espeletia schultzii* Wedd, a giant caulescent rosette species, along an altitudinal gradient in the Venezuelan Andes. Acta Oecol 19:73–79 [11]

Rada F, Goldstein G, Azocar A, Meinzer F (1985) Daily and seasonal osmotic changes in a tropical treeline species. J Exp Bot 36:989–1000 [12]

Rada F, Goldstein G, Azocar A, Torres F (1987) Supercooling along an altitudinal gradient in *Espeletia schultzii*, a caulescent giant rosette species. J Exp Bot 38:491–497 [1]

Rahn H, Carey C, Balmas K, Bhatia B, Paganelli C (1977) Reduction of pore area of the avian eggshell as an adaptation to altitude. Proc Natl Acad Sci USA 74:3095–3098 [3, 11]

Rahn H, Paganelli CV (1982) The role of gas-phase diffusion at altitude. In: Loeppky JA, Riedesel ML (eds) Oxygen transport to human tissues. Elsevier/North Holland, Amsterdam, pp 214–221 [11]

Ram J, Singh SP, Singh JS (1988) Community level phenology of grassland above treeline in Central Himalaya, India. Arct Alp Res 20:325–332 [13, 16]

Rau W, Hofmann H (1996) Sensitivity to UV-B of plants growing in different altitudes in the Alps. J Plant Physiol 148:21–25 [8]

Rauh W (1939) Über polsterförmigen Wuchs. Ein Beitrag zur Kenntnis der Wuchsformen der höheren Pflanzen. Nova Acta Leopold 7:272–505 [4, 9]

Rauh W (1940) Die Wuchsformen der Polsterpflanzen. Bot Arch 40:289–462 [2, 4]

Rauh W (1978) Die Wuchs- und Lebensformen der tropischen Hochgebirgsregionen und der Subantarktis, ein Vergleich. In: Troll C, Lauer W (eds) Geoecological relations between the southern temperate zone and the tropical mountains. Steiner, Wiesbaden, pp 62–92 [2]

Raven PH (1973) Evolution of subalpine and alpine groups in New Zealand. N Z J Bot 11:177–200 [2]

Rawat AS, Purohit AN (1991) CO_2 and water vapour exchange in four alpine herbs at two altitudes and under varying light and temperature conditions. Photosynth Res 28:99–108 [9, 11]

Rawat GS, Pangtey YPS (1987) Floristic structure of snowline vegetation in central Himalaya, India. Arct Alp Res 19: 195–201 [2, 6]

Read DJ, Haselwandter K (1981) Observations on the mycorrhizal status of some alpine plant communities. New Phytol 88:341–352 [10]

Rebetez M, Lugon M, Baeriswyl PA (1997) Climatic change and debris flows in high mountain regions: The case study of the Ritigraben torrent (Swiss Alps). Climatic change 36:371–389 [17]

Rehder H (1970) Zur Ökologie insbesondere Stickstoffversorgung subalpiner und alpiner Pflanzengesellschaften im Naturschutzgebiet Schachen (Wettersteingebirge). Diss Bot 6 [6, 10]

Rehder H (1976a) Nutrient turnover studies in alpine ecosystems. I. Phytomass and nutrient relations in four mat communities of the northern calcareous Alps. Oecologia 22:411–423 [10]

Rehder H (1976b) Nutrient turnover studies in alpine ecosystems. II. Phytomass and nutrient relations in the Caricetum firmae. Oecologia 23:49–62 [10]

Rehder H (1994) Soil nutrient dynamics in East African alpine ecosystems. In: Rundel PW, Smith AP, Meinzer FC (eds) Tropical alpine environments. Cambridge University Press, Cambridge, pp 223–228 [6]

Rehder H (1994) Soil nutrient dynamics in East African alpine ecosystems. In: Rundel PW, Smith AP, Meinzer FC (eds) Tropical alpine environments. Cambridge University Press, Cambridge, pp 223–228 [10]

Reich PB (1993) Reconciling apparent discrepancies among studies relating life span, structure and function of leaves in contrasting plant life forms and climates: "The blind men and the elephant retold". Funct Ecol 7:721–725 [13]

Reich PB, Walters MB, Ellsworth DS (1997) From tropics to tundra: global convergence in plant functioning. Proc Natl Acad Sci USA 94:13730–13734 [13]

Reisigl H, Pitschmann H (1958) Obere Grenzen von Flora und Vegetation in der Nivalstufe der zentralen Ötztaler Alpen (Tirol). Vegetatio 8:93–129 [2]

Resvoll TR (1917) Om planter som passer til kort o kold sommer. Johansen, Kristiania (Oslo), (in Norwegian) [16]

Retzer JL (1974) Alpine soils. In: Ives JD, Barry RG (eds) Arctic and alpine environments. Methuen, London, pp 771–802 [6]

Reynolds DN (1984a) Populational dynamics of three annual species of alpine plants in the Rocky Mountains. Oecologia 62:250–255 [16]

Reynolds DN (1984b) Alpine annual plants: phenology, germination, photosynthesis, and growth of three Rocky Mountain species. Ecology 65:759–766 [11, 16]

Richards AJ (1997) Plant breeding systems (2nd ed)

Richards JH (1985) Ecophysiological characteristics of seedling and sapling subalpine larch, Larix lyallii, in the winter environment. Ber Eidgenöss Anst Forstl Versuchswes 270:103–112 [7]

Richardson SG, Salisbury FB (1977) Plant responses to the light penetrating snow. Ecology 58:1152–1158 [5]

Rieger JB (1974) Geomorphic processes in the arctic. In: Ives JD, Barry RG (eds) Arctic and alpine environments. Methuen, London, pp 703–720 [6]

Rikhari HC, Negi GCS, Pant GB, Rana BS, Singh SP (1992) Phytomass and primary productivity in several communities of a central Himalayan alpine meadow, India. Arct Alp Res 24:334–351 [15]

Robberecht R, Caldwell MM (1983) Protective mechanisms and acclimation to solar ultraviolet-B radiation in Oenothera stricta. Plant Cell Environ 6:477–485 [8]

Robberecht R, Caldwell MM, Billings WD (1980) Leaf ultraviolet optical properties along a latitudinal gradient in the arctic-alpine life zone. Ecology 61:612–619 [5, 8]

Roberts J, Wareing PF (1975) A study of the growth of four provenances of Pinus contorta Dougl. Ann Bot 39:93–99 [7]

Rochefort RM, Little RL, Woodward A, Peterson DL (1994) Changes in sub-alpine tree distribution in western North-America: a review of climatic and other causal factors. Holocene 4:89–100 [7]

Rozema J, vandeStaaij J, Bjorn LO, Caldwell M (1997) UV-B as an environmental factor in plant life: Stress and regulation. Trend Ecol Evolut 12:22–28 [8]

Rübel E (1925) Alpine und arktische Flora und Vegetation. I, Alpenmatten-Überwinterungsstadien. Veröffentlichungen Geobot Inst (Stifung Rübel, Zürich) 3:37–53 [16]

Rundel PW (1994) Tropical alpine climates. In: Rundel PW, Smith AP, Meinzer FC (eds) Tropical alpine environments. Cambridge University Press, Cambridge, pp 21–44 [3, 7]

Rundel PW, Smith AP, Meinzer FC (1994) Tropical alpine environments. Cambridge University Press, Cambridge [1, 2]

Rundel PW, Witter MS (1994) Population dynamics and flowering in a Hawaiian alpine rosette plant, Argyroxiphium sandwicense. In: Rundel PW, Smith AP, Meinzer FC (eds) Tropical alpine environments. Cambridge University Press, Cambridge, pp 295–306 [12, 16]

Rusek J (1993) Air-pollution-mediated changes in alpine ecosystem and ecotones. Ecol Appl 3:409–416 [17]

Russell RS (1948) The effect of arctic and high mountain climates on the carbohydrate content of Oxyria digyna. J Ecol 36:91–95 [12]

Rusterholz H-P, Stöcklin J, Schmid B (1993) Populationsbiologische Studien an Geum reptans L. Verh Ges Ökol 22:337–346 [16]

Ruthsatz B (1977) Pflanzengesellschaften und ihre Lebensbedingungen in den Andinen Halbwüsten Nordwest-Argentiniens. Diss Bot 39 [1, 9, 10, 11]

Ruthsatz B, Hofmann U (1984) Die Verbreitung von C4-Pflanzen in den semiariden Anden NW-Argentiniens mit einem Beitrag zur Blattanatomie ausgewählter Beispiele. Phytocoenologia 12:219–249 [9, 11]

Ryser P (1996) The importance of tissue density for growth and life span of leaves and roots: a comparison of five ecologically contrasting grasses. Funct Ecol 10:717–723 [12]

Ryser P, Lambers H (1995) Root and leaf attributes accounting for the performance of fast- and slow-growing grasses at different nutrient supply. Plant Soil 170:251–265 [12]

Saetersdal M, Birks HJB (1997) A comparative ecological study of Norwegian mountain plants in relation to possible future climatic change. J Biogeogr 24:127–152 [17]

Sage RF, Sage TL (2002) Microsite characteristics of *Muhlenbergia richardsonis* (Trin.) Rydb., an alpine C$_4$ grass from the White Mountains, California [11]

Sage RF, Schäppi B, Körner C (1997) Effect of atmospheric CO$_2$ enrichment on Rubisco content in herbaceous species from high and low altitude. Acta Oecol 18:183–192 [11]

Sakai A, Larcher W (1987) Frost survival of plants. Responses and adaptation to freezing stress. Ecological studies, vol 62. Springer, Berlin Heidelberg New York [4, 5, 7, 8, 12]

Salgado-Labouriau ML (1986) Late Quaternary paleoecology of Venezuelan high mountains. In: Vuilleumier F, Monasterio M (eds) High altitude tropical biogeography. Oxford University Press, New York, pp 202–217 [2]

Salisbury E (1974) Seed size and mass in relation to environment. Proc R Soc Lond Biol 186:83–88 [16]

Salisbury FB, Spomer GG (1964) Leaf temperatures of alpine plants in the field. Planta 60:497–505 [4]

Sandvik SM, Totland O, Nylehn J (1999) Breeding system and effects of plant size and flowering time on reproductive success in the alpine herb *Saxifraga stellaris* L. Arct Antarct Alp Res 31:196–201 [16]

Sarapultsev IE (2001) The phenomenon of pseudoviviparity in alpine and arctomontane grasses (*Deschampsia* Beauv., *Festuca* L., and *Poa* L.). Russ J Ecol 32:170 [16]

Sarmiento G (1986) Ecological features of climate in high tropical mountains. In: Vuilleumier F, Monasterio M (eds) High altitude tropical biogeography. Oxford University Press, New York, pp 11–45 [3]

Sauberer F, Dirmhirn I (1958) Das Strahlungsklima. In: Steinhauser F, Eckel O, Lauscher F (eds) Klimatographie von Österreich. Springer, Wien, pp 13–102 [3]

Sawada S, Nakajima Y, Tsukuda M, Sasaki K, Hazama Y, Futatsuya M, Watanabe A (1994) Ecotypic differentiation of dry matter production processes in relation to survivorship and reproductive potential on *Plantago asiatica* populations along climatic gradients. Funct Ecol 8:400–409 [16]

Schachtschabel P, Blume HP, Hartge KH, Schwertmann U (1982) Lehrbuch der Bodenkunde. Enke, Stuttgart [6]

Schäppi B (1996) Growth dynamics and population development in an alpine grassland under elevated CO$_2$. Oecologia 106:93–99 [13]

Schäppi B, Körner Ch (1996) Growth responses of an alpine grassland to elevated CO$_2$. Oecologia 105:43–52 [13, 17]

Schäppi B, Körner Ch (1997) In situ effects of elevated CO2 on the carbon and nitrogen status of alpine plants. Funct Ecol 11:290–299 [12, 10, 17]

Schenk K, Härtel O (1937) Untersuchungen über den Wasserhaushalt von Alpenpflanzen am natürlichen Standort. Mit einem Beitrag zur Bodenkunde eines alpinen Standortes. Jahrb Wiss Bot 85:592–640 [9]

Scherff EJ, Galen C, Stanton ML (1994) Seed dispersal, seedling survival and habitat affinity in a snowbed plant: limits to the distribution of the snow buttercup, *Ranunculus adoneus*. Oikos 69:405–413 [16]

Schiechtl HM (1988) Hangsicherung mit ingenieurbiologischen Methoden im Alpenraum. In: Erosionsbekämpfung im Hochgebirge. Jahrb Ges Ingenieurbiol 3:50–77 [17]

Schimper AFW (1898) Pflanzen-Geographie auf physiologischer Grundlage. Fischer, Jena (Engl. transl). 1903: Plant geography upon a physiological basis. Clarendon Press, Oxford) [1, 3, 9]

Schimper AFW, von Faber FC (1935) Pflanzengeographie auf physiologischer Grundlage, 2nd edn. Fischer, Jena [1]

Schinner F (1982) CO$_2$-Freisetzung, Enzymaktivitäten und Bakteriendichte von Böden unter Spaliersträuchern und Polsterpflanzen in der alpinen Stufe. Acta Oecol 3:49–58 [4, 6, 10]

Schinner F (1982b) Soil microbial activities and litter decomposition related to altitude. Plant Soil 65:87–94 [6]

Schinner F (1983) Litter decomposition, CO$_2$-release and enzyme activities in a snowbed and on a windswept ridge in an alpine environment. Oecologia 59:288–291

Schinner F, Gstraunthaler G (1981) Adaptation of microbial activities to the environmental conditions in alpine soils. Oecologia 50:113–116 [6]

Schinner F, Margesin R, Pümpel T (1992) Extracellular protease-producing psychrotrophic bacteria from high alpine habitats. Arct Alp Res 24:88–92 [6]

Schmidt L (1977) Phytomassevorrat und Nettoprimarproduktivität alpiner Zwergstrauchbestände. Oecol Plant 12:195–213 [15]

Schönenberger W, Frey W (1988) Untersuchungen zur Ökologie und Technik der Hochlagenaufforstung. Forschungsergebnisse aus dem Lawinenanrissgebiet Stillberg. Schweiz Z Forstwes 139:735–820 [7]

Schröter C (1908/1926) Das Pflanzenleben der Alpen. Eine Schilderung der Hochgebirgsflora. Raustein, Zürich [2, 7, 8, 9]

Schulze ED (1982) Plant life forms and their carbon, water and nutrient relations. Encycl Plant Physiol N Ser 12B:616–676, Springer, Berlin Heidelberg New York [15]

Schulze ED, Beck E, Scheibe R, Ziegler P (1985) Carbon dioxide assimilation and stomatal response of afroalpine giant rosette plants. Oecologia 65:207–213 [1, 4, 9, 11]

Schulze ED, Chapin FS III (1987) Plant specialization to environments of different resource availability. Ecological studies, vol 61. Springer, Berlin Heidelberg New York, pp 120–148 [9]

Schulze ED, Mooney HA, Dunn EL (1967) Wintertime photosynthesis of bristlecone pine (*Pinus aristata*) in the White Mountains of California. Ecology 48:1044–1047 [7]

Schütz W (2002) Dormancy characteristics and germination timing in two alpine *Carex* species. Basic Appl Ecol 3:125–134 [16]

Schütz W, Stöcklin J (2001) Seed weight differences between alpine and lowland plants? Verh Ges Ökol (Göttingen) 31:55

Schwartz Clements E (1905) The relation of leaf structure to physical factors. Trans Am Microsc Soc 26:19–101 [9]

Schwarz W (1970) Der Einfluß der Photoperiode auf das Austreiben, die Frosthärte und die Hitzeresistenz von Zirben und Alpenrosen. Flora 159:258–285 [8]

Schwarzenbach FH (1956) Die Beeinflussung der Viviparie bei einer Grönländischen Rasse von *Poa alpina* L. durch den jahreszeitlichen Licht- und Temperaturwechsel. Ber Schweizer Bot Ges 66:204–223 [16]

Schweingruber F, Dietz H (2001) Annual rings in the xylem of dwarf shrubs and perennial dicotyledonous herbs. Dendrochronologia 19:115–126 [16]

Schweingruber FH (1987) Flächenhafte dendroklimatische Temperaturrekonstruktionen für Europa. Naturwissenschaften 74:205–212 [7]

Schweingruber FH, Bartholin T, Schaer E, Briffa KR (1988) Radiodensitometric-dendroclimatological conifer chronologies from Lapland (Scandinavia) and the Alps (Switzerland). Boreas 17:559–566 [7]

Scott D, Billings WD (1964) Effects of environmental factors on standing crop and productivity of an alpine tundra. Ecol Monogr 34:243–270 [5 10, 11, 12, 15]

Scott D, Hillier RD, Billings WD (1970) Correlation of CO_2 exchange with moisture regime and light in some Wyoming subalpine meadow species. Ecology 51:701–702 [7]

Scott PA, Bentley CV, Fayle DCF, Hansell RIC (1987) Crown forms and shoot elongation of white spruce at the treeline, Churchill, Manitoba, Canada. Arct Alp Res 19:175–186 [7]

Scuderi LA, Schaaf CB, Orth KU, Band LE (1993) Alpine treeline growth variability: simulation using an ecosystem process model. Arct Alp Res 25:175–182 [7]

Seastedt TR, Vaccaro L (2001) Plant species richness, productivity, and nitrogen and phosphorus limitations across a snowpack gradient in Alpine Tundra, Colorado, USA. Arct Antarct Alp Res 33:100–106 [10]

Seastedt TR, Walker MD, Bryant DM (2001) Controls on decomposition processes in alpine tundra. In: Bowman WD, Seastedt WD (eds) (2001) Structure and function of an alpine ecosystem - Niwot Ridge, Colorado. Oxford University Press, Oxford, pp 222–235 [6]

Semenova GV, Onipchenko VG (1994) Soil seed banks. In: Onipchenko VG, Blinnikov MS (eds) Experimental investigation of alpine plant communities in the northwestern Caucasus. Veröff Geobot Inst (Stiftung Rübel, Zürich) 115:69–82 [16]

Semichatowa OA (1965) About respiration of high mountain plants. Isdatelstwo Nauka Prob Bot 7:142–158 (in Russian) [1, 11]

Senn G (1922) Untersuchungen über die Physiologie der Alpenpflanzen. Verh Schweiz Naturforsch Ges 103:149–169 [1, 9, 13]

Seybold A, Egle K (1940) Über die Blattpigmente der Alpenpflanzen. Bot Arch 40:560–570 [8, 11]

Shanks REC (1956) Altitudinal and microclimatic relationships of soil temperature under natural vegetation. Ecology 37:1–7 [7]

Shaver GR, Billings WD (1975) Root production and root turnover in a wet tundra ecosystem, Barrow, Alaska. Ecology 56:401–409 [13]

Shaver GR, Chapin FS III (1986) Effect of fertilizer on production and biomass of tussock tundra, Alaska, USA. Arct Alp Res 18:261–268 [10]

Shibata O (1985) Altitudinal botany. Uchida Rokakuho, Tokyo (in Japanese) [1]

Shibata O, Arai T, Kinoshita T (1975) Photosynthesis in *Polygonum reynoutria* L. ssp. *asiatica* grown at different altitudes. J Fac Sci 10:27–34 [11]

Shibata O, Nishida T (1993) Seasonal changes in sugar and starch content of the alpine snowbed plants, *Primula cuneifolia* ssp. *hakusanensis* and *Fauria crista-galli*, in Japan. Arct Alp Res 25:207–210 [1, 12]

Simpson BB, Todzia CA (1990) Patterns and processes in the development of the high Andean flora. Am J Bot 77:1419–1432 [2]

Skre O (1985) Allocation of carbon and nitrogen in Norwegian alpine plants. Aquilo Ser Bot 23:23–35 [10, 11, 12]

Slatyer RO (1976) Water deficits in timberline trees in the Snowy Mountains of south-eastern Australia. Oecologia 24:357–366 [1, 7]

Slatyer RO (1978) Altitudinal variation in the photosynthetic characteristics of snow gum, *Eucalyptus pauciflora* Sieb. ex Spreng. VII. Relationship between gradients of field temperature and photosynthetic temperature optima in the Snowy Mountains area. Aust J Bot 26:111–121 [1, 7]

Slatyer RO, Ferrar PJ (1977) Altitudinal variation in the photosynthetic characteristics of snow gum, *Eucalyptus pauciflora* Sieb. ex Spreng. V. Rate of acclimation to an altered growth environment. Aust J Plant Physiol 4:595–609 [7]

Slatyer RO, Noble IR (1992) Dynamics of montane treelines. In: Hansen AJ, di Castri F (eds) Landscape boundaries. Consequences for biotic diversity and ecological flows. Ecological studies, vol 92. Springer, Berlin Heidelberg New York, pp 346–359 [7]

Smeets N (1980) Mineralstoffverhältnisse in einem Krummseggenrasen (Caricetum vurvulae) im Glocknergebiet. Flora 170:51–67 [10]

Smidt S, Mutsch F (1993) Messungen der nassen Freilanddeposition an alpinen Höhenprofilen. Proc Int Symp "Stoffeinträge aus der Atmosphäre und Waldbodenbelastung in den Ländern der ARGE ALP und ALPEN ADRIA". GSF-Bericht (Neuherberg – München) 39/93: 21–29 [17]

Smith AP (1980) The paradox of plant height in an Andean giant rosette species. J Ecol 68:63–73 [13]

Smith AP (1994) Introduction to tropical alpine vegetation. In: Rundel PW, Smith AP, Meinzer FC (eds) Tropical alpine environments. Cambridge University Press, Cambridge, pp 1–19 [2, 9, 13]

Smith AP, Young TP (1987) Tropical alpine plant ecology. Annu Rev Ecol Syst 18:137–158 [1]

Smith AP, Young TP (1994) Population biology of *Senecio keniodendron* (Asteraceae) – an afroalpine giant rosette plant. In: Rundel PW, Smith AP, Meinzer FC (eds) Tropical alpine environments. Cambridge University Press, Cambridge, pp 273–293 [7, 16]

Smith DJ (1987) Frost-heave activity in the Mount Rae area, Canadian Rocky Mountains. Arct Alp Res 19:155–166 [6]

Smith JD (1975) Resistance of turfgrasses to low-temperature-basidiomycete snow mold and recovery from damage. Can Plant Dis Surv 55:147–154 [5]

Smith JMB, Klinger LF (1985) Aboveground: belowground phytomass ratios in Venezuelan páramo vegetation and their significance. Arct Alp Res 17:189–198 [15]

Smith WK, Carter GA (1988) Shoot structural effects on needle temperatures and photosynthesis in conifers. Am J Bot 75:496–500 [4, 7]

Smith WK, Geller GN (1979) Plant transpiration at high elevations: theory, field measurements, and comparisons with desert plants. Oecologia 41:109–122 [4, 9]

Sobrevila C (1989) Effects of pollen donors on seed formation in *Espeletia schultzii* (Compositae) populations at different altitudes. Plant Syst Evol 166:45–67 [16]

Solhaug KA (1991) Long day stimulation of dry matter production in *Poa alpina* along a latitudinal gradient in Norway. Holarct Ecol 14:161–168 [13]

Sonesson M (1986) Photosynthesis in lichen populations from different altitudes in Swedish Lapland. Polar Biol 5:113–124 [11]

Sonesson M (1989) Water, light and temperature relations of the epiphytic lichens *Parmelia olivacea* and *Parmeliopsis ambigua* in northern Swedish Lapland. Oikos 56:402–415 [5]

Sonesson M, Callaghan TV, Carlsson BA (1996) Effects of enhanced ultraviolet radiation and carbon dioxide concentration on the moss *Hylocomium splendens*. Glob Change Biol 2:67–73 [17]

Sonesson M, Hoogesteger J (1983) Recent tree-line dynamics (*Betula pubescens* Ehrh. ssp. *tortuosa* (Ledeb.) Nyman) in northern Sweden. Collect Nord 47:47–54 [7]

Sonesson M, Schipperges B, Carlsson BA (1991) Seasonal patterns of photosynthesis in alpine and subalpine populations of the lichen *Nephroma arcticum*. Oikos 65:3–12 [1]

Sörensen T (1941) Temperature relations and phenology of the northeast Greenland flowering plants. Medd Grönl 125/9, Reitzels, Kopenhagen [8, 16]

Sowell JB, Koutnik DL, Lansing AJ (1982) Cuticular transpiration of whitebark pine (*Pinus albicaulis*) within a Sierra Nevadan timberline ecotone, USA. Arct Alp Res 14:97–103 [7]

Söyrinki N (1938) Studien über die generative und vegetative Vermehrung der Samenpflanzen in der alpinen Vegetation Petsamo-Lapplands. Ann Bot Soc Zool Bot Fenn Vanamo (Helsinki) 11:1–311 [16]

Spatz G, Mühlschlegel F, Jussel U, Weis GB (1989) Zur Futterqualität von Pflanzenbeständen entlang einem Höhengradienten an der Glocknerstraße. In: Cernusca A (ed) Struktur und Funktion von Graslandökosystemen im Nationalpark Hohe Tauern. Veröffentlichungen Österr MaB-Hochgebirgsprogramms, vol 13. Österr Akad Wiss (Wien) and Wagner, Innsbruck, pp 515–530 [12]

Spence JR (1990) Seed rain in grassland, herbfield, snowbank and fellfield in the alpine zone, Craigieburn Range, South Island, New Zealand. N Z J Bot 28:439–450 [16]

Spence JR, Shaw RJ (1981) A checklist of the alpine vascular flora of the Teton Range, Wyoming, with notes on biology and habitat preferences. Great Basin Nat 41:232–242 [2]

Spindler K, Wilfing H, Rastbichler-Zessernig E, ZurNedden D, Nothdurfter H (1995) Human mummies. The man in the ice 3. Springer, Vienna [17]

Spinner H (1936) Stomates et altitude. Ber Schweiz Bot Ges 46:12–27 [9]

Spomer GG (1964) Physiological ecology studies of alpine cushion plants. Physiol Plant 17:717–724 [2]

Squeo A, Rada F, Azocar A, Goldstein G (1991) Freezing tolerance and avoidance in high tropical Andean plants: is it equally represented in species with different plant height? Oecologia 86:378–382 [1, 7, 8]

Squeo FA, Rada F, Garcia C, Ponce M, Rojas A, Azocar A (1996) Cold resistance mechanisms in high desert Andean plants. Oecologia 105:552–555 [8]

Squeo FA, Veit H, Arancio G, Gutierrez JR, Arroyo MTK, Olivares N (1993) Spatial heterogeneity of high mountain vegetation in the Andean desert zone of Chile. Mountain Res and Dev 13:203–209 [5]

Stanton ML, Galen C (1993) Blue light controls solar tracking by flowers of an alpine plant. Plant Cell Environ 16:983–989 [16]

Stanton ML, Rejmanek M, Galen C (1994) Changes in vegetation and soil fertility along a predictable snowmelt gradient in the Mosquito Range, Colorado, USA. Arct Alp Res 26:364–374 [4, 5, 6]

Stehlik I, Holderegger R, Schneller JJ, Abbott RJ, Bachmann K (2000) Molecular biogeography and population genetics of alpine plant species. Bull Geobot Inst ETH 66:47–59 [2]

Steinger T, Körner Ch, Schmid B (1996) Long-term persistence in a changing climate: DNA analysis suggests very old ages of clones of alpine *Carex curvula*. Oecologia 105:94–99 [2, 5, 16, 17]

Steinhäusser H (1970) Gebietsverdunstung und Wasservorrat in verschiedenen Seehöhen Österreichs. Österr Wasserwirtsch 22:163–170 [9]

Steltzer H, Bowman WD (1998) Differential influence of plant species on soil nitrogen transformations within moist meadow Alpine tundra. Ecosystems 1:464–474 [16]

Stenström A, Jonsdottir IS (1997) Responses of the clonal sedge, *Carer bigelowii*, to two seasons of simulated climate change. Glob Change Biol 3:89–96 [17]

Stenström M, Gugerli F, Henry GHR (1997) Response of *Saxifraga oppositifolia* L. to simulated climate change at three contrasting latitudes. Glob Change Biol 3:44–54 [16, 17]

Stenström M, Molau U (1992) Reproductive biology in *Saxifraga oppositifolia*: phenology, mating system and reproductive success. Arct Alp Res 24:337–343 [16]

Stewart WS, Bannister P (1974) Dark respiration rates in *Vaccinium* spp. in relation to altitude. Flora 163:415–421 [11, 12]

Stocker O (1935) Assimilation und Atmung westjavanischer Tropenbaume. Planta 24:402–445 [11]

Stöcklin J (1992) Umwelt, Morphologie und Wachstumsmuster klonaler Pflanzen – eine Übersicht. Bot Helv 102:3–21 [16]

Stöcklin J, Bäumler E (1996) Seed rain, seedling establishment and clonal growth strategies on a glacier foreland. J Veg Sci 7:45–56 [16]

Stöcklin J, Favre P (1994) Effects of plant size and morphological constraints on variation in reproductive components in two related species of *Epilobium*. J Ecol 82:735–746 [16]

Streb P, Shang W, Feierabend J, Bligny R (1998) Divergent strategies of photoprotection in high-mountain plants. Planta 207:313–324 [11]

Sturges DL (1989) Response of mountain big sagebrush to induced snow accumulation. J Appl Ecol 26:1035–1041 [5]

Sullivan JH, Teramura AH, Ziska LH (1992) Variation in UV-B sensitivity in plants from a 3000-m elevational gradient in Hawaii. Am J Bot 79:737–743 [1, 8]

Sundblad LG, Andersson B (1995) No difference in frost hardiness between high and low altitude *Pinus sylvestris* (L.) offspring. Scand J For Res 10:22–26 [7]

Sundriyal RC (1992) Structure, productivity and energy flow in an alpine grassland in the Garhwal Himalaya. J Veg Sci 3:15–20 [17]

Sundriyal RC, Joshi AP (1992) Annual nutrient budget for an alpine grassland in the Garhwal Himalaya. J Veg Sci 3:21–26 [1, 10]

Suzuki S, Kudo G (1997) Short-term effects of simulated environmental change on phenology, leaf traits, and shoot growth of alpine plants on a temperate mountain, northern Japan. Glob Change Biol 3:108–115 [16, 17]

Sveshnikova VM (1973) Water regime of plants under the extreme conditions of high-mountain deserts of Pamirs. UNESCO (Ecology and conservation 5), Plant response to climatic factors. Proc Uppsala Symp (1970), pp 555–561 [1, 9]

Swan LW (1992) The aeolian biome. Ecosystems of the earth's extremes. Bioscience 42:262–270 [2, 15]

Szeicz JM, MacDonald GM (1993) A dendroecological analysis of white spruce stand dynamics at the subarctic alpine treeline. Bull Ecol Soc Am Suppl 74:452 [7]

Takahashi K (1944) Die Baum- und Waldgrenze im Hida-Gebirge (japanische Nordalpen). Ein Beitrag zur Baum- und Waldgrenze Ostasiens. Jpn J Bot 13:269–343 [7]

Tappeiner U, Cernusca A (1996) Microclimate and fluxes of water vapour, sensible heat and carbon dioxide in structurally differing subalpine plant communities in the central Caucasus. Plant Cell Environ 19:403–417 [1, 9, 11]

Tappeiner U, Cernusca A, Nakhutsrisvili GS (1989) Bestandesstruktur und Lichtklima ausgewählter Pflanzenbestände der subalpinen Stufe des Zentralkaukasus. Sitzungsber Österr Akad Wiss (Wien) Math Natwiss Kl Abt I 197:395–421 [15]

Tatewaki M (1968) Distribution of alpine plants in northern Japan. In: Wright HE Jr, Osburn WH (eds) Arctic and alpine environments. Indiana University Press, Bloomington, pp 119–136 [2]

Terashima I, Masuzawa T, Ohba H (1993) Photosynthetic characteristics of a giant alpine plant, *Rheum nobile* Hook. f. et Thoms. and of some other alpine species measured at 4300 m, in the eastern Himalaya, Nepal. Oecologia 95:194–201 [1, 11]

Terashima I, Masuzawa T, Ohba H, Yokoi Y (1995) Is photosynthesis suppressed at higher elevations due to low CO_2 pressure? Ecology 76:2663–2668 [11]

Tevini M, Thoma U, Iwanzik W (1983) Effects of enhanced UV-B radiation on germination, seedling growth, leaf anatomy and pigments of some crop plants. Z Pflanzenphysiol 109:435–448 [8]

Theodose TA, Bowman WD (1997) Nutrient availability, plant abundance and species diversity in two alpine tundra communities. Ecology 78:1861–1872 [10]

Theodose TA, Jaeger CH, Bowman WD, Schardt JC (1996) Uptake and allocation of N-15 in alpine plants: Implications for the importance of competitive ability in predicting community structure in a stressful environment. Oikos 75:59–66 [10, 16]

Theurillat J-P (1995) Climate change and the alpine flora: some perspectives. In: Guisan A, Holten JI, Spichiger R, Tessier L (eds) Potential ecological impacts of climate change in the alps and Fennoscandian mountains. Ed Conserv Jard Bot, Geneve, pp 121–127 [17]

Theurillat J-P, Guisan A (2001) Potential impact of climate change on vegetation in the European Alps: a review. Climatic Change 50:77–109 [17]

Theurillat JP, Schlüssel A (1996) L'écocline subalpin-alpin: diversite et phenologie des plantes vasculaires. Bull Murithienne 114:163–169 [2]

Thomas BD, Bowman WD (1998) Influence of N$_2$-fixing *Trifolium* on plant species composition and biomass production in alpine tundra. Oecologia 115:26–31 [16]

Thomas WH, Duval B (1995) Sierra Nevada, California, USA, snow algae: snow albedo changes, algal-bacterial interrelationships, and ultraviolet radiation effects. Arct Alp Res 27:389–399 [5]

Thompson K (1978) The occurrence of buried viable seed in relation to environmental gradients. J Biogeogr 5:425–430 [16]

Thompson K, Rabinowitz D (1989) Do big plants have big seeds? Am Nat 133:722–728 [16]

Tieszen LL, Bonde EK (1967) The influence of light intensity on growth and chlorophyll in arctic, subarctic, and alpine populations of *Deschampsia caespitosa* and *Trisetum spicatum*. Univ Color Stud Ser Biol 25:1–21 [11]

Tieszen LL, Senyimba MM, Imbamba SK, Troughton JH (1979) The distribution of C3 and C4 grasses and carbon isotope discrimination along an altitudinal and moisture gradient in Kenya. Oecologia 37:337–350 [9, 11]

Till-Bottraud I, Gaudeul M (2002) Intraspecific genetic diversity in alpine plants. In: Körner Ch, Spehn E (eds) Mountain biodiversity. A global assessment. Parthenon, New York, pp 23–34 [2]

Tissue DT, Oechel WC (1987) Response of *Eriophorum vaginatum* to elevated CO$_2$ and temperature in the Alaskan tussock tundra. Ecology 68:401–410 [17]

Todaria NP (1988) Ecophysiology of mountain plants: I. Photosynthesis. Acta Physiol Plant 10:199–226 [11]

Todaria NP, Thapliyal AP, Purohit AN (1980) Altitudinal effects on chlorophyll and carotenoid contents in plants. Photosynthetica 14:236–238 [8, 11]

Tosca C, Labroue L (1981) Calcicoles et calcifuges: composition minerale de quelques especes des pelouses d altitude. Oecol Plant 2:149–154 [10]

Totland O (1993) Pollination in alpine Norway: flowering phenology, insect visitors, and visitation rates in two plant communities. Can J Bot 71:1072–1079 [16]

Totland O (1996) Flower heliotropism in an alpine population of *Ranunculus acris* (Ranunculaceae): effects on flower temperature, insect visitation, and seed production. Am J Bot 83:452–458 [16]

Totland O (1997) Effects of flowering time and temperature on growth and reproduction in *Leontodon autumnalis* var. *taraxaci*, a late-flowering alpine plant. Arctic Alpine Res 29:285–290 [16]

Totland O (2001) Enviroment-dependent pollen limitation and selection on floral traits in an alpine species. Ecology 82:2233–2244 [16]

Tranquillini W (1960) Das Lichtklima wichtiger Pflanzengesellschaften. In: Ruhland W (ed) Handbuch der Pflanzenphysiologie, vol 5: Die CO$_2$-Assimilation. Springer, Berlin Göttingen Heidelberg, pp 304–317 [3]

Tranquillini W (1963) Climate and water relations of plants in the sub-alpine region. In: Rutler AJ, Whitehead FH (eds) The water relations of plants. Blackwell, London, pp 153–167 [7, 9]

Tranquillini W (1964) The physiology of plants at high altitudes. Annu Rev Plant Physiol 15:345–362 [1]

Tranquillini W (1976) Water relations and alpine timberline. In: Lange OL, Kappen L, Schulze ED (eds) Analysis and Synthesis. Ecological studies, vol 19. Springer, Berlin Heidelberg New York, pp 473–491 [8]

Tranquillini W (1979) Physiological ecology of the alpine timberline. Tree existence at high altitudes with special references to the European Alps. Ecological studies vol 31. Springer, Berlin Heidelberg New York [1, 4, 7]

Tranquillini W (1982) Frost drought and its ecological significance. Encycl Plant Physiology 12B:379–400 [7, 9]

Tranquillini W, Plank A (1989) Oekophysiologische Untersuchungen an Rotbuchen (*Fagus sylvatica* L.) in verschiedenen Höhenlagen Nord- und Südtirols. Centbl Gesamte Forstwes 106:225–246 [7]

Tranquillini W, Platter W (1983) Der winterliche Wasserhaushalt der Larche (*Larix decidua* Mill.) an der alpinen Waldgrenze. Verh Ges Ökol 11:433–443 [7]

Troll C (1944) Strukturböden, Solifluktion und Frostklimate der Erde. Geol Rundsch 34:545–694 [6]

Troll C (1959) Die tropischen Gebirge. Ihre dreidimensionale klimatische und pflanzengeographische Zonierung. Bonn Geogr Abh 25 [7]

Troll C (1961) Klima und Pflanzenkleid der Erde in dreidimensionaler Sicht. Naturwissenschaften 9:332–348 [2, 7]

Troll C (1968) The Cordilleras of the tropical Americas. In: Troll C (ed) Geo-ecology of the mountainous regions of the tropical Americas. Dümmler, Bonn, pp 15–55 [2]

Troll C (1973) The upper timberlines in different climatic zones. Arct Alp Res 5:A3–A18 [2, 7]

Troll C (1978) *Polylepis – Hagenia – Leucosidea*. In: Troll C, Lauer W (eds) Geoecological relations between the southern temperate zone and the tropical mountains. Steiner, Wiesbaden, pp 561–563 [7]

Troll C, Lauer W (eds) (1978) Geoökologische Beziehungen zwischen der temperierten Zone der Südhalbkugel und den Tropengebieten. Steiner, Wiesbaden [1, 2, 3]

Tschager A, Hilscher H, Franz S, Kull U, Larcher W (1982) Jahreszeitliche Dynamik der Fettspeicherung von *Loiseleuria procumbens* und anderen Ericaceen der alpinen Zwergstrauchheide. Oecol Plant 3:119–134 [12]

Tschurr FR (1992) Experimentelle Untersuchungen über das Regenerationsverhalten bei alpinen Pflanzen. Veröff Geobot Inst ETH (Stiftung Rübel, Zürich) 108:1–41, 84–121 [16]

Turesson G (1925) The plant species in relation to habitat and climate. Contributions to the knowledge of genecological units. Hereditas 6:147–236 [1]

Turesson G (1931) The geographical distribution of the alpine ecotype of some eurasiatic plants. Hereditas 15:329–346 [1]

Turesson G (1933) Untersuchungen über Grenzplasmolyse- und Saugkraftwerte in verschiedenen Ökotypen derselben Art. Jahrb Wiss Bot 66:721–746 [8, 9]

Turner H (1958a) Über das Licht- und Strahlungsklima einer Hanglage der Ötztaler Alpen bei Obergurgl und seine Auswirkung auf das Mikroklima und auf die Vegetation. Arch Meteor Geophys Bioklimatol, Ser B 8:273–325 [3]

Turner H (1958b) Maximaltemperaturen oberflächennaher Bodenschichten an der alpinen Waldgrenze. Wetter Leben 10:1–12 [4]

Turner H (1968) Über "Schneeschliff" in den Alpen. Wetter Leben 20:192–200 [7]

Turner H, Häsler R, Schönenberger W (1982) Contrasting microenvironments and their effects on carbon uptake and allocation by young conifers near alpine treeline in Switzerland. In: Waring RH (ed) Carbon uptake and allocation in subalpine ecosystems as a key to management. Proc IUFRO P 1.07-00, Ecology of subalpine zones. Oregon State University, Corvallis, pp 22–30

Tyree MT, Sperry JS (1989) Vulnerability of xylem to cavitation and embolism. Annu Rev Plant Physiol Plant Mol Biol 40:19–38 [7]

Ulmer W (1937) Ueber den Jahresgang der Frosthärte einiger immergrüner Arten der alpinen Stufe sowie der Zirbe und Fichte. Unter Berücksichtigung von osmotischem Wert, Zuckerspiegel und Wassergehalt. Jahrb Wiss Bot 84:553–592 [12]

Urbanska CM, Schütz M (1986) Reproduction by seed in alpine plants and revegetation research above timberline. Bot Helv 96/1:43–61 [16]

Urbanska KM (1988) High altitude revegetation research in the Swiss Alps: experimental establishment and performance of native plant populations in machine-graded ski runs above the timberline. Proc 8th HAR workshop. Color State Univ Info Ser 59:115–128 [17]

Van Bel AJE, Gamalei YV (1992) Ecophysiology of phloem loading in source leaves. Plant Cell Environ 15:265–270 [12]

van de Staaij JWM, Bolink E, Rozema J, Ernst WHO (1997) The impact of elevated UV-B (280–320 nm) radiation levels on the reproduction biology of a highland and a lowland population of Silene vulgaris. Plant Ecol 128:172–179 [17]

Van der Hammen T, Cleef AM (1986) Development of the high Andean páramo flora and vegetation. In: Vuilleumier F, Monasterio M (eds) High altitude tropical biogeography. Oxford University Press, New York, pp 153–201 [2]

Van Tatenhove F, Dikau R (1990) Past and present permafrost distribution in the Turtmanntal, Wallis, Swiss Alps. Arct Alp Res 22:302–316 [6]

Väre H, Vestberg M, Ohtonen R (1997) Shifts in mycorrhiza and microbial activity along an oroarctic altitudinal gradient in northern fennoscandia. Arct Alp Res 29:93–104 [10]

Vareschi V (1951) Zur Frage der Oberflächenentwicklung von Pflanzengesellschaften der Alpen und Subtropen. Planta 40:1–35 [1]

Vareschi V (1970) Flora de los Páramos de Venezuela. Universidad de los Andes, Merida, Venezuela [2]

Veit H (1993) Holocene solifluction in the Austrian and southern Tyrolean Alps: dating and climatic implications. In: Frenzel B (ed) Solifluction and climatic variation in the Holocene, vol 11. Fischer, Stuttgart, pp 25–32 [6]

Veit H (1996) Southern westerlies during the Holocene deduced from geomorphological and pedological studies in the Norte Chico, Chile (27–33°S). Palaeogeogr, Palaeoclimatol, Palaeoecol 123:107–119 [6]

Veit H, Höfner T (1993) Permafrost, gelifluction and fluvial sediment transfer in the alpine/subnival ecotone, central Alps, Austria: present, past and future. Z Geomorphol NF Suppl 92:71–84 [6]

Veit H, Stingl H, Emmerich KH, John B (1995) Zeitliche und räumliche Variabilität solifluidaler Prozesse und ihre Ursachen. Eine Zwischenbilanz nach acht Jahren Solifluktionsmessungen (1985–1993) an der Messtation "Glorer Hütte", Hohe Tauern. Oesterr Z Geomorphol NF Suppl 99: 107–122 [6]

Veit M, Bilger W, Mühlbauer T, Brummert W, Winter K (1996) Diurnal changes in flavonoids. J Plant Physiol 148:478–482 [8]

Villalba R, Leiva JC, Rubulls S, Suarez J, Lenzano L (1990) Climate, tree-ring, and glacial fluctuations in the Rio Frias Valley, Rio Negro, Argentina. Arct Alp Res 22:215–232 [7]

Virtanen R, Henttonen H, Laine K (1997) Lemming grazing and structure of a snowbed plant community – a long term experiment at Kilpisjärvi, Finnish Lapland. Oikos 79:155–166 [15]

Vitousek PM, Aplet G, Turner D, Lockwood JJ (1992) The Mauna Loa environmental matrix: foliar and soil nutrients. Oecologia 89:372–382 [6]

Volko P (1971) Das kurzwellige Strahlungsfeld der Atmosphäre – Richtwerte für Ingenieure und Architekten. Schweiz Blätter Heizung Lüftung 38:121–126 [3]

Voznesenskaya EV (1996) Structure of the photosynthetic apparatus in arborescent plants inhabiting the eastern Pamirs. Russ J Plant Physiol 43:342–348 (Transl. from Fiziol Rast 43:391–398) [11]

Vuilleumier BS (1971) Pleistocene changes in the fauna and flora of South America. Science 183:771–780 [2]

Vuilleumier F, Monasterio M (eds) (1986) High altitude tropical biogeography. Oxford University Press, Oxford [1]

Wagner A (1892) Zur Kenntnis des Blattbaues der Alpenpflanzen und dessen biologischer Bedeutung. Sitzungsber

Math Naturwiss Kl Kais Akad Wiss Wien 101:487–547 [1, 9, 11]

Wagner J, Larcher W (1981) Dependence of CO_2 gas exchange and acid metabolism of the Alpine CAM plant *Sempervivum montanum* on temperature and light. Oecologia 50:88–93 [11]

Wagner J, Mitterhofer E (1998) Phenology, seed development, and reproductive success of an alpine population of *Gentianella germanica* in climatically varying years. Bot Acta 111:159–166 [16]

Wagner J, Reichegger B (1997) Phenology and seed development of the alpine sedges *Carex curvula* and *Carex firma* in response to contrasting topoclimates. Arct Alp Res 29:291–299 [16]

Walker D (1968) A reconnaissance of the non-arboreal vegetation of the Pindaunde Catchment, Mount Wilhelm, New Guinea. J Ecol 56:445–465 [1]

Walker DA, Halfpenny JC, Walker MD, Wessman CA (1993) Long-term studies of snow-vegetation interactions. Bioscience 43:287–301 [4]

Walker MD (1995) Patterns and causes of arctic plant community diversity. In: Chapin FS III, Körner Ch (eds) Arctic and alpine biodiversity: patterns, causes and ecosystem consequences. Ecological studies vol 113. Springer, Berlin Heidelberg New York, pp 3–20 [2]

Walker MD, Ingersoll RC, Webber PJ (1995) Effects of interannual climate variation on phenology and growth of two alpine forbs. Ecology 76:1067–1083 [5, 16, 17]

Walker MD, Webber PJ, Arnold EH, Ebert-May D (1994) Effects of interannual climate variation on aboveground phytomass in alpine vegetation. Ecology 75:393–408 [15]

Wallace LL, Harrison AT (1978) Carbohydrate mobilization and movement in alpine plants. Am J Bot 65:1035–1040 [12]

Walter H (1931) Die Hydratur der Pflanze und ihre physiologisch-ökologische Bedeutung. Untersuchungen über den osmotischen Wert. Fischer, Jena [8, 9]

Walter H (1973) Vegetation of the earth in relation to climate and ecophysiological conditions. English University Press, London [7]

Walter H, Breckle SW (1991–1994) Ökologie der Erde, vols 1–4. Fischer, Stuttgart [1]

Walter H, Medina E (1969) Die Bodentemperatur als ausschlaggebender Faktor für die Gliederung der subalpinen und alpinen Stufe in den Anden Venezuelas. Ber Dtsch Bot Ges 82:275–281 [7]

Ward JK, Strain BR (1997) Effects of low and elevated CO_2 partial pressure on growth and reproduction of *Arabidopsis thaliana* from different elevations. Plant Cell Environ 20:254–260 [17]

Wardle P (1968) Engelmann spruce (*Picea engelmanii* Engel.) at its upper limits on the Front Range, Colorado. Ecology 49:483–495 [7]

Wardle P (1971) An explanation for alpine timberline. N Z J Bot 9:371–402 [7]

Wardle P (1974) Alpine timberlines. In: Ives JD, Barry RG (eds) Arctic and alpine environments. Methuen, London, pp 371–402 [1, 2, 7]

Wardle P (1981a) Is the alpine timberline set by physiological tolerance reproductive capacity, or biological interactions? Proc Ecol Soc Aust 11:53–66 [7]

Wardle P (1981b) Winter desiccation of conifer needles simulated by artificial freezing. Arct Alp Res 13:419–423 [7]

Wardle P (1993) Causes of alpine timberline: a review of the hypotheses. In: Alden J, Mastrantonio JL, Odum S (eds) Forest development in cold climates. Plenum Press, New York, pp 89–103 [7]

Wardle P (1998) Comparison of alpine timberlines in New Zealand and the Southern Andes. Roy Soc New Zealand Miscel Publ 48:69–90 [7]

Watson A, Miller GR, Green FHW (1966) Winter browning of heather (*Calluna vulgaris*) and other moorland plants. Trans Bot Soc Edinb 40(2):201–203 [5]

Waughman GJ, French JRJ, Jones K (1981) Nitrogen fixation in some terrestrial environments. In: Broughton WJ (ed) Nitrogen fixation, vol 1: Ecology. Clarendon Press, Oxford, pp 135–192 [10]

Webber PJ, Ebert May D (1977) The magnitude and distribution of belowground plant structures in the alpine tundra of Niwot Ridge, Colorado. Arct Alp Res 9:157–174 [6]

Webster GL (1961) The altitudinal limits of vascular plants. Ecology 42:587–590 [1, 2]

Wehrmeister RR, Bonde EK (1977) Comparative aspects of growth and reproductive biology in arctic and alpine populations of *Saxifraga cernua* L. Arct Alp Res 9:401–406 [16]

Weih M, Johanson U, Gwynn-Jones D (1998) Growth and nitrogen utilization in seedlings of mountain birch (*Betula pubescence ssp tortuosa*) as affected by ultraviolet radiation (UV-A and UV-B) under laboratory and outdoor conditions. Trees 12:201–207 [17]

Weih M, Karlsson PS (2001) Growth responses of mountain birch to air and soil temperature: is increasing leaf-nitrogen content an acclimation to lower air temperature? New Phytol 150:147–155 [10]

Weilenmann K (1981) Bedeutung der Keim- und Jungpflanzenphase für alpine Taxa verschiedener Standorte. Ber Geobot Inst ETH (Stiftung Rübel, Zürich) 48:68–119 [16]

Welker JM, Bowman WD, Seastedt TR (2001) Environmental change and future directions in alpine research. In: Bowman WD, Seastedt WD (eds) Structure and function of an alpine ecosystem – Niwot Ridge, Colorado. Oxford University Press, Oxford, pp 304–322 [17]

Welling P, Laine K (2002) Regeneration by seeds in alpine meadow and heath vegetation in sub-arctic Finland. J Veg Sci 13:217–226 [16]

Werner P (1981) Almen: Bäuerliches Wirtschaftsleben in der Gebirgsregion. Callwey, München [17]

Werner P (1988) La flore. Pillet, Martigny [2]

Wesche K (2000) The high-altitude environment of Mt. Elgon (Uganda, Kenya): climate vegetation, and the impact of fire. Ecotropical monographs 2, JF Carthaus, Bonn

Whitfield CJ (1932) Ecological aspects of transpiration. I. Pikes peak region: climatic aspects. Bot Gaz 93:436–452 [9]

Wielgolaski FE (ed) (1975) Fennoscandian tundra ecosystems. Springer, Berlin Heidelberg New York [1]

Wieser G (1997) Carbon dioxide gas exchange of cembran pine (Pinus cembra) at the alpine timberline during winter. Tree Physiol 17:473–477 [7]

Wieser G, Körner Ch, Cernusca A (1984) Die Wasserbilanz von Graslandökosystemen in den österreichischen Alpen. Verhandl Ges Oekologie (Bern) 12:89–99 [9]

Wijmstra TA (1978) Palaeobotany and climate change. In: Gribbin J (ed) Climatic change. Cambridge University Press, Cambridge, pp 25–45 [7]

Wildi B, Lütz C (1996) Antioxidant composition of selected high alpine plant species from different altitudes. Plant Cell Environ 19:138–146 [5, 8, 11]

Wildner-Eccher MT (1988) Keimungsverhalten von Gebirgspflanzen und Temperaturesistenz der Samen und Keimpflanzen. PhD Thesis, University of Innsbruck [16]

Williams LD, Barry RG, Andrews JT (1972) Application of computed global radiation for areas of high relief. J Appl Meteorol 11:526–533 [4]

Williams MW, Bales RC, Brown AD, Melack JM (1995) Fluxes and transformations of nitrogen on a high-elevation catchment, Sierra Nevada. Biogeochemistry 28:1–31 [17]

Williams RJ (1987) Patterns of air temperature and accumulation of snow in subalpine heathlands and grasslands on the Bogong High Plains, Victoria. Aust J Ecol 12:153–163 [4]

Wilson JW (1954) The influence of "midnight sun" conditions on certain diurnal rhythms in Oxyria digyna. J Ecol 42: 81–94 [12]

Winiger M (1981) Zur thermisch-hygrischen Gliederung des Mount Kenya. Erdkunde 35:248–263 [7]

Wohlgemuth T (1993) Der Verbreitungsatlas der Farn- und Blütenpflanzen der Schweiz (Welten und Sutter 1982) auf EDV: Die Artenzahlen und ihre Abhängigkeit von verschiedenen Faktoren. Bot Helv 103:55–71 [2]

Wojciechowski MF, Heimbrook ME (1984) Dinitrogen fixation in alpine tundra, Niwot Ridge, Front Range, Colorado, USA. Arct Alp Res 16:1–10 [6, 10]

Wolfsegger M, Posch A (1980) Der Wasserhaushalt von Böden am Südhang des Hochtores (Hohe Tauern). In: Franz H (ed) Untersuchungen an alpinen Böden in den Hohen Tauern 1974–1978, Stoffdynamik und Wasserhaushalt. Veröffentlichungen Österr MaB-Hochgebirgsprogramms Hohe Tauern, vol 3. Wagner, Innsbruck, pp 223–242 [9]

Woodward FI (1975) The climatic control of the altitudinal distribution of Sedum rosea (L.) Scop. and S. telephium L. II. The analysis of plant growth in controlled environments. New Phytol 74:335–348 [13]

Woodward FI (1979a) The differential temperature responses of the growth of certain plant species from different altitudes. I. Growth analysis of Phleum alpinum L., P. bertolonii D.C., Sesleria albicans Kit. and Dactylis glomerata L. New Phytol 82:385–395 [12, 13]

Woodward FI (1979b) The differential temperature responses of the growth of certain plant species from different altitudes. II. Analyses of the control and morphology of leaf extension and specific leaf area of Phleum bertolonii D.C. and P. alpinum L. New Phytol 82:397–405 [12, 13]

Woodward FI (1983) The significance of interspecific differences in specific leaf area to the growth of selected herbaceous species from different altitudes. New Phytol 95:313–323 [1, 12]

Woodward FI (1986) Ecophysiological studies on the shrub Vaccinium myrtillus L. taken from a wide altitudinal range. Oecologia 70:580–586 [10, 11]

Woodward FI (1987) Climate and plant distribution. Cambridge University Press, Cambridge [8]

Woodward FI, Friend AD (1988) Controlled environment studies on the temperature responses of leaf extension in species of Poa with diverse altitudinal ranges. J Exp Bot 39:411–420 [13]

Woodward FI, Körner Ch, Crabtree RC (1986) The dynamics of leaf extension in plants with diverse altitudinal ranges. I. Field observations on temperature responses at one altitude. Oecologia 70:222–226 [13]

Wookey PA, Welker JM, Parsons AN, Press MC, Callaghan TV, Lee JA (1994) Differential growth, allocation and photosynthetic responses of Polygonum viviparum to simulated environmental change at a high arctic polar semi-desert. Oikos 70:131–139 [17]

Woolhouse HW (1986) Adaptation of photosynthesis to stress: a critical appraisal of current approaches and future perspectives. In: Baker NR, Long SP (eds) Photosynthesis in contrasting environments. Elsevier, Amsterdam, pp 1–12 [11]

Wyka T (1999) Carbohydrate storage and use in an alpine population of the perennial herb, Oxytropis sericea. Oecologia 120:198–208 [12]

Wyka T (2000) Effect of nutrients on growth rate and carbohydrate storage in Oxytropis sericea: a test of the carbon accumulation hypothesis. Int J Plant Sci 161:381–386 [12]

Yoshimura Y, Koshima S (1997) A community of snow algae on a Himalayan glacier: change of algal biomass and community structure with altitude. Arct Alp Res 29:126–137 [2]

Yoshino MM (1975) Climate in a small area. University of Tokyo Press, Tokyo [3]

Young DR, Smith WK (1983) Effect of cloudcover on photosynthesis and transpiration in the subalpine understory species Arnica latifolia. Ecology 64:681–687 [11]

Young TP (1994) Population biology of Mount Kenya lobelias. In: Rundel PW, Smith AP, Meinzer FC (eds) Tropical alpine environments. Cambridge University Press, Cambridge, pp 251–272 [16]

Zachhuber K, Larcher W (1978) Energy contents of different alpine species of *Saxifraga* and *Primula* depending on their altitudinal distribution. Photosynthetica 12:436–439 [12]

Zalenskij OV (1955) Photosynthesis and frost resistance of plants under the conditions of high mountains. Exp Bot 10:194–227 (in Russian) [1]

Zimov SA, Zimova GM, Daviodov SP, Daviodova AI, Voropaev YV, Voropaeva ZV, Prosiannikoh SF, Prosiannikova OV, Semilatova IV, Semiletov IP (1993) Winter biotic activity and production of CO_2 in Siberian soils: a factor in the greenhouse effect. J Geophys Res 98:5017–5023 [5]

Ziska LH, Teramura AH, Sullivan JH (1992) Physiological sensitivity of plants along an elevational gradient to UV-B radiation. Amer J Bot 79:863–871 [1, 8]

Zoller H (1987) Zur Geschichte der Vegetation im Spätglazial und Holozän der Schweiz. Mitt Naturforsch Ges Luzern 29:123–149 [7]

Zoller H (2000) La découverte des Alpes de Pétrarque à Gessner. In: Pont JC, Lacki J (eds) Une cordée originale. Georg, Geneva, pp 417–428

Zoller H, Lenzin H, Erhardt A (2002) Pollination and breeding system of *Eritrichum nanum* (Boraginaceae). Plant Syst Evol 233:1–14

Zukrigl K (1975) Zur Geschichte der Hochlagenwälder in den Seetaler Alpen (Steiermark). Eine pollenanalytische Untersuchung des kleinen Moores im Winterleitenkessel. Centralbl Gesamte Forstwes 92:175–188 [7]

Zumbrunn R, Friedli HJ, Neftel A, Rauber D (1983) CO_2 measurements with an infrared laser spectrometer on flask samples collected at Jungfraujoch high-altitude research station (3500 meters asl) and with light aircraft up to 8000 meters over Switzerland. J Geophys Res 88:6853–6857 [3, 11]

Taxonomic index (genera)

Geographical index

Subject index

Color Plates

Plate 1. Characteristic **plant life forms** in the alpine zone. Although most alpine plants are small in stature, tall forbs (**a**) occur as well; here, *Senecio formosus* at 4000 m in the Venezuelan Andes. A key element of all alpine floras is tussock grasses (**b**); here, a tall assemblage on the slopes of Pico di Orizaba, 4100 m, Mexico (*Festuca* sp, *Muehlemberiga* sp.). **c** Dwarf shrubs (*Podocarpus lawrencii*, 2050 m, Kosciusco National Park, SE Australia) and **d** cushion plants (*Silene acaulis*, 2600 m, Mt. Glungezer, Tirolian Alps) profit from radiative warming. **e** *Linaria alpina* is a creeper on loose substrate (Oetztal Alps, 3000 m) and **f** *Espeletia* sp. (El Angel, 3600 m, Ecuador) exemplifies the giant form of rosettes; much smaller and sessile rosettes are key elements of alpine vegetation worldwide. Cryptogam pioneers: **g** crusts of algae and tiny bryophytes stabilize the ground, and **h** lichens can settle on most hostile places like bare rock (*Rhizocarpon* sp. and others)

Plate 2. The **alpine life zone** has many "faces" and is characterized by sharp environmental contrasts over short distances. The upper limits: **a** scree and rock (Tien Shan, Kasachstan, 3200–4000 m). **b** Scree vegetation in the Central Alps at 2600 m. Alpine grasslands, often co-dominated by sedges on wet (**c**) or well-drained (**d**) ground (Alps, 2500 m). **e** Alpine semi-desert in Tenerife, Canary Islands, 2500 m and **f** tussock grasslands, Cumbres Calchaquies, NW Argentina, 4250 m (*Festuca* species, with a giant *Azorella* cushion in front). The lowest belt of the alpine life zone is often covered by shrubs and dwarf shrubs: **g** Paramos, Venezula, 3800 m and **h** alpine heathland on Mt. Perisher, 2050 m, SE Australia

Plate 3. Alpine plants have to cope with many types of **environmental stress.** Dry microhabitats may heat above 60 °C, thus creating "mini deserts" dominated by CAM plants like *Sempervivum montanum* (**a**). A few meters away one may find moisture-soaken late snow beds, where flowers of *Soldanella pusilla* (**b**) manage to melt through snow in order to make maximum use of a short growing season (both examples from the Central Alps at 2300 m). It would be fatal if alpine plants (here, *Empetrum nigrum* in the northern Scandes) terminated their dormant state when experiencing warm spells under an icy greenhouse roof (**c**) during late winter. Because of deep dormancy and high freezing resistance in winter, it is the summer that bears the greatest risk of freezing damage, as happened here (**d**) to flowers of *Anemone sulphurea* at 2500 m in the Swiss Alps on 8 July 1998. **e** Late lying wet spring snow enhances the danger of snow mold (here, a ruined mat of *Calluna vulgaris* in the Tirolian Alps at 2200 m). **f** The sudden release from snow to full solar radiation needs very efficient protection of premature tissue from photodamage, involving carotenoids, flavonoids and anthocyanids (sprouting *Rumex alpinus*, 2500 m, Swiss Alps). **g** Mobile substrate often creates fatal mechanical damage – a niche for specialists (*Cerastium uniflorum*, Oetztal Alps, 3250 m). **h** Safety through controlled development: the same group of shoots (*Empetrum nigrum* and *Vaccinium vitis idea*) photographed in the subarctic-alpine belt of northern Sweden; *left* in late winter (12 May, dormant, with photosynthetic pigments screened from solar radiation) and *right* fully active on July 20

Plate 4. Millennia of **land use** shaped the alpine landscape in many parts of the world. **a** Central Asia, Tien Shan, Kasachstan, 2600 m, alpine pastures with *Kobresia* and *Leontopodium* ("Edelweiss"); **b** Lechtal Alps, Tirol, 1700 m, hayfields near treeline with *Poa*, *Nardus*, *Anthoxanthum*, *Alchemilla*, etc. Much of the **research presented in this book** was obtained from two areas: **c** Niwot Ridge, Front Range Colorado, 3500–3600 m and **d** alpine grasslands near Furka Pass, Swiss Alps, 2500 m. Understanding alpine plant life requires both field work and controlled environment work: **e, f** in situ measurements of photosynthesis on Mt. Glungezer, Tirol, 2600 m; **g, h** use of cooled solar domes (Innsbruck, Austria) and field devices such as open-top chambers (Furka Pass, Switzerland) to simulate global change (in this case, CO_2 enrichment) in alpine vegetation under growth conditions as natural as possible (see Chapters 11 and 17).